Grundlehren der mathematischen Wissenschaften 213

A Series of Comprehensive Studies in Mathematics

I. R. Shafarevich

Basic Algebraic Geometry

Translated from the Russian by K. A. Hirsch

With 19 Figures

Revised Printing

Springer-Verlag
Berlin Heidelberg New York 1977

Igor R. Shafarevich
Steklov Mathematical Institute of the Academy of Sciences of the USSR

Title of the Russian Original Edition: *Osnovy algebraicheskoi geometrii*,
Publisher: Nauka, Moscow, 1972

AMS Subject Classifications (1970): Primary 14—XX, Secondary 20 GXX

ISBN 3-540-06691-8 Springer-Verlag Berlin Heidelberg New York
ISBN 0-387-06691-8 Springer-Verlag New York Heidelberg Berlin

Library of Congress Catalog Card Number 74-1653. Printed in Germany.
Typesetting, printing, and binding: Brühlsche Universitätsdruckerei, Lahn-Gießen.
2141/3140-543210.

Preface

Algebraic geometry occupied a central place in the mathematics of the last century. The deepest results of Abel, Riemann, Weierstrass, many of the most important papers of Klein and Poincaré belong to this domain.

At the end of the last and the beginning of the present century the attitude towards algebraic geometry changed abruptly. Around 1910 Klein wrote:

"When I was a student, Abelian functions*—as an after-effect of Jacobi's tradition—were regarded as the undisputed summit of mathematics, and each of us, as a matter of course, had the ambition to forge ahead in this field. And now? The young generation hardly know what Abelian functions are." (Vorlesungen über die Entwicklung der Mathematik im XIX. Jahrhundert, Springer-Verlag, Berlin 1926, Seite 312).

The style of thinking that was fully developed in algebraic geometry at that time was too far removed from the set-theoretical and axiomatic spirit, which then determined the development of mathematics. Several decades had to lapse before the rise of the theory of topological, differentiable and complex manifolds, the general theory of fields, the theory of ideals in sufficiently general rings, and only then it became possible to construct algebraic geometry on the basis of the principles of set-theoretical mathematics.

Around the middle of the present century algebraic geometry had undergone to a large extent such a reshaping process. As a result, it can again lay claim to the position it once occupied in mathematics. The range of applicability of its ideas enlarged extraordinarily towards algebraic varieties over arbitrary fields and complex manifolds of the most general kind. Algebraic geometry, quite apart from many better achievements, succeeded in removing the charge of being "incomprehensible" and "unconvincing".

The basis for this rebuilding of algebraic geometry was algebra. In its first versions the use of a delicate algebraic apparatus often led

* From the present-day point of view, the theory of Abelian functions is the analytical aspect of the theory of projective algebraic group varieties.

to the disappearance of that vivid geometric style that was characteristic for the preceding period. However, the last two decades have brought many simplifications in the foundations of algebraic geometry, which have made it possible to come remarkably close to the ideal combination of logical transparency and geometrical intuitiveness.

The aim of the book is to set forth the elements of algebraic geometry to a fairly wide extent, so as to give a general idea of this branch of mathematics and to provide a basis for the study of the more specialist literature. The reader is not assumed to have any prior knowledge of algebraic geometry, neither of its general theorems nor of concrete examples. For this reason, side by side with the development of the general theory, applications and special cases take a prominent place, because they motivate new concepts and problems to be raised.

It seems to me that the logic of the subject will be clearer to the reader if in the spirit of the "biogenetic law" he repeats, in a very condensed way, the evolution of algebraic geometry. Therefore the very first section, for example, is devoted to the simplest properties of plane algebraic curves. Similarly, Part One of the book discusses only algebraic varieties situated in a projective space, and it is only in Part Two that the reader comes across schemes and the general concept of a variety.

Part Three is concerned with algebraic varieties over the complex field and their connections with complex analytic manifolds. In this part the reader needs some acquaintance with the elements of topology and the theory of analytic functions.

My sincere thanks are due to all who have helped me with their advice during the work on this book. It is based on notes of some courses I have given at the University of Moscow. Many members of the audience and many readers of these notes have made very useful comments to me. I am particularly indebted to the editor, B. G. Moishezon. Numerous conversations with him were very useful for me. A number of proofs spread over the book are based on his advice.

Table of Contents

Chapter II. Local Properties

Chapter III. Divisors and Differential Forms

Part II. Schemes and Varieties

Chapter V. Schemes

Chapter VI. Varieties

Part III. Algebraic Varieties over the Field of Complex Numbers and Complex Analytic Manifolds

Chapter VII. Topology of Algebraic Varieties

Chapter VIII. Complex Analytic Manifolds

Chapter IX. Uniformization

Advice to the Reader

The first two parts of the book assume very little knowledge on the part of the reader. This amounts to the contents of a university course in algebra and analytic geometry and the rudiments of the theory of fields, which the reader could find, for example, in any of the following books: van der Waerden, Algebra, Vol. I, Ch. V and VIII, Zariski and Samuel, Commutative algebra, Vol. I, Ch. II, or Lang, Algebra, Ch. VII and X. Apart from this, frequent use is made of Hilbert's Basis Theorem and Hilbert's Nullstellensatz. Proofs are contained in the book by Zariski and Samuel, Vol. I, Ch. IV, § 1 and Vol. II, Ch. II, § 3.

In addition, in a few places we use certain isolated results of commutative algebra, and for their proofs the reader is referred to the book by Zariski and Samuel. In all cases the matter concerns only a few pages, which can be understood independently of the remaining parts of the book.

The third part assumes more knowledge. Essentially this concerns topology. Singular homology theory is taken as known, as are properties of differential forms, and Stokes' theorem. In Ch. VII, § 1 the concept of a differentiable manifold is applied, also Poincaré's duality law, and some properties of intersections of cycles; in §§ 3 and 4 of the same chapter we use the combinatorial classification of surfaces, but these three sections are not necessary for an understanding of the rest of the book. In the last section of the book we use one result from Morse theory, which can, however, simply be taken on trust. Finally, in the third part the reader is assumed to be familiar with the elements of the theory of analytic functions—a standard university course is amply sufficient.

The second and third parts of the book are based on the first. However, there are passages in it that are not needed for an understanding of what follows. They are Ch. IV, §§ 2 and 3, Ch. I, §§ 6.4 and 6.5, Ch. II, § 1.5, Ch. II, § 4.5, Ch. II, § 5.5, Ch. III, §§ 5.6 and 5.7; Ch. III, § 3 is fairly isolated: it is connected only with Ch. VIII, § 1.3.

The reader who is interested only in varieties over the field of complex numbers, and altogether in the more "classical" aspects of algebraic geometry, might study Ch. V only superficially. Finally, there are a number of places where we report without proofs on further developments of the questions considered in the book. Of course, these passages are not essential for an understanding of what follows.

In conclusion I wish to indicate the literature that has a bearing on the problems treated in the book and can form the nucleus of a further intensive study of algebraic geometry.

Every reader who is interested in algebraic geometry simply has to study the cohomology theory of algebraic coherent sheaves. Within the framework of the theory of varieties one can become acquainted with it through Serre's paper "Faisceaux algébriques cohérents", Ann. of Math. (2) **61**, 197–278 (1955), or Zariski's "Algebraic sheaf theory", Bull. Amer. Math. Soc. **62**, 117–141 (1956). Within the framework of the theory of schemes there is an account of the theory in the notes of Manin's "Lectures on algebraic geometry", Moscow State University 1968. A natural continuation of this theory is the general Riemann-Roch theorem, which can be read up in the paper by Borel and Serre, "Le théorème de Riemann-Roch", Bull. Soc. Math. France **86**, 97–136 (1958), or Manin's "Lectures on the K-functor in algebraic geometry" [Uspekhi Mat. Nauk **24**:5, 3–86 (1969) = Russian Math. Surveys **24**:5, 1–89 (1969)].

In this book there are frequent references to the Riemann-Roch theorem for curves, but it is never proved. Of course, it follows from the general Riemann-Roch theorem, but it can also be easily derived directly from general properties of the cohomology of algebraic coherent sheaves. Such a proof can be found in the book by Serre "Groupes algébriques et corps de classes", Hermann et Cie., Paris 1959, Ch. II. One can become acquainted with the theory of algebraic surfaces in the book "Algebraic surfaces", Trudy Mat. Inst. Steklov **75** (1965).

The elements of the theory of algebraic groups can be found in Borel, "Groupes linéaires algébriques", Ann. of Math. (2) **64**, 20–82 (1956), or Mumford, "Abelian varieties", Oxford University Press, London 1970.

So far there are no accounts of the general theory of schemes having the character of a textbook. Mumford's mimeographed lecture notes "Introduction to algebraic geometry", Harvard University Notes, can serve as an excellent introduction, and a very full account is in Grothendieck and Dieudonné's many-volume work "Eléments de géometrie algébrique". (Vol. I, Springer-Verlag, Berlin-Heidelberg-New York 1971), which is not yet completed.

In our book the number-theoretical aspect of algebraic geometry is almost nowhere touched upon, although this aspect played a very important role in the development of this branch of mathematics and several of the most brilliant applications are connected with it. An idea of this circle of problems can be obtained from Lang's "Diophantine geometry", Interscience, New York-London 1962, and Cassel's paper "Diophantine equations with special reference to elliptic curves", J. London Math. Soc. **41**, 193–291 (1966).

One can become acquainted with the "analytic" direction in the theory of algebraic varieties and the closely related theory of analytic manifolds in the book by Weil "Introduction à l'étude des variétés Kählériennes", Hermann et Cie., Paris 1958, and Chern's "Complex manifolds", Univ. of Recife, 1959.

Finally, a great help in the understanding of algebraic geometry is familiarity with the works of the classical, above all the Italian geometers. Of the vast literature I only mention a few works that are least specialized: F. Enriques and O. Chisini, "Lezioni sulla teoria geometrica delle equazione e delle funzione algebriche, 3 vols., Bologna 1915–1924; G. Castelnuovo and F. Enriques, "Die algebraischen Flächen vom Gesichtspunkte der birationalen Transformationen aus", Enzykl. d. math. Wiss., III, 3; F. Severi "Vorlesungen über algebraische Geometrie", Leipzig 1921; O. Zariski, "Algebraic surfaces", second ed. Springer-Verlag, Berlin-Heidelberg-New York 1971 (the basic text of the book contains an account of the classical papers, and the appendix their translation into the language of present-day concepts).

Part I. Algebraic Varieties in a Projective Space

Chapter I. Fundamental Concepts

§ 1. Plane Algebraic Curves

The first chapter is concerned with a number of fundamental concepts of algebraic geometry. In the first section we analyse some examples, which prepare us for the introduction of these concepts.

1. Rational Curves. The curve given by the equation

$$y^2 = x^2 + x^3 \tag{1}$$

has the property that the coordinates of its points can be expressed as rational functions of a parameter. To derive this expression we observe that the line $y = tx$ through the origin of coordinates intersects the curve (1), apart from the origin, in a single point. For by substituting $y = tx$ in (1) we obtain $x^2(t^2 - x - 1) = 0$. The root $x = 0$ corresponds to the point $O = (0, 0)$. Apart from this we have one other root $x = t^2 - 1$. From the equation of the line we find that $y = t(t^2 - 1)$. Thus, we have the required parameterization

$$x = t^2 - 1, \quad y = t(t^2 - 1), \tag{2}$$

and we have also clarified its geometric meaning: t is the slope of the line passing through the points (x, y) and O, and x and y corresponding to t are the coordinates of the point of intersection, other than O, of the line $y = tx$ and the curve (1). We can represent this parameterization even more intuitively by drawing any line not passing through O (for example, the line with the equation $x = 1$) and associating with a point P the point of intersection Q of the line OP with the chosen line (projection of the curve from O) (Fig. 1). Here the coordinate on the chosen line plays the role of the parameter t. Both from this geometric interpretation and from (2) it is clear that the parameter t is uniquely determined (for $x \neq 0$) by the point (x, y).

Now we give a general definition of plane algebraic curves for which such a representation is possible. As a preliminary we introduce some concepts. We fix a field k. Henceforth we mean by points the points of the (x, y)-plane whose coordinates belong to k.

A plane algebraic curve is the set of all points whose coordinates satisfy an equation

$$f(x, y)=0, \tag{3}$$

where $f(x, y)$ is a polynomial with coefficients in k.

In what follows we assume that k is algebraically closed. This is due to the fact that otherwise an algebraic curve could have too few points. For example, if we restrict ourselves to points with real coordinates, we would have to say that the equations $x^2+2y^2=0$

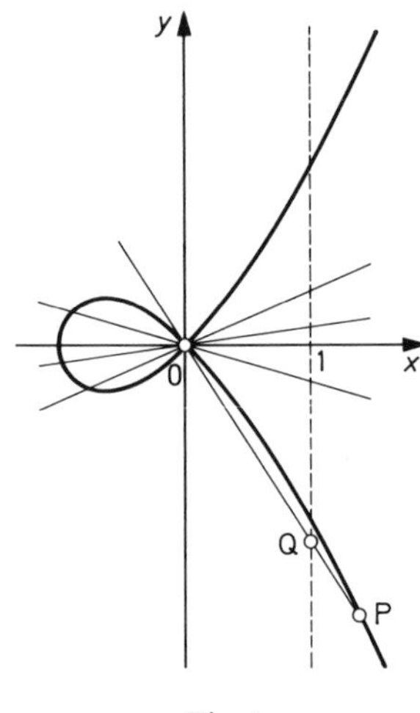

Fig. 1

and $x^4+y^4=0$ determine one and the same "curve"–the origin of coordinates. By considering points with complex coordinates we obtain two distinct curves. This does not mean that we exclude from the discussion points whose coordinates lie in a field k that is not algebraically closed. On the contrary, many problems lead to this situation.

Here are some examples.

If k is the field of real numbers, we come to the "ordinary" geometry of algebraic curves. If k is the field of rational numbers, then the problem of finding all the points of (3), that is, all the solutions in rational numbers of the corresponding equations, is one of the basic problems in the theory of Diophantine equations.

If k is the field with p elements (p a prime number), then we have the problem of solving the congruence $f(x, y)\equiv 0 \pmod p$.

In all these cases the investigation of points whose coordinates lie in the algebraic closure of the corresponding field k turns out to be very important.

If the polynomial $f(x, y)$ splits into two factors: $f = gh$, then the curve defined by it is the union of the two curves defined by the equations $g(x, y) = 0$ and $h(x, y) = 0$, respectively. If f is irreducible, then the curve defined by it is also called irreducible. Since every polynomial is a product of irreducible factors, every plane curve is the union of finitely many irreducible curves.

An irreducible plane algebraic curve X defined by the equation $f(x, y) = 0$ is said to be rational if there exist two rational functions $\varphi(t)$ and $\psi(t)$, of which at least one is not constant, such that

$$f(\varphi(t), \psi(t)) = 0 \tag{4}$$

identically in t. Clearly, if $t = t_0$ is a value of the parameter different from the finite set of values for which the denominators of φ and ψ vanish, then the point $(\varphi(t_0), \psi(t_0))$ lies on X. Later we shall show that for a suitably chosen parametrization φ, ψ this correspondence between values of the parameter t and points of a curve is one-to-one if certain finite sets are excluded from the values of the parameter as well as the points of the curve. Here the parameter t can be chosen as a rational function $\chi(x, y)$ of the coordinates x and y. If the coefficients of the rational functions φ and ψ belong to a subfield k_0 of k and $t_0 \in k_0$, then the coordinates of the point $(\varphi(t_0), \psi(t_0))$ also belong to k_0. This circumstance points to one of the possible applications of the idea of a rational curve. Suppose that the polynomial $f(x, y)$ has rational coefficients. If we know that the curve (3) is rational and that the coefficients of φ and ψ lie in the field of rational numbers, then the parametrization $x = \varphi(t)$, $y = \psi(t)$ gives us all the rational points of the curve, with the possible exception of finitely many, when t ranges over all rational numbers. For example, all the solutions of the indeterminate equation (1) can be obtained from (2) when t ranges over all the rational values.

Another application of rational curves is connected with the integral calculus. We assume that the Eq. (3) of a rational curve determines y as an algebraic function of x. Then any rational function $g(x, y)$ is a (compound) function of x. The rationality of the curve (3) implies the following important fact: for every rational function $g(x, y)$ the indefinite integral

$$\int g(x, y)\, dx \tag{5}$$

can be expressed in terms of elementary functions. For since the curve (3) is rational, it admits a parametrization $x = \varphi(t)$, $y = \psi(t)$, where φ and ψ are rational functions. Substituting this expression in the integral (5) we reduce it to the form $\int g(\varphi(t), \psi(t)) \varphi'(t)\, dt$, which is the integral of a rational function. Such an integral is known to be expressible in terms of elementary functions. Substituting the expression $t=\chi(x, y)$ of the parameter in terms of the coordinates, we obtain an expression of the integral (5) by elementary functions of the coordinates.

We now give some examples of rational curves. Curves of order 1, that is, straight lines, are, of course, rational.

Let us show that an irreducible curve of X of order 2 is rational. We take a point (x_0, y_0) on X and consider the line through (x_0, y_0) with the slope t. Its equation is of the form

$$y-y_0=t(x-x_0)\,. \tag{6}$$

Let us find the points of intersection of the curve with this line. It is sufficient to substitute the expression for y obtained from (6) in the equation of X. So we obtain an equation for x

$$f(x, y_0+t(x-x_0))=0\,, \tag{7}$$

which is easily seen to be of degree 2. We know one of the roots of the quadratic equation, namely $x=x_0$, because the point (x_0, y_0) is assumed to lie on the curve. We denote by A the coefficient of x in the equation obtained after dividing (7) by the coefficient of x^2. Then for the remaining root we have $x+x_0=-A$, $x=-x_0-A$. Since t occurs in the coefficients of (7), A is a rational function of t. Substituting this expression for x in (6) we obtain also for y an expression in the form of a rational function of t. These expressions, as is clear from the trend of the argument, satisfy the equation of the curve and hence show that the curve is rational.

This parametrization has an obvious geometric meaning: the point (x, y) is associated with the slope of the line joining it to the point (x_0, y_0), and the parameter t with the point of intersection of the curve with the line passing through (x_0, y_0) and having the slope t. This point is uniquely determined, because we are concerned with an irreducible curve of the second order. Just as we have done in connection with (1), we can interpret this parametrization as the projection of X from the point (x_0, y_0) onto a line that does not pass through this point (Fig. 2).

Observe that in the construction of the parametrization we have used the point (x_0, y_0) on X. If the coefficients of the polynomial $f(x, y)$ and the coordinates x_0, y_0 of this point belong to some subfield k_0 of k, then so do the coefficients of the functions giving the parametrization. For example, we can find the general form of the solution in rational numbers of a Diophantine equation of degree 2 if we know at least one solution.

The problem of the existence of at least one solution is fairly delicate. It is solved by the so-called Legendre theorem (see, for example, [7], Ch. I, § 7.2).

Now we consider another application of our parametrization. As we have seen, the equation $y^2 = ax^2 + bx + c$ of degree 2 defines a rational curve. From this it follows that, no matter what the rational function $g(x, y)$ is, the integral $\int g(x, \sqrt{ax^2 + bx + c})dx$ can be expressed in terms of elementary functions. Our parametrization gives an explicit form of a substitution that reduces this integral to one of a

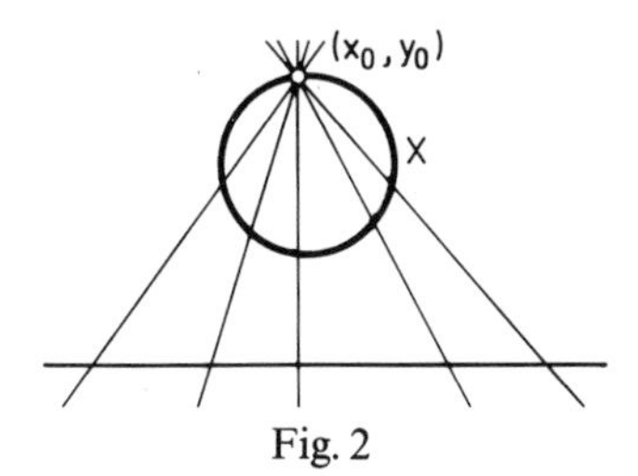

Fig. 2

rational function. It is easy to check that we arrive in this way at Euler's well known substitution.

Now we pass to curves of the third order. At the beginning of this section we have already had an example of a rational curve of order 3. We show next that there exist also non-rational curves of order 3. An example is the curve with the equation $x^3 + y^3 = 1$. This follows from the more general result that the curve with the equation

$$x^n + y^n = 1 \tag{8}$$

is non-rational for $n > 2$, if n is not divisible by the characteristic p of k.

Suppose that the curve (8) is rational and that $x = \varphi(t)$, $y = \psi(t)$ is a parametrization of it. We write the rational functions φ and ψ in the form

$$\varphi(t) = \frac{p(t)}{r(t)}, \qquad \psi(t) = \frac{q(t)}{r(t)},$$

where p, q and r are polynomials, which we may assume to be relatively prime. By hypothesis, the relation

$$(p(t))^n + (q(t))^n - (r(t))^n = 0 \tag{9}$$

must hold identically. Differentiating it and dividing the result by n (which is possible, because n is not divisible by the characteristic of k), we obtain the relation

$$p(t)^{n-1}\cdot p'(t)+q(t)^{n-1}\cdot q'(t)-r(t)^{n-1}\cdot r'(t)=0\,. \tag{10}$$

We consider (9) and (10) as a system of linear equations in p^{n-1}, q^{n-1}, $-r^{n-1}$, with the matrix

$$\begin{pmatrix} p & q & r \\ p' & q' & r' \end{pmatrix}.$$

Solving this system by the usual method we see that p^{n-1}, q^{n-1}, and $-r^{n-1}$ are proportional to $qr'-rq'$, $rp'-pr'$, and $pq'-qp'$, respectively. Since p, q and r are relatively prime, it follows that

$$p^{n-1}|(qr'-rq'),\ q^{n-1}|(rp'-pr'),\ r^{n-1}|(pq'-qp')\,.$$

We denote the degrees of p, q, and r by a, b, and c, with $a\geqslant b\geqslant c$, say. Then the first relation gives $(n-1)\,a\leqslant b+c-1$, which together with $b\leqslant a$, $c\leqslant a$, $n\geqslant 3$ leads to a contradiction.

These examples lead us to the following question. How can we recognize whether a given plane algebraic curve is rational? As we shall see later, this question is connected with rather subtle concepts of algebraic geometry.

2. Connections with the Theory of Fields. We now show that the question raised just above can be stated as a question in the theory of fields. For this purpose we associate with every irreducible plane algebraic curve a certain field, just as a field is associated with every polynomial, namely the least extension in which the polynomial has a root.

Let X be an irreducible curve, given by the equation $f(x,y)=0$, as in § 1.1 (3). We consider rational functions $u(x,y)=p(x,y)/q(x,y)$ (where p and q are polynomials with coefficients in k) such that $q(x,y)$ is not divisible by $f(x,y)$. We say that such functions are defined on X. Two functions $p(x,y)/q(x,y)$ and $p_1(x,y)/q_1(x,y)$ defined on X are said to be equal on X if the polynomial $p(x,y)q_1(x,y)-q(x,y)p_1(x,y)$ is divisible by $f(x,y)$. It is easy to verify that rational functions, considered to within equality on X, form a field. This is called the field of rational functions on X and is denoted by $k(X)$.

Obviously all the elements of $k(X)$ can be expressed as rational functions of x and y. Here x and y are algebraically dependent: they are connected by the relation $f(x,y)=0$. Starting out from this fact it is easy to check that the transcendence degree of $k(X)$ is 1.

If X is a line, for example $y=0$, then every rational function $\varphi(x,y)$ is equal on X to the rational function $\varphi(x,0)$ of the single variable x, therefore the field of rational functions on a line is the same as the field of rational functions of a single variable $x: k(X)=k(x)$.

Suppose now that X is rational and has the parametrization $x=\varphi(t)$, $y=\psi(t)$. With any rational function $u=p(x,y)/q(x,y)$ we associate the rational function $u(\varphi(t), \psi(t))$ of t obtained by substituting φ and ψ for x and y. First of all, let us verify that this substitution makes sense, in other words, that the denominator $q(\varphi(t), \psi(t))$ gives a function of t that is not identically zero. Suppose that $q(\varphi(t), \psi(t)) \equiv 0$. We compare this equation with § 1.1 (4). By giving t various values in k we see that the equations $f(x,y)=0$, $q(x,y)=0$ have infinitely many solutions in common. (We must recall that the field is algebraically closed, hence infinite.) But this is possible only when the polynomials f and q have a common factor, a simple result of elimination theory (see [33], §27).

Thus, our substitution gives a definite result for any function $u(x,y)$ that is defined on X. Furthermore, since φ and ψ satisfy § 1.1 (4), two functions u and u_1 that are equal on X give, after substitution, identical rational functions of t. Thus, to every element of $k(X)$ there corresponds a definite element of $k(t)$. This correspondence is, of course, an isomorphism of $k(X)$ with a subfield of $k(t)$. This isomorphism leaves the elements of k fixed.

At this place we make use of a theorem on rational functions, which is called Lüroth's theorem. It states that every subfield of the field of rational functions $k(t)$ containing k is of the form $k(g(t))$, where $g(t)$ is some rational function, in other words, that the subfield consists of all rational functions of $g(t)$. If the function $g(t)$ is not a constant, then the correspondence $f(u) \to f(g(t))$ defines evidently an isomorphism of the fields of rational functions $k(u)$ and $k(g(t))$. Therefore Lüroth's theorem can be put in the following form: a subfield of the field of rational functions $k(t)$ containing k, but different from k, is itself isomorphic to a field of rational functions. Lüroth's theorem can be proved by starting out from simple properties of extensions of fields (see [33], § 63). Applying Lüroth's theorem to our situation we see that if the curve X is rational, then the field $k(X)$ is isomorphic to the field of rational functions $k(t)$. Suppose, conversely, that for some curve X given by the equation $f(x,y)=0$ the field $k(X)$ is isomorphic to the field of rational functions $k(t)$, and that under this isomorphism x and y correspond to $\varphi(t)$ and $\psi(t)$. Since the relation $f(x,y)=0$ holds in $k(X)$, it is preserved under the isomorphism and gives $f(\varphi(t), \psi(t))=0$, which means that the curve X is rational.

It is easy to see that any field $K \supset k$ of transcendence degree 1 over k and generated by two elements x and y is isomorphic to $k(X)$, where X is an irreducible plane algebraic curve. For since the transcendence degree of K over k is 1, x and y must be connected by an algebraic relation. If this is $f(x, y)=0$, with an irreducible polynomial f, then clearly X can be taken to be the algebraic curve defined by this equation. Hence it follows that the question on rational curves raised at the end of § 1.1 is equivalent to the following question in the theory of fields: when is a field $K \supset k$ of transcendence degree 1 over k and generated over k by two elements isomorphic to the field of rational functions $k(t)$ of single variable? The condition that K should be generated over k by two elements is rather unnatural from the algebraic point of view. It would be more natural to consider extensions generated by any finite number of elements. However, we shall show later that this would not lead to a more general concept.

In conclusion we mention that the preceding arguments enable us to solve the problem of the uniqueness of the parametrization of a rational curve. Let X be a rational curve. By Lüroth's theorem the field $k(X)$ is isomorphic to the field of rational functions $k(t)$. Suppose that under this isomorphism x and y correspond to $\varphi(t)$ and $\psi(t)$. Then we obtain a parametrization of $X: x=\varphi(t)$, $y=\psi(t)$. Let us show that this parametrization has the following properties:

1) every point $(x_0, y_0) \in X$, with finitely many possible exceptions, can be represented in the form $x_0=\varphi(t_0)$, $y_0=\psi(t_0)$ for some t_0;

2) for all points, with finitely many possible exceptions, this representation is unique.

Suppose that under the isomorphism $k(X) \to k(t)$ the function $\chi(x, y)$ goes over into t. Then the inverse isomorphism $k(t) \to k(X)$ is given by the formula $u(t) \to u(\chi(x, y))$. Bearing in mind that the two correspondences are inverses of one another, we arrive at the relations

$$x=\varphi(\chi(x, y)), \qquad y=\psi(\chi(x, y)), \tag{1}$$

$$t=\chi(\varphi(t), \psi(t)). \tag{2}$$

The first relation gives 1). For if $\chi(x, y)=p(x, y)/q(x, y)$, and $q(x_0, y_0) \neq 0$ (there are only finitely many points $(x_0, y_0) \in X$ for which $g(x_0, y_0)=0$ because the polynomials $q(x, y)$ and $f(x, y)$ are relatively prime), then we can consider the value $\chi(x_0, y_0)$. Suppose that the point (x_0, y_0) is such that $\chi(x_0, y_0)$ is different from the roots of the denominators of $\varphi(t)$ and $\psi(t)$ (for the same reason there are only finitely many points (x_0, y_0) for which this is not so). Then the formula (1) gives the required representation for the point (x_0, y_0). Similarly, from (2) it follows that the parameter value t if it exists, is uniquely determined by (x_0, y_0), with the possible exception of the finitely many points for which $q(x_0, y_0)=0$.

Observe that we have proved 1) and 2) not for any parametrization of a rational curve, but for a specially constructed one. For an arbitrary parametrization 2) need not be true: for example, the curve $y^2 = x^2 + x^3$ of § 1.1 (1) has, apart from the parametrization given by § 1.1 (2), also the parametrization $x = t^4 - 1$, $y = t^2(t^4 - 1)$, which is obtained from the first by replacing t by t^2. Obviously the parameter values t and $-t$ now correspond to one and the same point on the curve.

3. Birational Isomorphism of Curves. If a plane algebraic curve is not rational, the coordinates of its points can nevertheless often be expressed in terms of the coordinates of the points of another, possibly simpler, curve. Consider, for example, a curve of the form

$$y^2 = f(x), \tag{1}$$

where $f(x)$ is a polynomial of even degree $2n$, and set $f(x) = g(x)(x - \alpha)$, where $g(x)$ is of degree $2n - 1$. We divide both sides of (1) by $(x - \alpha)^{2n}$ and set

$$x - \alpha = u^{-1}, \qquad \frac{y}{(x - \alpha)^n} = v. \tag{2}$$

Then $g(x)/(x - \alpha)^{2n-1} = f_1(u)$, where f_1 is of degree not greater than $2n - 1$, and (1) takes the form

$$v^2 = f_1(u). \tag{3}$$

It is clear that, conversely, the coordinates of a point of (1) can be expressed rationally in terms of coordinates of a point of (3). For it follows from (2) that

$$x = u^{-1} + \alpha, \qquad y = vu^{-n}. \tag{4}$$

The transformations (2) and (4) are inverses of one another. Thus, the curves (1) and (3) are transformed into each other, and this can sometimes turn out to be useful, because the degree of f_1 is smaller by at least 1 than that of f.

Here we come across a new type of connection that may exist between algebraic curves. The notion of a rational curve is included in this more general concept: the formulae (1) and (2) at the end of § 1.2 can be interpreted as saying that the curve X is transformed into the line $s = 0$ in the (s, t)-plane.

Here are the precise definitions. Let X and Y be irreducible curves given by the equations $f(x, y) = 0$ and $g(u, v) = 0$, respectively. A rational mapping of X into Y is defined as a pair of rational functions $\varphi(x, y)$ and $\psi(x, y)$ defined on X such that the function $g(\varphi(x, y), \psi(x, y))$ vanishes on X. It is easy to verify on the basis of the argument in § 1.2 that for all points $(x_0, y_0) \in X$, except finitely many, the values $\varphi(x_0, y_0)$ and $\psi(x_0, y_0)$ are defined and that $(\varphi(x_0, y_0), \psi(x_0, y_0)) \in Y$.

Two curves X and Y are called *birationally isomorphic* if there exist rational mappings of X into Y and of Y into X that are inverse to one another. In other words, there must exist a mapping of X into Y given by functions $\varphi(x, y)$ and $\psi(x, y)$, and a mapping of Y into X given by functions $\xi(u, v)$ and $\eta(u, v)$, such that

$$\left.\begin{aligned}\xi(\varphi(x, y), \psi(x, y)) &= x, \\ \eta(\varphi(x, y), \psi(x, y)) &= y\end{aligned}\right\} \quad \text{on } X,$$

$$\left.\begin{aligned}\varphi(\xi(u, v), \eta(u, v)) &= u, \\ \psi(\xi(u, v), \eta(u, v)) &= v\end{aligned}\right\} \quad \text{on } Y.$$

For example, the curves (1) and (3) are birationally isomorphic, and (2) and (4) are the corresponding mappings.

So we arrive at one of the central problems of algebraic geometry: how to classify plane algebraic curves to within birational isomorphism. Even today we can hardly claim that there is an exhaustive solution of this problem. Nevertheless, a number of strong results pointing in the direction of a solution will be proved later.

Clearly, from the algebraic point of view the simplest type of algebraic curves are curves birationally isomorphic to a line, that is, rational curves. Having constructed in § 1.2 an example of a non-rational curve we have shown incidentally that the solution of the problem raised above is not trivial: not all curves are birationally equivalent to each other.

Let X and Y be two irreducible birationally isomorphic plane algebraic curves and suppose that the mappings of one into the other are given by the formulae

$$(u, v) = (\varphi(x, y), \psi(x, y)), \qquad (x, y) = (\xi(u, v), \eta(u, v)).$$

As in the investigation of rational curves, we can establish a connection between the fields of rational functions $k(X)$ and $k(Y)$ on these curves. For this purpose we associate with any rational function $w(x, y)$ defined on X the function $w(\xi(u, v), \eta(u, v))$ on Y. It is easy to verify that in this way we obtain a mapping of $k(X)$ into $k(Y)$, which is an isomorphism of these fields. Conversely, if the fields $k(X)$ and $k(Y)$ are isomorphic, then to functions $x, y \in k(X)$ there must correspond under this isomorphism functions $\xi(u, v), \eta(u, v) \in k(Y)$, and to functions $u, v \in k(Y)$ functions $\varphi(x, y), \psi(x, y) \in k(X)$, and again a trivial verification shows that the pairs of functions φ, ψ and ξ, η determine a birational isomorphism of the curves X and Y. Thus, two curves are birationally equivalent if and only if their fields of rational functions are isomorphic.

We see that the problem of classifying algebraic curves to within birational isomorphism is the geometrical aspect of the natural algebraic problem of classifying (to within isomorphism) extensions of k, of transcendence degree 1 and generated by finitely many elements.

In this problem it is natural not to confine our attention to fields of transcendence degree 1, but to consider fields of arbitrary finite transcendence degree. We shall see later that this wider statement of the problem also has a geometrical interpretation. However, here we have to go beyond the limits of the theory of algebraic curves and have to consider algebraic varieties of arbitrary dimension.

Exercises

1. Calculate the area of the loop of the curve (1) in § 1.1, the so-called *folium Cartesii.*

2. Show that the lemniscate, defined by the equation $(x^2+y^2)^2=a^2(x^2-y^2)$, is rational. *Hint:* consider the points of intersection of the lemniscate with the pencil of circles $x^2+y^2=t(x-y)$.

3. Show that the curve of order 3 given by the equation $y^2=x^3+Ax+B$ is rational if and only if the polynomial x^3+Ax+B has multiple roots (the characteristic of the ground field is not equal to 2).

4. Find a rational parametrization of the circle $x^2+y^2=1$.

5. Find the general form of the solution of the equation in Exercise 4 in rational numbers, and derive from it the well-known formulae for Pythagorean numbers (see, for example, [32] Ch. 1, Exercise 9a) that is, the solutions of the equation $x^2+y^2=z^2$ in integers.

6. Show that the field of trigonometric functions, that is, the field of all rational functions of $\sin\varphi$ and $\cos\varphi$, is isomorphic to a field of rational functions. In particular, every trigonometric equation can be transformed into an algebraic equation.

7. Show that the curve given in polar coordinates by the equation $r=\sin 3\varphi$, and in Cartesian coordinates by $(x^2+y^2)^2=y(3x^2-y^2)$, is rational.

8. Given $2n$ numbers $\alpha_i (i=1, \ldots, 2n)$, $\alpha_i \neq 0$, we set $\alpha_i+\alpha_i^{-1}=a_i$. Show that the functions $u=x+x^{-1}$, $v=y/x^n$ determine a mapping of the curve

$$y^2=\prod_{i=1}^{2n}(x-\alpha_i)(x-\alpha_i^{-1})$$

into the curve

$$v^2=\prod_{i=1}^{2n}(u-a_i).$$

Is this mapping a birational isomorphism?

9. Prove that the curve given by the equation $f_{n-1}(x,y)+f_n(x,y)=0$ is rational if it is irreducible. Here f_{n-1} and f_n denote homogeneous polynomials of degree $n-1$ and n, respectively.

10. Show that the formulae $u=x/(1-y)$, $v=\sqrt{3}(1+y)/(1-y)$ determine a rational mapping of the curve $x^3+y^3=1$ into the curve $v^2=4u^3-1$. Is this mapping a birational isomorphism? The characteristic of the ground field is not equal to 2 or 3.

11. Prove that every curve of the third order is birationally equivalent to the curve in § 1.3 (1) with $n=2$. Here the coefficients of a parametrizing function belong to the field k_0

if the curve contains a point x with coordinates in k_0. *Hint:* draw the pencil of lines through x and consider the pairs of points of intersection other than x of the lines of the pencil with the curve. The characteristic of the ground field is not equal to 2.

12. Show that a curve of order 3 containing a point y with coordinates in k_0 is birationally isomorphic to a curve § 1.3 (3) with $n=2$ and that the coefficients of the parametrization formulae belong to k_0. *Hint:* draw the tangent at y and apply the construction of Exercise 11 to its point of intersection, other than y, with the curve. Show that the arguments at the beginning of § 1.3 are applicable to the curve so obtained.

§ 2. Closed Subsets of Affine Spaces

Throughout what follows we are concerned with one and the same algebraically closed field k, which we call the ground field.

1. Definition of Closed Subset. At various stages of the development of algebraic geometry the idea of what its basic object is—"the natural notion of an algebraic variety"—has changed. There have been projective and quasiprojective varieties, abstract algebraic varieties, schemes, algebraic spaces.

In this book algebraic geometry is treated in gradually increasing generality. In the first chapters the most general concept, comprising all the algebraic varieties to be studied there, is that of a quasiprojective variety. In the later chapters this role is played by *schemes.* We now define one class of algebraic varieties, which plays a fundamental role in all the subsequent definitions. Since the word variety is reserved for the more general concepts, we use another term.

We denote by $\mathbb{A}^n$ the n-dimensional affine space over k. Its points are therefore of the form $\alpha=(\alpha_1, \ldots, \alpha_n)$, $\alpha_i \in k$.

Definition. A *closed subset in* $\mathbb{A}^n$ is a subset $X \subset \mathbb{A}^n$ consisting of all common zeros of finitely many polynomials with coefficients in k. Occasionally we speak briefly of a *closed subset.*

Henceforth we write a polynomial in n variables $T_1, \ldots, T_n$ in the form $F(T)$, understanding by T the set of variables $T_1, \ldots, T_n$. If a closed subset X consists of all the common zeros of the polynomials $F_1(T), \ldots, F_m(T)$ then we call $F_1(T)=\cdots=F_m(T)=0$ the *equations of* X.

A set X defined by an infinite system of equations $F_\lambda(T)=0$ is also closed. For, the ideal $\mathfrak{A}$ in the ring of polynomials in $T_1, \ldots, T_n$ generated by all the polynomials $F_\lambda(T)$ has a finite basis: $\mathfrak{A}=(G_1, \ldots, G_m)$. It is easy to see that X is defined by the system of equations $G_1=\cdots=G_m=0$.

Hence it follows that the intersection of any number of closed sets is closed. For if the X_λ are closed, then to obtain a system of equations defining $X=\bigcap X_\lambda$ it is sufficient to take the union of the systems defining all the X_λ.

The union of finitely many closed sets is also closed. Clearly, it is sufficient to verify this for the case of two sets. If $X=X_1\cup X_2$, and if X_1 is determined by the system of equations $F_i(T)=0$ $(i=1,\ldots,m)$ and X_2 by the system $G_j(T)=0$ $(j=1,\ldots,l)$, then X, as is easy to verify, is determined by the system $F_i(T)\,G_j(T)=0\,(i=1,\ldots,m; j=1,\ldots,l)$.

Let X be a closed subset of an affine space. A set $U\subset X$ is called *open* if its complement $X-U$ is closed. Any open set U containing x is called a neighbourhood of x. The intersection of all closed subsets of X containing a given subset $M\subset X$ is closed. It is called the *closure* of M and is denoted by $\bar{M}$. A subset M is called *dense* in X if $\bar{M}=X$. This means that M is not contained in any closed subset $Y\subset X$, $Y\neq X$.

Example 1. The whole affine space $\mathbb{A}^n$ is closed: it is given by the empty set of equations, or by the equation $0=0$.

Example 2. The subset $X\subset\mathbb{A}^1$ consisting of all the points of $\mathbb{A}^1$ other than zero is not closed: every polynomial $F(T)$ that vanishes for all $T\neq 0$ must vanish identically.

Example 3. Let us determine all the closed subsets $X\subset\mathbb{A}^1$. Such a set is given by a system of equations $F_1(T)=0,\ldots,F_m(T)=0$ in a single variable T. If all the F_i vanish identically, then $X=\mathbb{A}^1$. If the polynomials $F_i(T)$ are relatively prime, then they have no common roots, and X has no points. But if all these polynomials have the greatest common divisor $D(T)$, then $D(T)=(T-\alpha_1)\cdots(T-\alpha_n)$, and X consists of the finite set of points $T=\alpha_1,\ldots,T=\alpha_n$.

Example 4. Let us determine the closed subsets $X\subset\mathbb{A}^2$. They are given by a system of equations

$$F_1(T)=0,\ldots,F_m(T)=0, \tag{1}$$

where now $T=(T_1,T_2)$. If all the F_i vanish identically, then $X=\mathbb{A}^2$. Suppose that this is not so. If the polynomials $F_1,\ldots,F_m$ have no common divisor, then, as we have shown in § 1.2, the system (1) has only a finite (possibly empty) set of solutions. Suppose, finally, that all the polynomials $F_i(T)$ have a greatest common divisor $D(T)$. Then $F_i(T)=D(T)\,G_i(T)$, where now the polynomials $G_i(T)$ have no common divisor. Evidently $X=X_1\cup X_2$, where X_1 is given by the system of equations $G_1=\cdots=G_m=0$, and X_2 by the single equation $D=0$. As we have seen, X_1 is a finite set of points. The closed sets given in $\mathbb{A}^2$ by one equation are plane algebraic curves. Thus, a closed set $X\subset\mathbb{A}^2$ either consists of a finite (possibly empty) set of points, or it is the union of a plane algebraic curve and a finite set of points, or it is the whole of $\mathbb{A}^2$.

Example 5. With a point $\alpha \in \mathbb{A}^r$ with the coordinates $(\alpha_1, \ldots, \alpha_r)$ and a point $\beta \in \mathbb{A}^s$ with the coordinates $(\beta_1, \ldots, \beta_s)$ we associate the point $(\alpha, \beta) \in \mathbb{A}^{r+s}$ with the coordinates $(\alpha_1, \ldots, \alpha_r, \beta_1, \ldots, \beta_s)$. Thus, $\mathbb{A}^{r+s}$ is identified with the set of pairs (α, β), $\alpha \in \mathbb{A}^r$, $\beta \in \mathbb{A}^s$. Let $X \subset \mathbb{A}^r$ and $Y \subset \mathbb{A}^s$ be closed sets. The set of pairs $(x, y) \in \mathbb{A}^{r+s}$, $x \in X$, $y \in Y$, is called the *product* of X and Y and is denoted by $X \times Y$. This is also a closed set. For if X is given by the equations $F_i(T) = 0$ and Y by the equations $G_j(U) = 0$, then $X \times Y$ is given in $\mathbb{A}^{r+s}$ by the equations $F_i(T) = 0, G_j(U) = 0$.

Example 6. A set $X \subset \mathbb{A}^n$ given by a single equation $F(T_1, \ldots, T_n) = 0$ is called a *hypersurface*.

2. Regular Functions on a Closed Set. Let X be a closed set in an affine space $\mathbb{A}^n$, and let k be the ground field.

Definition. A function f given on X and taking values in k is called *regular* if there exists a polynomial $F(T)$ with coefficients in k such that $f(x) = F(x)$ for all points $x \in X$.

For a given function f the polynomial F is, in general, not uniquely determined. For example, without changing f we can add to it any polynomial that occurs in the system of equations for X.

The collection of regular functions on a given closed set X forms a ring and even an algebra over k if the operations of addition, multiplication, and multiplication of elements by k are defined as in analysis, namely by means of the same operations on the values at each point $x \in X$. The ring so obtained is denoted by $k[X]$ and is called the coordinate ring of the closed set X.

We denote by $k[T]$ the ring of polynomials with coefficients in k in the variables $T_1, \ldots, T_n$. Obviously we can associate with every polynomial $F \in k[T]$ a function $f \in k[X]$, regarding F as a function on the set of points X. Thus, we obtain a ring homomorphism of $k[T]$ onto $k[X]$. The kernel of this homomorphism consists of all polynomials $F \in k[T]$ that vanish at all the points $x \in X$. Like every kernel of a homomorphism, this set is an ideal of the ring $k[T]$. It is called the ideal of X and is denoted by $\mathfrak{A}_X$. Clearly

$$k[X] = k[T]/\mathfrak{A}_X .$$

Thus, the ring $k[X]$ is determined by the ideal $\mathfrak{A}_X$.

Example 1. If X is a point, then $k[X] = k$.

Example 2. If $X = \mathbb{A}^n$, then $\mathfrak{A}_X = 0$, and $k[X] = k[T]$.

Example 3. Suppose that $X \subset \mathbb{A}^2$ is given by the equation $T_1 T_2 = 1$. Then $k[X] = k[T_1, T_1^{-1}]$ consists of all rational functions of T_1 of the form $G(T_1)/T_1^n$, where $n \geqslant 0$ and $G(T_1)$ is a polynomial.

Example 4. Let us show that $k[X \times Y] = k[X] \otimes_k k[Y]$ for any closed set X and Y. We define a homomorphism

$$\varphi : k[X] \otimes_k k[Y] \to k[X \times Y]$$

by the condition

$$\varphi\Big(\sum_i f_i \otimes g_i\Big)(x, y) = \sum_i f_i(x)\, g_i(y).$$

It is clear that in this way we do, in fact, obtain a regular function on the set $X \times Y$ and that φ is an epimorphism, because the functions α_i and β_j (in the notation of § 2.1, Example 5) belong to its image and generate the whole ring $k[X \times Y]$. To prove that it is a monomorphism it is sufficient to verify that if $\{f_i\}$ are linearly independent over k in $k[X]$, and $\{g_j\}$ in $k[Y]$, then $\varphi(f_i \otimes g_j)$ are linearly independent in $k[X \times Y]$. The equation

$$\sum_{i,j} c_{ij} f_i(x)\, g_j(y) = 0$$

implies for any fixed y that $\sum_j c_{ij} g_j(y) = 0$, from which it follows that $c_{ij} = 0$.

Since the ring $k[X]$ is a homomorphic image of the polynomial ring $k[T]$, the Hilbert basis theorem for ideals holds in it. The following analogue of Hilbert's Nullstellensatz is also valid in it: if the functions $f \in k[X]$ vanish at all the points $x \in X$ at which the functions $g_1, \ldots, g_m$ vanish, then $f^r \in (g_1, \ldots, g_m)$ for some $r > 0$. For suppose that f is given by a polynomial $F(T)$, and g_i by polynomials $G_i(T)$, and that $F_j = 0$ $(j = 1, \ldots, l)$ are the equations of X. Then the polynomial $F(T)$ vanishes at all points $\alpha \in \mathbb{A}^n$ at which the polynomials $G_1, \ldots, G_m$, $F_1, \ldots, F_l$ vanish. For since $F_j(\alpha) = 0$, we have $\alpha \in X$, and then $F(\alpha) = 0$ by hypothesis. Applying Hilbert's theorem to the ring of polynomials we see that $F^r \in (G_1, \ldots, G_m, F_1, \ldots, F_l)$ and hence that $f^r \in (g_1, \ldots, g_m)$ in $k[X]$.

How is the ideal $\mathfrak{A}_X$ of a closed set X connected with the system of equations $F_1 = \cdots = F_m = 0$ of this set? We know that $F_i \in \mathfrak{A}_X$ by definition of $\mathfrak{A}_X$, therefore $(F_1, \ldots, F_m) \subset \mathfrak{A}_X$. However, we do not always have $(F_1, \ldots, F_m) = \mathfrak{A}_X$. For example, if $X \subset \mathbb{A}_1$ is given by the equation $T^2 = 0$, that is, if it consists of the point $T = 0$, then $\mathfrak{A}_X$ consists of the polynomials without a constant term. Thus, $\mathfrak{A}_X = (T)$, but $(F_1, \ldots, F_m) = (T^2)$. However, we can always give the same set by a system of equations $G_1 = \cdots = G_l = 0$ such that $(G_1, \ldots, G_l) = \mathfrak{A}_X$. To see this it is sufficient to recall that every ideal in the ring $k[T]$ has a finite basis. Let $G_1, \ldots, G_l$ be a basis of the ideal $\mathfrak{A}_X$, so that $\mathfrak{A}_X = (G_1, \ldots, G_l)$. Then obviously the equations $G_1 = \cdots = G_l = 0$ determines the same set X and have the required property. Occasionally it is convenient even to assume that a closed set is given by the infinite system of equations $F=0$, where F are all the polynomials of the ideal $\mathfrak{A}_X$. For if $(F_1, \ldots, F_m) = \mathfrak{A}_X$, then all these equations are consequences of the equations $F_1 = \cdots = F_m = 0$.

Relations between closed sets are often reflected in their ideals. For example, if X and Y are closed sets in the affine space $\mathbb{A}^n$, then $X \supset Y$ if and only if $\mathfrak{A}_X \subset \mathfrak{A}_Y$. Hence it follows that with every closed set Y contained in X we can associate the ideal $\mathfrak{a}_Y$ of the ring $k[X]$ consisting of the images of the polynomials $F \in \mathfrak{A}_Y$ under the homomorphism $k[T] \to k[X]$. Conversely, every ideal $\mathfrak{a}$ of $k[X]$ determines an ideal $\mathfrak{A}$ in $k[T]$: $\mathfrak{A}$ consists of all the inverse images of the elements of $\mathfrak{a}$ under the homomorphism $k[T] \to k[X]$. It is clear that $\mathfrak{A} \supset \mathfrak{A}_X$. The equations $F=0$, where F are all the polynomials in $\mathfrak{A}$, determine a closed set $Y \subset X$.

From Hilbert's Nullstellensatz it follows that Y is empty if and only if $\mathfrak{a}_Y = k[X]$. Otherwise the ideal $\mathfrak{a}_Y \subset k[X]$ could be written as the collection of all functions $f \in k[X]$ that vanish at all the points of Y.

In particular, every point $x \in X$ is a closed subset and hence determines an ideal $\mathfrak{m}_x \subset k[X]$. By definition this ideal is the kernel of the homomorphism $k[X] \to k$ that associates with every function $f \in k[X]$ its value at x. Since $k[X]/\mathfrak{m}_x$ is a field, the ideal $\mathfrak{m}_x$ is maximal. Conversely, every maximal ideal $\mathfrak{m} \subset k[X]$ corresponds to some point $x \in X$. For it determines a closed subset $Y \subset X$. For every point $y \in Y$ we have $\mathfrak{m} \subset \mathfrak{m}_y$, and since $\mathfrak{m}$ is a maximal ideal, we see that $\mathfrak{m} = \mathfrak{m}_y$. If $u \in k[X]$, then the set of points $x \in X$ at which $u(x)=0$ is closed. It is denoted by $V(u)$ and is called a *hypersurface* in X.

3. Regular Mappings. Let $X \subset \mathbb{A}^n$ and $Y \subset \mathbb{A}^m$ be closed sets.

Definition. A mapping $f: X \to Y$ is called *regular* if there exist m regular functions $f_1, \ldots, f_m$ on X such that $f(x) = (f_1(x), \ldots, f_m(x))$ for all $x \in X$.

Thus, any regular mapping $f: X \to \mathbb{A}^m$ is given by m functions $f_1, \ldots, f_m \in k[X]$. To verify that we are concerned with a mapping $f: X \to Y$ (Y being a closed subset of $\mathbb{A}^m$) it is clearly sufficient to check that the functions $f_1, \ldots, f_m$ as elements of the ring $k[X]$ satisfy the equations of Y.

Example 1. The notion of a regular function on X is the same as that of a regular mapping of X into $\mathbb{A}^1$.

Example 2. A linear transformation is a regular mapping.

Example 3. The projection $f(x, y) = x$ determines a regular mapping of the curve given by the equation $xy = 1$ into $\mathbb{A}^1$.

Example 4. The preceding example can be generalized as follows: let $X \subset \mathbb{A}^n$ be a closed set and F a regular function on X. Consider the set $X' \subset \mathbb{A}^{n+1}$ given by the equation $F_i(T_1, \ldots, T_n) = 0$, where $F_i = 0$ are the equations of X in $\mathbb{A}^n$ and $T_{n+1}F(T_1, \ldots, T_n) = 1$. The projection $\varphi(x_1, \ldots, x_{n+1}) = (x_1, \ldots, x_n)$ is a regular mapping $\varphi: X' \to X$.

Example 5. The mapping $f(t) = (t^2, t^3)$ is a regular mapping of the line $\mathbb{A}^1$ into the curve given by the equation $x^3 = y^2$.

Example 6. Here is an example of great importance in number theory. We assume that the coefficients of the equation $F_i(T)$ of the closed set $X \subset \mathbb{A}^n$ belong to the prime field $\mathbb{F}_p$ of p elements.

As we said in § 1.1, the points of X whose coordinates lie in F_p correspond to solutions of the system of congruences $F_i(T) \equiv 0 \pmod{p}$. We consider the mapping φ of the space $\mathbb{A}^n$, defined by the formulae

$$\varphi(\alpha_1, \ldots, \alpha_n) = (\alpha_1^p, \ldots, \alpha_n^p).$$

This is obviously a regular mapping. It is important that φ carries X into itself. For if $\alpha \in X$, that is, $F_i(\alpha) = 0$, then by a property of fields of characteristic p and the fact that $F_i(T) \in \mathbb{F}_p[T]$ we have $F_i(\alpha_1^p, \ldots, \alpha_n^p) = (F_i(\alpha_1, \ldots, \alpha_n))^p = 0$. The mapping $\varphi: X \to X$ so obtained is called a *Frobenius mapping*. Its significance lies in the fact that the points of X whose coordinates are contained in $\mathbb{F}_p$ are characterized among all the points of X as the fixed points of φ. For the equation $\alpha_i^p = \alpha_i$ has as its solutions precisely all the elements of $\mathbb{F}_p$.

Let us clarify how a regular mapping acts on the ring of regular functions on closed set. We begin with a remark that refers to arbitrary sets and mappings. If $f: X \to Y$ is a mapping of a set X into a set Y, then we can associate with every function u on Y (with values in an arbitrary set Z) a function v on X as follows: $v(x) = u(f(x))$. Clearly the mapping $v: X \to Z$ determined by v is the product of the mappings $u: Y \to Z$ and $f: X \to Y$. We denote the function v by $f^*(u)$. Thus, f^* is a mapping of the functions on Y into the functions on X. Suppose now that f is a regular mapping $X \to Y$. The mapping f^* carries regular functions on Y into regular functions on X. For if u is given by a polynomial $F(T_1, \ldots, T_n)$ and f by polynomials $F_1, \ldots, F_m$, then $v = f^*(u)$ is obtained simply by substituting F_i for T_i in F, that is, it is given by the polynomial $F(F_1, \ldots, F_m)$. Furthermore, regular mappings can be characterized as mappings carrying regular functions into regular functions. For suppose that a mapping of closed sets $f: X \to Y$ is such that for any function u regular on Y the function $f^*(u)$ is also regular. Then this is so, in particular, for the functions t_i defined by the coordinates $T_i (i=1, \ldots, m)$ on Y. Consequently, the functions $f^*(t_i)$ are regular on X. But this means that f is a regular mapping.

We have seen that if a mapping f is regular, then f^* is the mapping $f^*: k[Y] \to k[X]$. From the definition of this mapping it easily follows that f^* is an algebra homomorphism of $k[Y]$ into $k[X]$. Let us show that, conversely, every algebra homomorphism $\varphi: k[Y] \to k[X]$ has the form $\varphi = f^*$, where f is a regular mapping of X into Y. Let $t_1, \ldots, t_m$ be coordinates in the space $\mathbb{A}^m$ in which Y is contained, regarded as functions on Y. Obviously $t_i \in k[Y]$, and hence $\varphi(t_i) \in k[X]$. We set $\varphi(t_i) = s_i$ and consider the mapping f given by the formulae $f(x) = (s_1(x), \ldots, s_m(x))$. Of course, it is regular. We show that $f(x) \in Y$. For if $H \subset \mathfrak{A}_Y$, then $H(t_1, \ldots, t_m) = 0$ in $k[Y]$, and hence also $\varphi(H) = 0$ on X. Let $x \in X$. Then $H(f(x)) = \varphi(H)(x) = 0$, and this means that $f(x) \in Y$.

Definition. A regular mapping $f: X \to Y$ of closed sets is called an *isomorphism* if it has an inverse, in other words, if there exists a regular mapping $g: Y \to X$, such that $fg = 1$ $gf = 1$.

In this case the closed sets X and Y are also called isomorphic. Obviously, an isomorphism is a one-to-one mapping.

From what we have said above it follows that if f is an isomorphism, then f^* is an algebra isomorphism between $k[X]$ and $k[Y]$. It is easy to check that the converse is also true, so that closed sets are isomorphic if and only if their rings of regular functions are isomorphic over k.

The facts just proved show that the correspondence $X \to k[X]$ determines an equivalence of the category of closed subsets of affine spaces, (and their regular mappings) and a certain subcategory of the category of commutative algebras over k (and their homomorphisms). What this category is, in other words, what algebras are of the form $k[X]$, is clarified in Exercises 1 and 2. (See also Theorem 5 in § 3.)

Example 7. The parabola given by the equation $y = x^l$ is isomorphic to a line, and the mappings $f(x, y) = x$, $g(t) = (t, t^l)$ determine an isomorphism.

Example 8. The projection $f(x, y) = x$ of the hyperbola $xy = 1$ into the x-axis is not an isomorphism, because this mapping is not one-to-one: there are no points (x, y) on the hyperbola for which $f(x, y) = 0$. See also Exercise 7.

Example 9. The mapping $f(t) = (t^2, t^3)$ of a line onto the curve given by the equation $x^3 = y^2$ is easily seen to be one-to-one. However, it is not an isomorphism, because the inverse mapping is of the form $g(x, y) = y/x$, and the function y/x is not regular at the origin of coordinates (see Exercise 5).

Example 10. Let X and Y be closed subsets of $\mathbb{A}^r$. Consider $X \times Y \subset \mathbb{A}^{2r}$ (Exercise 5 in § 1.1) and the linear subspace $\Delta \subset \mathbb{A}^{2r}$ given by the equations $t_1 = u_1, \ldots, t_r = u_r$, the so-called diagonal. With every point $z \in X \cap Y$ we associate the point $\varphi(z) = (z, z) \in \mathbb{A}^{2r}$, which obviously belongs to $(X \times Y) \cap \Delta$. It is easy to check that the mapping $\varphi: X \cap Y \to (X \times Y) \cap \Delta$ so obtained determines an isomorphism between $X \cap Y$ and $(X \times Y) \cap \Delta$. Making use of this fact we can always reduce the study of the intersection of two closed sets to that of the intersection of another closed set with a linear subspace.

Later we shall be mainly interested in concepts and properties of closed sets that are invariant under an isomorphism. The system of equations by which the set is determined does not necessarily have this property: sets given in distinct spaces $\mathbb{A}^r$ by distinct systems of equations may be isomorphic. Therefore it would be natural to look for an invariant definition of a closed set, independent of its realization in some affine space. Such a definition will be given in Ch. V in connection with the notion of a scheme.

Now let us find out when the kernel of the homomorphism $f^*: k[Y] \to k[X]$ corresponding to a regular mapping $f: X \to Y$ is trivial, in other words, when f^* is an isomorphic embedding of $k[Y]$ in $k[X]$. Let us see when $f^*(u) = 0$ for $u \in k[Y]$. This means that $u(f(x)) = 0$ for all points $x \in X$. In other words, u vanishes on all points of the image $f(X)$ of X under the mapping f. The set of points $y \in Y$ for which $u(y) = 0$ is obviously closed; therefore, if it contains $f(X)$, then it also contains its closure $\overline{f(X)}$. Repeating the same arguments in the reverse order we see that $f^*(u) = 0$ if and only if $u = 0$ on $\overline{f(X)}$ or, what is the same, if $u \in \mathfrak{a}_{\overline{f(X)}}$. In particular, it follows that the kernel of the homomorphism f^* is 0 if and only if $\overline{f(X)} = Y$, that is, if $f(X)$ is dense in Y.

This is necessarily so if $f(X) = Y$, but it can happen that $f(X) \neq Y$, yet $\overline{f(X)} = Y$ (see Example 3).

Exercises

1. Let $\mathfrak{a}$ be an ideal of a ring A. The set of elements $a \in A$ for each of which there exists an integer n_a such that $a^{n_a} \equiv 0 \pmod{\mathfrak{a}}$ is called the *radical* of $\mathfrak{a}$ and is denoted by $\sqrt{\mathfrak{a}}$. Show that $\sqrt{\mathfrak{a}}$ is also an ideal. Show that an ideal $\mathfrak{a} \in k[T_1, \ldots, T_n]$ is an ideal of a closed subset of an affine space $\mathbb{A}^n$ if and only if $\sqrt{\mathfrak{a}} = \mathfrak{a}$.

2. Show that an algebra A over a field k is of the form $k[X]$, where X is a closed set, if and only if it is finitely generated over k and has no nilpotent elements (that is, from $a^n = 0$, $a \in A$, it follows that $a = 0$).

3. A set $X \subset \mathbb{A}^2$ is defined by the equations $f: x^2+y^2=1$ and $g: x=1$. Find the ideal $\mathfrak{A}_X$. Is $\mathfrak{A}_X=(f,g)$?

4. Let $X \subset \mathbb{A}^2$ be the plane algebraic curve defined by the equation $y^2=x^3$. Show that all the elements of the ring $k[X]$ can be written uniquely in the form $P(x)+Q(x)y$, where $P(x)$ and $Q(x)$ are polynomials.

5. Let X be the curve of Exercise 4, and let $f(t)=(t^2,t^3)$ be a regular mapping $\mathbb{A}^1 \to X$. Show that f is not an isomorphism. *Hint:* Use the result of Exercise 4 in trying to construct a regular inverse mapping.

6. Let X be the curve defined by the equation $y^2=x^2+x^3$, and f the mapping $\mathbb{A}^1 \to X$ defined by the formula $f(t)=(t^2-1, t(t^2-1))$. Show that the corresponding homomorphism f^* maps the ring $k[X]$ isomorphically onto the subring of the polynomial ring $k[t]$ consisting of the polynomials $g(t)$ for which $g(1)=g(-1)$.

7. Show that the hyperbola defined by the equation $xy=1$ and the line $\mathbb{A}^1$ are not isomorphic.

8. Find $f(\mathbb{A}^2)$ for the regular mapping $f:\mathbb{A}^2 \to \mathbb{A}^2$ given by the formula $f(x,y)=(x,xy)$. Is this set open in $\mathbb{A}^2$? Is it dense? Is it closed?

9. The same as in Exercise 8 for the mapping $f:\mathbb{A}^3 \to \mathbb{A}^3$ given by $f(x,y,z)=(x,xy,xyz)$.

10. An isomorphism $f:X \to X$ of a closed set X into itself is called an automorphism. Show that all the automorphisms of the line $\mathbb{A}^1$ are of the form $f(x)=ax+b$, $a \neq 0$.

11. Show that the mapping $f(x,y)=(x,y+P(x))$, where $P(x)$ is an arbitrary polynomial in x, is an automorphism of $\mathbb{A}^2$. Show also that these automorphisms form a group.

12. Show that if $f(x_1,\dots,x_n)=(P_1(x_1,\dots,x_n),\dots,P_n(x_1,\dots,x_n))$ is an automorphism of $\mathbb{A}^n$, then the Jacobian $|\partial P_i/\partial x_j| \in k$. Denoting the value of this Jacobian by $J(f)$, show that the correspondence $f \to J(f)$ determines a homomorphism of the group of all automorphisms of $\mathbb{A}^n$ into the group of non-zero elements of k.

13. Suppose that X consists of two points. Show that the ring $k[X]$ is isomorphic to the direct sum of two copies of k.

14. Let $f:X \to Y$ be a regular mapping. The subset $T \subset X \times Y$ consisting of the points of the form $(x, f(x))$ is called the graph of f. Show a) that T is a closed subset of $X \times Y$ and b) that T is isomorphic to X.

15. The mapping $p_Y: X \times Y \to Y$ defined by the formula $p_Y(x,y)=y$, is called a projection. Show that for $Z \subset X$ and a regular mapping $f:X \to Y$ we have $f(Z)=p_Y((Z \times Y) \cap T)$, where T is the graph of f, and $Z \times Y$ consists of all points (z,y), $z \in Z$, $y \in Y$.

16. Show that for any regular mapping $f:X \to Y$ there exists a regular mapping $g:X \to X \times Y$ which is an isomorphism of X with a closed subset of $X \times Y$ and for which $f=p_Y g$ (any regular mapping can be split into an embedding and a projection).

17. Show that if $X=\bigcup U_\alpha$ is a covering of a closed set X by open sets, then there exist finitely many sets $U_{\alpha_1},\dots,U_{\alpha_r}$, such that $X=U_{\alpha_1} \cup \dots \cup U_{\alpha_r}$.

18. Show that the Frobenius mapping φ is one-to-one. Is it an isomorphism if, for example, $X=\mathbb{A}^1$?

§ 3. Rational Functions

1. Irreducible Sets. In § 1.1 we have come across the concept of an irreducible plane algebraic curve. Now we state an analogous concept in the general case.

Definition. A closed set X called *reducible* if there exist closed subsets $X_1 \subset X$, $X_2 \subset X$, $X_1 \neq X$, $X_2 \neq X$, such that $X=X_1 \cup X_2$. Otherwise X is called *irreducible*.

Theorem 1. *Every closed set is the union of finitely many irreducible ones.*

Proof. Suppose that the theorem is false for a closed set X. Then X is reducible: $X = X_1 \cup X_1'$, and the theorem is false for X_1 or X_1', say X_1. Then X_1 is reducible, and again one of the closed sets whose union it is must be reducible. So we construct an infinite sequence of closed sets $X \supset X_1 \supset X_2 \supset \ldots$, $X \neq X_1$, $X_1 \neq X_2, \ldots$. Let us show that such a sequence can not exist. If it did, then for the corresponding ideals we should have

$$\mathfrak{A}_X \subset \mathfrak{A}_{X_1} \subset \mathfrak{A}_{X_2}, \ldots, \mathfrak{A}_X \neq \mathfrak{A}_{X_1}, \quad \mathfrak{A}_{X_1} \neq \mathfrak{A}_{X_2}, \ldots.$$

But such a sequence cannot exist because in a polynomial ring every ideal has a finite basis, hence every ascending sequence of ideals breaks off. This proves the theorem.

If in a representation $X = \bigcup X_i$ we have $X_i \subset X_j$ for $i \neq j$, then we can discard X_i from this representation. Repeating this several times we arrive at a representation $X = \bigcup X_i$ in which $X_i \not\subset X_j$ for $i \neq j$. Such a representation is called an *incontractible* (or irredundant) decomposition of X into irreducible closed sets, and the X_i are called *irreducible components* of X.

Theorem 2. *The incontractible representation of a closed set is unique.*

Let $X = \bigcup_i X_i = \bigcup_j Y_j$ be two incontractible representations. Then $X_i = X_i \cap X = X_i \cap (\cup Y_j) = \bigcup_j (X_i \cap Y_j)$. Since X_i is irreducible by hypothesis, for some j we have $X_i \cap Y_j = X_i$, that is, $X_i \subset Y_j$. Interchanging the two decompositions we see that for j there exists an i' such that $Y_j \subset X_{i'}$. Consequently $X_i \subset Y_j \subset X_{i'}$, and since the decompositions are incontractible, $i' = i$ and $Y_j = X_i$. This proves the theorem.

Next we state the concept of irreducibility of a closed set X in terms of the ring $k[X]$. If X is reducible, $X = X_1 \cup X_2$, then since $X \supset X_1$, $X \neq X_1$, there exists a polynomial F_1 that vanishes on X_1, but not identically on X, and a similar polynomial F_2 for X_2. But then $F_1 \cdot F_2$ vanishes both on X_1 and on X_2, hence on X. The corresponding regular functions $f_1, f_2 \in k[X]$ have the property that $f_1 \neq 0, f_2 \neq 0, f_1 \cdot f_2 = 0$. In other words, f_1 and f_2 are divisors of zero in $k[X]$. Suppose, conversely, that the ring $k[X]$ has divisors of zero: $f_1 \cdot f_2 = 0, f_1 \neq 0, f_2 \neq 0$. We denote by X_1 and X_2 the closed subsets of X corresponding to the ideals (f_1) and (f_2) of $k[X]$. In other words, X_i consists of all those points $x \in X$ for which $f_i(x) = 0$ $(i = 1, 2)$. Obviously $X_i \neq X$, because $f_i \neq 0$ on X, and $X = X_1 \cup X_2$, because $f_1 \cdot f_2 = 0$ on X, hence at every point $x \in X$ either $f_1(x) = 0$ or $f_2(x) = 0$. Thus, a closed set X is irreducible if and only if the ring

$k[X]$ has no divisors of zero. This in turn is equivalent to the fact that $\mathfrak{A}_X$ is a prime ideal.

Theorem 3. *The product of irreducible closed sets is irreducible.*

Suppose that X and Y are irreducible, but $X \times Y = Z_1 \cup Z_2$, $Z_i \neq X \times Y$ $(i=1, 2)$. Then for every point $x \in X$ the closed set $x \times Y$ consisting of the points (x, y), where y is an arbitrary point of Y, is isomorphic to Y, hence irreducible. Since $x \times Y = ((x \times Y) \cap Z_1) \cup ((x \times Y) \cap Z_2)$, either $x \times Y \subset Z_1$ or $x \times Y \subset Z_2$. We consider the set $X_1 \subset X$ consisting of those points $x \in X$ for which $x \times Y \subset Z_1$, and we show that this set is closed. For any point $y \in Y$ the set X_y of those points $x \in X$ for which $x \times y \in Z_1$ is closed: it is characterized by the fact that $(X \times y) \cap Z_1 = X_y \times y$, and the intersection of the closed sets $X \times y$ and Z_1 is closed. Since $X_1 = \bigcap_{y \in Y} X_y$, we see that X_1 is also closed. Similarly, the set X_2 consisting of the points $x \in X$ for which $x \times Y \subset Z_2$ is closed. So we see that $X_1 \cup X_2 = X$, and since X is irreducible, it follows that $X_1 = X$ or $X_2 = X$. In the first case $X \times Y = Z_1$, in the second $X \times Y = Z_2$. This contradiction proves the theorem.

2. Rational Functions. Every commutative ring without divisors of zero can be embedded in a field, its field of fractions.

Definition. If a closed set X is irreducible, then the field of fractions of the ring $k[X]$ is called the *field of rational functions* on X. It is denoted by $k(X)$.

Recalling the definition of the field of fractions we can say that $k(X)$ consists of those rational functions $F(T)/G(T)$ for which $G(T) \notin \mathfrak{A}_X$, and we take it that $F/G = F_1/G_1$ if $FG_1 - F_1 G \in \mathfrak{A}_X$. This means that $k(X)$ can also be constructed as follows. Consider the subring $\mathcal{O}_X \subset k(T_1, \ldots, T_n)$ consisting of those rational functions $f = P/Q$, $P, Q \in k[T]$, for which $Q \notin \mathfrak{A}_X$. The functions f for which $P \in \mathfrak{A}_X$ form an ideal M_X, and $k(X) = \mathcal{O}_X / M_X$.

In contrast to a regular function on a closed set, a rational function does not always assume a definite value at a point of this set; for example, $1/x$ at 0 or x/y at $(0, 0)$. Let us clarify when this happens.

Definition. A rational function $\varphi \in k(X)$ is called *regular at a point* $x \in X$ if it can be written in the form $\varphi = f/g$, $f, g \in k[X]$, $g(x) \neq 0$. In this case the element $f(x)/g(x)$ of k is called the value of $\varphi(x)$ and is denoted by $\varphi(x)$.

Theorem 4. *A rational function φ that is regular at all points of a closed set is a regular function of this set.*

Let $\varphi \in k(X)$ be regular at all points $x \in X$. This means that for every point x there exist elements $f_x, g_x \in k[X]$, $g_x(x) \neq 0$, such that $\varphi = f_x/g_x$. Consider the ideal $\mathfrak{a}$ generated by all the functions g_x, $x \in X$. It has a finite base, so that there exist finitely many points $x_1, \ldots, x_N$, such that $\mathfrak{a} = (g_{x_1}, \ldots, g_{x_N})$. The functions g_{x_i} cannot have a common zero $x \in X$, because then all the functions of the ideal $\mathfrak{a}$ would vanish at x, whereas $g_x(x) \neq 0$. From the analogue to Hilbert's Nullstellensatz it follows that $\mathfrak{a} = (1)$, in other words, that there exist functions $u_1, \ldots, u_N \in k[X]$ such that $\sum_{i=1}^{N} u_i g_{x_i} = 1$. Multiplying both sides of this equality by φ and using the fact that $\varphi = f_{x_i}/g_{x_i}$, we find that $\varphi = \sum_{i=1}^{N} u_i f_{x_i}$, that is, $\varphi \in k[X]$. This proves the theorem.

The set of points at which a rational function φ on a closed set X is regular is non-empty and open. The first assertion follows from the fact that φ can be represented in the form $\varphi = f/g$, where $f, g \in k[X]$, $g \neq 0$. This means that there exists a point $x \in X$ for which $g(x) \neq 0$. Obviously φ is regular at this point. To prove the second assertion we consider all representations $\varphi = f_i/g_i$. For every regular function g_i the set $Y_i \subset X$ consisting of those points $x \in X$ for which $g_i(x) = 0$ is obviously closed, hence $U_i = X - Y_i$ is open. The set of points U at which φ is regular, by definition, is of the form $U = \bigcup U_i$ and is therefore open. This open set is called the *domain of definition* of φ. For any finite system of rational functions $\varphi_1, \ldots, \varphi_m$ the set of points $x \in X$ at which they are all regular is also open and non-empty. The first assertion follows from the fact that the intersection of finitely many open sets is open, and the second from the following useful property: the intersection of finitely many non-empty open sets of an irreducible closed set is non-empty. For let $U_i = X - Y_i$, $i = 1, \ldots, m$; $\bigcap U_i = \emptyset$. Then $Y_i \neq X$ and $\bigcup Y_i = X$. But the Y_i are closed sets, and we arrive at a contradiction to the fact that X is irreducible. Thus, any finite set of rational functions can be equated with a non-empty open set. This remark is useful in view of the fact that a rational function $\varphi \in k(X)$ is uniquely determined when it is specified on some non-empty open subset $U \subset X$. For if $\varphi(x) = 0$ at all $x \in U$ and $\varphi \neq 0$ on X, then by taking any one representation $\varphi = f/g$, $f, g \in k[X]$, we find that X is the union of two closed sets: $X = X_1 \cup X_2$, $X_1 = X - U$, and X_2 is determined by $f = 0$. This contradicts the fact that X is irreducible.

3. Rational Mappings. Let $X \subset \mathbb{A}^n$ be an irreducible closed set. A rational mapping $X \to \mathbb{A}^m$ is given by an arbitrary collection of m functions $\varphi_1, \ldots, \varphi_m \in k(X)$. Now we define the concept of a rational mapping $\varphi: X \to Y$, where Y is a closed subset of $\mathbb{A}^m$.

Definition. A *rational mapping* $\varphi: X \to Y \subset \mathbb{A}^m$ is a collection of m functions $\varphi_1, \ldots, \varphi_m \in k(X)$ such that $(\varphi_1(x), \ldots, \varphi_m(x)) \in Y$ for every point $x \in X$ at which all the functions φ_i are regular. This mapping φ is said to be *regular* at such a point x, and the point $(\varphi_1(x), \ldots, \varphi_m(x))$ is called the *image* of x and is denoted by $\varphi(x)$.

The set of points of the form $\varphi(x)$, where x ranges over those points X at which φ is regular, is called the *image* of X and is denoted by $\varphi(X)$. Thus, a rational mapping is not a mapping of the whole set X into Y, but it necessarily determines a mapping of some non-empty open subset $U \subset X$ into Y.

The study of functions and mappings that are not defined at all points is an essential difference between algebraic geometry and other branches of geometry, for example, topology.

As was shown at the end of the preceding subsection, all the functions φ_i, and hence the rational mapping $\varphi = (\varphi_1, \ldots, \varphi_m)$, are defined on a certain non-empty open set $U \subset X$. Therefore rational mappings can be regarded as mappings of open subsets; but it must be borne in mind that distinct mappings may have distinct domains of definition. The same applies, of course, to rational functions. To verify that functions $\varphi_1, \ldots, \varphi_m$ determine a rational mapping $\varphi: X \to Y$ we have to check that the functions $(\varphi_1, \ldots, \varphi_m)$, as elements of $k(X)$, satisfy the equations of the set Y. For if this property holds, then for any polynomial $u(T_1, \ldots, T_m) \in \mathfrak{A}_Y$ the function $u(\varphi_1, \ldots, \varphi_m)$ vanishes on X. Therefore, at every point x where all the φ_i are regular, $u(\varphi_1(x), \ldots, \varphi_m(x)) = 0$, that is, $(\varphi_1(x), \ldots, \varphi_m(x)) \in Y$. Conversely, if we have a mapping $\varphi: X \to Y$, then for every $u \in \mathfrak{A}_Y$ the function $u(\varphi_1, \ldots, \varphi_m) \in k(X)$ vanishes on some non-empty open set $U \subset X$, hence on X. It follows that $u(\varphi_1, \ldots, \varphi_m) = 0$ in $k(X)$.

Let us clarify how a rational mapping acts on rational functions on a closed set. We assume that for a rational mapping $\varphi: X \to Y$ the set $\varphi(X)$ is dense in Y. We regard φ as a mapping of sets $U \to \varphi(X)$, where U is the domain of definition of φ, and we construct its corresponding mapping of functions. For every function $f \in k[Y]$ the function $\varphi^*(f)$ is rational on X. For if $Y \subset \mathbb{A}^m$ and if f is given by a polynomial $u(T_1, \ldots, T_m)$, then $\varphi^*(f)$ is given by the rational function $u(\varphi_1, \ldots, \varphi_m)$. So we have a mapping $\varphi^*: k[Y] \to k(X)$, which is, of course, a ring homomorphism of $k[Y]$ into the field $k(X)$. This homomorphism is even an isomorphic embedding of $k[Y]$ in $k(X)$. For if $\varphi^*(u) = 0$ for $u \in k[Y]$, this means that $u = 0$ on $\varphi(X)$. But if $u \neq 0$ on Y, then the equation $u = 0$ determines a closed subset $V(u) \subset Y$, different from Y. Then $\varphi(X) \subset V(u)$, and this contradicts the fact that $\varphi(X)$ is dense in Y. Clearly, the embedding φ^* of $k[Y]$ into $k(X)$ can be extended to an isomorphic embedding of the field of fractions $k(Y)$ of $k[Y]$ into $k(X)$. Thus, if $\varphi(X)$ is dense in Y, then the rational mapping φ determines an isomorphic embedding φ^* of $k(Y)$ in $k(X)$. If $\varphi: X \to Y$ and $\psi: Y \to Z$ are two mappings and if $\varphi(X)$ is dense in Y, then, as is easy to see, the product $\psi\varphi: X \to Z$ can be defined, and if $\psi(Y)$ is dense in Z, then $(\psi\varphi)(X)$ is also dense in Z. For the embeddings of fields we then have the relation $(\psi\varphi)^* = \varphi^*\psi^*$.

Definition. A rational mapping $\varphi: X \to Y$ is called a *birational isomorphism* if it has an inverse. This means that there exists a rational mapping $\psi: Y \to X$ such that $\varphi(X)$ is dense in Y, and $\psi(Y)$ in X, and that $\psi\varphi = 1$, $\varphi\psi = 1$. In that case X and Y are called birationally isomorphic.

It is obvious that if a rational mapping $\varphi: X \to Y$ is a birational isomorphism, then the embedding $\varphi^*: k(Y) \to k(X)$ is an isomorphism. It is easy to verify that the converse also holds (for plane algebraic curves this was done in § 1). Thus, two closed sets X and Y are birationally isomorphic if and only if the fields $k(X)$ and $k(Y)$ are isomorphic over k.

Examples. In § 1 we have analysed a number of examples of birational isomorphism between plane algebraic curves. Obviously, isomorphic closed sets are birationally isomorphic. In Examples 8 and 9 of § 2.3 the mappings, although not isomorphisms, are birational isomorphisms.

Closed sets that are birationally isomorphic to an affine space are called rational. In § 1 we have come across rational algebraic curves. Here are some other examples.

Example 1. An irreducible hypersurface X determined in $\mathbb{A}^n$ by an equation $F(T_1, \ldots, T_n) = 0$ of degree 2 is rational. The proof given in § 1.1 for $n = 2$ works in the general case. The corresponding mapping can again be interpreted as a projection of X from some point $x \in X$ onto a hyperplane $l \subset \mathbb{A}^n$ that does not pass through x. We only have to take x so that it is not a "vertex" on X, that is, $(\partial F/\partial T_i)(x) \neq 0$ for at least one $i = 1, \ldots, n$.

Example 2. Consider the hypersurface X in $\mathbb{A}^3$ defined by the cubic equation $x^3 + y^3 + z^3 = 1$ and suppose that the characteristic of the ground field is not 3. There are some straight lines on X, for example, the lines L_1 and L_2 given by the systems of equations

$$L_1: \begin{array}{c} x + y = 0, \\ z = 1, \end{array} \qquad L_2: \begin{array}{c} x + \varepsilon y = 0, \\ z = \varepsilon, \end{array}$$

where ε is a cube root of unity, $\varepsilon \neq 1$. The lines L_1 and L_2 are skew.

We describe a rational mapping of X onto a plane geometrically and leave it to the reader to derive the formulae and to verify that we are dealing with a birational isomorphism. We take a plane E in $\mathbb{A}^3$ that does not contain L_1 or L_2. As is easy to verify, for $x \in X - L_1 - L_2$ there exists a unique line L passing through x and intersecting L_1 and L_2. We denote the point of intersection $L \cap E$ by $f(x)$. This is the required rational mapping $X \to E$.

In algebraic geometry we are concerned with two equivalence relations: isomorphism and birational isomorphism. Clearly, birational isomorphism is a coarser relationship than isomorphism, in other words, non-isomorphic closed sets may be birationally isomorphic. Therefore often the classification of closed sets from the point of view of birational isomorphism turns out to be simpler and more lucid than from the point of view of isomorphism. Isomorphism, being defined at all points, is close to such geometric concepts as homeomorphism or diffeomorphism and therefore more convenient. An important problem is the clarification of connections between these two equivalence relations. The point is: how much cruder is birational isomorphism than isomorphism, in other words, how many closed sets that are distinct from the point of view of isomorphism belong to one and the same type from the point of view of birational isomorphism? Later on we shall frequently come across this problem.

Both these equivalence relations can be defined purely algebraically: closed sets X and Y are isomorphic if and only if the rings $k[X]$ and $k[Y]$ are isomorphic, and they are birationally isomorphic if and only if the fields $k(X)$ and $k(Y)$ are isomorphic. In this context it is important to clarify what rings are of the form $k[X]$ and what fields of the form $k(X)$, where X is an irreducible closed set. The answer is very simple.

Theorem 5. *An algebra A over a field k is isomorphic to a ring $k[X]$, where X is an irreducible closed set, if and only if A has no divisors of zero and is finitely generated over k. An extension K of k is isomorphic to a field $k(X)$ if and only if it is finitely generated.*

The necessity of all these conditions is obvious. If an algebra A is generated by finitely many elements $t_1, \ldots, t_n$, then $A \simeq k[T_1, \ldots, T_n]/\mathfrak{A}$, where $\mathfrak{A}$ is an ideal of the polynomial ring $k[T_1, \ldots, T_n]$. Since A has no divisor of 0, $\mathfrak{A}$ is a prime ideal. Suppose that $\mathfrak{A} = (F_1, \ldots, F_m)$. Consider the closed set $X \subset \mathbb{A}^n$ defined by the equations $F_1 = \cdots = F_m = 0$; we show that $\mathfrak{A}_X = \mathfrak{A}$, and then $K[X] \simeq k[T_1, \ldots, T_n]/\mathfrak{A}_X \simeq A$.

If $F \in \mathfrak{A}_X$, then by Hilbert's Nullstellensatz $F^r \in \mathfrak{A}$ for some $r > 0$. Since $\mathfrak{A}$ is a prime ideal, we then have $F \in \mathfrak{A}$. Therefore $\mathfrak{A}_X \subset \mathfrak{A}$, and since the inclusion $\mathfrak{A} \subset \mathfrak{A}_X$ is obvious, we have $\mathfrak{A}_X = \mathfrak{A}$.

If the field K is generated over k by the finitely many elements $t_1, \ldots, t_n$, then the algebra $A = k[t_1, \ldots, t_n]$ satisfies the conditions of the theorem, and by what we have already proved, $A = k[X]$. Since K is the field of fractions of A, we have $K = k(X)$.

In conclusion we prove one result that illustrates the concept of a birational isomorphism.

Theorem 6. *Every irreducible closed set X is birationally isomorphic to a hypersurface in some affine space $\mathbb{A}^n$.*

Proof. The field $k(X)$ is finitely generated over k, say, by the elements $t_1, \ldots, t_n$, coordinates in $\mathbb{A}^n$ regarded as functions on X.

Suppose that $t_1, \ldots, t_d$ are algebraically independent over k, and that d is the maximal number. Then every element $y \in k(X)$ depends algebraically on $t_1, \ldots, t_d$, and there exists a relationship $f(t_1, \ldots, t_d, y) = 0$ for which the polynomial $f(T_1, \ldots, T_d, T_{d+1})$ is irreducible over k.

Let $f(T_1, \ldots, T_{d+1})$ be such a polynomial for $t_1, \ldots, t_{d+1}$. We claim that $f'_{T_i}(T_1, \ldots, T_{d+1}) \neq 0$ for at least one $i = 1, \ldots, d+1$. For if this were not the case, then all the T_i would occur in f with degrees that are multiples of the characteristic p of k, that is, f would be of the form $f = \Sigma a_{i_1 \ldots i_{d+1}} T_1^{pi_1} \ldots T_{d+1}^{pi_{d+1}}$. We set $a_{i_1 \ldots i_{d+1}} = b_{i_1 \ldots i_{d+1}}^p$,

$$g = \Sigma b_{i_1 \ldots i_{d+1}} T_1^{i} \ldots T_{d+1}^{i_{d+1}}$$

and find that $f = g^p$, a contradiction to the irreducibility of f.

If $f'_{T_i} \neq 0$, then the d elements $t_1, \ldots, t_{i-1}, t_{i+1}, \ldots, t_{d+1}$ are algebraically independent over k. For the element t_i is algebraic over the field $k(t_1, \ldots, t_{i-1}, t_{i+1}, \ldots, t_{d+1})$ because $f'_{T_i} \neq 0$, and hence T_i occurs in f. Therefore, if the elements $t_1, \ldots, t_{i-1}, t_{i+1}, \ldots, t_{d+1}$ were dependent, then the transcendance degree of the field $k(t_1, \ldots, t_{d+1})$ would be less than d, and this contradicts the independence of the elements $t_1, \ldots, t_d$.

Thus, we can always renumber $T_1, \ldots, T_n$ such that $t_1, \ldots, t_d$ are independent over k and that $f'_{T_{d+1}} \neq 0$. This shows that t_{d+1} is separable over $k(t_1, \ldots, t_d)$. Since t_{d+2} is algebraic over this field, by Abel's theorem on the primitive element we can find an element y such that $k(t_1, \ldots, t_{d+2}) = k(t_1, \ldots, t_d, y)$. Repeating the process of adjoining elements $t_{d+1}, \ldots, t_n$ we represent the field $k(X)$ in the form $k(z_1, \ldots, z_{d+1})$, where $z_1, \ldots, z_d$ are algebraically independent over k and

$$f(z_1, \ldots, z_d, z_{d+1}) = 0, \tag{1}$$

the polynomial f is irreducible over k, and $f'_{T_{d+1}} \neq 0$. Obviously the field of rational functions $k(Y)$ over the closed set Y defined by (1) is isomorphic to $k(X)$. But this means that X and Y are birationally isomorphic, and the theorem is proved.

Note 1. By virtue of the condition $f'_{T_{d+1}} \neq 0$ in (1) the element z_{d+1} is separable over $k(z_1, \ldots, z_d)$. Consequently, $k(X)/k(z_1, \ldots, z_d)$ is a finite separable extension.

Note 2. From the proof of Theorem 6 and Abel's theorem on the primitive element it follows that $z_1, \ldots, z_{d+1}$ can be chosen as linear combinations of the original coordinates $x_1, \ldots, x_n$: $z_i = \sum_{j=1}^{n} c_{ij} x_j$ $(i = 1, \ldots, d+1)$. The mapping $(x_1, \ldots, x_n) \to (z_1, \ldots, z_{d+1})$ given by these formulae is a projection of $\mathbb{A}^n$ parallel to the linear subspace defined by the equations $\sum_{j=1}^{n} c_{ij} x_j = 0$ $(i = 1, \ldots, d+1)$. This indicates the geometric meaning of the birational mapping whose existence is established in Theorem 6.

Exercises

1. Let k be a field of characteristic $\neq 2$. Decompose the closed set $X \subset \mathbb{A}^3$ defined by the equations $x^2 + y^2 + z^2 = 0$, $x^2 - y^2 - z^2 + 1 = 0$, into irreducible components.

2. Show that if X is the closed set of § 2, Exercise 4, then the elements of the field $k(X)$ have a unique representation in the form $u(x) + v(x)y$, where $u(x)$ and $v(x)$ are arbitrary rational functions.

3. Show that the mapping of f of Exercises 5, 6, and 8 in § 2 are birational isomorphisms.

4. Decompose the closed set X defined in $\mathbb{A}^3$ by the equations $y^2 = xz$, $z^2 = y^3$, into irreducible components. Show that all its irreducible components are birationally isomorphic to $\mathbb{A}^1$.

5. Show that if a closed set X is defined in $\mathbb{A}^n$ by a single equation $f_{n-1}(T_1, \ldots, T_n) + f_n(T_1, \ldots, T_n) = 0$, where f_{n-1} and f_n are homogeneous polynomials of degree $n-1$ and n, respectively, and X is irreducible, then it is birationally isomorphic to $\mathbb{A}^{n-1}$. (Such a closed set is called a monoid).

6. At what points of the circle given by the equation $x^2 + y^2 = 1$ is the rational function $(1-y)/x$ regular?

7. At what points of the curve X with the equation $y^2 = x^2 + x^3$ is the rational function $t = y/x$ regular? Show that $t \notin k[X]$.

§ 4. Quasiprojective Varieties

1. Closed Subsets of a Projective Space. Let $\mathbb{P}^n$ be an n-dimensional projective space, so that a point $\xi \in \mathbb{P}^n$ is given by $n+1$ elements $(\xi_0 : \ldots : \xi_n)$ of k and not all the ξ_i are 0. Two points $(\xi_0 : \ldots : \xi_n)$ and $(\eta_0 : \ldots : \eta_n)$ are taken to be identical if and only if there exists a $\lambda \neq 0$

such that $\eta_i = \lambda \xi_i (i=0, \ldots, n)$. For any collection $(\xi_0 : \ldots : \xi_n)$ determining ξ the ξ_i are called homogeneous coordinates of this point.

We say that a polynomial $f(S) \in k[S_0, \ldots, S_n]$ vanishes at a point $\xi \in \mathbb{P}^n$ if $f(\xi_0, \ldots, \xi_n) = 0$, no matter what coordinates ξ_i of ξ are chosen. It is clear that then $f(\lambda\xi_0, \ldots, \lambda\xi_n) = 0$ for all $\lambda \neq 0$, $\lambda \in k$. We write f in the form $f = f_0 + f_1 + \cdots + f_r$, where f_i is the sum of all the terms of degree i in f. Then

$$f(\lambda\xi_0, \ldots, \lambda\xi_n) = f_0(\xi_0, \ldots, \xi_n) + \lambda f_1(\xi_0, \ldots, \xi_n) + \cdots + \lambda^r f_r(\xi_0, \ldots, \xi_n).$$

Since k is infinite, the equality $f(\lambda\xi_0, \ldots, \lambda\xi_n) = 0$, which holds for all $\lambda \neq 0$, $\lambda \in k$, implies that $f_i(\xi_0, \ldots, \xi_n) = 0$. Thus, if a polynomial f vanishes at some point ξ, then all its homogeneous components vanish at that point.

Definition. A subset $X \subset \mathbb{P}^n$ is called *closed* if it consists of all points at which finitely many polynomials with coefficients in k vanish simultaneously.

The set of all polynomials $f \in k[S_0, \ldots, S_n]$ that vanish at all points $x \in X$ form an ideal in the ring $k[S]$, which is called the ideal of X and is denoted by $\mathfrak{A}_X$. By what was said above, $\mathfrak{A}_X$ has the property that if a polynomial f is contained in it, then so are all its homogeneous components. Ideals having this property are called homogeneous. Thus, the ideal of a closed subset of a projective space is homogeneous. From this it follows that it has a basis consisting of homogeneous polynomials: it is sufficient to take any basis and to consider the system of all homogeneous components of the polynomials of the basis. In particular, every closed subset of a projective space can be given by a system of homogeneous equations.

Thus, to every closed subset $X \subset \mathbb{P}^n$ there corresponds a homogeneous ideal $\mathfrak{A}_X \subset k[S_0, \ldots, S_n]$. Conversely, every homogeneous ideal $\mathfrak{A} \subset k[S]$ determines a closed subset $X \subset \mathbb{P}^n$. For if $F_1, \ldots, F_n$ is a homogeneous basis of $\mathfrak{A}$, then X is determined by the system of equations $F_1 = 0, \ldots, F_n = 0$. If this system has no solutions in the field other than the null solution, then naturally X is taken to be given by the empty set.

In the case of closed subsets of affine spaces an ideal $\mathfrak{A} \subset k[T]$ determines the empty set only if $\mathfrak{A} = (1)$: this is the content of Hilbert's Nullstellensatz. In the case of closed subsets of a projective space this need not be so: for example, clearly the empty set is also determined by the ideal $(S_0, \ldots, S_n)$. We denote by I_s the ideal of $k[S]$ consisting of those polynomials in which only terms of degree at least s occur. Obviously the ideal I_s determines the empty set – it contains, for example, the polynomials S_i^s, which vanish simultaneously only at the origin.

Lemma. *A homogeneous ideal $\mathfrak{A} \subset k[S]$ determines the empty set if and only if it contains the ideal I_s for some $s > 0$.*

We have already seen that I_s determines the empty set. This is true a fortiori for any ideal containing it. Suppose now that a homogeneous ideal $\mathfrak{A} \subset k[S]$ determines the empty set. Let $F_1, \ldots, F_m$ be a homogeneous basis of $\mathfrak{A}$ and $\deg F_i = n$. Then by hypothesis the polynomials $F_i(1, T_1, \ldots, T_n)$, where $T_j = S_j/S_0$. have no common roots. For a common root $(\alpha_1, \ldots, \alpha_n)$ would give a common root $(1, \alpha_1, \ldots, \alpha_n)$ of $F_1, \ldots, F_m$. By Hilbert's theorem, there must then exist polynomials $G_i(T_1, \ldots, T_n)$ such that $\sum_i F_i(1, T_1, \ldots, T_n) G_i(T_1, \ldots, T_n) = 1$. Substituting in these equations $T_j = S_j/S_0$ and multiplying by the common denominator, which is of the form $S_0^{m_0}$, we find that $S_0^{m_0} \in \mathfrak{A}$. Similarly, for every $i = 1, \ldots, n$ we can find an integer $m_i > 0$ such that $S_i^{m_i} \in \mathfrak{A}$. If now $m = \max(m_0, \ldots, m_n)$ and $s = (m-1)(n+1)+1$, then in every term $S_0^{a_0} \ldots S_n^{a_n}$ with $a_0 + \cdots + a_n \geqslant s$ at least one S_i must occur with an exponent $a_i \geqslant m \geqslant m_i$, and since $S_i^{m_i} \in \mathfrak{A}$, this term is contained in $\mathfrak{A}$. This shows that $I_s \subset \mathfrak{A}$.

Later we shall consider simultaneously closed subset of affine and projective spaces. We call them affine and projective closed sets.

For projective closed sets the same terminology as for affine sets is applicable, namely if X and Y are two closed sets and $Y \subset X$, then $X - Y$ is called *open* in X. As before, the union of any number and the intersection of a finite number of open sets are open, and the union of a finite number and the intersection of any number of closed sets are closed. The set $\mathbb{A}_0^n$ of points $\xi = (\xi_0 : \ldots : \xi_n)$ for which $\xi_0 \neq 0$ is obviously open. Its points can be put into one-to-one correspondence with the points of the n-dimensional affine space, by setting $\alpha_i = \xi_i/\xi_0 (i = 1, \ldots, n)$ and assigning to the point $\xi \in \mathbb{A}_0^n$ the point $(\alpha_1, \ldots, \alpha_n) \in \mathbb{A}^n$. Therefore we call $\mathbb{A}_0^n$ an *affine open subset*. Similarly the sets $\mathbb{A}_i^n (i = 0, \ldots, n)$ consist of the points for which $\xi_i \neq 0$. Clearly $\mathbb{P}^n = \bigcup \mathbb{A}_i^n$.

For every projective closed set $X \subset \mathbb{P}^n$ the sets $U_i = X \cap \mathbb{A}_i^n$ are open in X. As subsets of $\mathbb{A}_i^n$ they are closed. For if X is given by the system of homogeneous equations $F_0 = \cdots = F_m = 0$ and $\deg F_i = n_i$, then, for example, U_0 is given by the system of equations

$$S_0^{-n_j} F_j = F_j(1, T_1, \ldots, T_n) = 0 \; (j = 1, \ldots, m),$$
$$T_i = S_i/S_0 (i = 1, \ldots, n).$$

We call the U_i *affine open subsets* of X. Clearly $X = \bigcup U_i$. A closed subset $U \subset \mathbb{A}_0^n$ determines a closed projective set $\bar{U}$, which is called its closure and is the intersection of all projective closed sets containing U. It is easy to check that homogeneous equations for $\bar{U}$ are obtained by the process inverse to the one just described: if $F(T_1, \ldots, T_n)$ is any polynomial of $\mathfrak{A}$ and if $\deg F = l$. then the equations of $\bar{U}$ are of the form $S_0^l F(S_1/S_0, \ldots, S_n/S_0) = 0$. Hence it follows that

$$U = \bar{U} \cap \mathbb{A}_0^n \, . \tag{1}$$

So far we have considered two objects that can lay claim to be called algebraic varieties: affine and projective closed sets. It is natural to try and find a single concept of which these two types of varieties would be particular cases. This will be done more fully in Ch. V in connection with the concept of a scheme. Here we introduce a more special concept, which combines projective and affine closed sets.

Definition. A *quasiprojective variety* is an open subset of a closed projective set.

Obviously a closed projective set is quasiprojective. For affine closed sets this follows from (1).

A *closed subset* of a quasiprojective variety is defined as its intersection with a closed set of a projective space. An open set and a neighbourhood of a point are defined similarly. The notion of an irreducible variety and the theorem on the decomposition of a variety into irreducible components carries over verbatim from the case of affine closed sets.

A *subvariety* Y of a quasiprojective variety $X \subset \mathbb{P}^n$ is now defined as any subset $Y \subset X$ that is itself a quasiprojective variety in $\mathbb{P}^n$. Obviously this is equivalent to the fact that $Y = Z - Z_1$, where $Z \supset Z_1$ and where Z and Z_1 are closed in X.

2. Regular Functions. Passing to the investigation of functions on quasiprojective varieties we begin with the projective space $\mathbb{P}^n$. Here we come across an important difference between functions of homogeneous and of inhomogeneous coordinates: a rational function of homogeneous coordinates

$$f(S_0, \ldots, S_n) = \frac{P(S_0, \ldots, S_n)}{Q(S_0, \ldots, S_n)} \tag{1}$$

cannot be regarded as a function of a point $x \in \mathbb{P}^n$ even when $Q(x) \neq 0$, because the value $f(\alpha_0, \ldots, \alpha_n)$ changes when all the homogeneous coordinates are multiplied by a common factor. However, homogeneous functions of degree 0, that is, functions $f = P/Q$, where P and Q are homogeneous of the same degree, can be regarded as functions of a point.

If X is a quasiprojective variety, $X \subset \mathbb{P}^n$, $x \in X, f = P/Q$ is a homogeneous function of degree 0, and $Q(x) \neq 0$, then f determines in some neighbourhood of x a function with values in k. This function is called *regular* in a neighbourhood of x or simply at x. A function given on X

and regular at all points $x \in X$ is called regular on X. All functions that are regular on X form a ring, which is denoted by $k[X]$.

Let us show that for a closed subset X of an affine space our definition of a regular function is the same as that given in §2. If X is irreducible, then this is the content of Theorem 4 in §3. In general, it is sufficient to make insignificant modifications of the arguments by which this theorem was proved. In it we understand regularity of a function in the sense of the definition given in §2.

By hypothesis, every point $x \in X$ has a neighbourhood U_x in which $f = p_x/q_x$, where p_x and q_x are regular functions on X, and $q_x \neq 0$ on U_x. Therefore

$$q_x f = p_x \tag{2}$$

on U_x. But we may assume that (2) holds on the whole of X. To see this it is sufficient to choose a regular function that vanishes on $X - U_x$ but not at x, and to multiply it by p_x and q_x. Then (2) also holds outside U_x, because both sides of the equality vanish. As in the proof of Theorem 4 of §3, we can find points $x_1, \ldots, x_N$ and regular functions $h_1, \ldots, h_N$ such that $\sum_{i=1}^{N} q_{x_i} h_i = 1$. Multiplying (2) for $x = x_i$ by h_i and adding up we see that

$$f = \sum_{i=1}^{N} p_{x_i} h_i ,$$

that is, f is a regular function.

In contrast to the case of affine closed sets, the ring $k[X]$ can consist of constants only. In §5 we shall show that this happens always when X is a closed projective set. This can easily be verified directly when $X = \mathbb{P}^n$. For if $f = P/Q$, where P and Q are forms of the same degree, we may assume that P and Q are relatively prime. Then the function f is non-regular at points x where $Q(x) = 0$. On the other hand, $k[X]$ may turn out to be unexpectedly large. Namely, if X is an affine closed set, then as a ring $k[X]$ is finitely generated over k. Rees and Nagata have constructed examples of quasiprojective varieties for which this is not so. This shows that only for affine closed sets is the ring $k[X]$ a natural invariant.

Now we pass to mappings. Every mapping of a quasiprojective variety X into an affine space $\mathbb{A}^n$ is given by n functions on X with values in k. If these functions are regular on X, then the mapping is called regular.

Definition. Let $f: X \to Y$ be a mapping of quasiprojective varieties and $Y \subset \mathbb{P}^m$. This mapping is called *regular* if for every point $x \in X$ and every open affine set $\mathbb{A}_i^m$ containing the point $f(x)$ there exists a neighbourhood U of x such that $f(U) \subset \mathbb{A}_i^m$ and the mapping $f: U \to \mathbb{A}_i^m$ is regular.

Let us verify that the property of regularity does not depend on the particular open affine set $\mathbb{A}_i^m$ containing $f(x)$ we have used. If $f(x)=(y_0, \ldots, \underset{i}{\hat{1}}, \ldots, y_m) \in \mathbb{A}_i^m$ is also contained in $\mathbb{A}_j^m$, then $y_j \neq 0$ and the coordinates of this point in $\mathbb{A}_j^m$ are of the form

$$(y_0/y_j, \ldots, \underset{i}{1/y_j}, \ldots, \underset{j}{\hat{1}}, \ldots, y_m/y_j).$$

Therefore, if the mapping $f: U \to \mathbb{A}_i^m$ is given by the functions $(f_0, \ldots, \underset{i}{\hat{1}}, \ldots, f_m)$, then $f: U \to \mathbb{A}_j^m$ is given by the functions

$$(f_0/f_j, \ldots, 1/f_j, \ldots, \underset{j}{\hat{1}}, \ldots, f_m/f_j).$$

By hypothesis, $f_j(x) \neq 0$ and the set U' of points of U at which $f_j \neq 0$ is open. On U' the functions $f_1/f_j, \ldots, 1/f_j, \ldots, f_m/f_j$ are regular, and hence the mapping $f: U' \to \mathbb{A}_j^m$ is regular.

As for affine closed sets, a regular mapping $f: X \to Y$ determines a mapping $f^*: k[Y] \to k[X]$.

Now let us see by what formulae a regular mapping of an irreducible variety is given in homogeneous coordinates. Suppose, for example, that $f(x) \in \mathbb{A}_0^m$ and that the mapping $f: U \to \mathbb{A}_0^m$ is given by regular functions $f_1, \ldots, f_m$. By definition, $f_i = P_i/Q_i$, where P_i and Q_i are forms of the same degree in homogeneous coordinates of the point x and $Q_i(x) \neq 0$. Taking these fractions to the least common denominator we find that $f_i = F_i/F_0$, where all the $F_0, \ldots, F_m$ are forms of the same degree and $F_0(x) \neq 0$. In other words. $f(x) = (F_0(x): \cdots : F_m(x))$, as a point in $\mathbb{P}^m$. In such a substitution we must bear in mind that the representation of a regular function as a ratio of two forms is not unique. Therefore two formulae

$$\begin{aligned} f(x) &= (F_0(x): \cdots : F_m(x)), \\ g(x) &= (G_0(x): \cdots : G_m(x)) \end{aligned} \tag{3}$$

can give one and the same mapping. This is so if and only if

$$F_i G_j = F_j G_i \quad \text{on } X, \quad 0 \leqslant i, j \leqslant m. \tag{4}$$

So we arrive at a second version of the definition of a regular mapping.

A regular mapping $f: X \to \mathbb{P}^m$ of an irreducible quasiprojective variety is given by a collection of forms

$$(F_0 : \cdots : F_m) \tag{5}$$

of the same degree in homogeneous coordinates of a point $x \in \mathbb{P}^n$. Two mappings (3) are called identical if the conditions (4) hold. It is required that for every point $x \in X$ there exists an expression (5) for f such that

$F_i(x) \neq 0$ for at least one i. Then the point $(F_0(x): \cdots : F_m(x))$ is denoted by $f(x)$.

The importance of considering all the expressions (5) of a regular mapping is illustrated by the example of the projection of a conic onto a line. If the curve is the circle $x_1^2 + x_2^2 = 1$ and the centre of the projection is the point $(1, 0)$, then the mapping is given by the formula $t = x_2/(x_1 - 1)$. We introduce projective coordinates: $x_1 = u_1/u_0$, $x_2 = u_2/u_0$, $t = v_1/v_0$. Then the mapping can be written in the form $f(u_0:u_1:u_2) = (u_1 - u_0:u_2)$. Both the forms u_2 and $u_1 - u_0$ vanish at the point $(1:1:0)$. But on the circle $u_2^2 = (u_0 - u_1)(u_0 + u_1)$, and therefore the same mapping can be given by the formula $f(u_0:u_1:u_2) = (-u_2:u_1 + u_0)$. The form $u_1 + u_0$ does not vanish at $(1:1:0)$, which shows that f is regular.

Having defined a regular mapping of quasiprojective varieties, it is natural to define an isomorphism as a regular mapping having a regular inverse.

A quasiprojective variety X' isomorphic to a closed subset of an affine space is called an *affine variety*. Here it can happen that X lies, but is not closed, in $\mathbb{A}^n$. For example, the quasiprojective set $X = \mathbb{A}^1 - 0$, which is not closed in $\mathbb{A}^1$, is isomorphic to a hyperbola, which is closed in $\mathbb{A}^2$ (Example 3 of §2.3). Thus, the concept of a closed affine set is not invariant under isomorphism, whereas that of an affine variety is invariant by definition.

Similarly, a quasiprojective variety isomorphic to a closed projective set is called *projective variety*. We shall show in §5 that if $X \subset \mathbb{P}^n$ is projective, then it is closed in $\mathbb{P}^n$, so that the concept of a closed projective set and a projective variety are the same and are invariant under isomorphism.

There exist quasiprojective varieties that are neither affine nor projective (see Exercise 5 in §4 and Exercise 4, 5, and 6 in §5).

Later we shall come across properties of a variety X that need only be checked in an arbitrary neighbourhood U of any point $x \in X$. In other words, if $X = \bigcup U_\alpha$, where the U_α are any open sets, then it is enough to check such a property for each of the U_α. Such properties are called *local*. Here is an example.

Lemma 1. *The property of a subset $Y \subset X$ of being closed in a quasiprojective variety X is local.*

This proposition means that if $X = \bigcup U_\alpha$, where U_α is open and $Y \cap U_\alpha$ is closed in every U_α, then Y is itself closed. By the definition of open sets, $U_\alpha = X - Z_\alpha$, where Z_α is closed, and by the definition of closed sets, $U_\alpha \cap Y = U_\alpha \cap T_\alpha$, where the T_α are closed.

Let us verify that $Y = \bigcap (Z_\alpha \cup T_\alpha)$, from which it follows, of course, that Y is closed. If $y \in Y$ and $y \in U_\alpha$, then $y \in U_\alpha \cap Y \subset T_\alpha$, and if $y \notin U_\alpha$, then $y \in X - U_\alpha = Z_\alpha$, so that $y \in Z_\alpha \cup T_\alpha$ for all α. Conversely, let $x \in Z_\alpha \cup T_\alpha$ for all α. From the fact that $X = \bigcup U_\alpha$ it follows that $x \in U_\beta$ for some β. Then $x \notin Z_\beta$, and hence $x \in T_\beta$, $x \in T_\beta \cap U_\beta \subset Y$.

In studying local properties we confine our attention to affine varieties.

Lemma 2. *Every point $x \in X$ has a neighbourhood that is isomorphic to an affine variety.*

By hypothesis, $X \subset \mathbb{P}^n$. If $x \in \mathbb{A}_0^n$ (that is, the coordinate u_0 of x is not 0), then $x \in X \cap \mathbb{A}_0^n$, and by definition of a quasiprojective variety $X \cap \mathbb{A}_0^n = Y - Y_1$, where Y and $Y_1 \subset Y$ are closed subsets of $\mathbb{A}_0^n$. Since $x \in Y$, there exists a polynomial F in the coordinates in $\mathbb{A}_0^n$ for which $F = 0$ on Y_1, $F(x) \neq 0$. We denote by (F) the set of points of the variety Y where $F = 0$. Obviously $D(F) = Y - (F)$ is a neighbourhood of x. We show that this neighbourhood is isomorphic to an affine variety. Let $F_1 = 0, \ldots, F_m = 0$ be the equations of Y in $\mathbb{A}_0^n$. We define a variety Z in $\mathbb{A}^{n+1}$ by the equations

$$\begin{gathered} F_1(T_1, \ldots, T_n) = \cdots = F_m(T_1, \ldots, T_n) = 0\,, \\ F(T_1, \ldots, T_n) \cdot T_{n+1} = 1\,. \end{gathered} \tag{6}$$

The mapping $\varphi: (x_1, \ldots, x_{n+1}) \to (x_1, \ldots, x_n)$ clearly determines a regular mapping of Z into $D(F)$, and $\psi: (x_1, \ldots, x_n) \to (x_1, \ldots, x_n, F(x_1, \ldots, x_n)^{-1})$ is a regular mapping of $D(F)$ into Z inverse of φ. This proves the lemma.

If $Y = \mathbb{A}^1$, $F = T$, then Z is a hyperbola, and the isomorphism we have constructed coincides with the mapping considered in Example 3 of § 2.3.

Definition. An open set $D(f) = X - V(f)$ consisting of points of an affine variety X for which $f(x) \neq 0$ ($f \in k[X]$), is called a *principal open set.*

The significance of these sets lies in the fact that, as we have seen, they are affine and that it is easy to indicate their rings $k[D(f)]$. For by construction, $f \neq 0$ on $D(f)$, so that $f^{-1} \in k[D(f)]$, and the Eqs. (6) show that $k[D(f)] = k[X][1/f]$.

Lemmas 1 and 2 show, for example, that the images of closed subsets under an isomorphism are closed. We can even show that under any regular mapping $f: X \to Y$ the inverse image $f^{-1}(Z)$ of any closed set $Z \subset Y$ is closed in X.

By definition of a regular mapping any points $x \in X$ and $f(x) \in Y$ have neighbourhoods U of x and V of $f(x)$ such that $f(U) \subset V \subset \mathbb{A}^m$ and that $f: U \to V$ is a regular mapping. By Lemma 2 we may take U to be an affine variety. By Lemma 1 it is sufficient to verify that $f^{-1}(Z) \cap U = f^{-1}(Z \cap V) \cap U$ is closed in U. Since $Z \cap V$ is closed in V, it is defined by equations $g_1 = \cdots = g_m = 0$, where g_i are regular functions on V. But then $f^{-1}(Z \cap V) \cap U$ is determined by the equations $f^*(g_1) = \cdots = f^*(g_m) = 0$ and is therefore also closed.

From what we have shown it follows that the inverse image of an open set is open.

It is easy to verify that a regular mapping can be defined as a mapping $f: X \to Y$ such that the inverse image of every open set is open (continuity) and that for every point $x \in X$ and every function φ that is regular in a neighbourhood of the point $f(x) \in Y$, the function $f^*(\varphi)$ is regular in a neighbourhood of x.

3. Rational Functions. In defining rational functions on quasiprojective varieties we are up against the fact that the general case is entirely different from the case of affine varieties. Namely, for an affine variety X we have defined rational functions on X as ratios of functions that are regular on the whole of X. As we have seen, in the general case it can happen that on a variety there are no functions other than constants that are regular everywhere, and then there are no non-constant rational functions. Therefore we define rational functions on a quasiprojective variety $X \subset \mathbb{P}^n$ as functions that can be defined on X by homogeneous functions on $\mathbb{P}^n$. More accurately, we consider an irreducible quasiprojective variety $X \subset \mathbb{P}^n$ and denote (by analogy with §3.2) by $\mathcal{O}_X$ the set of rational functions in homogeneous coordinates $S_0, \ldots, S_n$ of the form $f = P/Q$, where P and Q are forms of the same degree and $Q \notin \mathfrak{A}_X$. As for affine varieties, the irreducibility of X implies that $\mathcal{O}_X$ is a ring without divisors of zero. We denote by M_X the set of functions $f \in \mathcal{O}_X$ for which $P \in \mathfrak{A}_X$. Clearly $\mathcal{O}_X/M_X$ is a field, which is called the *field* of *rational functions* on the variety X and is denoted by $k(X)$. Since a form vanishes on an irreducible quasiprojective variety X precisely when it vanishes on an open subset U, we have $k(X) = k(U)$. In particular, $k(X) = k(\bar{X})$, where $\bar{X}$ is the closure of X in the projective space. Therefore, in studying fields of rational functions we may restrict our attention, if we wish, to affine or projective varieties.

It is easy to verify that if X is an affine variety, then the definition above is the same as that in §3. For when we divide numerator and denominator of a homogeneous function of degree zero $f = P/Q$, $\deg P = \deg Q = m$, by S_0^m, we can write it in the form of a rational function of $T_i = S_i/S_0 (i = 1, \ldots, n)$. In this way an isomorphism is established between the field of homogeneous rational functions of degree zero in $S_0, \ldots, S_n$ and the field $k(T_1, \ldots, T_n)$. A trivial verification shows that the ring and the ideal of $k(T_1, \ldots, T_n)$, which were denoted in §3.2 by $\mathcal{O}_X$ and M_X, then correspond to the objects we have here denoted by the same letters.

Just as in the preceding subsection, where we have used the rational functions on the space $\mathbb{P}^n$ to define regular functions, so we call a function $f \in k(X)$ *regular* at a point $x \in X$ if it can be represented in the form $f = F/G$, where F and G are homogeneous of the same degree, and $G(x) \neq 0$. Then $f(x) = F(x)/G(x)$ is called its *value* at x. As in the case of affine varieties, the set of points at which a given rational function f is regular forms a non-empty open subset U of X. The set U is called the *domain of definition* of f. Obviously rational functions can also be defined as functions that are regular on open sets $U \subset X$.

A rational mapping $f: X \to \mathbb{P}^m$ is determined (by analogy to the second definition of a regular mapping given in §4.2) by specifying $m+1$ forms $(F_0 : \cdots : F_m)$ of $n+1$ homogeneous coordinates of the projective space $\mathbb{P}^n$ containing X. Here at least one of the forms must not vanish on X. Two mappings $(F_0 : \cdots : F_m)$ and $(G_0 : \cdots : G_m)$ are called *equal* if $F_i G_j = F_j G_i$ on X. When we divide all the forms F_i by one that is different from zero, we can give a rational mapping by $m+1$ rational functions on X with the same concept of equality of mappings. If a rational mapping f can be given by functions $(f_0 : \cdots : f_m)$ for which all the f_i are regular at $x \in X$ and not all vanish at this point, then the mapping is regular at x. It also determines a regular mapping of some neighbourhood of x into $\mathbb{P}^m$.

The set of points at which a rational mapping is regular is open. Therefore we can also define a rational mapping as a mapping of some open set $U \subset X$. If $Y \subset \mathbb{P}^m$ is a quasiprojective variety and $f: X \to \mathbb{P}^m$ a rational mapping, then we say that f maps X into Y if there exists an open set $U \subset X$ in which f is regular with $f(U) \subset Y$. The union $\tilde{U}$ of all such open sets is called the domain of regularity of f, and $f(\tilde{U})$ the image of X in Y.

As in the case of affine varieties, if the image of a rational mapping $f: X \to Y$ is dense in Y, then it determines a field embedding $f^*: k(Y) \to k(X)$. If a rational mapping $f: X \to Y$ has an inverse rational mapping, then f is called a birational isomorphism, and X and Y birationally isomorphic. In this case the embedding $f^*: k(Y) \to k(X)$ is an isomorphism.

Now we can make the connection between the concepts of an isomorphism and a birational isomorphism more explicit.

Proposition. *Two irreducible varieties X and Y are birationally isomorphic if and only if they contain isomorphic open subsets $U \subset X$ and $V \subset Y$.*

For let $f: X \to Y$ be a birational isomorphism, let $g: Y \to X$, $g = f^{-1}$, be a rational mapping, and let $U_1 \subset X$ and $V_1 \subset Y$ be the domains of regularity of f and g. Since by hypothesis $f(U_1)$ is dense in Y, we see that $f^{-1}(V_1) \cap U_1$ is not empty and, as was shown in §4.2, is open. We set

$U=f^{-1}(V_1)\cap U_1$, $V=g^{-1}(U_1)\cap V_1$. A simple verification shows that $f(U)=V$, $g(V)=U$, $fg=1$, $gf=1$, that is, U and V are isomorphic.

4. Examples of Regular Mappings

1. Projection. Let E be a d-dimensional subspace of a projective space $\mathbb{P}^n$, determined by $n-d$ linearly independent linear equations $L_1=L_2=\cdots=L_{n-d}=0$, where L_i are linear forms. The mapping $\pi(x)=(L_1(x):\cdots:L_{n-d}(x))$ is called a *projection* with centre at E. This mapping is regular on $\mathbb{P}^n-E$, because at the points of this set the forms $L_i(i=1,\ldots,n-d)$ do not vanish simultaneously. Therefore $\pi(x)$ determines a regular mapping $\pi:X\to\mathbb{P}^{n-d-1}$, where X is any closed subset of $\mathbb{P}^n$ disjoint from E. The geometrical meaning of a projection is the following. As a model of $\mathbb{P}^{n-d-1}$ we take any $(n-d-1)$-dimensional subspace $H\subset\mathbb{P}^n$ disjoint from E. Through any point $x\in\mathbb{P}^n-E$ and E there passes a unique $(d+1)$-dimensional subspace E_x. This subspace intersects H in a unique point, namely $\pi(x)$. If X intersects E but is not contained in it, then the projection is a rational mapping. The case $d=0$, that is, a projection from a point, has already occurred several times.

2. Veronese Mapping. Consider all homogeneous polynomials F of degree m in the variables $S_0,\ldots,S_n$. They form a linear space whose dimension is easily calculated to be $\binom{n+m}{m}$.

We are interested in the varieties determined in $\mathbb{P}^n$ by an equation $F=0$. Such varieties are called *projective hypersurfaces*. Since proportional polynomials determine one and the same hypersurface, the hypersurfaces correspond to the points of a projective space $\mathbb{P}^{\nu_{n,m}}$ of dimension $\nu_{n,m}=\binom{n+m}{m}-1$. We denote homogeneous coordinates in $\mathbb{P}^{\nu_{n,m}}$ by $v_{i_0\cdots i_n}$, where $i_0,\ldots,i_n$ are any non-negative numbers such that $i_0+\cdots+i_n=m$. Consider the mapping v_m of $\mathbb{P}^n$ into $\mathbb{P}^{\nu_{n,m}}$ defined by the formulae

$$v_{i_0\cdots i_n}=u_0^{i_0}\ldots u_n^{i_n},\quad i_0+\cdots+i_n=m. \tag{1}$$

Obviously it is regular, because among the monomials on the right-hand side of (1) there are, in particular, u_i^m, which vanish only when all the $u_i=0$. This v_m is called a *Veronese mapping*, and $v_m(\mathbb{P}^n)$ a *Veronese variety*. From (1) it follows that on $v_m(\mathbb{P}^n)$

$$v_{i_0\cdots i_n}v_{j_0\cdots j_n}=v_{k_0\cdots k_n}v_{l_0\cdots l_n} \tag{2}$$

if $i_0+j_0=k_0+l_0,\ldots,i_n+j_n=k_n+l_n$. Conversely, from (2) it is easy to derive that at least one coordinate of the form $v_{0\ldots m\ldots 0}$ is different from zero and that, for example, in the open set $v_{m0\ldots 0}\neq 0$ the mapping

$$u_0=v_{m0\ldots 0},\, u_1=v_{m-1,1,0,\ldots},\ldots,u_n=v_{m-1,0\ldots 1}$$

is the inverse of v_m. Therefore $v_m(\mathbb{P}^n)$ is determined by (2) and v_m is an isomorphic embedding of $\mathbb{P}^n$ into $\mathbb{P}^{\nu_{n,m}}$.

The significance of a Veronese mapping lies in the fact that if $F=\sum a_{i_0\dots i_n}u_0^{i_0}\dots u_n^{i_n}$ is a form of degree m in the homogeneous coordinates of the point $x\in\mathbb{P}^n$ and if H is the hypersurface defined by the equation $F=0$ in $\mathbb{P}^n$, then $v_m(H)$ is the intersection of $v_m(\mathbb{P}^n)$ and the hyperplane with the equation $\Sigma a_{i_0\dots i_n}v_{i_0\dots i_n}=0$ in $\mathbb{P}^{\nu_{n,m}}$. Therefore a Veronese mapping makes it possible to reduce the study of certain problems connected with hypersurfaces to the case of hyperplanes.

Exercises

1. Show that an affine variety U is irreducible if and only if its closure $\bar{U}$ in a projective space is irreducible.
2. Assign to any affine variety U contained in $\mathbb{A}_0^n$ its closure $\bar{U}$ in the projective space $\mathbb{P}^n$. Show that this gives a one-to-one correspondence between affine varieties in $\mathbb{A}_0^n$ and those projective varieties in $\mathbb{P}^n$ that have no component contained in the hyperplane $S_0=0$.
3. Show that the subvariety in $\mathbb{P}^3$ defined by the equations $x_1x_3=x_2^2$, $x_0x_2=x_1^2$, $x_0x_3=x_1x_2$ (which is called a twisted cubic curve) is irreducible.
4. Decompose into irreducible components the variety defined by two of the equations in Exercise 3.
5. Show that the variety $X=\mathbb{A}^2-x$, where $x=(0,0)$, is not isomorphic to an affine variety. *Hint:* Work out $k[X]$ and use the fact that for an affine variety any proper ideal $\mathfrak{A}\subset k[X]$ determines a non-empty subvariety.
6. Show that any quasiprojective variety is open in its closure in a projective space.
7. Show that any rational mapping $\varphi:\mathbb{P}^1\to\mathbb{P}^n$ is regular.
8. Show that any regular mapping $\varphi:\mathbb{P}^1\to\mathbb{A}^n$ maps $\mathbb{P}^1$ into a single point.
9. Determine a birational isomorphism f between an irreducible quadric X in $\mathbb{P}^3$ and a plane $\mathbb{P}^2$, by analogy with Example 1 in §3.3 (stereographic projection). At what points is f not regular? At what point is f^{-1} not regular?
10. In Example 9, find open isomorphic sets $U\subset X$ and $V\subset\mathbb{P}^2$.
11. Show that the mapping $y_0=x_1x_2$, $y_1=x_0x_2$, $y_2=x_0x_1$ defines a birational isomorphism f of the plane $\mathbb{P}^2$ with itself. At what point is f and at what points is f^{-1} not regular? Find open sets between which f determines an isomorphism.
12. Show that the variety $v_m(\mathbb{P}^n)$ is not contained in any linear subspace of $\mathbb{P}^{\nu_{n,m}}$.
13. Show that the variety of Exercise 3 coincides with $v_3(\mathbb{P}^1)$.
14. Show that the variety $\mathbb{P}^2-X$, where X is a conic in $\mathbb{P}^2$, is affine. *Hint:* Use a Veronese mapping v_2.

§ 5. Products and Mappings of Quasiprojective Varieties

1. Products. The definition of a product of affine varieties (Example 5 of §2.1) was so natural that it did not require any explanations. For arbitrary quasiprojective varieties the matter is somewhat more complicated. Therefore we begin by discussing quasiprojective subvarieties of affine spaces. If $X \subset \mathbb{A}^n$ and $Y \subset \mathbb{A}^m$ are such varieties, then the set $X \times Y = \{(x, y); x \in X, y \in Y\}$ is a quasiprojective subvariety of $\mathbb{A}^n \times \mathbb{A}^m = \mathbb{A}^{n+m}$. For if $X = X_1 - X_0$, $Y = Y_1 - Y_0$, where X_1, X_0 and Y_1, Y_0 are closed subvarieties of the spaces $\mathbb{A}^n$ and $\mathbb{A}^m$, respectively, then the representation $X \times Y = X_1 \times Y_1 - (X_1 \times Y_0 \cup Y_1 \times X_0)$ shows that $X \times Y$ is quasiprojective. We call this quasiprojective variety the *direct product* of X and Y. At this place we have to verify that if X and Y are replaced by isomorphic varieties, then $X \times Y$ is also replaced by an isomorphic variety. This is easy to check. Let $\varphi: X \to X' \subset \mathbb{A}^p$ and $\psi: Y \to Y' \subset \mathbb{A}^q$ be isomorphisms. Then $(\varphi \times \psi): X \times Y \to X' \times Y'$, where $(\varphi \times \psi)(x, y) = (\varphi(x), \psi(y))$, is a regular mapping, and $(\varphi^{-1}, \psi^{-1})$ is its inverse.

We now turn to quasiprojective varieties and clarify what we expect from the concept of a product. Let $X \subset \mathbb{P}^n$ and $Y \subset \mathbb{P}^m$ be two quasiprojective varieties. We denote by $X \times Y$ the set of pairs (x, y), $x \in X$, $y \in Y$. We wish to regard this set as a quasiprojective variety, and for this purpose we have to specify an embedding φ of it in a projective space $\mathbb{P}^N$ such that $\varphi(X \times Y)$ is a quasiprojective subvariety of $\mathbb{P}^N$. Here it is natural to ask that the definition should be local, in other words, that arbitrary points $x \in X$ and $y \in Y$ should have affine neighbourhoods $U \subset X$ of x and $V \subset Y$ of y such that $\varphi(U \times V)$ is open in $\varphi(X \times Y)$ and that φ defines an isomorphism between the direct product of the affine varieties U and V (whose definition is already known to us) and the variety $\varphi(U \times V) \subset \varphi(X \times Y)$. It is easy to see that by the property of being local the embedding φ is in essence uniquely determined; more accurately, if $\psi: X \times Y \to \mathbb{P}^M$ is another such embedding, then $\psi\varphi^{-1}$ determines an isomorphism between $\varphi(X \times Y)$ and $\psi(X \times Y)$. For it is sufficient to show that arbitrary $x \in X$ and $y \in Y$ have neighbourhoods $W_1 \subset \varphi(X \times Y)$ of $\varphi(x, y)$ and $W_2 \subset \psi(X \times Y)$ of $\psi(x, y)$ such that $\psi\varphi^{-1}$ defines an isomorphism between W_1 and W_2. With this aim we consider affine neighbourhoods $U \subset X$ of x and $V \subset Y$ of y, whose existence is guaranteed by the property of being local. We can even assume that $U \times V$ is isomorphic to both $\varphi(U \times V)$ and $\psi(U \times V)$, going over, if necessary, to smaller affine neighbourhoods. Then $\varphi(U \times V) = W_1$ and $\psi(U \times V) = W_2$ are the required affine neighbourhoods, because according to the assumption we have made both are isomorphic to the direct product $U \times V$ of the affine varieties U and V.

Let us proceed to the construction of the embedding φ having the requisite property. Here we can at once confine ourselves to the case when $X = \mathbb{P}^n$, $Y = \mathbb{P}^m$; if the embedding $\varphi(\mathbb{P}^n \times \mathbb{P}^m) \to \mathbb{P}^N$ has already been constructed, then a simple verification shows that its restriction to $X \times Y \subset \mathbb{P}^n \times \mathbb{P}^m$ has all the necessary properties.

To construct φ we consider the space $\mathbb{P}^{(n+1)(m+1)-1}$ in which homogeneous coordinates $w_{i,j}$ are numbered by double suffixes i and j

$(i=0, \ldots, n; j=0, \ldots, m)$. If $x=(u_0 : \ldots : u_n) \in \mathbb{P}^n$, $y=(v_0 : \ldots : v_m) \in \mathbb{P}^m$, then we set

$$\varphi(x, y) = (w_{ij}), \ w_{ij} = u_i v_j (i=0, \ldots, n; j=0, \ldots, m). \tag{1}$$

Clearly, multiplication of the homogeneous coordinates of x (or y) by a common non-zero factor does not change the point

$$\varphi(x, y) \in \mathbb{P}^{(n+1)(m+1)-1}.$$

In order to show that $\varphi(\mathbb{P}^n \times \mathbb{P}^m)$ is a closed set in $\mathbb{P}^{(n+1)(m+1)-1}$, we write down its equations,

$$w_{ij} w_{kl} = w_{kj} w_{il} \quad (i, k=0, \ldots, n; j, l=0, \ldots, m). \tag{2}$$

Substitution shows that w_{ij} as defined in (1) satisfies (2). Conversely, if the w_{ij} satisfy (2) and if $w_{00} \neq 0$ say, then setting $k=l=0$ in (2) we see that $(\ldots : w_{ij} : \ldots) = \varphi(x, y)$, where $x=(w_{00} : \ldots : w_{n0})$, $y=(w_{00} : \ldots w_{0m})$. This argument shows at the same time that $\varphi(x, y)$ is uniquely determined by x and y, so that φ is an embedding of $\mathbb{P}^n \times \mathbb{P}^m$ in $\mathbb{P}^{(n+1)(m+1)-1}$. We consider the open sets $\mathbb{A}_0^n (u_0 \neq 0)$ and $\mathbb{A}_0^m (v_0 \neq 0)$ in $\mathbb{P}^n$ and $\mathbb{P}^m$, respectively. It is clear that

$$\varphi(\mathbb{A}_0^n \times \mathbb{A}_0^m) = W_{00} = \mathbb{A}_{00}^{(n+1)(m+1)-1} \cap \varphi(\mathbb{P}^n \times \mathbb{P}^m),$$

where $\mathbb{A}_{00}^{(n+1)(m+1)-1} = \{w_{00} \neq 0\}$. If $(w_{ij}) = \varphi(x, y) \in W_{00}$ and $z_{ij} = w_{ij}/w_{00}$, $x_i = u_i/u_0$, $y_j = v_j/v_0$ are inhomogeneous coordinates, then as we have just found, $z_{i0} = x_i$, $z_{0j} = y_j$, $z_{ij} = x_i y_j = z_{i0} z_{0j}$, for $i>0$, $j>0$. Hence it follows that W_{00} is isomorphic to the affine space $\mathbb{A}^{n+m}$ with the coordinates $(x_1, \ldots, x_n, y_1, \ldots, y_m)$ and that φ determines an isomorphism $\mathbb{A}_0^n \times \mathbb{A}_0^m \to W_{00} = \varphi(\mathbb{A}_0^n \times \mathbb{A}_0^m)$. This proves that our construction has the property of being local.

Note 1. The points (w_{ij}) can be interpreted as $(n+1)$-by-$(m+1)$ matrices; the Eqs. (2) can be written in the form $\begin{vmatrix} w_{ij} & w_{il} \\ w_{kj} & w_{kl} \end{vmatrix} = 0$ and indicate that the rank of the matrix (w_{ij}) is 1, and the Eqs. (1) show that this matrix is the product of a column of type $(n+1, 1)$ and a row of type $(1, m+1)$.

Note 2. The simplest case case $n=m=1$ has a simple geometric meaning. We have one Eq. (2) $w_{11} w_{00} = w_{01} w_{10}$ so that $\varphi(\mathbb{P}^1 \times \mathbb{P}^1)$ is a non-degenerate quadric Q in $\mathbb{P}^3$. The set $\varphi(\alpha \times \mathbb{P}^1)$, where $\alpha = (\alpha_0 : \alpha_1)$, is given in $\mathbb{P}^3$ by the equations $\alpha_1 w_{00} = \alpha_0 w_{10}$, $\alpha_1 w_{01} = \alpha_0 w_{11}$ and determines a line in $\mathbb{P}^3$. Similarly $\varphi(\mathbb{P}^1 \times \beta)$, where $\beta \in \mathbb{P}^1$, is a line. When α ranges over the whole of $\mathbb{P}^1$, the lines of the first kind give all the lines of a family of generators of Q. The lines of the second kind give the second family.

After having given the definition of a direct product by means of the embedding φ of $\mathbb{P}^n \times \mathbb{P}^m$ in $\mathbb{P}^{(n+1)(m+1)-1}$, it is convenient to interpret some concepts that were first defined by means of this embedding in terms of the set $\mathbb{P}^n \times \mathbb{P}^m$. For example, let us find out what subsets of $\mathbb{P}^n \times \mathbb{P}^m$ go over under φ into algebraic varieties. Subvarieties $X \subset \mathbb{P}^{(n+1)(m+1)-1}$ are defined by equations $F_i(w_{00}:\ldots:w_{nm})=0$, where the F_i are homogeneous polynomials. After the substitution (1) this can be written in the coordinates u_i and v_j in the form $G_i(u_0:\ldots:u_n; v_0:\ldots:v_m)=0$, where the G_i are homogeneous both in $u_0, \ldots, u_n$ and in $v_0, \ldots, v_m$, of the same degree of homogeneity with respect to the two systems of variables. Conversely, as is easy to verify, polynomials with this property of homogeneity can always be represented as polynomials in the products $u_i v_j$. However, if the equations are homogeneous both in the u_i and v_j, then they always determine in $\mathbb{P}^n \times \mathbb{P}^m$ an algebraic subvariety even if the degrees of homogeneity are different. If the polynomial

$$G(u_0:\ldots u_n; v_0:\ldots:v_m)$$

is of degree r in the u_i and s in the v_j and if $r>s$, say, then the equation $G=0$ is equivalent to the system $v_i^{r-s}G=0 (i=0, \ldots, m)$ of which we know already that it determines an algebraic variety.

Later we shall be faced with a similar problem for the product $\mathbb{P}^n \times \mathbb{A}^m$. Let $\mathbb{A}^m = \mathbb{A}_0^m \subset \mathbb{P}^m$ be given by the condition $v_0 \neq 0$. The equations of a closed set are of the form $G_i(u_0:\ldots:u_n; v_0:\ldots:v_m)=0$. Suppose that the G_i are homogeneous of degree r_i in $v_0, \ldots, v_m$. When we divide the equations by $v_0^{r_i}$ and set $y_j = v_j/v_0$, we obtain equations $g_i(u_0:\ldots:u_n; y_1, \ldots, y_m)=0$, where the g_i are homogeneous in $u_0, \ldots, u_n$, but in general inhomogeneous in $y_1, \ldots, y_m$. So we have proved the following result.

Theorem 1. *A subset $X \subset \mathbb{P}^n \times \mathbb{P}^m$ is closed if and only if it is given by a system of equations*

$$G_i(u_0:\ldots:u_n; v_0:\ldots:v_m)=0 \ (i=1, \ldots, t),$$

homogeneous in each system of variables u_i and v_j separately. Every closed subset of $\mathbb{P}^n \times \mathbb{A}^m$ is given by a system of equations

$$g_i(u_0:\ldots:u_n; y_1, \ldots, y_m)=0 \ (i=1, \ldots, t), \tag{3}$$

homogeneous in the variables $u_0, \ldots, u_n$.

Of course, the matter is similar for a product of any number of spaces. For example, a variety in $\mathbb{P}^{n_1} \times \cdots \times \mathbb{P}^{n_l}$ is given by a system of equations that are homogeneous in each of the l groups of variables.

2. Closure of the Image of a Projective Variety. The image of an affine variety under a regular mapping need not be a closed set. For mappings of an affine variety into another this is shown by Examples 3 and 4 of §2.3. For a mapping of an affine variety into a projective variety this is even more obvious: an example is the embedding of $\mathbb{A}^n$ in $\mathbb{P}^n$ as an open set $\mathbb{A}_0^n$. In this respect projective varieties differ radically from affine varieties.

Theorem 2. *The image of a projective variety under a regular mapping is closed.*

The proof makes use of a concept which will also occur later. Let $f: X \to Y$ be a regular mapping of arbitrary quasiprojective varieties. The subset Γ_f of $X \times Y$ consisting of the points of the form $(x, f(x))$ is called the graph of f.

Lemma 1. *The graph of a regular mapping is closed in $X \times Y$.*

First of all, it is sufficient to take Y as a projective space. For if $Y \subset \mathbb{P}^m$, then $X \times Y \subset X \times \mathbb{P}^m$, f determines a mapping $\bar{f}: X \to \mathbb{P}^m$, and $\Gamma_f = \Gamma_{\bar{f}} \cap (X \times Y)$. Therefore we set $Y = \mathbb{P}^m$. Let i be the identity mapping of $\mathbb{P}^m$ onto itself. We consider the regular mapping $(f, i): X \times \mathbb{P}^m \to \mathbb{P}^m \times \mathbb{P}^m$, $(f, i)(x, y) = (f(x), y))$. Clearly Γ_f is the inverse image of Γ_i under the regular mapping (f, i). In § 4.2 we have verified that the inverse image of a closed set under a regular mapping is closed. Therefore everything reduces to a verification that Γ_i is closed in $\mathbb{P}^m \times \mathbb{P}^m$. But Γ_i consists of the points $(x, y) \in \mathbb{P}^m \times \mathbb{P}^m$, $x = (u_0 : \ldots : u_m)$, $y = (v_0 : \ldots : v_m)$, for which $(u_0 : \ldots : u_m)$ is proportional to $(v_0 : \ldots : v_m)$. This can be written in the form $u_i v_j = u_j v_i$, $w_{ij} = w_{ji}$ $(i, j = 0, \ldots, m)$. The fact that Γ_i is closed follows from Theorem 1, and consequently the Lemma is proved.

We now return to the proof of the theorem. Let Γ_f be the graph of f and $p: X \times Y \to Y$ the projection defined by $p(x, y) = y$. Obviously $f(X) = p(\Gamma_f)$. By Lemma 1, Theorem 2 is a consequence of the following more general proposition.

Theorem 3. *If X is a projective and Y a quasiprojective variety, then the projection $p: X \times Y \to Y$ carries closed sets into closed sets.*

The proof of the theorem can be reduced to a very simple case. First of all, if X is a closed subset of $\mathbb{P}^n$, then by proving the theorem for $\mathbb{P}^n$ we also prove it for X because $X \times Y$ is closed in $\mathbb{P}^n \times Y$, and if Z is closed in $X \times Y$, then it is closed in $\mathbb{P}^n \times Y$. Therefore we may assume that $X = \mathbb{P}^n$. Secondly, since the concept of being closed is local, we may cover Y by affine open sets U_i and prove the theorem for each of them. Therefore we may assume that Y is an affine variety. Finally, if Y is closed in $\mathbb{A}^m$, then $\mathbb{P}^n \times Y$ is closed in $\mathbb{P}^n \times \mathbb{A}^m$, so that it is sufficient for us to prove the theorem when $X = \mathbb{P}^n$, $Y = \mathbb{A}^m$. What does the theorem mean in this case? According to Theorem 1, every closed subset $Z \subset \mathbb{P}^n \times \mathbb{A}^m$ is given by a system of equations (3) of § 5.1, which we write in the form $g_i(u; y) = 0 (i = 1, \ldots, t)$. Clearly if $y_0 \in \mathbb{A}^m$, then $p^{-1}(y_0)$ consists of all non-zero solutions of the system $g(u, y_0) = 0$, and hence $y_0 \in p(Z)$ if and only if the system of equations $g_i(u, y_0) = 0$ has a non-zero solution. Thus, Theorem 3 asserts that for any system (3) of § 5.1 the set T of those $y_0 \in \mathbb{A}^m$

for which the system $g_i(u, y_0)=0$ has a non-zero solution is closed. By Lemma 1 in § 4.1 the system $g_i(u, y_0)=0$ $(i=1, \ldots, t)$ has a non-zero solution if and only if $(g_i(u, y_0), \ldots, g_t(u, y_0)) \not\supset I_s$ for all $s=1, 2, \ldots$. We now verify that for any given $s \geqslant 1$ the points $y_0 \in \mathbb{A}^m$ for which $(g_1(u, y_0), \ldots, g_t(u, y_0)) \not\supset I_s$ form a closed set T_s. Then $T=\bigcap T_s$ is also closed. We denote by k_i the degree of the homogeneous polynomial $g_i(u, y)$ in the variables $u_0, \ldots, u_n$. Let $M^{(\alpha)}$ be an arbitrary numbering of the monomials of degree s in the variables $u_0, \ldots, u_n$. The condition $(g_1(u, y_0), \ldots, g_t(u, y_0)) \supset I_s$ means that all the $M^{(\alpha)}$ can be represented in the form

$$M^{(\alpha)}=\sum_{i=1}^{t} g_i(u, y_0) F_{i,\alpha}(u). \tag{1}$$

Comparing the homogeneous components of degree s we can see that there must be a similar equation in which $\deg F_{i,\alpha}=s-k_i$ (and $F_{i,\alpha}=0$ if $k_i>s$). We denote by $N_i^{(\beta)}$ the monomials of degree $s-k_i$, numbered arbitrarily. We see that the relation (1) is equivalent to the fact that all the monomials $M^{(\alpha)}$ are linear combinations of the polynomials $g_i(u, y_0) N_i^{(\beta)}$. Of course, this is equivalent to the fact that the monomials $g_i(u, y_0) N_i^{(\beta)}$ generate the whole linear vector space S of homogeneous polynomials of degree s in $u_0, \ldots, u_n$. Conversely, the condition

$$(g_1(u, y_0), \ldots, g_t(u, y_0)) \not\supset I_s$$

means that the polynomials $g_i(u, y_0) N_i^{(\beta)}$ together do not generate the space S. In order to write down these conditions, we have to express the coefficients of the polynomials $g_i(u, y_0) N_i^{(\beta)}$ as a rectangular matrix and equate to zero all the minors of this matrix of order $\sigma=\dim S$. Clearly, these minors are polynomials in the coefficients of the $g_i(u, y_0)$, hence are polynomials in the coordinates of y_0. They give the equations of the set T_s. Theorem 3, and hence also Theorem 2, are now proved.

Corollary 1. *If a function φ is regular on an irreducible projective variety, then $\varphi \in k$, that is, φ is a constant.*

Proof. We can regard φ as a mapping $f: X \rightarrow \mathbb{A}^1$, hence also as a mapping $\bar{f}: X \rightarrow \mathbb{P}^1$. Since φ is regular, so is f, and a fortiori $\bar{f}$. Hence, by Theorem 2 its image is closed. But since f is regular and $f(X)=\bar{f}(X)$, the set $\bar{f}(X)$ is closed and is contained in $\mathbb{A}^1$, that is, it does not contain the point at infinity $x_{\infty} \in \mathbb{P}^1$. From this it follows that either $f(X)=\mathbb{A}^1$ or that $f(X)$ is a finite set S (Example 3 of § 2.1). The first case is impossible, because $f(X)$ must be closed in $\mathbb{P}^1$, whereas $\mathbb{A}^1$ is not. Hence $f(X)=S$. If S consists of the points $\alpha_1, \ldots, \alpha_t$, then $X=\bigcup f^{-1}(\alpha_i)$, and if $t>1$, this contradicts the irreducibility of X. Therefore S consists of a single point, which means that φ is constant.

Corollary 1 and Theorem 4 of § 3 are examples of the fact that affine and projective varieties have diametrically opposite properties. On an affine variety there is a wealth of regular functions: they form the whole ring $k[X]$, but the only regular functions on an irreducible variety are constants. Here is another example of the contrast between affine and projective varieties.

Corollary 2. *A regular mapping $f: X \to Y$ of a projective irreducible variety X into an affine variety Y maps X into a point.*

Let $Y \subset \mathbb{A}^m$. The mapping f is given by m functions

$$f(x) = (\varphi_1(x), \ldots, \varphi_m(x)).$$

Each of the functions $\varphi_i(x)$ is constant, by Corollary 1, $\varphi_i = \alpha_i \in k$. Therefore $f(X) = (\alpha_1, \ldots, \alpha_m)$.

We give a further example of an application of Theorem 2. For this purpose we make use of a representation of forms of degree m in $n+1$ variables by points of the space $\mathbb{P}^{\nu_{n,m}}$ (§ 4.4).

Proposition. *Points $\xi \in \mathbb{P}^{\nu_{n,m}}$ corresponding to reducible polynomials F form a closed set.*

The proposition asserts that the condition of reducibility of a homogeneous polynomial can be written in the form of algebraic relations between its coefficients. For conics, that is, for the case $m = n = 2$, this relation is well-known from analytic geometry: if $F = \sum_{i,j=0}^{2} a_{ij} U_i U_j$, then F is reducible if and only if $|a_{ij}| = 0$.

Proceeding to the proof of the proposition we denote by X the set of points $\xi \in \mathbb{P}^{\nu_{n,m}}$ corresponding to reducible polynomials, and by $X_j (j = 1, \ldots, m-1)$ the set of all points corresponding to polynomials X that split into factors of degree j and $m-j$. Clearly $X = \bigcup X_j$, and we need only prove that each X_j is closed.

Consider the projective spaces $\mathbb{P}^{\nu_{n,j}}$ and $\mathbb{P}^{\nu_{n,m-j}}$. The multiplication of two polynomials of degree j and $m-j$ determines a mapping

$$f: \mathbb{P}^{\nu_{n,j}} \times \mathbb{P}^{\nu_{n,m-j}} \to \mathbb{P}^{\nu_{n,m}},$$

which is regular, as is easy to see. Obviously $X_j = f(\mathbb{P}^{\nu_{n,j}} \times \mathbb{P}^{\nu_{n,m-j}})$. As we have seen in § 5.1, the product of projective spaces is a projective variety, hence it follows from Theorem 2 that X_j is closed.

3. Finite Mappings. The projection mapping introduced in §4.4 has an important property, which we can state after recalling some algebraic concepts. Let B be a ring containing a ring A. An element $b \in B$ is said to be integral over A if it satisfies an equation $b^m + a_1 b^{m-1} + \cdots + a_m = 0$, $a_i \in A$. The ring B is called integral over A if each of its elements is integral over A. It is easy to prove (see, for example, [37], Vol. 1, Ch. V, § 1) that a ring B having finitely many generators over A is integral over A if and only if it is a module of finite type over A.

Let X and Y be affine varieties and $f: X \to Y$ a regular mapping such that $f(X)$ is dense in Y. Then f^* determines an isomorphic embedding of $k[Y]$ into $k[X]$. Making use of this we may regard $k[Y]$ as a subring of $k[X]$.

Definition 1. A mapping f is called *finite* if $k[X]$ is integral over $k[Y]$.*

From the above property of integral rings it follows that the compositum of two finite mappings is itself finite. A typical instance of a mapping that is not finite is Example 3 in §2.3.

If f is a finite mapping, then every point $y \in Y$ has only finitely many inverse images. For suppose that $X \subset \mathbb{A}^n$ and that $t_1, \ldots, t_n$ are coordinates in $\mathbb{A}^n$ as functions on X. It is enough to prove that every coordinate t_j assumes only finitely many values on the set $f^{-1}(y)$. By definition t_j satisfies an equation $t_j^m + b_1 t_j^{m-1} + \cdots + b_m = 0$, $b_i \in k[Y]$. For $x \in f^{-1}(y)$, $y \in Y$, we obtain the equation

$$(t_j(x))^m + b_1(y)(t_j(x))^{m-1} + \cdots + b_m(y) = 0\,, \tag{1}$$

which has finitely many roots.

The meaning of the concept of finiteness lies in the fact that when y ranges over Y, no root of (1) tends to infinity because the coefficient of the leading term does not vanish. Therefore, as y ranges over Y, the points $f^{-1}(y)$ can merge, but they cannot disappear. A more precise form of this remark is the following theorem.

Theorem 4. *A finite mapping is epimorphic.*

Let $f: X \to Y$ be a finite mapping, X and Y affine varieties, $y \in Y$. We denote by $\mathfrak{m}_y$ the ideal of $k[Y]$ consisting of the functions that vanish at y. If $t_1, \ldots, t_n$ are coordinates as functions on Y and if $y = (\alpha_1, \ldots, \alpha_n)$, then $\mathfrak{m}_y = (t_1 - \alpha_1, \ldots, t_n - \alpha_n)$. The equations of the variety $f^{-1}(y)$ are of the form $f^*(t_1) = \alpha_1, \ldots, f^*(t_n) = \alpha_n$, and the set $f^{-1}(y)$ is empty if and only if $(f^*(t_1) - \alpha_1, \ldots, f^*(t_n) - \alpha_n) = k[X]$. From now on we do not distinguish between functions $u \in k[Y]$ and $f^*(u) \in k[X]$, regarding $k[Y]$ as a subring of $k[X]$. Then the preceding condition can be written in the form $(t_1 - \alpha_1, \ldots, t_n - \alpha_n) = k[X]$ or $\mathfrak{m}_y k[X] = k[X]$. From the fact that $k[X]$ is integral over $k[Y]$ it follows that $k[X]$ is a module of finite type over $k[Y]$. In view of this, Theorem 4 follows from a purely algebraic proposition.

Lemma 2. *If a ring B is a module of finite type over a subring A (with unit element), then $\mathfrak{a}B \neq B$ for any proper ideal $\mathfrak{a} \subset A$.*

Let $B = A\omega_1 + \cdots + A\omega_n$: then from $\mathfrak{a}B = B$ it would follow that $\omega_i \in \mathfrak{a}\omega_1 + \cdots + \mathfrak{a}\omega_n$, that is, $\omega_i = \sum \alpha_{ij}\omega_j, \alpha_{ij} \in \mathfrak{a}$. Therefore $\sum_j (\alpha_{ij} - \delta_{ij})\omega_j = 0$ and hence $d\omega_j = 0$ $(j = 1, \ldots, n)$, $d = |\alpha_{ij} - \delta_{ij}|$. Therefore

* Such a mapping is sometimes called *finite and dominant*.

$dB=0$ and hence $d=0$. Since $\alpha_{ij} \in \mathfrak{a}$, it follows from $|\alpha_{ij}-\delta_{ij}|=0$ that $1 \in \mathfrak{a}$ and hence $\mathfrak{a}=A$. This contradiction proves the lemma.

Corollary. *A finite mapping carries closed sets into closed sets.*

It is sufficient to verify this for irreducible closed sets. If $f: X \to Y$ is a finite mapping and $Z \subset X$ is irreducible, then we have to apply Theorem 4 to the mapping $\bar{f}: Z \to \overline{f(Z)}$, the restriction of f to Z. It is obviously finite, hence $f(Z)=\overline{f(Z)}$, that is, $f(Z)$ is closed.

The property of being finite is local.

Theorem 5. *If $f: X \to Y$ is a regular mapping of affine varieties and if every point $x \in Y$ has an affine neighbourhood U such that $V=f^{-1}(U)$ is affine and $f: V \to U$ finite, then f is also finite.*

We set $k[X]=B$, $k[Y]=A$. In §4.2 we have given the definition of a principal open set. For every point we can take a neighbourhood U being a principal open set and satisfying the conditions of the theorem (see Exercise 11).

Let $D(g_\alpha)$ be a system of such open sets, which we may take to be finite in number. Then $Y=\bigcup D(g_\alpha)$, that is, the ideal generated by all the g_α is equal to A. In our case, $V_\alpha=f^{-1}(D(g_\alpha))=D(f^*(g_\alpha))$, $k[D(g_\alpha)]=A[1/g_\alpha]$, $k[V_\alpha]=B[1/g_\alpha]$. By hypothesis, $B[1/g_\alpha]$ has a finite basis $\omega_{i,\alpha}$ over $A[1/g_\alpha]$. Here we may assume that $\omega_{i,\alpha} \in B$; if the basis consisted of elements $\omega_{i,\alpha}/g_\alpha^{m_i}$, $\omega_{i,\alpha} \in B$, then the elements $\omega_{i,\alpha}$ would also be a basis. We consider the union of all the basis elements $\omega_{i,\alpha}$ and show that they form a basis of B over A.

Every element $b \in B$ has a representation

$$b=\sum_i \frac{a_{i,\alpha}}{g_\alpha^{n_\alpha}}\,\omega_{i,\alpha}$$

for each α. Since the elements $g_\alpha^{n_\alpha}$ generate the unit ideal of A, there exist $h_\alpha \in A$ such that $\sum_\alpha g_\alpha^{n_\alpha} h_\alpha=1$. Therefore

$$b=b\sum_\alpha g_\alpha^{n_\alpha} h_\alpha=\sum_i\sum_\alpha a_{i,\alpha}h_\alpha\omega_{i,\alpha},$$

which proves the theorem.

Definition 2. A regular mapping $f: X \to Y$ of quasiprojective varieties is called *finite* if every point $y \in Y$ has an affine neighbourhood V such that the set $U=f^{-1}(V)$ is affine and the mapping of affine varieties $f: U \to V$ is finite.

Obviously, for every finite mapping f the set $f^{-1}(y)$ is finite for every $y \in Y$.

From Theorem 4 it follows that every finite mapping is epimorphic.

This property leads to important consequences concerning arbitrary mappings.

Theorem 6. *If $f: X \to Y$ is a regular mapping and $f(X)$ is dense in Y, then $f(X)$ contains a set that is open in Y.*

The assertion of the theorem easily reduces to the case when X and Y are irreducible and affine, which we shall now assume. Then $k[Y] \subset k[X]$. We denote the transcendence degree of the extension $k(X)/k(Y)$ by r, and we choose r elements $u_1, \ldots, u_r \in k[X]$ that are algebraically independent over $k(Y)$. Then $k[X] \supset k[Y][u_1, \ldots, u_r] \supset k[Y]$, and $k[Y][u_1, \ldots, u_r] = k[Y \times \mathbb{A}^r]$. Thus, the mapping f can be represented as the composition of two mappings: $f = g \circ h$, $h: X \to Y \times \mathbb{A}^r$ and $g: Y \times \mathbb{A}^r \to Y$, where g is simply the projection onto the first factor. Every element $v \in k[X]$ is algebraic over $k[Y \times \mathbb{A}^r]$, hence we can find for it an element $a \in k[Y \times \mathbb{A}^r]$ such that αv is integral over $k[Y \times \mathbb{A}^r]$. We choose such elements $a_1, \ldots, a_m$ for some system of generators $v_1, \ldots, v_m$ of the ring $k[X]$, and we set $F = a_1 \ldots a_m$. Since in the open set $D(F) \subset Y \times \mathbb{A}^r$ the functions a_i are invertible, the functions v in $D(h^*(F)) \subset X$ are integral, that is, the restriction

$$h: D(h^*(F)) \to D(F)$$

is finite. By Theorem 4, $D(F) \subset h(X)$. It remains to show that $g(D(F))$ contains a set that is open in Y.

Let

$$F(y, T) = \sum F_\alpha(y) T^{(\alpha)},$$

where the $T^{(\alpha)}$, monomials in the variables $T_1, \ldots, T_r$, are coordinates in $\mathbb{A}^r$. For points $y \in Y$ for which not all the $F_\alpha(y) = 0$ there exist values $T_i = \tau_i$ such that $F(y, \tau) \neq 0$. Therefore $g(D(F)) \supset \bigcup D(F_\alpha)$.

Theorem 6 shows how much simpler regular mappings of algebraic varieties are than continuous or differentiable mappings.

The famous everywhere dense twisting of the torus, that is, the mapping

$$f: \mathbb{R}^1 \to T, \; T = (\mathbb{R}/\mathbb{Z})^2, \; f(x) = (x, x\sqrt{2}) \bmod \mathbb{Z}^2$$

gives an example of a situation that cannot occur for algebraic varieties by virtue of Theorem 6.

Theorem 7. *If X is closed in $\mathbb{P}^n$ and $X \subset \mathbb{P}^n - E$, where E is a d-dimensional linear subspace, then the projection $\pi: X \to \mathbb{P}^{n-d-1}$ with centre at E determines a finite mapping $X \to \pi(X)$.*

Proof. Let $y_0, \ldots, y_{n-d-1}$ be homogeneous coordinates in $\mathbb{P}^{n-d-1}$ and let π be given by the formulae $y_j = L_j(x)$, $x \in X$ $(j = 0, \ldots, n-d-1)$.

Obviously, $U_i = \pi^{-1}(\mathbb{A}_i^{n-d-1}) \cap X$ is given by the condition $L_i(x) \neq 0$ and is an affine open subset of X. We show that $\pi: U_i \to \mathbb{A}_i^{n-d-1} \cap \pi(X)$ is a finite mapping. Every function $g \in k[U_i]$ is of the form $g = G_i(x_0, \ldots, x_n)/L_i^m$, where G_i is a form of degree m. Consider the mapping $\pi_1: X \to \mathbb{P}^{n-d}: z_j = L_j^m(x)$ $(j = 0, \ldots, n-d-1)$, $z_{n-d} = G_i(x)$, where $z_0, \ldots, z_{n-d}$ are homogeneous coordinates in $\mathbb{P}^{n-d}$. It is a regular mapping and its image $\pi_1(X)$ is closed in $\mathbb{P}^{n-d}$, by Theorem 2. Let $F_1 = \cdots = F_s = 0$ be its equations. Since $X \subset \mathbb{P}^n - E$, the forms $L_i (i = 0, \ldots, n-d-1)$ do not vanish simultaneously on X. This means that the point $0 = (0 : \ldots : 1)$ is not contained in $\pi_1(X)$, in other words, that the equations $z_0 = \cdots = z_{n-d-1} = F_1 = \cdots = F_s = 0$ have no solution in $\mathbb{P}^{n-d}$. By Lemma 1 of §4.1 it follows that

$$(z_0, \ldots, z_{n-d-1}, F_1, \ldots, F_s) \supset T_l$$

for some $l > 0$. In particular, $z_{n-d}^l \in (z_0, \ldots, z_{n-d-1}, F_1, \ldots, F_s)$. This means that

$$z_{n-d}^l = \sum_{j=0}^{n-d-1} z_j H_j + \sum_{j=1}^{s} F_j P_j,$$

where H_j and P_j are polynomials. Denoting by $H^{(q)}$ the homogeneous component of H of degree q, we deduce that

$$\Phi(z_0, \ldots, z_{n-d}) = z_{n-d}^l - \sum z_j H_j^{(l-1)} = 0 \quad \text{on} \quad \pi_1(X). \tag{2}$$

The homogeneous polynomial Φ is of degree l and as a polynomial in z_{n-d} it has the leading coefficient 1:

$$\Phi = z_{n-d}^l - \sum_{j=0}^{l-1} A_{l-j}(z_0, \ldots, z_{n-d-1}) z_{n-d}^j. \tag{3}$$

Substituting the transformation formulae for π_1 in (2) we find that $\Phi(L_0^m, \ldots, L_{n-d-1}^m, G_i) = 0$ on X. with Φ of the form (3). Dividing this relation by L_i^{ml} we obtain the required relation

$$g^l + \sum_{j=0}^{l-1} A_{l-j}(x_0^m, \ldots, 1, \ldots, x_{n-d-1}^m) g^j = 0,$$

where $x_r = y_r/y_i$ are coordinates in $\mathbb{A}^{n-d-1}$. This proves the theorem.

An application of a Veronese mapping makes it possible to generalize this result substantially.

Theorem 8. *Let $F_0, \ldots, F_s$ be linearly independent forms of degree m on $\mathbb{P}^n$ that do not vanish simultaneously on a closed variety $X \subset \mathbb{P}^n$. Then $\varphi(x) = (F_0(x) : \ldots : F_s(x))$ determines a finite mapping $X \to \varphi(X)$.*

Let $v_m: \mathbb{P}^n \to \mathbb{P}^{\nu_{n,m}}$ be a Veronese mapping, and L_i linear forms on $\mathbb{P}^{\nu_{n,m}}$ corresponding to the forms F_i on $\mathbb{P}^n$. It is clear that then $\varphi = \pi \circ v_m$, where π is the projection determined by the forms $L_0, \ldots, L_s$. Since v_m is an isomorphism between X and $v_m(X)$, the theorem follows from Theorem 7.

4. Normalization Theorem. We consider an irreducible projective variety $X \subset \mathbb{P}^n$ other than the whole of $\mathbb{P}^n$. Then there exists a point $x \in \mathbb{P}^n$, $x \notin X$, and the mapping φ projecting X from x is regular. The variety $\varphi(X) \subset \mathbb{P}^{n-1}$ is projective, by Theorem 2, and the mapping $\varphi: X \to \varphi(X)$ is finite, by Theorem 7. If $\varphi(X) \neq \mathbb{P}^{n-1}$, then we can apply the same arguments to it. In the end we arrive at a mapping $X \to \mathbb{P}^m$, which is finite as the composition of finite mappings. The result we have proved is called the normalization theorem:

Theorem 9. *For every irreducible projective variety X there exists a finite mapping $\varphi: X \to \mathbb{P}^m$ onto a projective space.*

The analogous fact is true for affine varieties. To prove this we consider an affine variety $X \subset \mathbb{A}^n$. We assume that $\mathbb{A}^n$ is open in $\mathbb{P}^n$ and denote by $\bar{X}$ the closure of X in $\mathbb{P}^n$. Let $X \neq \mathbb{A}^n$. We choose a point $x \in \mathbb{P}^n - \mathbb{A}^n$, $x \notin \bar{X}$, and we consider the projection $\varphi: \bar{X} \to \mathbb{P}^{n-1}$ from this point. Here X is projected into "finite" points of $\mathbb{P}^{n-1}$, that is, into points of the affine space $\mathbb{A}^{n-1} = \mathbb{P}^{n-1} \cap \mathbb{A}^n$. We may continue this process as long as $X \neq \mathbb{A}^n$, and as a result we obtain a projection $\varphi: \bar{X} \to \mathbb{P}^m$ for which $\varphi(X) = \mathbb{A}^m$. So we have proved:

Theorem 10. *For every irreducible affine variety X there exists a finite mapping $\varphi: X \to \mathbb{A}^m$ onto an affine space.*

Theorems 9 and 10 allow us to reduce the study of certain (rather crude) properties of projectives of affine varieties to the case of a projective and affine space. For $m = 1$ this was the point of view of Riemann, who considered algebraic curves as covering of the Riemann sphere ($\mathbb{P}^1$ over the field of complex numbers).

Theorem 10 indicates that a ring A without divisors of zero that is finitely generated over a field k is integral over a ring isomorphic to a polynomial ring. This result can also be proved directly. It is easy to derive Hilbert's Nullstellensatz from it (see [37], Vol. II, Ch. 7, § 3).

Exercises

1. Show that a variety $\varphi(\mathbb{P}^r \times \mathbb{P}^s)$ does not lie in any linear subspace of $\mathbb{P}^{(r+1)(s+1)-1}$ other than the whole of $\mathbb{P}^{(r+1)(s+1)-1}$.

2. Consider the mapping of varieties $\mathbb{P}^1 \times \mathbb{P}^1 \to \mathbb{P}^1$: $p_1(x, y) = x$, $p_2(x, y) = y$. Show that $p_1(X) = p_2(X) = \mathbb{P}^1$ for every closed irreducible subvariety $X \subset \mathbb{P}^1 \times \mathbb{P}^1$, provided that X does not belong to one of the following types:
   a) a point $(x_0, y_0) \in \mathbb{P}^1 \times \mathbb{P}^1$;
   b) a set $x_0 \times \mathbb{P}^1$, where x_0 is a fixed point of $\mathbb{P}^1$;
   c) a set $\mathbb{P}^1 \times y_0$.

3. Verify directly Corollary 1 to Theorem 2 for the case $X = \mathbb{P}^n$.

4. Let $X = \mathbb{A}^2 - x$, where x is a point. Show that X is not isomorphic to an affine nor to a projective variety (see Exercise 5 in §4).

5. The same as in Exercise 4, for $\mathbb{P}^2 - x$.

6. The same as in Exercise 4 and 5 for $\mathbb{P}^1 \times \mathbb{A}^1$.

7. Is the mapping $f: \mathbb{A}^1 \to X$, where X is given by the equations $y^2 = x^3$, $f(t) = (t^2, t^3)$, finite?

8. Let X be a hypersurface in $\mathbb{P}^r$, L a line passing through the origin of coordinates, and φ_L the projection of X parallel to L onto an $(r-1)$-dimensional subspace not containing L. Denote by S the set of all those lines L for which φ_L is not finite. Show that S is an algebraic variety. Find S when $r=2$ and X is given by the equation $xy=1$.

9. Show that the intersection of affine open subsets is affine. *Hint:* Use Example 10 of §2.3.)

10. Show that forms of degree $m = l'l$ in $n+1$ variables that are l-th powers of forms correspond to points of some closed subset of $\mathbb{P}^{\nu_{m,n}}$.

11. Let $f: X \to Y$ be a regular mapping of affine varieties. Show that the inverse image of a principal affine open set is a principal affine open set.

§ 6. Dimension

1. Definition of Dimension. In §2 we have seen that the closed algebraic subvarieties $X \subset \mathbb{A}^2$ are finite sets of points, plane algebraic curves, or $\mathbb{A}^2$ itself. This division into three types corresponds to the intuitive notion of dimension: we have varieties of dimension 0, 1 and 2. Presently we shall give a definition of dimension for an arbitrary algebraic variety.

How can we arrive at such a definition? First of all, the dimension of an n-dimensional projective or affine space must, of course, be n. Secondly, if there is a finite mapping $X \to Y$, then naturally we assume that X and Y have the same dimension. Since by the normalization theorems (Theorem 9 and 10 of §5) any projective or affine variety X has a finite mapping onto some space $\mathbb{P}^m$ or $\mathbb{A}^m$, this m is naturally taken as the definition of the dimension of the variety. But here the question arises whether this is a good definition: could there perhaps exist two finite mappings $f: X \to \mathbb{A}^n$ and $g: X \to \mathbb{A}^m$, with $m \neq n$? Suppose that X is irreducible. Then from the finiteness of a mapping $f: X \to \mathbb{A}^n$ it follows that the field of rational functions $k(X)$ is a finite extension of the field $f^* k(\mathbb{A}^n)$, which in turn is isomorphic to the field $k(T_1, \ldots, T_n)$. Therefore $k(X)$ has the transcendence degree n, and this is a characterization of the number n, independent of the choice of the finite mapping $f: X \to \mathbb{A}^n$. In this way we have motivated to a certain extent the following definition of dimension.

Definition. The *dimension* of a quasiprojective reducible variety X is the transcendence degree of the field $k(X)$.

The dimension of reducible varieties is the maximum of the dimensions of its irreducible components.

The dimension of a variety X is denoted by $\dim X$.

If Y is a closed subvariety of X, then the number $\dim X - \dim Y$ is called the *codimension* of Y in X and is denoted by codim Y or $\operatorname{codim}_X Y$.

Observe that if X is an irreducible variety and U is open in X, then $k(U)=k(X)$, hence $\dim U = \dim X$.

Example 1. $\dim \mathbb{A}^n = \dim \mathbb{P}^n = n$, because $k(\mathbb{A}^n)$ is the same as the field of rational functions in n variables. Since by definition the dimension is invariant under a birational isomorphism, we see that $\mathbb{A}^n$ and $\mathbb{A}^m$ are not birationally isomorphic when $n \neq m$.

Example 2. A plane irreducible algebraic curve is of dimension 1, as we have seen in § 1.

Example 3. If X consists of a single point, then obviously $\dim X = 0$, hence the same is true when X is a finite set. Conversely, if $\dim X = 0$, then X is a finite set. It is enough to verify this for an irreducible affine variety X. Let $X \subset \mathbb{A}^n$ and $t_1, \ldots, t_n$ be coordinates in $\mathbb{A}^n$ as functions on X, that is, as elements of $k[X]$. By hypothesis, t_i is algebraic over k, hence assumes only finitely many values. From this it follows that X is finite.

Example 4. Let us show that if X and Y are irreducible varieties, then $\dim(X \times Y) = \dim X + \dim Y$.

It is sufficient to consider the case when X and Y are affine varieties, $X \subset \mathbb{A}^N$, $Y \subset \mathbb{A}^M$. Let $\dim X = n$ and $\dim Y = m$, and let $t_1, \ldots, t_N$ and $u_1, \ldots, u_M$ be coordinates in $\mathbb{A}^N$ and $\mathbb{A}^M$, regarded as functions on X and Y, respectively, with $t_1, \ldots, t_n$ algebraically independent in $k(X)$, and $u_1, \ldots, u_m$ in $k(Y)$. By definition $k[X \times Y]$ is generated by the elements $t_1, \ldots, t_N, u_1, \ldots, u_M$, and by the assumptions we have made all these elements are algebraically dependent on $t_1, \ldots, t_n, u_1, \ldots, u_m$. It remains for us to show that these latter elements are independent. Suppose that there exists a relation between them, $F(T; U) = F(T_1, \ldots, T_n, U_1, \ldots, U_m) = 0$ on $X \times Y$. Then for any point $x \in X$ we have $F(x; U_1, \ldots, U_m) = 0$ on Y. Since $u_1, \ldots, u_m$ are independent on Y, each coefficient $a_i(x)$ of the polynomial $F(x, U)$ is 0. This means that the corresponding polynomial $a_i(T_i, \ldots, T_n) = 0$ on X. Now we use the independence of the elements $t_1, \ldots, t_n$ on X to deduce that all the $a_i(T_1, \ldots, T_n) = 0$, so that $F(T; U) = 0$ identically.

One-dimensional and two-dimensional algebraic varieties are called curves and surfaces, respectively.

Theorem 1. *If* $Y \subset X$, *then* $\dim Y \leqslant \dim X$. *If* X *is irreducible,* Y *is closed in* X, *and* $\dim Y = \dim X$, *then* $X = Y$.

It is enough to prove the assertion for the case when X and Y are affine and irreducible.

Let $Y \subset X \subset \mathbb{A}^N$ and $\dim X = n$. Then among the coordinates $t_1, \ldots, t_N$ any $n+1$ are algebraically dependent as elements of $k[X]$, that is, they are connected by a relation $F(t_{i_1}, \ldots, t_{i_{n+1}})=0$ on X. A fortiori this holds on Y. But this means that the transcendence degree of $k(Y)$ is not greater than n, that is, $\dim Y \leqslant \dim X$.

Let $\dim Y = \dim X = n$. Then some n among the coordinates $t_1, \ldots, t_N$ are independent on Y, say $t_1, \ldots, t_n$. A fortiori they are independent on X. Let $u \in k[X]$, $u \neq 0$ on X. Then u depends algebraically on $t_1, \ldots, t_n$ on X, that is, it satisfies a relation

$$a_0(t_1, \ldots, t_n)u^l + \cdots + a_l(t_1, \ldots, t_n) = 0 \tag{1}$$

on X.

We can choose the polynomial on the left-hand side of (1) to be irreducible, and then $a_l(t_1, \ldots, t_n) \neq 0$ on Y. A fortiori the relation (1) holds on Y. Suppose that $u=0$ on Y. Then it follows from (1) that $a_l(t_1, \ldots, t_n)=0$ on Y, and since $t_1, \ldots, t_n$, by hypothesis, are independent on X, we have $a_l(T_1, \ldots, T_n)=0$ on the whole of $\mathbb{A}^N$. This contradicts the fact that $a_l(t_1, \ldots, t_n) \neq 0$ on X. Thus, if $u=0$ on Y, then $u=0$ on X.

As we have seen, an irreducible plane algebraic curve is of dimension 1. A generalization of this fact is the following result.

Theorem 2. *All irreducible components of a hypersurface in $\mathbb{A}^N$ or $\mathbb{P}^N$ are of codimension* 1.

Proof. It is sufficient to consider the case of a hypersurface in $\mathbb{A}^N$. Suppose that a variety $X \subset \mathbb{A}^N$ is given by the equation $F(T)=0$. To the decomposition $F = F_1 \ldots F_{l'}$ into irreducible factors there corresponds a representation $X = X_1 \cup \ldots \cup X_{l'}$, where X_i defined by the equation $F_i=0$. Clearly it is enough to prove the theorem for the varieties X_i. We show that they are irreducible. If X_i were reducible, then there would exist polynomials G and H such that $GH=0$ on X_i, $G \neq 0$, $H \neq 0$ on X_i. By Hilbert's Nullstellensatz it follows that $F_i|(GH)^l$ for some $l>0$. Owing to the irreducibility of F_i it follows that $F_i|G$ or $F_i|H$, and this contradicts the condition $G \neq 0$, $H \neq 0$ on X_i.

We assume that the variable T_N actually occurs in $F_i(T)$ and show that the coordinates $t_1, \ldots, t_{N-1}$ are algebraically independent on X_i. In fact, from a relation $G(t_1, \ldots, t_{N-1})=0$ on X_i it would follow that $F_i|G^l$ for some $l>0$, which is impossible, because G does not contain T_N. Thus, $\dim X_i \geqslant N-1$, and since $X \neq \mathbb{A}^N$, it follows from Theorem 1 that $\dim X_i = N-1$.

Theorem 3. *Any variety $X \subset \mathbb{A}^N$ whose components all have codimension 1 is a hypersurface, and the ideal $\mathfrak{A}_X$ is principal.*

It is enough to consider the case when X is irreducible. Since $X \neq \mathbb{A}^N$ (because $\dim X = N-1$), there exists a non-zero polynomial F that vanishes on X. By the irreducibility of X some irreducible factor H of

F also vanishes on X. Then $\mathbb{A}^n_H \supset X$, and since we have seen in the proof of Theorem 1 that $\mathbb{A}^n_H$ is irreducible, by Theorem 2 $X = \mathbb{A}^n_H$. If $G \in \mathfrak{A}_X$, by Hilbert's theorem $H \mid G^l$, and since H is irreducible, we see that $G \in (H)$, that is, $\mathfrak{A}_X = (H)$.

The following variant of Theorem 3 is proved similarly.

Theorem 3'. *Any subvariety $X \subset \mathbb{P}^{N_1} \times \cdots \times \mathbb{P}^{N_l}$ whose components all have codimension 1 is given by a single equation, homogeneous in each of the l groups of variables.*

Instead of the uniqueness of the decomposition of a polynomial into irreducible factors we need only use the uniqueness of the decomposition of a polynomial that is homogeneous in each group of variables into factors of the same kind. This can be derived from the fact that if $F(x_0, \ldots, x_{N_1}, y_0, \ldots, y_{N_2}, \ldots, u_0, \ldots, u_{N_l})$ is homogeneous in each of the groups of variables $(x_0, \ldots, x_{N_1}), \ldots, (u_0, \ldots, u_{N_l})$ and $F = G \cdot H$, then G and H have the same property.

2. Dimension of an Intersection with a Hypersurface. If we attempt to investigate the varieties defined by more than one equation, then we come at once to the problem of the dimension of the intersection of a variety with a hypersurface. We investigate this problem first for projective varieties. If X is closed in $\mathbb{P}^N$ and $F \neq 0$ is a form on X, then we denote by X_F the closed subset of X defined by the condition $F = 0$.

For any projective variety $X \subset \mathbb{P}^N$ we can find a form $G(U_0, \ldots, U_N)$ of any preassigned degree m that does not vanish on any of the components X_i. It is sufficient to choose in each component X_i of X a point $x_i \in X_i$ and to find a linear form L that does not vanish at any of these points. For G we can take a suitable power of L. Let X be closed in $\mathbb{P}^N$ and assume that the form F does not vanish on any component of X. By Theorem 1, $\dim X_F < \dim X$. We set $X_F = X^{(1)}$, and applying to it the same arguments we find a form F_1, $\deg F_1 = \deg F$, that does not vanish on any component of $X^{(1)}$. So we obtain a sequence of varieties $X^{(i)}$ and of forms $F_i (i = 0, \ldots)$ such that

$$X = X^{(0)} \supset X^{(1)} \supset \ldots, X^{(i+1)} = X^{(i)}_{F_i}, F_0 = F. \tag{1}$$

By Theorem 1, $\dim X^{(i+1)} < \dim X^{(i)}$. Hence when $\dim X = n$, then $X^{(n+1)}$ is empty. In other words, the forms $F_0 = F, F_1, \ldots, F_n$ do not vanish simultaneously on X.

Now let X be an irreducible variety. We consider the mapping $\varphi : X \to \mathbb{P}^n$ given by:

$$\varphi(x) = (F_0(x) : \ldots : F_n(x)). \tag{2}$$

This mapping satisfies the conditions of Theorem 8 in §5, and by this theorem the mapping $X \to \varphi(X)$ is finite. But if $X \to Y$ is a finite mapping, then as we have seen, $\dim X = \dim Y$. Therefore $\dim \varphi(X) = \dim X = n$, and since $\varphi(X)$ is closed in $\mathbb{P}^n$, by Theorem 2 of § 5, we see that $\varphi(X) = \mathbb{P}^n$, by Theorem 1 of §6.1. Suppose now that $\dim X^{(1)} = \dim X_F < n-1$. Then in the sequence (1) already $X^{(n)}$ is empty. In other words, the forms $F_0, \ldots, F_{n-1}$ do not vanish simultaneously on X. This means that the point $(0:0:\ldots 0:1)$ is not contained in $\varphi(X)$, but this contradicts the fact that $\varphi(X) = \mathbb{P}^n$. So we have proved the following result.

Theorem 4. *If a form F does not vanish on an irreducible projective variety X, then* $\dim X_F = \dim X - 1$.

Corollary 1. *On a projective variety X there exist subvarieties of any dimension $s < \dim X$.*

Corollary 2. *(Inductive definition of dimension.) For an irreducible projective variety X we have* $\dim X = 1 + \sup \dim Y$, *where Y ranges over all the proper subvarieties of X.*

Corollary 3. *The dimension of a projective variety X can be defined as the largest number n for which there exists a chain of irreducible subvarieties $Y_0 \supset Y_1 \supset \ldots \supset Y_n$, $Y_i \neq Y_{i+1}$.*

Corollary 4. *The dimension n of a projective variety X can be defined as $N-s-1$, where s is the maximum dimension of the linear subspaces of $\mathbb{P}^N$ that are disjoint from X.*

Let $E \subset \mathbb{P}^N$ be a linear subspace and $\dim E = s$. If $s \geqslant N-n$, then E can be given by not more than n equations, and a subsequent application of Theorem 4 shows that $\dim(X \cap E) \geqslant 0$, hence $X \cap E$ is not empty (the dimension of the empty set is -1!). Setting $m=1$ in the construction of the sequence (1), we obtain $n+1$ linear forms $L_0, \ldots, L_n$ that do not vanish simultaneously on X. If E is the subspace defined by them, then $\dim E = N-n-1$, and $X \cap E$ is empty.

Corollary 5. *The dimension of the set of zeros of r forms $F_1, \ldots, F_r$ on an n-dimensional variety is not less than $n-r$.*

This is proved by applying Theorem 4 in succession $r-1$ times.

Note. Corollary 5 yields a fairly strong existence theorem: if $r \leqslant n$, then r forms have a common zero on an n-dimensional variety. For example (with $X = \mathbb{P}^n$), n homogeneous equations in $n+1$ unknowns have a non-zero solution. From this existence theorem we can deduce a number of important results.

1°. On $\mathbb{P}^2$ any two curves intersect (since any curve is given by a single homogeneous equation). On a quadric $Q \subset \mathbb{P}^3$ there are non-intersecting curves, for example, lines of one and the same family of generators. Therefore $\mathbb{P}^2$ and Q are not isomorphic. Since they are birationally isomorphic (Example 1 in §3.3), here we have an example of birationally isomorphic, but not isomorphic, projective varieties. This example will occur again later.

2°. Theorem 3 is false even for curves on a quadric Q. Not every curve $C \subset Q$ can be defined by equating a single form of $\mathbb{P}^3$ to zero. In fact, supposing that the two non-intersecting curves which we have found above on Q are given by the equations $F_1 = 0$ and $F_2 = 0$, we come to a contradiction to Corollary 5 according to which the system $F_1 = 0$, $F_2 = 0$, $G = 0$ (the equation of Q) has a solution.

Theorem. 5. *Under the assumptions of Theorem* 4 *all the components of the variety* X_F *have the same dimension* $\dim X - 1$.

We consider the finite mapping $\varphi: X \to \mathbb{P}^n$ ($n = \dim X$) that was constructed in the proof of Theorem 4. Let $\mathbb{A}_i^n$ be open affine sets covering $\mathbb{P}^n$; then $\varphi^{-1}(\mathbb{A}_i^n) = U_i$ are affine open sets in X, as we can easily see by applying a Veronese mapping with $m = \deg F$. Obviously it is enough to show that all the components of the affine variety $X_F \cap U_i$ have the dimension $n-1$ for all $i = 1, \ldots, n$. From here on our arguments refer to a fixed U_i which we denote by U. Clearly $X_F \cap U = V(f)$, where $f = F/F_i$, so that it coincides on U with a set of zeros of the regular function $f \in k[U]$. We have constructed a finite mapping $\varphi: U \to \mathbb{A}^n$, given by n regular functions $f_1, \ldots, f_n$, with $f_1 = f$.

To prove that all the components of $V(f)$ have the dimension $n-1$ it is sufficient to show that their dimensions are not less than $n-1$. We show that the functions $f_2, \ldots, f_n$ are algebraically independent on each of these components. Let $P \in k[T_2, \ldots, T_n]$. To show that $R = P(f_2, \ldots, f_n)$ does not vanish on any component of $V(f)$, we need only show that from a relation $R \cdot Q = 0$ on $V(f)$, $Q \in k[U]$, it follows that $Q = 0$ on $V(f)$. For if $V(f) = U^{(1)} \cup \ldots \cup U^{(t)}$ is an incontractible decomposition into irreducible components and $R = 0$ on $U^{(1)}$, we can take for Q any function that vanishes on $U^{(2)} \cup \ldots \cup U^{(t)}$, but not on $U^{(1)}$. Then $R \cdot Q = 0$ on $V(f)$, but $Q \neq 0$ on $V(f)$. By Hilbert's Nullstellensatz our proposition can be rephrased as follows: if $f \mid (R \cdot Q)^l$ for some $l > 0$, then $f \mid Q^m$ for some $m > 0$.

Thus, Theorem 5 is a consequence of the following purely algebraic fact.

Lemma. *Let* $B = k[T_1, \ldots, T_r]$, *let* $A \supset B$ *be a ring without divisors of zero that is integral over* B, $x = T_1$, $y = P(T_2, \ldots, T_r) \neq 0$, $u \in A$. *If* $x \mid (yu)^l$ *in* A *for some* $l > 0$, *then* $x \mid u^m$ *for some* $m > 0$.

The only property of the polynomials x and y that we make use of is that they are relatively prime in the ring $k[T_1, \ldots, T_r]$. Observe that we may replace y^l by z and u^l by v, and we need only show that if x and z

are relatively prime in $k[T_1, \ldots, T_r]$, then from $x|zv$ in A it follows that $x|v^m$ in A for some $m>0$. Thus, the lemma asserts that in a certain sense the property of the polynomials z and x in B of being relatively prime is preserved on transition to the ring A, which is integral over B.

We denote by K the field of fractions of B. If an element $t \in A$ is integral over B, then it is clearly algebraic over K. We denote by $F(T) \in K[T]$ the polynomial of least degree with leading coefficient 1 such that $F(t)=0$, the so-called minimal polynomial of t. Division with remainder shows that any polynomial $G(T) \in K[T]$ for which $G(t)=0$ is divisible by F in $K[T]$. Hence we can conclude that an element t is integral over B if and only if $F[T] \in B[T]$. For if t is integral and $G(t)=0$, where $G \in B[T]$ has the leading coefficient 1, then $G(T)= F(T) \cdot H(T)$ in $K(T)$. But from the fact that in B the decomposition into prime factors is unique (remember that $B=k[T_1, \ldots, T_r]$) it follows that then $F(T) \in B[T]$ and $H(T) \in B[T]$, a simple consequence of Gauss's lemma.

Now it is easy to complete the proof of the lemma. Let $zv=xw$, $v, w \in A$, and let $F(T)=T^l+b_1 T^{l-1}+ \cdots +b_l$ be the minimal polynomial of w. Since w is integral over B, we have $b_i \in B$. It is easy to see that the minimal polynomial $G(T)$ of v is of the form $(x^l/z^l)F((z/x)T)$. Therefore

$$\begin{aligned} G(T) &= T^l+(xb_1/z)T^{l-1}+ \cdots +(x^l b_l/z^l), \\ & v^l+(xb_1/z)\, v^{l-1}+ \cdots +(x^l b_l/z^l)=0. \end{aligned} \tag{3}$$

Since v is integral over B, we have $x^i b_i/z^i \in B$, and as z and x are relatively prime, $z^i | b_i$. From (3) it then follows that $x|v^l$. This proves the lemma and with it Theorem 5.

Corollary 1. *If $X \subset \mathbb{P}^N$ is a quasiprojective irreducible variety, F is a form that does not vanish identically on X, and X_F is not empty, then each of its components has codimension 1.*

Proof. By definition, X is open in some closed subset $\bar{X}$ of $\mathbb{P}^N$. Since X is irreducible, so is $\bar{X}$, and consequently $\dim \bar{X} = \dim X$. By Theorem 5, $(\bar{X})_F = \bigcup Y_i$, $\dim Y_i = \dim X - 1$. But, as is easy to see, $X_F=(\bar{X})_F \cap X$; hence it follows that $X_F = \bigcup (Y_i \cap X)$, and $Y_i \cap X$ is either empty or open in Y_i, therefore $\dim (Y_i \cap X) = \dim X - 1$.

Usually one meets the special case of this corollary when $X \subset \mathbb{A}^n$ is an affine variety. If $\mathbb{A}^n \subset \mathbb{P}^n$, $\mathbb{A}^n = \mathbb{A}^n_0$, then $X_F = V(f)$, where $f=F/u_0^m$, $m = \deg F$. Thus, X_F coincides with the set of zeros of some regular function $f \in k[X]$.

Corollary 2. *If $X \subset \mathbb{P}^N$ is a quasiprojective irreducible n-dimensional variety and Y is the set of zeros of m forms on X and is not empty, then each of its components is of dimension not less than $n-m$.*

The proof by induction on m is obvious. Again, in the case of an affine variety we can speak of the set of zeros of m regular functions on X.

If X is projective and $n \geqslant m$, then we can claim that Y is not empty.

Theorem 6. *If X and Y are quasiprojective irreducible varieties in* $\mathbb{P}^N$, $\dim X = n$, $\dim Y = m$, $N \leqslant n+m$, *and* $X \cap Y \neq \emptyset$, *then*

$$\dim Z \geqslant n+m-N$$

for each component Z of $X \cap Y$.

Clearly the theorem is of local character, so that we need only prove it for affine varieties. Let $X, Y \subset \mathbb{A}^N$. Then $X \cap Y$ is isomorphic to $(X \times Y) \cap \Delta \subset \mathbb{A}^{2N}$ (Example 10 of §2.3). The theorem now follows from Corollary 2 to Theorem 5, because Δ is defined by N equations. For projective varieties, as above, the set $X \cap Y$ is not empty as long as $N \leqslant n+m$. Theorem 6 can be stated in a more symmetrical form in which it generalizes at once to an arbitrary number of subvarieties:

$$\operatorname{codim}_X \bigcap_{i=1}^{r} Y_i \leqslant \sum_{i=1}^{r} \operatorname{codim}_X Y_i . \tag{4}$$

3. A Theorem of the Dimension of Fibres. If $f: X \to Y$ is a regular mapping of quasiprojective varieties and $y \in Y$, then the set $f^{-1}(y)$ is called a fibre over the point y. Clearly, a fibre is a closed subvariety.

This terminology is justified by the fact that X is stratified into the disjoint fibres of distinct points $y \in f(X)$.

Theorem 7. *If $f: X \to Y$ is a regular mapping of irreducible varieties,* $f(X) = Y$, $\dim X = n$, $\dim Y = m$, *then $m \leqslant n$ and*

1) $\dim f^{-1}(y) \geqslant n-m$ *for every point $y \in Y$;*

2) *in Y there exists a non-empty open set U such that* $\dim f^{-1}(y) = n-m$ *for $y \in U$.*

Proof of 1). Clearly, this is a local property relative to y, so that we need only prove it by taking for Y any open set $U \subset Y$ containing y and for X the variety $f^{-1}(U)$. Therefore we may assume that Y is an affine variety. Let $Y \subset \mathbb{A}^N$. In the sequence (1) of §6.2 for Y we find that $Y^{(m)}$ is a finite set: $Y^{(m)} = Y \cap Z$, where Z is defined by m equations and $y \in Z$. We can choose U such that $Z \cap U \cap Y = y$, and we assume therefore that $Z \cap Y = y$. The subspace Z can be defined by m equations $g_1 = 0, \ldots, g_m = 0$. Thus, the system of equations $g_1 = 0, \ldots, g_m = 0$ defines the point y on Y. This means that on X the system of equations $f^*(g_1) = 0, \ldots, f^*(g_m) = 0$ defines the subvariety $f^{-1}(y)$. Now 1) follows from Corollary 2 to Theorem 5 (the affine case).

Proof of 2). We may replace Y by an open affine subset W of it, and X by an open affine subset $V \subset f^{-1}(W)$. Since V is dense in $f^{-1}(W)$,

$f(V)$ is dense in W. Therefore f determines an embedding $f^*: k[W] \to k[V]$. We now assume that $k[W] \subset k[V]$, and hence that $k(W) \subset k(V)$. Let $k[W] = k[w_1, \dots, w_M]$, $k[V] = k[v_1, \dots, v_N]$. Since $\dim W = m$, $\dim V = n$, the field $k(V)$ has over $k(W)$ the transcendence degree $n - m$. Suppose that $v_1, \dots, v_{n-m}$ are algebraically independent over $k(W)$ and that v_i $(i = n-m+1, \dots, N)$ are connected with them by the relations $F_i(v_i; v_1, \dots, v_{n-m}; w_1, \dots, w_M) = 0$. We denote by $\bar{v}_i$ the restriction of v_i to $f^{-1}(y) \cap V$. Then

$$k[f^{-1}(y) \cap V] = k[\bar{v}_1, \dots, \bar{v}_N]. \tag{1}$$

We regard F_i as polynomials in $v_i, v_1, \dots, v_{n-m}$ taking $w_1, \dots, w_M$ into the coefficients, and we denote by Y_i the subvariety of W defined by the vanishing of the leading coefficient of this polynomial. We set $Y_0 = \bigcup Y_i$, $U = W - Y_0$. Clearly U is open and not empty. If $y \in U$, then none of the polynomials $F_i(T_i; T_1, \dots, T_{n-m}, w_1(y), \dots, w_M(y))$ vanishes, that is, all the $\bar{v}_i$ are algebraically dependent on $\bar{v}_1, \dots, \bar{v}_{n-m}$. In conjunction with (1) this shows that $\dim f^{-1}(y) \leqslant n - m$, and 2) now follows by virtue of 1).

This proves the theorem. In the next subsection we shall see that 2) need not hold for all y, in other words, that the dimension of a fibre can actually jump.

Corollary. *The sets* $Y_l = \{y \in Y, \dim f^{-1}(y) \geqslant l\}$ *are closed in* Y.

By Theorem 7, $Y_{n-m} = Y$, and there exists a closed subset $Y' \subset Y$, $Y' \neq Y$ such that $Y_l \subset Y'$ for $l > n - m$. If Z_i are the irreducible components of Y', then $\dim Z_i < \dim Y$ and we can apply induction on $\dim Y$ to the mapping $f^{-1}(Z_i) \to Z_i$.

Theorem 7 leads to a criterion for irreducibility of varieties which is often useful.

Theorem 8. *If* $f: X \to Y$ *is a regular mapping of projective varieties,* $f(X) = Y$, *and if* Y *and all the fibres* $f^{-1}(y)$ *are irreducible and of the same dimension, then* X *is irreducible.*

We set $\dim f^{-1}(y) = n$. Suppose that X is reducible and that $X = \bigcup X_i$ is an incontractable decomposition into irreducible components. By Theorem 2 of §5 all the $f(X_i)$ are closed. Since $Y = \bigcup f(X_i)$ and Y is irreducible we have $f(X_i) = Y$ for certain X_i. From Y we discard the union of those closed sets $f(X_i)$ that are different from Y, and we denote the remaining open set by Y'. We set $f^{-1}(Y') = X'$ and $X' = \bigcup X'_j$, where the X'_j are open subsets of those X_j for which $f(X_j) = Y$. Let $f_j: X'_j \to Y'$ be the restriction of f, and m_j the minimum of the numbers $\dim_{y \in Y'} f_j^{-1}(y)$.

By Theorem 7, this minimum is attained on some open set $U \subset Y'$, and since $\bigcup_j f_j^{-1}(y) = f^{-1}(y)$ is irreducible and of dimension n, we see that that $\max m_j = n$ and for a certain value $j = j_0$, $\dim f_{j_0}^{-1}(y) = n$ for $y \in U$, hence for all $y \in Y$. But then $f^{-1}(y) = \bigcup f_j^{-1}(y)$ for every $y \in Y$, $\dim f_j^{-1}(y) \leqslant n$, $\dim f_{j_0}^{-1}(y) = n$, and from the irreducibility of $f^{-1}(y)$ if follows that $f^{-1}(y) = f_{j_0}^{-1}(y)$. But this means that $X_{j_0} = X'$ and consequently $X_{j_0} = X$.

A very special case of Theorem 8 is the irreducibility of the direct product of irreducible varieties, which was proved in §3.

4. Lines on Surfaces. After the hard work spent on the proof of the theorems on the dimensions of intersections, naturally we wish to see some applications of these theorems. As an example, we now treat the simple problem of the disposition of lines on surfaces in $\mathbb{P}^3$.

As a rule, the concept of dimension turns out to be useful if we have to give a rigorous meaning to the fact that an element of some set depends on a given number of parameters. For this purpose we must identify the set with some algebraic variety and apply our concept of dimension.

For example, we have seen that a hypersurface in $\mathbb{P}^n$ given by an equation of degree m can be put into correspondence with points of the projective space $\mathbb{P}^{\nu_{n,m}}$ where $\nu_{n,m} = \binom{n+m}{m} - 1$ (see §4.4).

We now turn to subvarieties that are not hypersurfaces; the simplest of these are the lines in $\mathbb{P}^3$.

A point of $\mathbb{P}^3$ corresponds to a line of a four-dimensional vector space E, and lines in $\mathbb{P}^3$ to two-dimensional linear subspaces, that is, planes in E. In an n-dimensional space E every plane $F \subset E$ has a basis of two vectors x, y which in turn determine a bivector $\omega = x \wedge y \in \Lambda^2 E$. If the vectors x and y have in some basis the coordinates $(x_1, \ldots, x_n)$ and $(y_1, \ldots, y_n)$, then the coordinates of the bivector $x \wedge y$ are $p_{ij} = x_i y_j - y_i x_j$ $(i, j = 1, \ldots, n)$. The bivector $x \wedge y$ uniquely determines a plane F, and under a change of basis x and y are multiplied by a non-zero element of the field. Therefore the point of projective space with the coordinates p_{ij} is uniquely determined by, and uniquely determines, the plane F.

In our case $n = 4$ and the bivector p_{ij} has sixteen coordinates p_{ij} $(i, j = 0, \ldots, 3)$, which are connected by the relations $p_{ii} = 0$, $p_{ij} = -p_{ji}$. Thus, they are six independent coordinates $p_{01}, p_{02}, p_{03}, p_{12}, p_{13}, p_{23}$. Every line $L \subset \mathbb{P}^3$ determines a point $(p_{01} : p_{02} : p_{03} : p_{12} : p_{13} : p_{23}) \in \mathbb{P}^5$; p_{ij} $(0 \leqslant i < j \leqslant 3)$ are called the Plücker coordinates of L.

Not every point $(p_{ij})(0 \leqslant i \leqslant j \leqslant 3)$ corresponds to a line. For this it is necessary and sufficient that the relation $p_{01}p_{23} - p_{02}p_{13} + p_{03}p_{12} = 0$ holds. Points satisfying this relation form a hypersurface of the second order in $\mathbb{P}^5$, which we denote by Π. Clearly $\dim \Pi = 4$. Thus, we have set up a one-to-one correspondence between the lines $L \in \mathbb{P}^3$ and the points of Π.

For the study of lines lying on surfaces the following result is very important.

Lemma. *The conditions for a line L with the Plücker coordinates p_{ij} to belong to a surface X given by the equation $F=0$ are algebraic relations between the p_{ij} and the coefficients of F that are homogeneous both in the p_{ij} and in the coefficients of F.*

We can write down a parametric representation of the coordinates of the points of L in terms of its Plücker coordinates.

Let x and y be a basis of the plane $\Lambda \subset E$. Then the set of vectors of the form

$$xf(y) - yf(x) \tag{1}$$

coincides, as is easy to verify, with Λ as f ranges over all linear forms on E. If the coordinates of f have the form $(\alpha_0, \alpha_1, \alpha_2, \alpha_3)$, that is, $f(x) = \Sigma \alpha_i x_i$, then the vector (1) has the coordinates $z_i = \sum_j \alpha_j p_{ij}$, $p_{ij} = x_i y_j - y_i x_j$. Therefore a point of L having the Plücker coordinates p_{ij} has the coordinates $u_i = \sum_j \alpha_j p_{ij}$. Substituting these expressions in the equation $F(u_0, u_1, u_2, u_3) = 0$ and equating to zero the coefficients of all monomials in α_i, we obtain the condition for $L \subset X$ to hold in the form of algebraic relations between the coefficients of F and the coordinates p_{ij}.

Now we pass to the interesting question of the lines on surfaces on $\mathbb{P}^3$. For a given m we consider the space P^ν, $\nu = \nu_{3,m} = (m+1)(m+2)(m+3)/6 - 1$ whose points correspond one-to-one to surfaces of degree m in $\mathbb{P}^3$ (that is, those given by a homogeneous equation of degree m). We denote by Γ_m the subset of pairs $(\xi, \eta) \in \mathbb{P}^\nu \times \Pi$ for which the line L corresponding to the point $\eta \in \Pi$ is contained in the surface X corresponding to the point $\xi \in \mathbb{P}^\nu$. By the lemma Γ_m is a projective variety. Let us find the dimension of Γ_m. For this purpose we consider the projections $\varphi(\mathbb{P}^\nu \times \Pi) \to \mathbb{P}^\nu$ and $\psi(\mathbb{P}^\nu \times \Pi) \to \Pi : \varphi(\xi, \eta) = \xi$, $\psi(\xi, \eta) = \eta$. Obviously φ and ψ are regular mappings. In what follows, we only consider them on Γ_m. Observe that $\psi(\Gamma_m) = \Pi$. This means simply that through any line there passes at least one surface of degree m, but perhaps a reducible one.

We determine the dimension of the fibres $\psi^{-1}(\eta)$ of ψ. By making a projective transformation we may assume that the line corresponding to η is defined by the equations $u_0 = 0$, $u_1 = 0$. Points $\xi \in \mathbb{P}^\nu$ such that $(\xi, \eta) \in \psi^{-1}(\eta) \subset \Gamma_m$, correspond to surfaces of degree m passing through this line. The equation has the form $F = 0$, where $F = u_0 G + u_1 H$, and where G and H are arbitrary forms of degree $m-1$. The set of these forms corresponds, of course, to a linear subspace of $\mathbb{P}^\nu$, whose dimension is easy to compute. It is

$$\mu = \frac{m(m+1)(m+5)}{6} - 1 . \tag{2}$$

Thus,

$$\dim \psi^{-1}(\eta) = \frac{m(m+1)(m+5)}{6} - 1 .$$

From Theorem 8 it follows that Γ_m is irreducible. Applying Theorem 7 we see that

$$\dim \Gamma_m = \dim \psi(\Gamma_m) + \dim \psi^{-1}(\eta) = \frac{m(m+1)(m+5)}{6} + 3 . \qquad (3)$$

We now consider the mapping $\varphi: \Gamma_m \to \mathbb{P}^\nu$. According to Theorem 2 of §5, its image is a closed subset of $\mathbb{P}^\nu$. Clearly $\dim \varphi(\Gamma_m) \leqslant \dim \Gamma_m$. Therefore, if $\dim \Gamma_m < \nu$, then $\varphi(\Gamma_m) \neq \mathbb{P}^\nu$, and this means that not every surface of degree m contains a line. The inequality $\dim \Gamma_m < \nu$ by (3) means that $m(m+1)(m+5)/6 + 3 < (m+1)(m+2)(m+3)/6 - 1$. It holds when $m > 3$. So we have obtained the following result.

Theorem 9. *For every $m > 3$ there exist surfaces of degree m that contain no line. Furthermore, such surfaces correspond to an open set in $\mathbb{P}^\nu$.*

Thus, there is a non-trivial algebraic relation between the coefficients of a form $F(u_0, u_1, u_2, u_3)$ of degree $m > 3$ that are necessary and sufficient for the surface given by the equation $F = 0$ to contain at least one line.

Of the remaining cases $m = 1, 2, 3$ the case $m = 1$ is trivial.

We consider the case $m = 2$, but the answer is well known from analytic geometry.

From $m = 2$, $\nu = 9$, $\dim \Gamma_m = 10$. From Theorem 7 it follows that $\dim \varphi^{-1}(\xi) \geqslant 1$. This is a standard fact: every quadric contains infinitely many lines.

Without going into the details of the proofs we note that here we come across the phenomenon of the dimension of fibres jumping of which we spoke in §6.3. For if a quadric is irreducible, then for the points corresponding to it $\dim \varphi^{-1}(\xi) = 1$, but if it degenerates into two planes, then, of course, $\dim \varphi^{-1}(\xi) = 2$.

We now consider the case $m = 3$. Then $\dim \Gamma_m = \nu = 19$. It is easy to construct a cubic surface $X \subset \mathbb{P}^3$ on which there are only finitely many lines. For example, if X is given in homogeneous coordinates by the equation

$$T_1 T_2 T_3 = 1 \qquad (4)$$

then in $\mathbb{A}^3$ there are no lines on X. For by writing the equation of a line in the parametric form $T_i = a_i t + b_i (i = 1, 2, 3)$ and substituting in (4) we come to a contradiction. The intersection of X with the plane at infinity contains three lines. Thus, in $\mathbb{P}^{19}$ there exist points ξ for which $\dim \varphi^{-1}(\xi) = 0$. By Theorem 7 this is only possible when $\dim \varphi(\Gamma_3) = 19$. Applying Theorem 1 we see that $\varphi(\Gamma_3) = \mathbb{P}^{19}$.

So we have proved the following result.

Theorem 10. *On every cubic surface there lies at least one line. In the space $\mathbb{P}^{19}$, whose points correspond to all cubic surfaces, there is an open subset such that on surfaces corresponding to its points there are finitely many lines.*

Cubic surfaces on which there are infinitely many lines actually exist, for example, cubic cones. Here, too, the dimensions of fibres can jump.

5. The Chow Coordinates of a Projective Variety. One of the most important applications of the theorems on the dimension of intersections lies in the fact that it enables us to specify subvarieties $X \subset \mathbb{P}^N$ of a fixed dimension n by coordinates. We have already seen this in the case of hypersurfaces: hypersurfaces determined by forms of degree m in $\mathbb{P}^n$ correspond to points of the projective space $\mathbb{P}^{\nu_{m,n}}$. Another example is the classification of lines in $\mathbb{P}^3$ by their Plücker coordinates.

Naturally, we try to reduce somehow the case of an arbitrary variety X to that of a hypersurface, and for this purpose we attempt to associate with X a hypersurface. Suppose, for example, that X is a curve in $\mathbb{P}^3$. We consider the set Y of all lines in $\mathbb{P}^3$ that intersect X. It is not hard to verify that Y corresponds to an algebraic subvariety $\tilde{Y}$ on the variety Π which describes all the lines in $\mathbb{P}^3$. Since the set Y_x of all lines passing through a given point $x \in X$ corresponds, as is easy to see, to a two-dimensional subvariety $\tilde{Y}_x \subset \Pi$ and $\dim X = 1$, it is easy to deduce from Theorem 7 that $\dim \tilde{Y} = 3$. Therefore $\tilde{Y}$ has codimension 1 in Π. If we knew that in Π, as in $\mathbb{P}^n$, any subvariety of codimension 1 is given by a single equation, then the coefficients of this equation could be taken as coordinates of X. The difficulty arising here can be avoided if we consider instead of lines intersecting X pairs of planes E_1, E_2 such that $E_1 \cap E_2 \cap X$ is not empty. Since planes in $\mathbb{P}^3$ correspond to points of $\mathbb{P}^3$ (because $\nu_{3,1} = 3$), the pair E_1, E_2 corresponds to points of the variety $\mathbb{P}^3 \times \mathbb{P}^3$ of dimension 6. The set of pairs for which $E_1 \cap E_2 \cap X \neq \emptyset$ corresponds to a subvariety $Y \subset \mathbb{P}^3 \times \mathbb{P}^3$ of dimension 5, and now we can apply Theorem 3′. This plan will be carried out in detail for arbitrary varieties $X \subset \mathbb{P}^n$.

The set of all hyperplanes $\xi \subset \mathbb{P}^N$ corresponds one-to-one to the points of an N-dimensional projective space, which we denote by $\tilde{\mathbb{P}}^N$, to distinguish it from the original $\mathbb{P}^N$. A hyperplane ξ and the point of $\tilde{\mathbb{P}}^N$ corresponding to it are denoted by one and the same letter.

We consider a projective irreducible variety $X \subset \mathbb{P}^N$, $\dim X = n$. In the product $\underbrace{\tilde{\mathbb{P}}^N \times \cdots \times \tilde{\mathbb{P}}^N}_{n+1} \times X$, which we abbreviate by $(\tilde{\mathbb{P}}^N)^{n+1} \times X$, we consider the subset Γ consisting of those systems $(\xi^{(0)}, \ldots, \xi^{(n)}, x)$ for which the hyperplanes $\xi^{(i)} (i=0, \ldots, n)$ contain the point $x \in X$. This Γ is a closed subset, and there are two regular mappings defined on it, $\varphi:\Gamma\to(\tilde{\mathbb{P}}^N)^{n+1}$ and $\psi:\Gamma\to X$.

Clearly $\psi(\Gamma)=X$. Let us determine the dimension of $\psi^{-1}(x_0)$, where $x_0\in X$. The set $\psi^{-1}(x_0)$ consists of the systems $(\xi^{(0)},\dots,\xi^{(n)},x_0)$ for which $x_0\in\xi^{(i)}$. All the $\xi\in\tilde{\mathbb{P}}^N$ for which $x_0\in\xi$ form a hyperplane $\tilde{\mathbb{P}}^{N-1}\subset\tilde{\mathbb{P}}^N$. Therefore $\psi^{-1}(x_0)=(\tilde{\mathbb{P}}^{N-1})^{n+1}$, and $\dim\psi^{-1}(x_0)=(N-1)\times(n+1)$. It follows from Theorem 7 that $\dim\Gamma=(N-1)\times(n+1)+n=N(n+1)-1$, and from Theorem 8 that Γ is irreducible.

There exist points $y\in(\tilde{\mathbb{P}}^N)^{n+1}$ such that $y\in\varphi(\Gamma)$ and $\varphi^{-1}(y)$ is a single point. This is immediately obvious from the construction process for the sequence (1) in §6.2, if we take for the F_i linear forms that vanish at some point $x\in X$. Now we can apply Theorem 7 and deduce that $\dim\varphi(\Gamma)=\dim\Gamma=N(n+1)-1$. Since $\varphi(\Gamma)\subset(\tilde{\mathbb{P}}^N)^{n+1}$ and $\dim(\tilde{\mathbb{P}}^N)^{n+1}=N(n+1)$, we can apply Theorem 3′. It follows that $\varphi(\Gamma)$ is defined by a single form F_X in $n+1$ variables. The form F_X is homogeneous in each system of variables. Obviously we can choose it so that it does not contain multiple factors. Then it is determined by X uniquely to within a constant factor. F_X is called an *associated form*, and its coordinates the *Chow coordinates* of X. Let us show that the variety X, in turn, is uniquely determined by the form F_X. For this purpose it is sufficient to verify that a point $x\in\mathbb{P}^N$ is contained in X if and only if any $n+1$ hyperplanes $\xi^{(0)},\dots,\xi^{(n)}$ containing it satisfy the relation

$$F_X(\xi^{(0)},\dots,\xi^{(n)})=0\,. \tag{1}$$

For if $x\in X$, then (1) holds by the definition of F_X. But if $x\notin X$, then we can find $n+1$ hyperplanes $\xi^{(0)},\dots,\xi^{(n)}$ containing x such that $\xi^{(0)}\cap\dots\cap\xi^{(n)}\cap X=\emptyset$, which again follows immediately from the construction of the sequence (1) in §6.2. Such $\xi^{(0)},\dots,\xi^{(n)}$ do not satisfy the relation (1).

The forms F_X have a "discrete" invariant, namely the degree, and for a given degree "continuous" invariants, the coefficients. First of all, let us explain the meaning of the degree of F_X. More accurately, it has $n+1$ degrees $d_0,\dots,d_n$ in each system of variables. But all these degrees are equal. In fact, since (1) is symmetrical with respect to $\xi^{(0)},\dots,\xi^{(n)}$ and this condition determines the set of zeros of F_X, under a permutation of the $\xi^{(i)}$ the form F_X can only be multiplied by a constant (which is easily seen to be equal to 1 or -1). Therefore all the numbers d_i are the same, and we denote their common value by d.

We choose n hyperplanes $\eta^{(1)},\dots,\eta^{(n)}\subset\tilde{\mathbb{P}}^N$ such that their intersection with X consists of finitely many points; This is always possible by virtue of the sequence (1) in §6.2. Let

$$\eta^{(1)}\cap\dots\cap\eta^{(n)}\cap X=\{x^{(1)},\dots,x^{(c)}\}\ \text{ and }\ x^{(j)}=(U_0^{(j)}:\dots:U_N^{(j)})\ (j=1,\dots,c)\,.$$

If we write the equation of $\xi^{(0)}$ in the form $\sum_{i=0}^{N} v_i u_i = 0$, then we see that $F_X(\xi^{(0)}, \eta^{(1)}, \ldots, \eta^{(n)})$ is a form of degree d in $v_0, \ldots, v_N$ and that this form vanishes if and only if for at least one $j = 1, \ldots, c$

$$\sum_{i=0}^{N} v_i u_i^{(j)} = 0 .$$

Hence it follows that

$$F_X(\xi^{(0)}, \eta^{(1)}, \ldots, \eta^{(n)}) = \alpha \prod_{j=1}^{c} (\sum v_i u_i^{(j)})^{r_j} , \tag{2}$$

where $r_j \geqslant 1$ are integers. So we see that $c \leqslant d$, and if $F_X(\xi^{(0)}, \eta^{(1)}, \ldots, \eta^{(n)})$ does not contain multiple factors, then $c = d$. Let us show that for a suitable choice of the hypersurfaces $\eta^{(1)}, \ldots, \eta^{(n)}$ the form $F_X(\xi^{(0)}, \eta^{(1)}, \ldots, \eta^{(n)})$ has no multiple factors.

Lemma. *If a polynomial $F(x, Y)$, $Y = (y_1, \ldots, y_m)$, does not have multiple factors, then either there exists a Y_0 such that $F(x, Y_0)$ does not have multiple factors, or $F'_x(x, Y) = 0$.*

The latter case is possible only if the characteristic p of k is positive, and then $F(x, y) = G(x^p, y)$.

We leave the simple proof of this lemma to the reader.

From the lemma and the fact that $F_X(\xi^{(0)}, \ldots, \xi^{(n)})$ has no multiple factors it follows that either for some choice of $\eta^{(1)}, \ldots, \eta^{(n)}$ the form $F_X(\xi^{(0)}, \eta^{(1)}, \ldots, \eta^{(n)})$ does not have multiple factors, or that all the variables $v_i^{(0)} (i = 0, \ldots, N)$ occur in F with exponents divisible by p. But then by virtue of the symmetry of F_X in the various groups of variables all the variables have this property, and hence F_X is the p-th power of a polynomial, contrary to our assumption.

By arguments like those used in the proof of Theorem 4 it is easy to verify that the points $\eta^{(1)}, \ldots, \eta^{(n)} \in (\tilde{\mathbb{P}}^N)^n$ for which the form $F_X(\xi^{(0)}, \eta^{(1)}, \ldots, \eta^{(n)})$ does not have multiple factors form a non-empty open set in $(\tilde{\mathbb{P}}^N)^n$. The points for which $F_X(\xi^{(0)}, \eta^{(1)}, \ldots, \eta^{(n)}) \neq 0$ have the same property. Hence in (2) we have $r_j = 1$ and $c = d$ for points of a certain non-empty open set.

The result we have obtained leads to the following characterization of the degree d of the associated form of a variety X: d is the maximum number of points of intersection in $X \cap E$, where E is a linear subspace, $\dim E = N - n$, and $X \cap E$ is finite. The number d is called the degree of X is denoted by $\deg X$.

The set of all forms $F(\xi^{(0)}, \ldots, \xi^{(n)})$ in $n+1$ groups of $N+1$ variables of degree d in each group, form a projective space $\mathbb{P}^{\nu_{N,n,d}}$, provided that we consider forms to within a constant factor. With a variety $X \subset \mathbb{P}^N$

of dimension n and degree d we have associated a form F_X and hence a point $c(X)$ in $\mathbb{P}^{\nu_{N,n,d}}$. We denote by $C_{N,n,d} \subset \mathbb{P}^{\nu_{N,n,d}}$ the set of points so obtained. The main problem is how to describe this set. We mention without proof a relevant result. It asserts that $C_{N,n,d}$ is a quasiprojective variety, and furthermore, it specifies the connection between X and F_X.

To begin with we introduce an important concept. Let S be a quasi-projective variety, Γ a closed subvariety of $S \times \mathbb{P}^N$, and $\varphi: \Gamma \to \mathbb{P}^N$, $\psi: \Gamma \to S$ its natural projections.

If for all $s \in S$ the subvarieties $X_s = \varphi\psi^{-1}(s)$ have one and the same dimension, then the family of subvarieties $\{X_s; s \in S\}$ is called *algebraic*. We say that S and Γ determine this family.

Proposition. *The family of all closed subvarieties* $X \subset \mathbb{P}^N$ *with* $\dim X = n$, $\deg X = d$, *is algebraic. There exists a quasiprojective subvariety* $C_{N,n,d} \subset \mathbb{P}^{\nu_{N,n,d}}$ *and a closed subvariety* $\Gamma \subset C_{N,n,d} \times \mathbb{P}^N$ *determining this family such that* $y \in C_{N,n,d}$ *is the associated form for* $\varphi\psi^{-1}(y)$.

Obviously, a form $F \in \mathbb{P}^{\nu_{N,n,d}}$ *is contained in* $C_{N,n,d}$ *if and only if* $F = F_X$, $\dim X = n$, $\deg X = d$.

A proof can be found in [18], Vol. 2, Ch. X, §8 or in [27], Ch. 1, §9.

Generally speaking, the set $C_{N,n,d}$ is not closed in $\mathbb{P}^{\nu_{N,n,d}}$. For example, it is easy to see that the quadratic form $F(x_0, x_1, x_2)$ is the associated form of a conic $F=0$ if and only if F is not the square of a linear form. However, the definition can be modified somewhat so as to arrive at closed varieties. For this purpose we introduce the concept of an n-dimensional cycle, by which we understand a formal linear combination $D = m_1 X_1 + \cdots + m_l X_l$ of subvarieties $X_i \subset \mathbb{P}^N$ of dimension n with integers $m_i > 0$. We set

$$\deg D = m_1 \deg X_1 + \cdots + m_l \deg X_l,\ F_D = F_{X_1}^{m_1} \cdots F_{X_l}^{m_l}.$$

The associated form of a so-defined cycle has all the properties of the associated form a subvariety, but in addition, the set $\overline{C}_{N,n,d}$ of all associated forms of cycles $D \subset \mathbb{P}^N$, $\dim D = n$, $\deg D = d$, is closed.

The concept of an associated form enables us to approach the classification problem for subvarieties and cycles of a projective space. Generally speaking, the algebraic variety $C_{N,n,d}$ is reducible. Its irreducible components, their numbers and dimension, give a representation of the collection of subvarieties of a given dimension and degree in $\mathbb{P}^N$. We emphasize that we consider here subvarieties in $\mathbb{P}^N$ not to within isomorphism, but that we regard them as distinct if they are distinct as sets.

In conclusion we give some examples. The case of a hypersurface is very simple (Exercise 13). Therefore the first non-trivial examples are curves in $\mathbb{P}^3$, that is, the case $N=3$, $n=1$. We list the results for $d=1, 2, 3$.

$d=1$; $C_{3,1,1}=\bar{C}_{3,1,1}=\Pi$ is a Plücker hypersurface of the second order. It is irreducible and of dimension 4 (see also Exercise 14).

$d=2$; $\bar{C}_{3,1,2}$ is reducible, $\bar{C}_{3,1,2}=C'\cup C''$. The components C' and C'' are irreducible and $\dim C'=\dim C''=8$. The points of C' correspond to plane conics, the points of C'' to pairs of lines (generally speaking skew):

$d=3$; $\bar{C}_{3,1,3}=C'\cup C''\cup C'''\cup C^{\mathrm{IV}}$,

$$\dim C'=\dim C''=\dim C'''=\dim C^{\mathrm{IV}}=12\,.$$

The points of C' correspond to triples of lines, the points of C'' to reducible curves consisting of a plane conic and a line (generally speaking, in another plane), and the points of C''' to plane cubic curves. Finally, the points C^{IV} correspond to twisted cubic curves. It can be shown that all these curves are obtained from the curve $v_3(\mathbb{P}^1)$ (see Exercise 18) by distinct linear transformations.

In all the cases we have been able to compute so far the irreducible components of the variety $C_{3,1,d}$ have not been too complicated. More accurately, they are all rational varieties, that is, birationally isomorphic to projective spaces. Whether this is so, in general, is an apparently very difficult but very fundamental problem.

Exercises

1. Let L be an $(n-1)$-dimensional linear subspace of $\mathbb{P}^n$, $X\subset L$ an irreducible closed variety, and $y\notin L$. We join y by lines to all the points $x\in X$. We denote by Y the set of points lying on all these lines. Show that Y is an irreducible projective variety and that $\dim Y=\dim X+1$.

2. Let $X\subset\mathbb{A}^3$ be a reducible curve whose components are the three coordinates axes. Show that the ideal $\mathfrak{A}_X$ cannot be generated by two elements.

3. Let $X\subset\mathbb{P}^2$ be a reducible zero-dimensional variety whose components are three non-collinear points. Show that the ideal $\mathfrak{A}_X$ cannot be generated by two elements.

4. Show that any finite set of points $S\subset\mathbb{A}^2$ can be determined by two equations (*Hint*: Choose a system of coordinates x, y in $\mathbb{A}^2$ so that the points of S have distinct x-coordinates. Then define S by the equations $y=f(x)$, $\prod(x-\alpha_i)=0$, where $f(x)$ is a polynomial.)

5. Show that any finite set of points $S\subset\mathbb{P}^2$ can be given by two equations.

6. Let $X\subset\mathbb{A}^3$ be an algebraic curve, and x, y, z coordinates in $\mathbb{A}^3$. Show that there exists a non-zero polynomial $f(x,y)$ that vanishes at all points of X. Show that all such polynomials form a principal ideal $(g(x,y))$ and that the curve $g(x,y)=0$ is the closure of the projection of X onto the (x,y)-plane parallel to the z-axis.

7. We use the notation of Exercise 6. Let $h(x,y,z)=g_0(x,y)z^n+\dots+g_n(x,y)$ be the polynomial of smallest positive degree in z in the ideal $\mathfrak{A}_X$. Show that if $f\in\mathfrak{A}_X$ and if the degree of f in z is m, then $f\cdot g_0^m=h\cdot U+v(x,y)$ and $v(x,y)$ is divisible by $g(x,y)$. Deduce that the equations $h=0$, $g=0$ determine a reducible curve consisting of X and finitely many lines parallel to the z-axis and determined by the equations $g_0(x,y)=0$, $g(x,y)=0$.

8. Using Exercises 6 and 7 show that any curve $X \subset \mathbb{A}^3$ can be determined by three equations.

9. By analogy with Exercises 6–8 show that any curve $X \subset \mathbb{P}^3$ can be determined by three equations.

10. Can any irreducible curve $X \subset \mathbb{A}^3$ (or $X \subset \mathbb{P}^3$) be determined by two equations? The answer to this question is not known.*

11. Let $F_0(x_0, \dots, x_n), \dots, F_n(x_0, \dots, x_n)$ be forms of degree $m_0, \dots, m_n$. Denote by Γ the subset of $\mathbb{P}^n \times \prod_i \mathbb{P}^{\nu_{n,m_i}}$ consisting of those systems $(F_0, \dots, F_n, x)$ for which $F_0(x) = \dots = F_n(x) = 0$. By considering the projections $\varphi: \Gamma \to \Pi \mathbb{P}^{\nu_{n,m_i}}$ and $\psi: \Gamma \to \mathbb{P}^n$ show that $\dim \Gamma = \dim \varphi(\Gamma) = \sum_i \nu_{n,m_i} - 1$. Deduce that there exists a polynomial $R(F_0, \dots, F_n)$ in the coefficients of the forms $F_0, \dots, F_n$ such that the equation $R=0$ is necessary and sufficient for the system of $n+1$ equations in $n+1$ unknowns $F_0 = \dots = F_n = 0$ to have a non-zero solution. What is the polynomial R if the forms $F_0, \dots, F_n$ are linear?

12. Show that the associated form of a point $(u_0 : \dots : u_N)$ is of the form $\sum_{i=0}^{N} v_i u_i$.

13. Show that if X is a hypersurface and $\mathfrak{A}_X = (G(u_0, \dots, u_N))$, then $F_X = G(\Delta_0, \dots, \Delta_N)$, where $(-1)^i \Delta_i$ is the minor of the matrix $(v_j^{(i)})$ obtained by deleting the i-th column.

14. Let $X \subset \mathbb{P}^3$ be a line with the Plücker coordinates $p_{ij} (0 \leqslant i \leqslant j \leqslant 3)$. Show that $F_X = \sum_{i,j} p_{ij} v_i^{(0)} \cdot v_j^{(1)}$.

15. Show that the degree of a hypersurface X defined by an equation $G=0$ is equal to the degree of the form G, provided that G has no multiple factors.

16. Show that subvarieties $X \subset \mathbb{P}^n$ of degree 1 are linear subspaces.

17. Let $v_m: \mathbb{P}^1 \to \mathbb{P}^m$ be a Veronese mapping, $X = v_m(\mathbb{P}^1)$. Show that the degree of X is m. *Hint:* Use the connection mentioned in §4.4 between hyperplane sections of $v_m(\mathbb{P}^1)$ and forms on $\mathbb{P}^1$.

18. Show that a curve X in $\mathbb{P}^n$ of degree 2 either lies in a plane and is given there by an equation of degree 2, or degenerates into two lines.

19. Let $X = X_1 \cup \dots \cup X_t$ be a decomposition into irreducible components assumed to be of equal dimension. Show that $\deg X = \sum \deg X_i$. How is the form F_X connected with the forms F_{X_i}?

20. Show that an irreducible curve $X \subset \mathbb{P}^n$ of degree d is contained in some d-dimensional linear subspace.

21. Show that on a Plücker hypersurface Π there lie two systems of 2-dimensional linear subspaces. The subspace of the first system is determined by a point $\xi \in \mathbb{P}^3$ and consists of all points of Π that correspond to lines $L \subset \mathbb{P}^3$ passing through ξ. The subspace of the second system is determined by a plane $\Xi \subset \mathbb{P}^3$ and consists of all those points on Π that corresponds to line $L \subset \mathbb{P}^3$ lying in Ξ. There are no other 2-dimensional linear subspaces on Π.

22. Let $F(x_0, x_1, x_2, x_3)$ be an arbitrary form of degree 4. Show that there exists a polynomial Φ in the coefficients of F such that the condition $\Phi = 0$ is necessary and sufficient for the surface determined by the equation $F=0$ to contain a line.

23. Let $X \subset \mathbb{P}^3$ be a non-degenerate quadric and $\Lambda_X \subset \Pi$ the set of points on the hypersurface Π corresponding to the lines on X. Show that Λ_X consists of two non-intersecting lines.

24. Show that the points of the space $\mathbb{P}^{\nu_{3,2}}$ corresponding to degenerate quadrics form a hypersurface.

---

* The question about curves in Λ^3 has been answered positively under some mild assumptions.

Chapter II. Local Properties

§ 1. Simple and Singular Points

1. The Local Ring of a Point. In this chapter we study local properties of points of algebraic varieties, that is, properties of points $x \in X$ that are preserved when X is replaced by any affine neighbourhood of x. Since every point has an affine neighbourhood, we may confine ourselves in the study of local properties to points on affine varieties.

The basic local invariant of a point x of a variety X is the local ring $\mathcal{O}_x$ of this point. This ring consists of all the functions that are regular in some neighbourhood of x. However, since distinct functions are regular in distinct neighbourhoods, this definition needs some care.

If X is irreducible, then $\mathcal{O}_x$ is a subring of the field $k(X)$ and consists of all the functions $f \in k(X)$ that are regular at x. Recalling the definition of $k(X)$ as the field of fractions of the coordinate ring $k[X]$, we see that $\mathcal{O}_x$ consists of the ratios $f/g, f, g \in k[X]$, $g(x) \neq 0$.

This construction becomes clearer when we draw attention to its general and purely algebraic character. It can be applied to any commutative ring A and a prime ideal $\mathfrak{p}$ in it. But here a new difficulty arises owing to the fact that A may have divisors of zero.

Consider the set of pairs (f, g), $f, g \in A$, $g \notin \mathfrak{p}$, which we identify according to the rule

$$(f, g) = (f', g')$$

if there exists an element $h \in A$, $h \notin \mathfrak{p}$, such that

$$h(fg' - gf') = 0. \tag{1}$$

The operations in this set are defined as follows:

$$(f, g) + (f', g') = (fg' + gf', gg'), \tag{2}$$

$$(f, g)\,(f', g') = (ff', gg'). \tag{3}$$

It is easy to verify that in this way we obtain a ring, which is called the *local ring* of the prime ideal $\mathfrak{p}$ and is denoted by $A_{\mathfrak{p}}$.

The mapping $\varphi: A \to A_{\mathfrak{p}}$, $\varphi(h) = (h, 1)$ is a homomorphism. The elements $\varphi(g)$, $g \notin \mathfrak{p}$, are invertible in $A_{\mathfrak{p}}$, and every element $u \in A_{\mathfrak{p}}$ can be written in the form $u = \varphi(f)/\varphi(g)$, $g \notin \mathfrak{p}$. Occasionally the somewhat inaccurate notation $u = f/g$ is used. The elements of the form $\varphi(f)/\varphi(g)$, $f \in \mathfrak{p}$, $g \notin \mathfrak{p}$, form an ideal $\mathfrak{m} \subset A_{\mathfrak{p}}$, and every element $u \in A_{\mathfrak{p}}$, $u \notin \mathfrak{m}$, has an inverse. Therefore $\mathfrak{m}$ contains all other ideals of $A_{\mathfrak{p}}$.

Here we come across one of the most fundamental concepts of commutative algebra:

A ring $\mathcal{O}$ is said to be *local* if it has an ideal $\mathfrak{m} \subset \mathcal{O}$, $\mathfrak{m} \neq \mathcal{O}$, containing all other ideals.

Lemma. *If A is a Noetherian ring, then every local ring $A_{\mathfrak{p}}$ is also Noetherian.*

We set $\bar{\mathfrak{a}} = \varphi^{-1}(\mathfrak{a})$ for an ideal $\mathfrak{a} \subset A_{\mathfrak{p}}$. This is an ideal in A, which by hypothesis has a finite basis: $\bar{\mathfrak{a}} = (f_1, \ldots, f_r)$. If $u \in \mathfrak{a}$, then $u = \varphi(f)/\varphi(g)$, $g \notin \mathfrak{p}$, $f, g \in A$. Hence it follows that $f \in \bar{\mathfrak{a}}$, and since $1/\varphi(g) \in A_{\mathfrak{p}}$, we see that $u \in \varphi(\bar{\mathfrak{a}})\, A_{\mathfrak{p}} = (\varphi(f_1), \ldots, \varphi(f_r))$. Therefore $\mathfrak{a} = (\varphi(f_1), \ldots, \varphi(f_r))$, that is, $\mathfrak{a}$ has a finite basis.

If $A = k[X]$, where X is an affine variety, and $\mathfrak{p} = \mathfrak{m}_x$, $x \in X$, then $A_{\mathfrak{p}}$ is called the *local ring of the point* x and is denoted by $\mathcal{O}_x$. According to the lemma, it is Noetherian.

For every pair (f, g) that determines an element of $\mathcal{O}_x$, the function f/g is regular in a neighbourhood $D(g)$ of x. The rule (1) indicates that we identify in $\mathcal{O}_x$ functions that coincide in some neighbourhood of x (in our case $D(h)$). Thus, $\mathcal{O}_x$ can also be defined as the ring whose elements are regular functions in various neighbourhoods of x, with the given rule of identification. Now this definition no longer depends on the choice of any affine neighbourhood U of x.

Let us choose, in particular, a variety U whose irreducible components all pass through x. Then a function f that vanishes on some neighbourhood $V \subset U$ of x vanishes on the whole of U. Therefore the homomorphism $\varphi: k[U] \to \mathcal{O}_x$ is an embedding, and we identify $k[U]$ with a subring of $\mathcal{O}_x$. In that case we can ignore the factor h in the rule of identification (1). In other words, $\mathcal{O}_x$ consists of functions on U without any identifications, and all the functions $\varphi \in \mathcal{O}_x$ are of the form f/g, $f, g \in k[U]$, $g(x) \neq 0$.

2. The Tangent Space. We define the tangent space at a point x of an affine variety X as the totality of lines passing through x and touching X. To define what it means to say that a line $L \subset \mathbb{A}^N$ touches the variety $X \subset \mathbb{A}^N$ we assume that the system of coordinates in $\mathbb{A}^N$ is chosen so that $x = (0, \ldots, 0) = O$. Then $L = \{ta, t \in k\}$, where a is a fixed point other than O. To investigate the intersection of X with L

we assume that X is given by a system of equations $F_1 = \dots = F_m = 0$, with $\mathfrak{A}_X = (F_1, \dots, F_m)$.

The set $X \cap L$ is then determined by the equations $F_1(ta) = \dots = F_m(ta) = 0$. Since we are now concerned with polynomials in a single variable t, their common roots are the roots of their greatest common divisor. Let

$$f(t) = \text{g.c.d.}\,(F_1(ta), \dots, F_m(ta)),$$
$$f(t) = c \prod (t - \alpha_i)^i . \tag{1}$$

The values $t = \alpha_i$ correspond to points of intersection of L with X. Observe that in (1) the values $t = \alpha_i$ are endowed with a multiplicity l_i, which we naturally interpret as multiplicities of the intersections of L with X. In particular, since $O \in L \cap X$, the root $t = 0$ occurs in (1). We arrive at the following definition.

Definition 1. The *intersection multiplicity* at a point O of a line L and a variety X is the multiplicity of the root $t = 0$ in the polynomial $f(t) = \text{g.c.d.}\,(F_1(ta), \dots, F_m(ta))$.

Thus, this multiplicity is the highest power of t that divides all the polynomials $F_i(ta)$. By definition, it is at least 1.

If the polynomials $F_i(ta)$ vanish identically, then the intersection multiplicity is taken to be $+\infty$.

Clearly, $f(t) = \text{g.c.d.}\,(F(ta), F \in \mathfrak{A}_X)$, so that the multiplicity does not depend on the choice of the generators F_i of $\mathfrak{A}_X$.

Definition 2. A line L *touches* the variety X at O if its intersection multiplicity at this point is greater than 1.

Let us write down conditions for tangency of L and X. Since $O \in X$, the constant terms of all the polynomials $F_i(T)$ are 0. We denote by L_i their linear parts, so that $F_i = L_i + G_i$ $(i = 1, \dots, m)$, where the G_i contain only terms of degree at least 2. Then $F_i(at) = tL_i(a) + G_i(ta)$, and $G_i(ta)$ is divisible by t^2. Therefore $F_i(at)$ is divisible by t^2 if and only if $L_i(a) = 0$. The conditions for tangency have the form

$$L_i(a) = \dots = L_m(a) = 0 . \tag{2}$$

Definition 3. The locus of points on lines touching X at x is called the *tangent space at the point* x. It is denoted by Θ_x or, when we have to emphasize the variety X in question, by $\Theta_{x,X}$.

Thus, (2) are the equations of the tangent space. They show that Θ_x is a linear subspace.

Example 1. The tangent space to $\mathbb{A}^n$ at each of its points is the whole of $\mathbb{A}^n$.

Example 2. Let $X \subset \mathbb{A}^n$ be a hypersurface and $\mathfrak{A}_X = (F)$. If $O \in X$ and $F = L + G$ (in the previous notation), then Θ_0 is defined by the single equation $L(T_1, \ldots, T_n) = 0$. Therefore, if $L \neq 0$, then $\dim \Theta_0 = n - 1$, and if $L = 0$, then $\Theta_0 = \mathbb{A}^n$ and $\dim \Theta_0 = n$. An example of the second case (with $n = 2$) is the curve with the equation $x^2 - y^2 + x^3 = 0$.

3. Invariance of the Tangent Space. Definition 3 above is given in terms of the equations of the variety X. Therefore it is not obvious that under an isomorphism $f: X \to Y$ the tangent spaces of the points x and $f(x)$ are isomorphic (that is, have the same dimension). We show that this is so, and for this purpose we reformulate the concept of the tangent space so that it only depends on the algebra $k[X]$.

We recall some definitions. A polynomial $F(T_1, \ldots, T_N)$ at a point $x = (x_1, \ldots, x_N)$ has a Taylor expansion $F(T) = F(x) + F_1(T) + F_2(T) + \cdots + F_l(T)$, where the F_i are homogeneous polynomials of degree i in the variables $T_j - x_j$. The linear form F_1 is called the *differential polynomial* of F at x and is denoted by dF or $d_x F$. It is

$$d_x F = \sum_{i=1}^{N} \left(\frac{\partial F}{\partial T_i} \right) (x)\, (T_i - x_i)\,.$$

From the definition it follows that

$$d_x(F + G) = d_x F + d_x G\,, \qquad d_x(FG) = F(x) d_x G + G(x) d_x F\,. \tag{1}$$

In the same notation we can write down the Eq. (2) above for the tangent space of X at x in the form

$$d_x F_1 = \cdots = d_x F_m = 0 \tag{2}$$

or

$$\sum_{i=1}^{N} \left(\frac{\partial F_j}{\partial T_i} \right) (x)\, (T_i - x_i) = 0 \qquad (j = 1, \ldots, m)\,, \tag{3}$$

where $\mathfrak{A}_X = (F_1, \ldots, F_m)$. Suppose that $g \in k[X]$ is determined by some polynomial G restricted to X. If we were to set $d_x g = d_x G$, the result would depend on the choice of G, more accurately, it would be determined only to within a term $d_x F$, $F \in \mathfrak{A}_X$. Since $\mathfrak{A}_X = (F_1, \ldots, F_m)$, we have $F = G_1 F_1 + \cdots + G_m F_m$, and by (1) and the fact that $F_i(x) = 0$ we obtain $d_x F = G_1(x) d_x F_1 + \cdots + G_m(x) d_x F_m$. Taking (2) into account, we see that the linear forms $d_x F$, $F \in \mathfrak{A}_X$, vanish on Θ_x; therefore, if we denote by $d_x g$ the restriction of $d_x G$ to Θ_x:

$$d_x g = d_x G_{\Theta_x}\,, \tag{4}$$

we associate with every function $g \in k[X]$ a uniquely determined linear form on Θ_x.

Definition. The linear function $d_x g$ defined by (4) is called the *differential of g at x.*

Obviously,

$$d_x(f+g)=d_x f+d_x g,\ d_x(fg)=f(x)\,d_x g+g(x)\,d_x f. \tag{5}$$

Thus, we have a homomorphism $d_x: k[X]\to\Theta_x^*$, where Θ_x^* is the space of linear forms on Θ_x. Since $d_x\alpha=0$ for $\alpha\in k$, we can replace the investigation of this homomorphism by that of $d_x:\mathfrak{m}_x\to\Theta_x^*$, where $\mathfrak{m}_x=\{f\in k[X];\ f(x)=0\}$. Clearly $\mathfrak{m}_x$ is an ideal of the ring $k[X]$.

Theorem 1. *The homomorphism d_x determines an isomorphism of the spaces $\mathfrak{m}_x/\mathfrak{m}_x^2$ and Θ_x^*.*

We have to prove that $\operatorname{Im} d_x=\Theta_x^*$, $\operatorname{Ker} d_x=\mathfrak{m}_x^2$. The first is obvious: every linear form φ on Θ_x is induced by some linear function f on $\mathbb{A}^n$ and $d_x f=\varphi$. To prove the second assertion we assume that $x=10,\dots,01$, $d_x g=0$, $g\in\mathfrak{m}_x$. Suppose that g is induced by the polynomial $G\in k[T_1,\dots,T_N]$. By hypothesis, the linear form $d_x G$ vanishes on Θ_x, and hence is a linear combination of the left-hand sides of the Eq. (2) of this subspace:

$$d_x G=\lambda_1 d_x F_1+\cdots+\lambda_m d_x F_m.$$

We set $G_1=G-\lambda_1 F_1-\cdots-\lambda_m F_m$. We see that G_1 does not contain terms of degree 0 or 1 in $T_1,\dots,T_N$, hence that $G_1\in(T_1,\dots,T_N)^2$. Further, $G_1|_X=G|_X=g$, and hence $g\in(t_1,\dots,t_N)^2$, where $t_i=T_i|_X$. Since obviously $\mathfrak{m}_x=(t_1,\dots,t_N)$, this proves the theorem.

It is standard knowledge that if L is a linear space and $M=L^*$ the space of all linear functions on L, then L can be identified with the space of all linear functions on M, that is, $L=M^*$. Applying this device to our situation we obtain the following corollary.

Corollary 1. *The tangent space at x is isomorphic to the space of linear functions on $\mathfrak{m}_x/\mathfrak{m}_x^2$.*

From this we can deduce a result on the behaviour of the tangent space under regular mappings of varieties. Let $f:X\to Y$ be such a mapping and $f(x)=y$. This determines a mapping $f^*:k[Y]\to k[X]$, and obviously $f^*(\mathfrak{m}_y)\subset\mathfrak{m}_x$, $f^*(\mathfrak{m}_y^2)\subset\mathfrak{m}_x^2$, so that the mapping

$$f^*:\mathfrak{m}_y/\mathfrak{m}_y^2\to\mathfrak{m}_x/\mathfrak{m}_x^2$$

is well-defined. Linear functions, like arbitrary functions, are mapped in the opposite direction, and since by Corollary 1 the spaces $\Theta_{x,X}$ and $\Theta_{y,Y}$ are isomorphic to the spaces of linear functions on $\mathfrak{m}_x/\mathfrak{m}_x^2$ and $\mathfrak{m}_y/\mathfrak{m}_y^2$, respectively, we arrive at a mapping $\Theta_{x,X}\to\Theta_{y,Y}$. This is called the differential mapping of f and is denoted by $d_x f$.

It is easy to verify that if $g: Y \to Z$ is another regular mapping and $z = g(y)$, then for the mapping $d(g \circ f): \Theta_{x,X} \to \Theta_{z,Z}$ the relation $d(g \circ f) = dg \circ df$ holds. If f is the identity mapping $X \to X$, then $d_x f$ is the identity mapping of Θ_x for every point $x \in X$. From these remarks we derive the following corollary.

Corollary 2. *Under an isomorphism of varieties the tangent spaces at corresponding points are mapped isomorphically. In particular, the dimension of a tangent space is invariant under isomorphisms.*

Theorem 2. *The tangent space $\Theta_{x,X}$ is a local invariant of x in X.*

We show how to define Θ_x in terms of the local ring $\mathcal{O}_x$ of x. We recall that the differential of a rational function F/G, $F, G \in k[T_1, \dots, T_n]$, is defined as

$$d_x(F/G) = (G(x) d_x F - F(x) d_x G)/G^2(x), \quad G(x) \neq 0.$$

We can regard a function $f \in \mathcal{O}_x$ as the restriction to X of a rational function F/G and define the differential as $d_x f = d_x(F/G)_{|\Theta_x}$. All the arguments preceding Theorem 1, and also its proof, remain valid, and we find that d_x determines an isomorphism $d_x: \mathfrak{m}_x/\mathfrak{m}_x^2 \to \Theta_x^*$, where now $\mathfrak{m}_x$ denotes the maximal ideal of $\mathcal{O}_x: \mathfrak{m}_x = \{f \in \mathcal{O}_x; f(x) = 0\}$. This proves Theorem 2.

We define the tangent space Θ_x at a point x of any quasiprojective variety X as $(\mathfrak{m}_x/\mathfrak{m}_x^2)^*$, where $\mathfrak{m}_x$ is the maximal ideal of the local ring $\mathcal{O}_x$ of x. By Theorem 2 it is also the tangent space at x of any of its affine neighbourhoods.

Thus, the tangent space is defined as an "abstract" vector space, not specified in the form of a subspace of some larger space. However, if X is affine and $X \subset \mathbb{A}^N$, then the embedding i of X in $\mathbb{A}^N$ determines an embedding di of $\Theta_{x,X}$ in $\Theta_{x,\mathbb{A}^N}$. Since $\Theta_{x,\mathbb{A}^N}$ can be identified with $\mathbb{A}^N$, we can regard $\Theta_{x,X}$ as embedded in $\mathbb{A}^N$ and revert to the definition given in § 1.2.

If X is projective and $X \subset \mathbb{P}^N$, $x \in X$ and $x \in \mathbb{A}_i^N$, then $\Theta_{x,X}$ is contained in $\mathbb{A}_i^N$. The closure of $\Theta_{x,X}$ in $\mathbb{P}^N$ does not depend on the choice of the open affine set $\mathbb{A}_i^N$. Although here one and the same term refers to two distinct objects, $\bar{\Theta}_{x,X} \subset \mathbb{P}^N$ is also occasionally called the tangent space to X at x.

The invariance of the tangent space permits us to answer some questions on embeddings of varieties in affine spaces. For example, if a point $x \in X$ is such that $\dim \Theta_x = N$, then X is not isomorphic to any subvariety of the affine space $\mathbb{A}^n$ with $n < N$: an isomorphism $f: X \to Y \subset \mathbb{A}^n$ would carry Θ_x into an isomorphic space $\Theta_{f(x)} \subset \mathbb{A}^n$. Starting out from this we can construct for any $n > 1$ an example of a curve $X \subset \mathbb{A}^n$ that is not isomorphic to any curve $Y \subset \mathbb{A}^m$ with $m < n$.

Indeed, let X be the image of $\mathbb{A}^1$ under the mapping

$$x_1 = t^n, \; x_2 = t^{n+1}, \ldots, x_n = t^{2n-1}. \tag{6}$$

It is sufficient to prove that $\Theta_{x,X} = \mathbb{A}^n$ for $x = (0, \ldots, 0)$. This means that none of the polynomials $F \in \mathfrak{A}_X$ contains linear terms in $T_1, \ldots, T_n$. Let $F \in \mathfrak{A}_X$ and $F = \sum_{i=1}^{n} a_i T_i + G$, $G \in (T_1, \ldots, T_n)^2$. Substituting (6) in F we see that $\sum_{i=1}^{n} a_i t^{n+i-1} + G(t^n, t^{n+1}, \ldots, t^{2n-1}) = 0$, identically in t. But this is impossible if at least one $a_i \neq 0$, because the terms $a_i t^{n+i-1}$ are of degree $\leqslant 2n-1$, whereas terms arising from $G(t^n, \ldots, t^{2n-1})$ are of degree $\geqslant 2n$ and cannot cancel.

From the preceding proof it follows that no neighbourhood of a point x on a curve X is isomorphic to a quasiprojective subvariety of $\mathbb{A}^m$ with $m < n$.

4. Singular Points. We now explain what can be said about the dimensions of tangent spaces of points of an irreducible quasiprojective variety X. Our result is of local character, and therefore we restrict our attention to affine varieties.

Let $X \subset \mathbb{A}^N$ be an irreducible variety. In the direct product $\mathbb{A}^N \times X$ we consider the set Θ of those pairs (a, x), $a \in \mathbb{A}^N$, $x \in X$, for which $a \in \Theta_x$. The Eq. (2) of § 1.3 show that Θ is closed in $\mathbb{A}^N \times X$. We denote by π the projection $\Theta \to X : \pi(a, x) = x$. Obviously $\pi(\Theta) = X$, $\pi^{-1}(x) = \{(a, x); a \in \Theta_x\}$. Thus, Θ is stratified into the tangent spaces to X at various points $x \in X$. The variety Θ is called the *tangent fibering* of X. By § 1.3 (3) the dimension of Θ_x is $N - r$, where r is the rank of the matrix $((\partial F_j/\partial T_i)(x))$. We denote by ϱ the rank of the matrix $(\partial F_j/\partial T_i)$ whose elements belong to $k[X]$. Then all the minors of this matrix of order greater than ϱ vanish, and there exist non-zero minors Δ_α of order ϱ. Hence it follows that in the matrix $(\partial F_j/\partial T_i)(x)$ all minors of order greater than ϱ vanish, so that $r \leqslant \varrho$, and $r < \varrho$ for precisely those points x for which all $\Delta_\alpha(x) = 0$. Therefore there exists a number s such that $\dim \Theta_x \geqslant s$ and that points $y \in X$ for which $\dim \Theta_y > s$ form a proper closed subset of X, in other words, a subvariety of smaller dimension.

Definition. Points x of an irreducible variety X for which $\dim \Theta_x = s = \min \dim \Theta_y$ are called *simple points*; the remaining points are called *singular*.

A variety for which a point x is simple is called non-singular at this point. A variety is called *smooth* if all its points are simple. As we have just seen, simple points form an open non-empty subset, and singular points a closed proper subset, of X.

Let us consider the example of a hypersurface (Example 2 of § 1.2). If $\mathfrak{A}_X=(F)$, then the equation of the tangent space at x is of the form

$$\sum_{i=1}^{n} (\partial F/\partial T_i)(x)(T_i-x_i)=0.$$

We show that in this case $s=\min \dim \Theta_y=n-1$. Clearly this is equivalent to the fact that $\partial F/\partial T_i$ do not vanish simultaneously on X. For characteristic 0 this would mean that F is constant, and for characteristic $p>0$ that all the variables occur in F with exponents that are multiples of p. But then (since k is algebraically closed) $F=F_1^p$, and this contradicts the fact that $\mathfrak{A}_X=(F)$. Thus, in our example $\dim \Theta_x=\dim X=n-1$ for simple points $x\in X$. Next we show that the same happens with an arbitrary irreducible variety and that the general case can be reduced to that of a hypersurface.

Theorem 3. *The dimension of the tangent space at a simple point is equal to the dimension of the variety.*

By virtue of the definition of a simple point the theorem asserts that $\dim \Theta_x \geqslant \dim X$ for all points x of an irreducible variety X, and that the set of points x for which $\dim \Theta_x=\dim X$ is open and non-empty. Clearly, this is a local statement, and it is enough for us to consider the case of an affine variety. We have seen that there exists an integer s such that $\dim \Theta_x\geqslant s$ for all $x\in X$ and that the set of points for which $\dim \Theta_x=s$ is open and non-empty. It remains to show that $s=\dim X$. Now we use Theorem 6 of Ch. I, § 3, which states that X is birationally isomorphic to a hypersurface Y.

Let $\varphi: X\to Y$ be this birational isomorphism. According to the proposition in Ch. I, § 4.3 there exist open and non-empty sets $U\subset X$ and $V\subset Y$ such that φ determines an isomorphism between them. By the remarks made before the statement of the theorem, the set W of simple points of Y is open, and $\dim \Theta_y=\dim Y=\dim X$ for $y\in W$. The set $W\cap V$ is also open and non-empty, hence so is $\varphi^{-1}(W\cap V)\subset U$. Since the dimension of a tangent space is invariant under an isomorphism, $\dim \Theta_x=\dim X$ for $x\in\varphi^{-1}(W\cap U)$, and the theorem is proved.

Now we turn to reducible varieties. For them even the inequality $\dim \Theta_x\geqslant \dim X$ ceases to be true. For example, if $X=X_1\cup X_2$, $\dim X_1=1$, $\dim X_2=2$, and if x lies in X_1 but not in X_2 and is a simple point of X_1, then $\Theta_x=1$ and $\dim X=2$. This is quite natural: the components of X that do not pass through x influence the dimension of X, but do not influence the space Θ_x. Therefore it is natural to introduce the following concept: the *dimension* $\dim_x X$ *of a variety* X *at a point* x is the maximum of the dimensions of the irreducible components of X that pass through x. Obviously, $\dim X=\max\limits_{x\in X}\dim_x X$.

Definition. A point x of an affine variety X is called *simple* if $\dim \Theta_x = \dim_x X$.

From Theorem 3 it follows that $\dim \Theta_x \geqslant \dim_x X$ for any point $x \in X$. For if X^i $(i=1,\dots,s)$ are the irreducible components of X passing through x, and Θ_x^i the tangent space to X^i at this point, then $\dim \Theta_x^i \geqslant \dim X^i$, $\Theta_x^i \subset \Theta_x$, and therefore $\dim \Theta_x \geqslant \max_i \dim \Theta_x^i \geqslant \max \dim X^i = \dim_x X$.

In exactly the same way it follows from Theorem 3 that the singular points are contained in a subvariety of smaller dimension than X.

5. The Tangent Cone. The simplest invariant that distinguishes a singular point from simple points is the dimension of its tangent space. However, there is a much more delicate invariant: the tangent cone to X at a singular point x. We shall not need this concept later, therefore we leave the detailed execution of the subsequent arguments to the reader as a (very simple) exercise.

Let X be an affine irreducible variety. The tangent cone to X at a point $x \in X$ consists of lines passing through x, which we define as an analogue to the limiting position of secants in differential geometry.

Suppose that $X \subset \mathbb{A}^N$, $x=(0,\dots,0)$ and that $\mathbb{A}^N$ is turned into a vector space by choosing x as the origin of coordinates. In $\mathbb{A}^{N+1} = \mathbb{A}^N \times \mathbb{A}^1$ we consider the set $\tilde{X}$ of pairs (a,t), $a \in \mathbb{A}^N$, $t \in \mathbb{A}^1$, for which $a \cdot t \in X$. As always, we have two projections, namely $\varphi : \tilde{X} \to \mathbb{A}^1$ and $\psi : \tilde{X} \to \mathbb{A}^N$. Clearly, $\tilde{X}$ is closed in A^{N+1}. It is easy to see that it is reducible (if $X \neq \mathbb{A}^N$) and that it consists of two components: $\tilde{X} = \tilde{X}_1 \cup \tilde{X}_2$; $\tilde{X}_2 = \{(a,0);\ a \in \mathbb{A}^N\}$, $\tilde{X}_1$ is the closure of $\varphi^{-1}(\mathbb{A}^1 - (0))$ in $\tilde{X}$. We denote by φ_1 and ψ_1 the restrictions of φ and ψ to $\tilde{X}_1$. The set $\psi_1(\tilde{X}_1)$ is the closure of the set of points on all secants of X that pass through x. The set $T_x = \psi_1 \varphi_1^{-1}(0)$ is called the *tangent cone* to X at x.

It is easy to write down the equations of the tangent cone. The equations of $\tilde{X}$ have the form

$$F(at) = 0\,, \qquad F \in \mathfrak{A}_X.$$

Let $F = F_l + F_{l+1} + \dots + F_m$, where F_j is a form of degree j, $F_l \neq 0$. Then $F(at) = t^l F_l(a) + \dots + t^m F_m(a)$. Since $F(0)=0$, we always have $l \geqslant 1$, and the equation of the component $\tilde{X}_2$ is $t=0$. It is easy to see that the equations of T_x are of the form $F_l = 0$, $F \in \mathfrak{A}_X$. Here F_l is called the *initial form of the polynomial* F. Thus, T_x is defined by equating to zero all initial forms of the polynomials of the ideal $\mathfrak{A}_X$. Since T_x is determined by homogeneous equations, it is a cone with its vertex at x. It is easy to see that $T_x \subset \Theta_x$, and $T_x = \Theta_x$ if x is a simple point.

Let us consider the example of a plane algebraic curve $X \subset \mathbb{A}^2$. If $\mathfrak{A}_X = (F(x, y))$ and F_l is the initial form of F, then the equation of T_x is $F_l(x, y) = 0$. Since F_l is a form in two variables and k is algebraically closed, F_l splits into a product of linear forms $F_l(x, y) = \prod (\alpha_i x + \beta_i y)^{l_i}$. Therefore T_x splits in this case into several lines $\alpha_i x + \beta_i y = 0$. These lines are called tangents to X at x, and l_i are the multiplicities of these tangents. If $l > 1$, then $\Theta_x = \mathbb{A}^2$. The number l is called the *multiplicity of the singular point* x. For $l = 2$ it is called a double point, for $l = 3$ a triple point.

For example, if $F = x^2 - y^2 + x^3$, $x = (0, 0)$, then T_x consists of two lines: $x + y = 0$, $x - y = 0$; if $F = x^2 y - y^3 + x^4$, $x = (0, 0)$, then T_x consists of three lines: $y = 0$, $x + y = 0$, $x - y = 0$; if $F = y^2 - x^3$, $x = (0, 0)$, then $y = 0$ is a double tangent.

Like our original definition of a tangent space, this definition of a tangent cone uses concepts that are not invariant under isomorphisms. It can be shown, however, that the tangent cone T_x is invariant under isomorphisms and is a local invariant of x.

Exercises

1. Show that the local ring of a point x of an irreducible variety X is the union (in $k(X)$) of all rings $k[U]$, where the U are neighbourhoods of x.

2. The mapping $\varphi(t) = (t^2, t^3)$ determines a birational isomorphism of the curve $y^2 = x^3$ and the line $\mathbb{A}^1$. What rational functions of t correspond to functions of the local ring $\mathcal{O}_x$ of the point $(0, 0)$?

3. The same for the birational isomorphism between $\mathbb{A}^1$ and the curve (1) of Ch. I, § 1.1.

4. Show that the local ring $\mathcal{O}_x$ of the point $(0, 0)$ of a curve $xy = 0$ is isomorphic to the subring $\Omega \subset \mathcal{O}' \oplus \mathcal{O}'$ ($\mathcal{O}'$ is the local ring of the point O on $\mathbb{A}^1$) consisting of those functions (f, g), $f, g \in \mathcal{O}'$ for which $f(0) = g(0)$.

5. Determine the local ring of the point $(0, 0, 0)$ of the curve consisting of the three coordinate axes in $\mathbb{A}^3$.

6. Determine the local ring of the point $(0, 0)$ of the curve $xy(x - y) = 0$.

7. Show that if $x \in X$, $y \in Y$ are simple points, then the point $(x, y) \in X \times Y$ is simple.

8. Show that if $X = X_1 \cup X_2$, $x \in X_1 \cap X_2$, and if $\Theta_{x,X}$, Θ_{x,X_1}, and Θ_{x,X_2} are the tangent spaces, then $\Theta_{x,X} \supset \Theta_{x,X_1} + \Theta_{x,X_2}$. Does equality always hold?

9. Show that a hypersurface of order 2 having a singular point is a cone.

10. Show that if a hypersurface of order 3 has two singular points, then the line containing them lies on the hypersurface.

11. Show that if a plane curve of order 3 has three singular points, then it splits into three lines.

12. Show that the singular points of the hypersurface in $\mathbb{P}^n$ given by the equation $F(x_0, \ldots, x_n) = 0$ are determined by the system of equations $F(x_0, \ldots, x_n) = 0$, $F_{x_i}(x_0, \ldots, x_n) = 0$, $(i = 0, \ldots, n)$. If the degree of the form F is not divisible by the characteristic of the field, then the first equation is a consequence of the others.

13. Show that if a hypersurface X in $\mathbb{P}^n$ contains a linear subspace L of dimension $r \geqslant n/2$, then it has singular points. *Hint:* Choose a coordinate system so that L is given by the equations $x_{r+1} = 0, \ldots, x_n = 0$, and look for singular points contained in L.

14. For what values of a has the curve $x_0^3+x_1^3+x_2^3+a(x_0+x_1+x_2)^3=0$ a singular point? What is then the nature of the singular points? Is the curve reducible?

15. Determine the singular points of the Steiner surface in $\mathbb{P}^3$:

$$x_1^2x_1^2+x_2^2x_0^2+x_0^2x_1^2-x_0x_1x_2x_3=0.$$

16. Determine the singular points of the dual Steiner surface in $\mathbb{P}^3$:

$$x_0x_1x_2+x_0x_1x_3+x_0x_2x_3+x_1x_2x_3=0.$$

17. Show that over a field of characteristic 0 the points of the space $\mathbb{P}^{\nu_{n,m}}$ (see Ch. I, § 5.2) corresponding to hypersurfaces having a singular point form a hypersurface in $\mathbb{P}^{\nu_{n,m}}$. *Hint:* Use the results of Exercise 11 in Ch. I, § 6.

18. Show that if a curve of order 3 has a singular point, then it is rational. *Hint:* Use the projection from this point.

19. Let $F(x_0, x_1, x_2)=0$ be the equation of an irreducible curve $X\subset\mathbb{P}^2$. Consider the rational mapping $\varphi: X\to\mathbb{P}^2$ given by the formulae $u_i=\partial F/\partial x_i(x_0,x_1,x_2)$, $i=0,1,2$. Show that a) $\varphi(X)$ is a point if and only if X is a line; b) if X is not a line, then φ is regular at $x\in X$ if and only if it is a simple point. The curve $\varphi(X)$ is called *dual* to X.

20. Show that if X is a conic, then so is $\varphi(X)$.

21. Find the dual to the curve $x_0^3+x_1^3+x_2^3=0$.

22. Show that if a hypersurface $X\subset\mathbb{P}^n$ does not have singular points and is not a hyperplane, then the set of linear subspaces Θ_x, $x\in X$, forms a hypersurface in the dual space $\tilde{\mathbb{P}}^n=\mathbb{P}^{\nu_{n,1}}$.

23. Let φ be a regular mapping of a variety $X\subset\mathbb{A}^n$ consisting of a projection onto some subspace. Determine the mapping $d\varphi$ of the linear space Θ_x, $x\in X$.

24. Show that for any integer $t>0$ the group $\mathfrak{m}_x^t/\mathfrak{m}^{t+1}\ x$ is a finite-dimensional vector space over k.

§ 2. Expansion in Power Series

1. Local Parameters at a Point. We investigate a simple point x of an n-dimensional variety X.

Definition. Functions $u_1,\dots,u_n\in\mathcal{O}_x$ are called *local parameters* at x if $u_i\in\mathfrak{m}_x$ and $u_1,\dots,u_n$ form a basis of the space $\mathfrak{m}_x/\mathfrak{m}_x^2$.

By virtue of the isomorphism $d_x:\mathfrak{m}_x/\mathfrak{m}_x^2\to\Theta_x^*$ we see that $u_1,\dots,u_n$ form a system of local parameters if and only if the linear forms $d_xu_1,\dots,d_xu_n$ are linearly independent on Θ_x. Since $\dim\Theta_x^*=n$, this in turn is equivalent to the fact that the equations

$$d_xu_1=\dots=d_xu_n=0 \tag{1}$$

have only the trivial solution in Θ_x.

We may replace X by an affine neighbourhood X' of x on which the functions $u_1,\dots,u_n$ are regular. We denote by X_i' the hypersurface determined in X' by the equation $u_i=0$. Let U_i be the polynomial that determines on X' the function u_i, and let $\mathfrak{A}_i=\mathfrak{A}_{X_i'}$, $\mathfrak{A}=\mathfrak{A}_{X'}$. Then $\mathfrak{A}_i\supset(\mathfrak{A},U_i)$, and from the definition of the tangent space it follows that $\Theta_i\subset L_i$, where Θ_i is the tangent space to X_i' at x, and $L_i\subset\Theta_x$ is determined by the equation $d_xU_i=0$. From the fact that the system (1) has only the trivial solution it follows that $L_i\neq\Theta_x$, that is, $\dim L_i=n-1$, and from the theorem on the dimension of an intersection and the inequality $\dim \Theta_i \geqslant \dim X_i'$ it follows that $\dim \Theta_i \geqslant n-1$. Therefore $\dim \Theta_i = n-1$, and this means that x is a simple point on X_i'. The intersection of the varieties X_i' in some neighbourhood of it is just the point x; if there were a component Y of the intersection $\bigcap X_i'$, $\dim Y > 0$, passing through x, then the tangent space to Y at x would be contained in all Θ_i, and this again contradicts the fact that the system (1) only has the trivial solution.

So we have proved the following proposition.

Theorem 1. *If $u_1, \ldots, u_n$ are local parameters at a point x, $u_1, \ldots, u_n$ are regular on X, and $X_i = V(u_i)$, then the point x is simple on each of the X_i and $\bigcap \Theta_i = 0$, where Θ_i is the tangent space to X_i at x.*

Here we encounter a general property of subvarieties, which we shall meet frequently later on.

Definition. Subvarieties $Y_1, \ldots, Y_r$ of a smooth variety X are said to be *transversal* at a point $x \in \bigcap Y_i$ if

$$\operatorname{codim}_\Theta\left(\bigcap_{i=1}^r \Theta_{x,Y_i}\right) = \sum_{i=1}^r \operatorname{codim}_X Y_i, \qquad \Theta = \Theta_{x,X}. \tag{2}$$

Using the inequality (4) in Ch. I, § 6.2 for the subspaces $\Theta_{x,Y_i} \subset \Theta$ and the inequality $\operatorname{codim}_\Theta \Theta_{x,Y_i} \leqslant \operatorname{codim}_X Y_i$, we see that (2) leads to the equality $\dim \Theta_{x,Y_i} = \dim Y_i$, which indicates that all the Y_i are smooth at x, and to the equality

$$\operatorname{codim}_\Theta \bigcap_{i=1}^r \Theta_{x,Y_i} = \sum_{i=1}^r \operatorname{codim}_\Theta \Theta_{x,Y_i},$$

which indicates that the linear spaces Θ_{x,Y_i} are transversal: they have the smallest possible intersection compatible with their dimensions. From the inclusion $\bigcap_{i=1}^r \Theta_{x,Y_i} \supset \Theta_{x,Y}$, where $Y = \bigcap Y_i$, we obtain similarly that Y is also smooth at x.

For example, two smooth curves on a surface having distinct tangents at a point of intersection are transversal (Fig. 3).

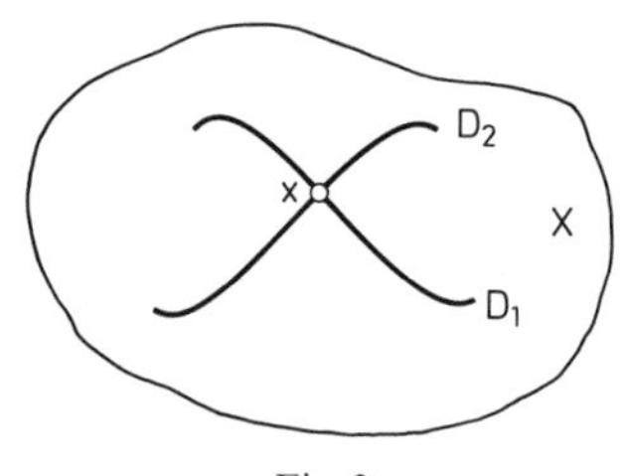

Fig. 3

Thus, Theorem 1 asserts that the subvarieties $V(u_i)$ are transversal.

Let X' be an affine neighbourhood of x in which $\bigcap X_i = x$. Then x is determined by equations $t_1 = \cdots = t_N = 0$ if $X' \subset \mathbb{A}^N$ and the t_i are coordinates, and $\bigcap X_i$ by the equations $u_1 = \cdots = u_n = 0$. From Hilbert's Nullstellensatz it now follows that $(t_1, \ldots, t_N)^l \subset (u_1, \ldots, u_n)$ for some $l > 0$. Here $(t_1, \ldots, t_N)$ and $(u_1, \ldots, u_n)$ are ideals of the ring $k[X']$. All the more this is true for the ideals $(t_1, \ldots, t_N)$ and $(u_1, \ldots, u_n)$ in $\mathcal{O}_x$. Observe that $(t_1, \ldots, t_N) = \mathfrak{m}_x$, so that $\mathfrak{m}_x^l \subset (u_1, \ldots, u_n)$. We now prove a more precise proposition.

Theorem 2. *The local parameters generate the maximal ideal* $\mathfrak{m}_x$ *of the local ring* $\mathcal{O}_x$.

As we have seen, $\mathfrak{m}_x = (t_1, \ldots, t_N)$. What we have to show is that all the $t_i \in (u_1, \ldots, u_n)$. By induction on $N - i$ we show that $t_i \in (u_1, \ldots, u_n, t_1, \ldots, t_{i-1})$. Suppose that this is true for $i = N, \ldots, l+1$. By hypothesis,

$$\mathfrak{m}_x = (u_1, \ldots, u_n, t_1, \ldots, t_N) = (u_1, \ldots, u_n, t_1, \ldots, t_l) . \tag{3}$$

From the definition of local parameters it follows that

$$t_l \equiv \sum_{j=1}^{n} \alpha_j u_j (\mathfrak{m}_x^2) , \qquad \alpha_j \in k . \tag{4}$$

By virtue of (3) every element of $\mathfrak{m}_x^2$ can be written in the form

$$\mu_1 u_1 + \cdots + \mu_n u_n + \mu'_1 t_1 + \cdots + \mu'_l t_l , \qquad \mu_j, \mu'_j \in \mathfrak{m}_x .$$

Therefore (4) means that

$$t_l = \sum_{j=1}^{n} \alpha_j u_j + \sum_{j=1}^{n} \mu_j u_j + \sum_{s=1}^{l} \mu'_s t_s$$

or

$$(1 - \mu'_l) t_l = \sum_{j=1}^{n} \alpha_j u_j + \sum_{j=1}^{n} \mu_j u_j + \sum_{s=1}^{l-1} \mu'_s t_s \in (u_1, \ldots, u_n, t_1, \ldots, t_{l-1}) . \tag{5}$$

Since $\mu'_l \in \mathfrak{m}_x$, we have $1 - \mu'_l \notin \mathfrak{m}_x$, and hence $(1 - \mu'_l)^{-1} \in \mathcal{O}_x$. Therefore it follows from (5) that $t_l \in (u_1, \ldots, u_n, t_1, \ldots, t_{l-1})$.

Note. The preceding argument proves the following general fact about local rings.

Nakayama's Lemma. *Let* M *be a module of finite type over a local ring* $\mathcal{O}$ *with maximal ideal* $\mathfrak{m}$. *If the elements* $u_1, \ldots, u_n \in M$ *are such that their images in* $M/\mathfrak{m}M$ *generate this module, then* $u_1, \ldots, u_n$ *generate* M.

It is important to mention that the property of a point x of being simple can be characterized by a purely algebraic property of its local ring $\mathcal{O}_x$. By definition, $x \in X$ is a simple point if and only if

$\dim_k \mathfrak{m}_x/\mathfrak{m}_x^2 = \dim_x X$. The left-hand side of this equality is defined for every Noetherian local ring $\mathcal{O}$. The right-hand side can also be expressed as a property of the local ring $\mathcal{O}_x$. For by Corollary 1 to Theorem 5 in Ch. I, § 6, the dimension of a variety X at x can be defined as the smallest integer r for which there exist r functions $u_1, \ldots, u_r \in \mathfrak{m}_x$ such that the set determined by the equations $u_1 = 0, \ldots, u_r = 0$ consists in a certain neighbourhood of x of this point only. By Hilbert's Nullstellensatz this property is equivalent to the fact that $(u_1, \ldots, u_r) \supset \mathfrak{m}_x^l$ for some $l > 0$.

For an arbitrary Noetherian local ring $\mathcal{O}$ with maximal ideal $\mathfrak{m}$ the smallest number of elements $u_1, \ldots, u_r \in \mathfrak{m}$ for which $(u_1, \ldots, u_r) \supset \mathfrak{m}^l$ for some $l > 0$ is called the dimension of $\mathcal{O}$ and is denoted by $\dim \mathcal{O}$. By Nakayama's lemma, $\mathfrak{m}$ itself is generated by n elements, where $n = \dim_{\mathcal{O}/\mathfrak{m}}(\mathfrak{m}/\mathfrak{m}^2)$. Therefore

$$\dim \mathcal{O} \leqslant \dim_{\mathcal{O}/\mathfrak{m}}(\mathfrak{m}/\mathfrak{m}^2) .$$

If $\dim \mathcal{O} = \dim_{\mathcal{O}/\mathfrak{m}}(\mathfrak{m}/\mathfrak{m}^2)$, then the local ring is called regular. We see that a point x is simple if and only if its local ring $\mathcal{O}_x$ is regular. And this is the algebraic meaning of simplicity of a point.

2. Expansion in Power Series. The method of associating power series with the elements of a local ring $\mathcal{O}_x$ is based on the following arguments. For every function $f \in \mathcal{O}_x$ we set $f(x) = \alpha_0$, $f_1 = f - \alpha_0$. Then $f_1 \in \mathfrak{m}_x$. Let $u_1, \ldots, u_n$ be a system of local parameters at x. By definition, the elements $u_1, \ldots, u_n$ generate the whole vector space $\mathfrak{m}_x/\mathfrak{m}_x^2$. Hence there exist $\alpha_1, \ldots, \alpha_n \in k$ such that $f_1 - \sum_{i=1}^{n} \alpha_i u_i \in \mathfrak{m}_x^2$. We set $f_2 = f_1 - \sum_{i=1}^{n} \alpha_i u_i = f - \alpha_0 - \sum_{i=1}^{n} \alpha_i u_i$. Since $f_2 \in \mathfrak{m}_x^2$, we have $f_2 = \Sigma g_j h_j$, g_j, $h_j \in \mathfrak{m}_x$. As above, there exist β_{ji}, $\gamma_{ji} \in k$ such that

$$g_j - \sum_{i=1}^{n} \beta_{ji} u_i \in \mathfrak{m}_x^2 , \quad h_j - \sum_{i=1}^{n} \gamma_{ji} u_i \in \mathfrak{m}_x^2 .$$

We set $\sum_j \left(\sum_i \beta_{ji} u_i\right)\left(\sum_i \gamma_{ji} u_i\right) = \sum_{1 \leq l, m \leq n} \alpha_{lm} u_l u_m$. Then $f_2 - \sum \alpha_{lm} u_l u_m \in \mathfrak{m}_x^2$, and hence $f - \alpha_0 - \sum \alpha_i u_i - \sum \alpha_{lm} u_l u_m \in \mathfrak{m}_x^2$. Continuing like this we can obviously find forms $F_i \in k[T_1, \ldots, T_n]$, $\deg F_i = i$, such that $f - \sum_{i=0}^{l} F_i(u_i, \ldots, u_n) \in \mathfrak{m}_x^{l+1}$.

Definition. The *ring of formal power series* in the variables $(T_1, \ldots, T_n) = T$ is the ring whose elements are infinite expressions of the form

$$\Phi = F_0 + F_1 + F_2 + \cdots , \tag{1}$$

where $F_i \in k[T]$ is a form of degree i, and the operations are defined by the rules: if $\Psi = G_0 + G_1 + G_2 + \cdots$, then

$$\Phi + \Psi = (F_0 + G_0) + (F_1 + G_1) + (F_2 + G_2) + \cdots,$$
$$\Phi \cdot \Psi = H_0 + H_1 + H_2 + \cdots, \qquad H_i = \sum_{j+l=i} G_j F_l.$$

The ring of formal power series is denoted by $k[[T]]$. It contains the field k (power series in which $F_i = 0$ for $i > 0$). If i is the first suffix for which $F_i \neq 0$, then F_i is called the initial form of (1). The initial form of a product is the product of the initial forms, therefore $k[[T]]$ has no divisors of the zero.

The preceding arguments enable us to assign to a function $f \in \mathcal{O}_x$ a power series $\Phi = F_0 + F_1 + F_2 + \cdots$.

So we arrive at the following definition.

Definition. A formal power series Φ is called a *Taylor series* of the function $f \in \mathcal{O}_x$ if for all $l \geqslant 0$

$$f - S_l(u_1, \ldots, u_n) \in \mathfrak{m}_x^{l+1}, \qquad S_l = \sum_{i=0}^{l} F_i. \tag{2}$$

Example. Let $X = \mathbb{A}^1$ and let x be the point corresponding to the coordinate value $t = 0$. Then $\mathfrak{m}_x = (t)$, and the power series $\sum_{m=0}^{\infty} \alpha_m t^m$ associated with the rational function $f(t) = P(t)/Q(t)$, $Q(0) \neq 0$, is such that

$$P(t)/Q(t) - \sum_{m=0}^{l} \alpha_m t^m \equiv 0(t^{l+1}),$$

that is,

$$P(t) - Q(t)\left(\sum_{m=0}^{l} \alpha_m t^m\right) \equiv 0(t^{l+1}).$$

This is the usual way of finding the coefficients of the power series of a rational function by the method of undetermined coefficients.

For example, $1/(1-t) = \sum_{m=0}^{\infty} t^m$, because

$$\frac{1}{1-t} - \sum_{m=0}^{l} t^l = \frac{t^{l+1}}{1-t} \equiv O(t^{l+1}).$$

The correspondence $f \to \Phi$ essentially depends on the choice of the local system of parameters $u_1, \ldots, u_n$.

The arguments just given prove the following proposition.

Theorem 3. *Every function f has at least one Taylor series.*

So far we have nowhere used the fact that x is a simple point. For $u_1, \ldots, u_n$ we can choose any system of elements of $\mathcal{O}_x$ whose images generate $\mathfrak{m}_x/\mathfrak{m}_x^2$. Now we make use of the fact that x is a simple point.

Theorem 4. *If x is a simple point, then the function has a unique Taylor series.*

Clearly it is sufficient to show that any Taylor series of the function $f=0$ is zero. By (2) this is equivalent to the statement: if $F_l(T_1, \dots, T_n)$ is a form of degree l, $u_1, \dots, u_n$ are local parameters of a simple point x, and

$$F_l(u_1, \dots, u_n) \in \mathfrak{m}_x^{l+1}, \tag{3}$$

then

$$F_l(T_1, \dots, T_n) = 0.$$

Suppose that this is not so. By a non-singular linear transformation we can achieve that the coefficient of T_n^l in the form F_l is different from 0. This coefficient is equal to $F_l(0, \dots, 0, 1)$, and if $F_l(\alpha_1, \dots, \alpha_n) \neq 0$ (such $\alpha_1, \dots, \alpha_n$ exist as $F_l \neq 0$), then we need only carry out a linear transformation that takes the vector $(\alpha_1, \dots, \alpha_n)$ into $(0, \dots, 0, 1)$. Thus, we may assume that

$$F_l(T_1, \dots, T_n) = \alpha T_n^l + G_1(T_1, \dots, T_{n-1})\, T_n^{l-1} + \dots + G_l(T_1, \dots, T_{n-1}),$$

where $\alpha \neq 0$ and G_i is a form of degree i.

From Theorem 2 of § 2.1 it follows easily that every element of the ideal $\mathfrak{m}_x^{l+1}$ can be written as a form of degree l in $u_1, \dots, u_n$ with coefficients in $\mathfrak{m}_x$. Therefore (3) can be written in the form

$$\begin{aligned} &\alpha u_n^l + G_1(u_1, \dots, u_{n-1})\, u_n^{l-1} + \dots + G_l(u_1, \dots, u_{n-1}) \\ &\quad = \mu u_n^l + H_1(u_1, \dots, u_{n-1})\, u_n^{l-1} + \dots + H_l(u_1, \dots, u_{n-1}), \end{aligned} \tag{4}$$

where $\mu \in \mathfrak{m}_x$, and H_i is a form of degree i. Hence it follows that $(\alpha - \mu) u_n^l \in (u_1, \dots, u_{n-1})$. Since $\alpha \neq 0$, we have $\alpha - \mu \notin m_x$ and $(\alpha - \mu)^{-1} \in \mathcal{O}_x$, therefore $u_n^l \in (u_1, \dots, u_{n-1})$. So we see that $V(u_n) \supset V(u_1) \cap \dots \cap V(u_{n-1})$. Hence $\Theta_n \supset \Theta_1 \cap \dots \cap \Theta_{n-1}$ (where Θ_i is the tangent space to $V(u_i)$ at x), and hence $\Theta_1 \cap \dots \cap \Theta_n = \Theta_1 \cap \dots \cap \Theta_{n-1}$. Therefore $\dim(\Theta_1 \cap \dots \cap \Theta_n) \geqslant 1$, which contradicts Theorem 1 of § 2.1. This proves the theorem.

Thus, we have a uniquely determined mapping $\tau: \mathcal{O}_x \to k[[T]]$, which associates with every function its Taylor series. A simple check, based on the definition (2) of τ, shows that τ is a homomorphism. We leave this check to the reader.

What is the kernel of τ? If $\tau(f) = 0$ for a function $f \in \mathcal{O}_x$, then by (2) this means that $f \in \mathfrak{m}_x^{l+1}$ for all l. In other words, $f \in \bigcap \mathfrak{m}_x^l$. Therefore we are concerned with functions analogous to those functions in analysis whose derivatives all vanish at a certain point. We prove that in our case such a function must be equal to zero. This is a consequence of the following more general theorem, if we bear in mind that, as we have proved in § 1.1, the ring $\mathcal{O}_x$ is Noetherian.

Theorem 5. *Let A be a Noetherian ring and $\mathfrak{a} \subset A$ any ideal for which the elements $1+\alpha$, $\alpha \in \mathfrak{a}$, are not divisors of zero in A. Then* $\bigcap_l \mathfrak{a}^l = 0$.

Proof. Let $\alpha \in \mathfrak{a}^l$ for any $l > 0$ and $\mathfrak{a} = (u_1, \ldots, u_n)$. This means that α can be represented as $\alpha = F_l(u_1, \ldots, u_n)$, where $F_l \in A[T_1, \ldots, T_n]$ is a form of degree l.

We consider the ideal generated by all the forms $F_l(T)$ $(l = 1, 2, \ldots)$ in the ring $A[T]$. Since A and hence $A[T]$ is Noetherian, this ideal has a finite basis, and we can choose a finite set of forms F_l, say $F_1, \ldots, F_m$ generating it. Then

$$F_{m+1}(T) = \sum_{i=1}^{m} G_i(T)\, F_i(T), \tag{5}$$

where $G_i \in A[T]$ is a form of degree $m+1-i$. We substitute in this equation $T_1 = u_1$, $T_2 = u_2, \ldots, T_n = u_n$. Since the degree of the form G_i is positive, we have $\mu_i = G_i(u_1, \ldots, u_n) \in \mathfrak{a}^{m+1-i} \subset \mathfrak{a}$. From (5) we find that $\alpha = \mu\alpha$, $\mu = \sum_{i=1}^{m} \mu_i \in \mathfrak{a}$. Hence it follows that $(1-\mu)\,\alpha = 0$, and since $\mu \in a$, by the condition on $\mathfrak{a}$ we have $\alpha = 0$. This completes the proof.

Corollary. *A function $f \in \mathcal{O}_x$ is uniquely determined by any one of its Taylor series. In other words, the mapping τ is an isomorphic embedding of the local ring $\mathcal{O}_x$ in the ring of formal power series $k[[T]]$.*

Nowhere in this section have we used the fact that X is an irreducible variety. On the contrary, from Theorem 5 and its corollary we can draw some conclusions on irreducibility.

Theorem 6. *If x is a simple point, then one and only one component of X passes through it.*

We replace X by a neighbourhood U of x, $X' = X - \bigcup Z_i$, where the Z_i are all the components of X that do not pass through x. Then $k[X'] \subset \mathcal{O}_x$. According to the corollary to Theorem 5, $\mathcal{O}_x$ is isomorphic to a subring of the ring of formal power series $k[[T]]$. Since $k[[T]]$ has no divisors of zero, this is also true of the ring $k[X']$, which is isomorphic to a subring of it. Therefore X' is irreducible, as was asserted by the theorem.

Corollary. *The set of singular points of an algebraic variety X is closed.*

Let $X = \bigcup X_i$ be the decomposition into irreducible components. From Theorem 6 it follows that the set of singular points of the variety is the union of the sets $X_i \cap X_j$ $(i \neq j)$ and the set of singular points of the X_i. Being the union of finitely many closed sets, it is closed.

3. Varieties over the Field of Real and the Field of Complex Numbers. Assuming that k is the field of real or complex numbers, we show that then the formal Taylor series of functions $f \in \mathcal{O}_x$ converge for small values of $T_1, \ldots, T_n$.

Let $\mathfrak{A}_X = (F_1, \ldots, F_m)$, $X \subset \mathbb{A}^N$ and $\dim_x X = n$. If $x \in X$ is a simple point, then the rank of the matrix

$$((\partial F_i/\partial T_j)(x)) \quad (i=1, \ldots, m; j=1, \ldots, N)$$

is $N-n$. Suppose that

$$|(\partial F_i/\partial T_j)(x)| \neq 0 \quad (i=1, \ldots, N-n; j=n+1, \ldots, N). \tag{1}$$

Let x be the origin of coordinates. Then $t_1, \ldots, t_n$ (coordinates restricted to X) form a system of local parameters of x. We denote by X' the set of components passing through x of the variety determined by the equations

$$F_1 = 0, \ldots, F_{N-n} = 0. \tag{2}$$

By (1) the dimension of its tangent space Θ' at x is n, and by the theorem on the dimension of an intersection $\dim_x X' \geqslant n$. Since $\dim \Theta' \geqslant \dim_x X'$, we see that $\dim_x X' = n$, and x is a simple point on X'. Hence by Theorem 6, X' is irreducible. Clearly $X' \supset X$, and since $\dim X' = \dim X$, it follows that $X' = X$.

So we see that X can be determined in some neighbourhood of x by the $N-n$ Eq. (2), with (1) holding. By the implicit function theorem (see, for example, [16], § 185) there exist a system of power series $\Phi_1, \ldots, \Phi_{N-n}$ in the n variables $T_1, \ldots, T_n$ and $\varepsilon > 0$ such that $\Phi_j(T_1, \ldots, T_n)$ converges for all T_i with $|T_i| < \varepsilon$, and

$$F_i(T_1, \ldots, T_n, \Phi_1(T), \ldots, \Phi_{N-n}(T)) = 0, \tag{3}$$

where the coefficients of the power series $\Phi_1, \ldots, \Phi_{N-n}$ are uniquely determined from the relations (3).

But the formal power series $\tau(t_{n+1}), \ldots, \tau(t_N)$ (if $t_1, \ldots, t_n$ are chosen as local parameters) also satisfy (3) and must therefore coincide with $\Phi_1, \ldots, \Phi_{N-n}$, so that it follows that $\tau(t_i)$ $(i=n+1, \ldots, N)$ converges for $|T_j| < \varepsilon$ $(j=1, \ldots, n)$.

Any function $f \in \mathcal{O}_x$ can be represented in the form

$$f = P(t_1, \ldots, t_N)/Q(t_1, \ldots, t_N), \quad Q(x) \neq 0$$

$$\text{and} \quad \tau(f) = P(\tau(t_1), \ldots, \tau(t_N))/Q(\tau(t_1), \ldots, \tau(t_N)).$$

The convergence of the series $\tau(f)$ follows therefore from standard theorems on convergence of series.

Similarly it can be shown that if $u_1, \ldots, u_n$ is any other system of local parameters, then

$$|(\partial \tau(u_i)/\partial T_j)(0, \ldots, 0)| \neq 0 \quad (i=1, \ldots, n; j=1, \ldots, n),$$

the Taylor series of $t_1, \ldots, t_n$ in terms of the local parameters $u_1, \ldots, u_n$ are obtained by inverting the series $\tau(u_i) = \Phi_i(T_1, \ldots, T_n)$ $(i = 1, \ldots, n)$, and therefore also have a positive radius of convergence. Hence it follows that the series $\tau(f), f \in \mathcal{O}_x$ has a positive radius of convergence for any choice of local parameters.

The implicit function theorem asserts not only the existence of convergent series $\Phi_1, \ldots, \Phi_{N-n}$, but also the fact that for some $\eta > 0$ any point $(t_1, \ldots, t_N) \in X$, $|t_i| < \eta$ $(i = 1, \ldots, N)$, has the form $t_{n+i} = \Phi_i(t_1, \ldots, t_n)$ $(i = 1, \ldots, N-n)$. It follows that the mapping $(t_1, \ldots, t_N) \to (t_1, \ldots, t_n)$ carries the set $(t_1, \ldots, t_N) \in X$, $|t_i| < \eta$, one-to-one and bicontinuously onto a domain of the n-dimensional space.

The space $\mathbb{P}^N$ over k (when k is the field of real or of complex numbers) is a topological space. An algebraic variety X in this space is also a topological space. We call the relevant topology in X *real* or *complex* according as k is the field of real or complex numbers. It must not be confused with the topological terms such as closure, openness, ..., which we have used earlier.

The preceding arguments show that in the real topology of an n-dimensional variety X any simple point has a neighbourhood homeomorphic to a domain of the real n-dimensional space. Therefore, if all the points of X are simple, then X is an n-dimensional manifold in the topological sense. If k is the field of complex numbers, then a simple point $x \in X$ has in the complex topology a neighbourhood homeomorphic to a domain in the n-dimensional complex, and so in the $2n$-dimensional real space. Therefore, if all the points of X are simple, then X is a $2n$-dimensional manifold.

As is easy to show, the space P^N is compact both in the real and complex topology. Therefore, if X is projective, then it is compact. If k is the field of complex numbers, the converse is also true: a quasi-projective variety X that is compact in its complex topology is a projective variety. See Ch. VII, § 2, Exercise 1.

In conclusion we mention that everything we have said here (excluding the last paragraph) carries over word-for-word to the case when k is a field of p-adic numbers.

Exercises

1. Show that the set of points in which n given functions on an n-dimensional variety X do not form a system of local parameters is closed.

2. Show that a polynomial $f \in k[T] = k[\mathbb{A}^1]$ is a local parameter at a point $T = \alpha$ if and only if α is a simple root of it.

3. Show that a formal power series $\Theta = F_0 + F_1 + \ldots$ has an inverse in $k[[T]]$ if and only if $F_0 \neq 0$.

4. Consider the ring $k\{T\}$ consisting of the expressions of the form $\alpha_{-n}T^{-n}+\alpha_{-n+1}T^{-n+1}+\cdots+\alpha_0+\alpha_1 T+\ldots$, where T is a variable and n an arbitrary integer. Show that $k\{T\}$ is a field isomorphic to the field of fractions of the ring $k[[T]]$.

5. Let $X\subset\mathbb{A}^2$ be the circle given by the equation $X^2+Y^2=1$, and x the point $(0,1)$. Show that X is a local parameter at x and that

$$\tau(Y)=\sum_{n=0}^{\infty}(-1)^n\frac{1}{n!}\frac{1}{2}\left(\frac{1}{2}-1\right)\cdots\left(\frac{1}{2}-n+1\right)X^{2n}.$$

The characteristic of the ground field is 0.

6. Show that if x is a singular point, then any function $f\in\mathcal{O}_x$ has infinitely many distinct Taylor series.

7. Let $X=\mathbb{A}^1$, $x\in X$. Show that $\tau(\mathcal{O}_x)$ does not coincide with the whole ring $k[[T]]$.

§ 3. Properties of Simple Points

1. Subvarieties of Codimension 1. The theory of local rings enables us to establish an important property of a smooth variety similar to Theorem 3 in Ch. I, § 6. It concerns the question whether a subvariety $Y\subset X$ of codimension 1 can be determined by a single equation. Generally speaking, this is not so. (Note 2 after Corollary 5 in Ch. I, § 6.2). We show, however, that on non-singular varieties it is true locally. To state this result we introduce the following definition.

Functions $f_1,\ldots,f_m\in\mathcal{O}_x$ are called *local equations* of a subvariety $Y\subset X$ in a neighbourhood of x if there exists an affine neighbourhood X' of x such that $\mathfrak{a}_{Y'}=(f_1,\ldots,f_m)$ in $k[X']$, where $Y'=Y\cap X'$, $f_i\in k[X']$. It is convenient to reformulate this concept in terms of the local ring $\mathcal{O}_x$ of x. For this purpose we consider the ideal $\mathfrak{a}_{x,Y}\subset\mathcal{O}_x$ consisting of the functions $f\in\mathcal{O}_x$ that vanish on Y in some neighbourhood of x.

Clearly, for an affine variety X

$$\mathfrak{a}_{x,Y}=\{f=u/v;\ u,\ v\in k[X],\ v(x)\neq 0,\ u\in\mathfrak{a}_Y\},$$

and if all the components of Y pass through x, then $\mathfrak{a}_Y=\mathfrak{a}_{x,Y}\cap k[X]$.

Lemma. *Functions $f_1,\ldots,f_m$ are local equations of Y in a neighbourhood of x if and only if $\mathfrak{a}_{x,Y}=(f_1,\ldots,f_m)$.*

It is obvious that if $\mathfrak{a}_{Y'}=(f_1,\ldots,f_m)$ in $k[X']$, then also $\mathfrak{a}_{x,Y}=(f_1,\ldots,f_m)$ in $\mathcal{O}_x$. Let $\mathfrak{a}_{x,Y}=(f_1,\ldots,f_m)$, $f_i\in\mathcal{O}_x$, $\mathfrak{a}_Y=(g_1,\ldots,g_s)$, $g_i\in k[X]$.

Since $g_i\in\mathfrak{a}_{x,Y}$, we have

$$g_i=\sum_{j=1}^{m}h_{ij}f_j\quad(i=1,\ldots,s),\ h_{ij}\in\mathcal{O}_x.\tag{1}$$

The functions f_i and h_{ij} are regular in some principal affine neighbourhood U of x. Let $U=X-V(g)$, $g\in k[X]$. The ring $k[U]$ consists of the elements of the form u/g^l, $u \in k[X]$, $l \geqslant 0$. Then by (1) $(g_1, \ldots, g_s) = \mathfrak{a}_Y \cdot k[U] \subset (f_1, \ldots, f_m)$.

We show that $\mathfrak{a}_Y k[U] = \mathfrak{a}_{Y'}$. It then follows that $\mathfrak{a}_{Y'} \subset (f_1, \ldots, f_m)$, and since $f_i \in \mathfrak{a}_{Y'}$, this will prove the lemma.

It remains to verify that $\mathfrak{a}_Y k[U] = \mathfrak{a}_{Y'}$. The inclusion $\mathfrak{a}_Y k[U] \subset \mathfrak{a}_{Y'}$ is obvious. Let $v \in \mathfrak{a}_{Y'}$. Then $v = u/g^l$, $u \in k[X]$, and hence $u = vg^l$; consequently, $u \in \mathfrak{a}_Y$, and since $1/g^l \in k[U]$, we have $v = u/g^l \in \mathfrak{a}_Y k[U]$. Our aim is to prove the following result.

Theorem 1. *An irreducible subvariety $Y \subset X$ of codimension 1 has one local equation in a neighbourhood of any non-singular point $x \in X$.*

The proof follows precisely the lines of the proof of Theorem 3 in Ch. I, § 6. However, there we used the unique factorization in $k[T]$. Here the role analogous to this ring is played by $\mathcal{O}_x$, which has a similar property.

Theorem 2. *In the local ring of a simple point decomposition into prime factors is unique.*

We now prove Theorem 1, assuming Theorem 2 to be true. In § 3.3 we return to the proof of Theorem 2.

As we said above, the proof of Theorem 1 is the same as that of Theorem 3 in Ch. I, § 6. Since the proposition has local character, we may take X to be affine. Let f be any function in $\mathcal{O}_x$ that vanishes on Y. We decompose it into prime factors in $\mathcal{O}_x$. Owing to the irreducibility of Y one of the prime factors must also vanish on Y. We denote it by g and show that it is a local equation of Y. Replacing X by a smaller affine neighbourhood, if necessary, we may assume that g is regular on X.

Since $V(g) \supset Y$ and both subvarieties are of codimension 1, we have $V(g) = Y \cup Y'$. If $x \in Y'$, then there exist functions h and h' such that $h \cdot h' = 0$ on $V(g)$, while $h \neq 0$ and $h' \neq 0$ on $V(g)$. This means that $(hh')^r$ for some $r > 0$ is divisible by g in $k[X]$, and a fortiori in $\mathcal{O}_x$. From the unique factorization in $\mathcal{O}_x$ it follows that then h or h' are divisible by g in $\mathcal{O}_x$. Hence h or h' vanish on $V(g)$ in some neighbourhood of x, and after going over to a smaller neighbourhood on the whole of $V(g)$. This contradicts the condition. Thus, $x \notin Y'$, and again after replacing X by a sufficiently small affine neighbourhood of x we may assume that $V(g) = Y$. If now u vanishes on Y, then u^s for some $s > 0$ is divisible by g in $k[X]$, and hence also in $\mathcal{O}_x$. From this it follows that u is divisible by g in $\mathcal{O}_x$. Thus, $\mathfrak{a}_{x,Y} = (g)$, and the theorem is proved.

Theorem 1 has many applications. Here is the first of them.

Theorem 3. *If X is a smooth variety and $\varphi: X \to \mathbb{P}^n$ a rational mapping of it into a projective space, then the set of points at which φ fails to be regular, has codimension not less than two.*

We recall that the set of points of non-regularity of a rational mapping is closed. The assertion of the theorem is of local character, and it is sufficient to verify it for some neighbourhood of a simple point $x \in X$. We may write φ in the form $\varphi = (f_0 : \ldots : f_n)$, $f_i \in k(X)$, and without changing φ we can multiply f_i by a common factor such that all the f_i lie in, but have no common factor, in $\mathcal{O}_x$. Here φ can fail to be regular only at points where $f_0 = f_1 = \cdots = f_n = 0$. But no variety Y of codimension 1 is contained in the set defined by these equations. For by Theorem 1, $\mathfrak{a}_{x,Y} = (g)$, and all the f_i would have the common factor g in $\mathcal{O}_x$, against the hypothesis. This proves the theorem.

Corollary 1. *Every rational mapping of a smooth curve into a projective space is regular.*

Corollary 2. *If two smooth projective curves are birationally isomorphic, then they are isomorphic.*

Let k be the field of complex numbers. From Corollary 2 it follows that the set of points of two curves X' and X'' are homeomorphic in their complex topology if X' and X'' are birationally isomorphic. For regular functions, and hence also regular mappings, are determined in this case by convergent power series and are therefore necessarily continuous.

The same is true for the sets of real points of curves defined by equations with real coefficients if a birational isomorphism $\varphi: X \to X'$ is defined over the field of real numbers, in other words, is given by formulae with real coefficients. From this it is sometimes easy to conclude that two curves are not birationally isomorphic over the field of real numbers. For example, the curve $y^2 = x^3 - x$ has a graph (Fig. 4) consisting of two components. Therefore it is not rational (over the field of real numbers): $\mathbb{P}^1$ is homeomorphic to a circle and consists of a single component.

It can be shown on the basis of a similar idea that the curve X with the equation $y^2 = x^3 - x$ is non-rational even over the field of complex numbers. To see this we have to compare the topological spaces of complex points on X and on $\mathbb{P}^1$ in their complex topology and to show that they are not homeomorphic. In fact, the first space is homeomorphic to a torus, and the second to a sphere. This is a special case of results that will be proved in Ch. VII, § 3. Figure 5 shows how the real points of X are situated among its complex points.

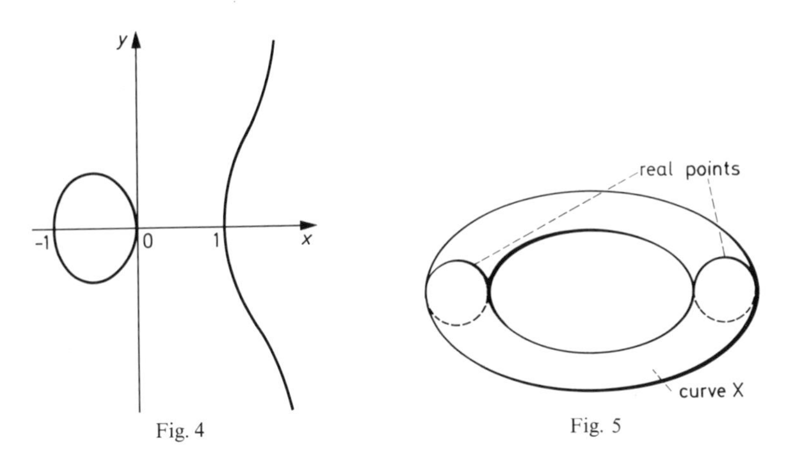

Fig. 4

Fig. 5

2. Smooth Subvarieties. Theorem 1 does not generalize to subvarieties of codimension greater than 1 (see, for example, Exercise 2 in Ch. I, § 6). But for subvarieties that are not singular at x, an analogous proposition is true. We prove a somewhat more precise fact and begin with an auxiliary proposition.

Theorem 4. *Let X be an affine variety, x a simple point of it, and $u_1, \dots, u_n$ regular functions on X forming a system of local parameters at x. Then the subvariety Y defined by the equations $u_1 = \dots = u_m = 0$ $(m \leqslant n)$ is non-singular at x and in some neighbourhood of x, $\mathfrak{a}_Y = (u_1, \dots, u_m)$, and $u_{m+1}, \dots, u_n$ form a system of local parameters at x on Y.*

The proof is by induction on m. For $m=1$ Theorem 1 shows that $\mathfrak{a}_Y = (f)$ in some affine neighbourhood of x. Let $u_1 = fv$. Then $d_x u_1 = v(x) d_x f$. Since u_1 occurs in a system of local parameters at x, we have $d_x u_1 \neq 0$. Therefore $v(x) \neq 0$, and hence $\mathfrak{a}_Y = (u_1)$ in a smaller open set. Since $d_x u_1 \neq 0$, we see that x is a simple point of Y.

Clearly, the tangent space $\Theta_{x,Y}$ to Y at x is obtained from $\Theta_{x,X}$ by imposing the condition $d_x u_1 = 0$. Therefore $d_x u_2, \dots, d_x u_n$ is a basis of $\Theta^*_{x,Y}$, that is, $u_2, \dots, u_n$ are local parameters at x on Y.

In the general case we set $X' = X_{u_1}$. Then Y is defined on X' by the equations $u_2 = \dots = u_n = 0$, and we can apply the induction.

Now we show that any subvariety Y that is non-singular at x can be obtained by the process described in Theorem 4 in some neighbourhood of a simple poiint.

Theorem 5. *Let X be a variety, $Y \subset X$, and x a simple point on Y and on X. Then we can choose a system of local parameters $u_1, \dots, u_n$ at x on X and an affine neighbourhood U of x such that $\mathfrak{a}_Y = (u_1, \dots, u_m)$.*

Proof. To an embedding of the tangent spaces $\Theta_{x,Y} \to \Theta_{x,X}$ there corresponds an epimorphism of the associated spaces $\varphi: \mathfrak{m}_{x,X}/\mathfrak{m}_{x,X}^2 \to \mathfrak{m}_{x,Y}/\mathfrak{m}_{x,Y}^2$, which is determined by restricting the functions from X to Y. We can choose a basis $u_1, \ldots, u_n$ in $\mathfrak{m}_{x,X}/\mathfrak{m}_{x,X}^2$ such that $u_1, \ldots, u_m \in \mathfrak{a}_Y$, and $u_{m+1}, \ldots, u_n$, restricted to Y, form a basis in $\mathfrak{m}_{x,Y}/\mathfrak{m}_{x,Y}^2$. We now consider an affine neighbourhood of x in which all the u_i are regular, and in it the subvariety Y' defined by the equations $u_1 = \cdots = u_m = 0$. By construction $Y' \supset Y$. We show that $Y' = Y$, from which the theorem follows by virtue of Theorem 4.

According to Theorem 4, Y' is non-singular at x, hence by § 2 Theorem 6 Y' is irreducible in a neighbourhood of x. From Theorem 4 it follows that $\dim Y' = n - m$. From the construction it is clear that $\dim \Theta_{x,Y} = n - m$, and hence $\dim Y = n - m$. Therefore $Y = Y'$, and since by Theorem 4 $\mathfrak{a}_{Y'} = (u_1, \ldots, u_m)$, we also have $\mathfrak{a}_Y = (u_1, \ldots, u_m)$ in some neighbourhood of x. This completes the proof.

In the special case $X = \mathbb{A}^m$, when k is the field of real or complex numbers, we have already proved the analogous fact in §2.3.

3. Factorization in the Local Ring of a Simple Point. Our proof of Theorem 2 is based on the embedding $\tau: \mathcal{O}_x \to k[[T]]$, where $k[[T]]$ is the ring of formal power series in the n variables $(T_1, \ldots, T_n) = T$.

To begin with we mention some properties of the ring $k[[T]]$ and the embedding τ. A formal power series must not be regarded as the sum of its terms if the structure of $k[[T]]$ only is to be used: in this ring the sum of infinitely many terms is not defined. To make it possible, we introduce a notion of convergence or, what is the same, a topology in a ring of formal power series. We denote by M the ideal of $k[[T]]$ consisting of the series Φ (in the notation of § 2.2, (1)) for which $F_0 = 0$. Clearly, $M = (T_1, \ldots, T_n)$ and M^l consists of all series Φ for which $F_i = 0$ for $i < l$. A topology is defined in $k[[T]]$ by taking as a system of neighbourhoods of 0 the ideals M^l. In other words, a sequence of power series Φ_m *converges* to Φ if the degree of the initial form of the series $\Phi_m - \Phi$ increases beyond all bounds together with m. This can be written as $\Phi_m \to \Phi$ or $\Phi = \lim \Phi_m$. It is easy to verify that in this topology $k[[T]]$ becomes a topological ring (for the definition and simplest properties of topological rings see [25], § 25).

The series $\sum_{m=0}^{\infty} \Phi_m$, $\Phi_m \in k[[T]]$, is said to converge to the sum Φ if $S_l \to \Phi$, where $S_l = \sum_{m=0}^{l} \Phi_m$. In this case we write $\Phi = \sum_{m=0}^{\infty} \Phi_m$. For example, in § 2.2 (1) $S_l = F_0 + \cdots + F_l$ and $S_l \to \Phi$, because the degree of the initial form of the series $\Phi - S_l$ is $l+1$. Therefore every formal power series is in this sense the sum of its terms.

The image $\tau(\mathcal{O}_x)$ of $\mathcal{O}_x$ is everywhere dense in $k[[T]]$. For if $u_1, \ldots, u_n$ is a system of generators of the ideal $\mathfrak{m}_x$ by means of which we have defined the mapping τ, then $\tau(u_i) = T_i$ and $\tau(p(u_i, \ldots, u_n)) = p(T_1, \ldots, T_n)$, where p is a polynomial. Since for every power series $\Phi = \lim S_l$, $S_l \in k[T]$, we have $\Phi = \lim S_l(\tau(u_1), \ldots, \tau(u_n))$, that is, Φ is the limit of a sequence of elements in $\tau\mathcal{O}_x$.

The proof of Theorem 2 is based on the unique factorization in $k[[T]]$, which has first to be established. This is a fairly elementary fact, analogous to the corresponding result for rings of polynomials. We only indicate the main steps of the proof. A completely elementary proof (which does not depend on the rest of the book) can be found in [37], Vol. 2, Ch. VII, § 1.

A power series $\Phi(T_1, \ldots, T_n)$ is called regular relative to the variable T_n if its initial form (of degree m, say) contains the term $c_m T_n^m$, $c_m \neq 0$.

A linear transformation of the variables $T_1, \ldots, T_n$ evidently induces a ring automorphism of $k[[T]]$. In particular, we can carry out a linear transformation under which the given series becomes regular relative to T_n.

Lemma 1 (Weierstrass' Preparation Theorem). *If a power series $\Phi \in k[[T]]$ is regular relative to the variable T_n and if the degree of its initial form is m, then there exists a series $U \in k[[T]]$ whose constant term does not vanish, such that the series ΦU is a polynomial in T_n over the ring $k[[T_1, \ldots, T_{n-1}]]$:*

$$\Phi U = T_n^m + R_1(T_1, \ldots, T_{n-1})\, T_n^{m-1} + \cdots + R_m(T_1, \ldots, T_{n-1}),$$
$$R_i(T_1, \ldots, T_{n-1}) \in k[[T_1, \ldots, T_{n-1}]].$$

For a proof see [37], Vol. 2, p. 139.

Lemma 2. *In a ring of formal power series decomposition of elements into prime factors is unique.*

Lemma 1 makes it possible to prove this proposition by induction on the number of variables $T_1, \ldots, T_n$, by reducing it to the analogous proposition on polynomials in T_n with coefficients in $k[[T_1, \ldots, T_{n-1}]]$. The reader can find a detailed proof in [37], Vol. 2, Ch. VII, § 1, Theorem 6 (p. 148).

Now we move on to the proof of Theorem 2. The usual proof of the unique factorization for integers is based on the existence of a greatest common divisor and is valid in any ring without divisors of zero in which any two elements a and b have a greatest common divisor d. Instead of the greatest common divisor one can also prove the existence of a least common multiple m, because $d = (a \cdot b)/m$. Next, the fact that m is the least common multiple of a and b means that $(a)\cap(b)=(m)$. Therefore we need only prove that in the ring $\mathcal{O}_x$ the intersection of principal ideals is a principal ideal. For this purpose we use the fact that this property is known to us in $k[[T]]$ from Lemma 2. We can establish a connection between the ideals in $\mathcal{O}_x$ and in $k[[T]]$ by using the embedding τ. From now on we identify the elements $f\in\mathcal{O}_x$ with $\tau(f)$. For every ideal $\mathfrak{a}\subset\mathcal{O}_x$ its *closure* $\bar{\mathfrak{a}}$ in $k[[T]]$ is defined as the set of power series Φ that are limits of sequences $\tau(f_l)$, $f_l\in\mathfrak{a}$.

By what we have said above, Theorem 2 is a consequence of the following relationships between the ideals $\mathfrak{a}$ and $\bar{\mathfrak{a}}$.

1. $\overline{\mathfrak{a}\cap\mathfrak{b}}=\bar{\mathfrak{a}}\cap\bar{\mathfrak{b}}$.

2. The ideal $\bar{\mathfrak{a}}$ is principal in $k[[T]]$ if and only if $\mathfrak{a}$ is principal in $\mathcal{O}_x$.

A proof of these propositions can be found in [37], Vol. 2, Ch. VIII. To help the reader with the analysis of the proof we give here a brief sketch, referring to the book for a detailed account of the arguments (the references are to [37], Vol. 2, Ch. VIII). We denote the ring $\mathcal{O}_x$ by A, and $k[[T]]$ by $\hat{A}$.

The main thing is the notion of *completion* of a module E over a ring A. This operation is similar to the transition from A to $\hat{A}$. For every A-module E a topology is defined, in which the submodules M^lE are neighbourhoods of zero, and we construct a topological $\hat{A}$-module $\hat{E}$ in which E is embedded as an everywhere dense set and in which the Cauchy convergence criterion holds: if $\alpha_n\in\hat{E}$ is a sequence of elements such that $\alpha_n-\alpha_m\to 0$ for $n, m\to\infty$, then α_n converges to a limit in $\hat{E}$ (this property is called completeness). The module $\hat{E}$ is constructed from the Cauchy sequences, that is, the sequences $\{\alpha_n\}$, $\alpha_n\in E$ for which $\alpha_n-\alpha_m\to 0$ as $n, m\to\infty$. Here two sequences $\{\alpha_n\}$ and $\{\beta_n\}$ can be identified if $\alpha_n-\beta_n\to 0$. Every homomorphism $f: E\to F$ extends to a homomorphism $\hat{E}\xrightarrow{f}\hat{F}$ of the completions. In the special case $E=A(=\mathcal{O}_x)$, the completion $\hat{E}$ coincides with $\hat{A}(=k[[T]])$. If E is an ideal $\mathfrak{a}\subset A$, then $\hat{E}$ coincides with its closure $\bar{\mathfrak{a}}$ in $\hat{A}$. If E is a module of finite type over A, then $\hat{E}$ is generated as an $\hat{A}$-module by the subset E:

$$\hat{E}=E\cdot\hat{A} \tag{1}$$

([37], § 2, Theorem 5).

The following is a fundamental property of completions: if a sequence $E\xrightarrow{f}F\xrightarrow{g}G$ of finite A-modules and mappings is exact, then the same is true for the sequence $\hat{E}\to\hat{F}\to\hat{G}$ ([37], § 4, Theorem 11). This property is expressed by saying that the ring $\hat{A}$ is *flat* over A.

To prove 1 we begin by verifying the analogous property

$$\overline{\mathfrak{a}+\mathfrak{b}}=\bar{\mathfrak{a}}+\bar{\mathfrak{b}}\,. \tag{2}$$

It follows at once from (1). Now we consider the exact sequences

$$0 \to \mathfrak{a} \cap \mathfrak{b} \to \mathfrak{b} \to \mathfrak{b}/\mathfrak{a} \cap \mathfrak{b} \to 0\,,$$
$$0 \to \mathfrak{a} \to \mathfrak{a} + \mathfrak{b} \to (\mathfrak{a} + \mathfrak{b})/\mathfrak{a} \to 0\,.$$

Since $\hat{A}$ is flat over A, we obtain from them the exact sequences

$$0 \to \overline{\mathfrak{a} \cap \mathfrak{b}} \to \overline{\mathfrak{b}} \to \widehat{\mathfrak{b}/\mathfrak{a} \cap \mathfrak{b}} \to 0\,,$$
$$0 \to \overline{\mathfrak{a}} \to \overline{\mathfrak{a} + \mathfrak{b}} \to ((\widehat{\mathfrak{a} + \mathfrak{b})/\mathfrak{a}}) \to 0\,,$$

The isomorphism theorem $\mathfrak{b}/\mathfrak{a}\ \mathfrak{b} \simeq \mathfrak{a} + \mathfrak{b}/\mathfrak{a}$ shows that also

$$\widehat{\mathfrak{b}/(\mathfrak{a} \cap \mathfrak{b})} \simeq \widehat{(\mathfrak{a}+\mathfrak{b})/\mathfrak{a}}\,, \qquad (\overline{\mathfrak{a} + \mathfrak{b}})/\overline{\mathfrak{a}} \simeq \overline{\mathfrak{b}}/\overline{\mathfrak{a} \cap \mathfrak{b}}\,.$$

The isomorphism theorem applied now to $\overline{\mathfrak{a}}$ and $\overline{\mathfrak{b}}$ and (2) yield the isomorphism $\overline{\mathfrak{b}}/\overline{\mathfrak{a} \cap \mathfrak{b}} \simeq \overline{\mathfrak{b}}/\overline{\mathfrak{a}} \cap \overline{\mathfrak{b}}$, from which it follows very easily that $\overline{\mathfrak{a}} \cap \overline{\mathfrak{b}} = \overline{\mathfrak{a} \cap \mathfrak{b}}$.

To prove 2 we assume that $\overline{\mathfrak{a}} = (\alpha)$. Since $\overline{\mathfrak{a}} = \mathfrak{a}\hat{A}$ by (1), we have

$$\alpha = \Sigma\, a_i \xi_i\,, \qquad a_i \in \mathfrak{a}\,, \qquad \xi_i \in \hat{A}\,, \tag{3}$$

and further, since $a_i \in \mathfrak{a}$,

$$a_i = \eta_i \alpha\,, \qquad \eta_i \in \hat{A}\,. \tag{4}$$

Substituting (4) in (3) and cancelling α we find that $\Sigma\, \xi_i \eta_i = 1$. Therefore not all the η_i are contained in M, and if $\eta_i \notin M$, then $\eta_i^{-1} \in \hat{A}$ and $\overline{\mathfrak{a}} = (a_i)$. Thus, $\overline{\mathfrak{a}} = \overline{a_i A}$, and from the fact that $\hat{A}$ is flat over A it now follows easily that also $\mathfrak{a} = a_i A$.

Exercises

1. Show that if t is a local parameter of a simple point of an algebraic curve, then every function $f \in \mathcal{O}_x$ can be represented uniquely in the form $f = t^n u$, where $n \geqslant 0$ and u is an invertible element in $\mathcal{O}_x$. Hence derive Theorem 2 for curves.

2. Prove the converse of Theorem 1 in § 2: if subvarieties $D_1, \ldots, D_n$ of codimension 1 intersect transversally at a point x and if $u_1, \ldots, u_n$ are their local equations in a neighbourhood of this point, then $u_1, \ldots, u_n$ form a system of local parameters at x.

3. Is Corollary 2 to Theorem 3 true without the assumption of smoothness? Is Theorem 3 true without this assumption?

4. Show that a point x of an algebraic curve X is simple if and only if it has a local equation.

5. A cone $X \subset \mathbb{A}^3$ is given by the equation $x^2 + y^2 - z^2 = 0$. Show that its generator L given by the equations $x = 0$, $y = z$ does not have a local equation in any neighbourhood of the point $(0, 0, 0)$.

6. A rational mapping $\varphi : \mathbb{P}^2 \to \mathbb{P}^2$ is given by the formula $(x_0 : x_1 : x_2) = (x_1 x_2 : x_0 x_2 : x_0 x_1)$. Let $x = (1 : 0 : 0)$ and $C \subset \mathbb{P}^2$ be a curve that is non-singular at x. According to Theorem 3 the mapping φ restricted to C is regular at x and therefore carries x into some point, which we denote by $\varphi_C(x)$. Show that $\varphi_{C_1}(x) = \varphi_{C_2}(x)$ if and only if the curves C_1 and C_2 touch at x, that is, $\Theta_{x,C_1} = \Theta_{x,C_2}$.

7. Show that if $\varphi=f/g$ is a rational function, if f and g are regular at a simple point x, and if the power series $\tau(f)$ is divisible by $\tau(g)$, then φ is regular at x. *Hint:* Use the connections between ideals $\mathfrak{a}$ in $\mathcal{O}_x$ and their closures $\bar{\mathfrak{a}}$ in the ring of formal power series.

8. Let $X\subset\mathbb{A}^n$ and $Y\subset\mathbb{A}^m$ be affine varieties passing through the origins of coordinates $0\subset\mathbb{A}^n$ and $0'\subset\mathbb{A}^m$. A system of formal power series $\Phi_1(T),\ldots,\Phi_m(T)$ $(T=(T_1,\ldots,T_n))$ is called a formal mapping of X into Y in a neighbourhood of 0 and 0′ if $\Phi_i(0)=0$ and $F(\Phi_1,\ldots,\Phi_m)\in\bar{\mathfrak{a}}_X\subset k[[T]]$ for all $F\in\mathfrak{a}_Y$. A composition of formal mappings is defined by substitution of the series. Two formal mappings $(\Phi_1,\ldots,\Phi_m)$ and $(\psi_1,\ldots,\psi_m)$ are called equal if $\Phi_i-\Psi_i\in\bar{\mathfrak{a}}_X$ $(i=1,\ldots,m)$; X and Y are called formally isomorphic in a neighbourhood of 0 and 0′ if there exist formal mappings $\varphi=(\Phi_1,\ldots,\Phi_m)$ of X into Y and $\psi=(\psi_1,\ldots,\psi_n)$ of Y into X such that $\psi\varphi$ and $\varphi\psi$ are equal to the identity mappings. Show that if the origin of coordinates is a simple point of X, then X is formally isomorphic to an affine space.

9. Show that a formal isomorphism of $\mathbb{A}^n$ with itself (an automorphism) in a neighbourhood of 0 is given by series $\Phi_1,\ldots,\Phi_n$ without constant terms such that the determinant formed from the linear terms does not vanish.

10. Show that two plane curves with the equations $F=0$ and $G=0$ passing through the origin of coordinates $0\in\mathbb{A}^2$ are formally isomorphic in a neighbourhood of 0 if and only if there exists a formal automorphism of $\mathbb{A}^2$ given by series Φ_1 and Φ_2 such that $F(\Phi_1,\Phi_2)=G\cdot U$, where U is a power series with a non-vanishing constant term.

11. Show that all plane algebraic curves having the origin of coordinates 0 as a double point with distinct tangents are formally isomorphic in a neighbourhood of 0 to the curve with the equation $xy=0$. *Hint:* Use Exercise 10. Look in Φ_1 and Φ_2 for the highest power of the ideal (x,y).

12. Give a formal classification of double points of plane algebraic curves over a field k of characteristic 0.

13. Let X be a hypersurface in $\mathbb{A}^n$ with the equation $F=F_2(T)+F_3(T)+\cdots+F_l(T)$, where $F_2(T)$ is a quadratic form of rank n. Show that X is formally isomorphic in a neighbourhood of 0 to the cone $T_1^2+\cdots+T_n^2=0$.

14. Denote by $k[[X]]$ the ring $k[[T]]/\bar{\mathfrak{a}}_X$. Show that $k[[X]]$ is the completion of the local ring at x. Show also that X and Y are formally isomorphic if and only if $k[[X]]$ and $k[[Y]]$ are isomorphic.

15. Construct an embedding $\tau:\mathcal{O}_x\to k[[X]]$ and show that the connections between $\mathcal{O}_x$ and $k[[T]]$ introduced in § 3.3 also hold between $\mathcal{O}_x$ and $k[[X]]$, even if x is a singular point.

16. Construct infinitely many smooth projective curves that are pairwise non-isomorphic to each other over the field of real numbers.

§ 4. The Structure of Birational Isomorphisms

1. The σ-Process in a Projective Space. In the preceding section we have shown (Corollary 2 to Theorem 3) that a birational isomorphism between smooth projective curves is an isomorphism. For varieties of higher dimension this is no longer true: for example, stereographic projection, which establishes a birational isomorphism between a non-degenerate quadric and a projective plane, is not a regular mapping (Exercise 9 of Ch. I, § 4, and the remark after Corollary 5 to Theorem 4 in Ch. I, § 6). In this section we define and investigate the simplest and most typical birational, but not regular, isomorphism: the σ-process.

We consider the projective spaces $\mathbb{P}^n$ with the homogeneous coordinates $x_0, \ldots, x_n$ and $\mathbb{P}^{n-1}$ with the homogeneous coordinates $y_1, \ldots, y_n$.

In the space $\mathbb{P}^n \times \mathbb{P}^{n-1}$ we denote a point $x \times y$, $x = (x_0 : \ldots : x_n)$, $y = (y_1 : \ldots : y_n)$ also by $(x_0 : \ldots : x_n; \; y_1 : \ldots : y_n)$. We consider the closed subvariety $\Pi \subset \mathbb{P}^n \times \mathbb{P}^{n-1}$ defined by the equations

$$x_i y_j = y_i x_j, \qquad 1 \leqslant i, j \leqslant n. \tag{1}$$

Definition 1. The mapping $\sigma : \Pi \to \mathbb{P}^n$ defined by the projection $\mathbb{P}^n \times \mathbb{P}^{n-1} \to \mathbb{P}^n$ is called the *σ-process*.

Let ξ denote the point $(1 : 0 : \ldots : 0) \in \mathbb{P}^n$. If $(x_0 : \ldots : x_n) \neq \xi$, then it follows from (1) that $(y_1 : \ldots : y_n) = (x_1 : \ldots : x_n)$, so that the mapping

$$(x_0 : \ldots : x_n) \to (x_0 : \ldots : x_n; x_1 : \ldots : x_n) \tag{2}$$

is inverse to σ. But if $(x_0 : \ldots : x_n) = \xi$, then arbitrary values of y_i satisfy the equations. Thus, $\sigma^{-1}(\xi) = \xi \times \mathbb{P}^{n-1}$, and σ determines an isomorphism between $\mathbb{P}^n - \xi$ and $\Pi - (\xi \times \mathbb{P}^{n-1})$. The point ξ is called the *centre* of the σ-process.

Now we describe the structure of Π in a neighbourhood of points of the form $(\xi; \; y_1 : \ldots : y_n)$. We have $y_i \neq 0$ for some i, consequently the chosen point lies in the open set U_i determined by the conditions $x_0 \neq 0$, $y_i \neq 0$. In this set we may even assume that $x_0 = 1$, $y_i = 1$. The Eqs. (1) then take the form $x_j = y_j x_i$, $1 \leqslant j \neq i \leqslant n$. Hence it follows that U_i is isomorphic to an affine space with the coordinates $y_1, \ldots, x_i, \ldots, y_n$.

In particular, we see that Π is not singular and hence, by Theorem 6 of § 2, is irreducible in a neighbourhood of each of its points. We shall soon see that Π is irreducible.

To give a clearer picture of the action of a σ-process, we consider it on some line L passing through ξ. Let $x_j = \alpha_j x_i$ $(j = 1, \ldots, n, \; j \neq i, \; i \neq 0)$ be the equations of this line. On L the mapping (2) takes the form $\sigma^{-1}(x_0 : \ldots : x_n) = (x_0 : \ldots : x_n; \; \alpha_1 : \ldots : \underset{i}{1} : \ldots : \alpha_n)$. We see that σ^{-1} is regular on L and carries it into a curve $\sigma^{-1}(L)$, which intersects $\xi \times \mathbb{P}^{n-1}$ in the point $(\xi; \; \alpha_1 : \ldots : \underset{i}{1} : \ldots : \alpha_n)$. We can interpret this result as follows. The mapping σ^{-1} is not regular at ξ, but by regarding it on the line L we obtain a regular mapping $\sigma^{-1} \colon L \to \Pi$. By using it we can define σ^{-1} also at ξ (over the field of real or of complex numbers this would mean that we define $\sigma^{-1}(x)$ for $x \in L$ and let x tend to ξ in the direction of L). However, the result depends on the choice of L (the limit process depends on the directions in which we carry it out). By choosing various lines L we obtain all possible points on $\xi \times \mathbb{P}^{n-1}$. Thus, although σ^{-1} is not regular at ξ, by resolving the resulting indeterminacy we obtain not arbitrary points of Π but only points of

$\xi \times \mathbb{P}^{n-1}$. Having this picture in mind we can say that σ^{-1} blows up ξ to $\xi \times \mathbb{P}^{n-1}$.

Observe that at the same time we have proved that Π is irreducible. For

$$\Pi = (\xi \times \mathbb{P}^{n-1}) \cup (\Pi - (\xi \times \mathbb{P}^{n-1})).$$

Since $\Pi - (\xi \times \mathbb{P}^{n-1})$ is isomorphic to $\mathbb{P}^n - \xi$, it is irreducible, and hence so is $\overline{\Pi - (\xi \times \mathbb{P}^{n-1})}$. What we have to verify is that

$$\xi \times \mathbb{P}^{n-1} \subset \overline{\Pi - (\xi \times \mathbb{P}^{n-1})}.$$

But necessarily

$$\sigma^{-1}(L) \subset \overline{\Pi - (\xi \times \mathbb{P}^{n-1})},$$

hence

$$\sigma^{-1}(L) \cap (\xi \times \mathbb{P}^{n-1}) \subset \overline{\Pi - (\xi \times \mathbb{P}^{n-1})}.$$

We have already seen that for a suitable choice of L we can obtain any point of $\xi \times \mathbb{P}^{n-1}$ on the left-hand side.

For $n=2$ we can give a clear illustration of the mapping $\sigma: \Pi \to \mathbb{P}^2$ and its action on lines L. The curves $\sigma^{-1}(L)$ intersect the line $\xi \times \mathbb{P}^1$ at points that do not change as L rotates about ξ in $\mathbb{P}^2$. Thus, Π looks like one loop of a helix (Fig. 6).

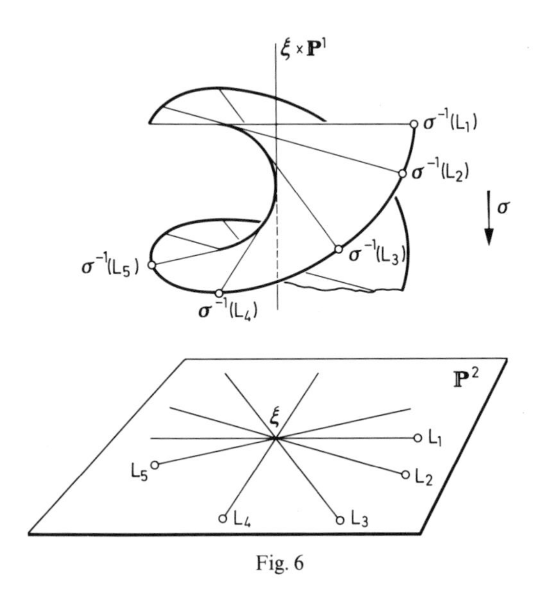

Fig. 6

2. The Local σ-Process. Now we construct for any quasiprojective variety X and a simple point x of it a variety Y and a mapping $\sigma: Y \to X$, analogous to that constructed above for $X = \mathbb{P}^n$.

We begin with an auxiliary construction.

Let X be a quasiprojective irreducible variety, ξ a simple point of it, and $u_1, \ldots, u_n$ functions regular on the whole of X and such that:

a) the equations $u_1 = \cdots = u_n = 0$ have on X the unique solution ξ;

b) the functions $u_1, \ldots, u_n$ forms a system of local coordinates at ξ.

We consider the product $X \times \mathbb{P}^{n-1}$ and in it the subvariety Y consisting of those points $(x; t_1 : \ldots : t_n)$, $x \in X$, $(t_1 : \ldots : t_n) \in \mathbb{P}^{n-1}$ such that $u_i(x)\, t_j = u_j(x)\, t_i$, $1 \leqslant i, j \leqslant n$. The regular mapping $\sigma : Y \to X$ which is the restriction to Y of the projection $X \times \mathbb{P}^{n-1} \to X$ is called the *local σ-process centred at ξ.*

Note that this construction, generally speaking, is not applicable when X is projective: we require the existence on X of non-constant everywhere regular functions $u_1, \ldots, u_n$. Therefore the new concept does not comprise the earlier introduced concept of a σ-process for the case $X = \mathbb{P}^n$. The connection between them is as follows.

We denote by X the affine subset defined in $\mathbb{P}^n$ by the condition $x_0 \neq 0$, and we set $Y = \sigma^{-1}(X)$. Then the mapping $\sigma : Y \to X$ induced on Y by the σ-process $\Pi \to \mathbb{P}^n$ is a local σ-process.

The following properties, which we have proved in § 4.1 for the σ-process, can be proved word-for-word in the same way for a local σ-process: the mapping $\sigma : Y \to X$ is regular and determines an isomorphism

$$Y - (\xi \times \mathbb{P}^{n-1}) \to X - \xi .$$

At a point $y \in \sigma^{-1}(\xi)$ we have $t_i \neq 0$ for some i, and we can set $s_j = t_j/t_i$, $j \neq i$. Then the equations of Y take the form $u_j = u_i s_j$ $(j = 1, \ldots, n,\ j \neq i)$. Hence we see that the ideal of y has the form

$$\begin{aligned} \mathfrak{m}_y &= (u_1 - u_1(y), \ldots, u_n - u_n(y), s_1 - s_1(y), \ldots, s_n - s_n(y)) \\ &= (s_1 - s_1(y), \ldots, u_i - u_i(y), \ldots, s_n - s_n(y)) . \end{aligned}$$

Therefore $\dim \Theta_{y, Y} \leqslant n$, and since $\dim \sigma^{-1}(X - \xi) = n$, the variety Y is smooth at every point $y \in \sigma^{-1}(X - \xi)$. Since

$$Y = \sigma^{-1}(X - \xi) \cup (\xi \times \mathbb{P}^{n-1}) ,$$

Y is either irreducible and therefore coincides with the closure $\overline{\sigma^{-1}(X - \xi)}$ of $\sigma^{-1}(X - \xi)$, or it has one component isomorphic to $\mathbb{P}^{n-1}$. In the second case the two components intersect: otherwise $\sigma^{-1}(X - \xi)$ would be closed, but then, by Theorem 3 of Ch. I, § 5, its image $X - \xi$ would also be closed. The point of intersection of the two components would be simple, and this contradicts Theorem 6 of § 2. Thus, Y is irreducible and smooth, and $s_1 - s_1(y), \ldots, u_i - u_i(y), \ldots, s_n - s_n(y)$ are local parameters at a point $y \in \sigma^{-1}(\xi)$ at which $t_i \neq 0$.

Now we establish a property that could be called the independence of a local σ-process of the choice of the functions $u_1, \ldots, u_n$.

Lemma. *If $v_1, \dots, v_n$ is another system of functions on X satisfying the conditions* a) *and* b), *if Y' is the resulting variety, and $\sigma': Y' \to X$ the corresponding local σ-process, then Y' and Y are isomorphic.*

There even exists an isomorphism $\varphi: Y \to Y'$ such that the diagram

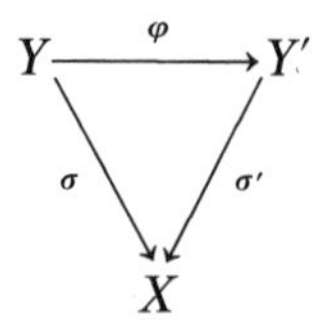

commutes.

Proof. Let $Y' \subset X \times \mathbb{P}^{n-1}$ and let $t'_1, \dots, t'_n$ be homogeneous coordinates in $\mathbb{P}^{n-1}$. In the open sets $Y - \sigma^{-1}(\xi)$ and $Y' - \sigma'^{-1}(\xi)$ we set

$$\begin{aligned} \varphi(x; t_1 : \dots : t_n) &= (x; v_1(x) : \dots : v_n(x)), \\ \psi(x; t'_1 : \dots : t'_n) &= (x; u_1(x) : \dots : u_n(x)). \end{aligned} \tag{1}$$

From property a) of the functions u_i it follows that φ and ψ are regular and that $\varphi(Y - \sigma^{-1}(\xi)) \subset Y'$, $\psi(Y' - \sigma'^{-1}(\xi)) \subset Y$.

We now consider an open set in which $t_i \neq 0$, and we set $s_j = t_j/t_i$. Since $v_l(\xi) = 0$ and $u_1, \dots, u_n$ is a basis of the ideal $\mathfrak{m}_\xi$, we have

$$v_l = \sum_{j=1}^{n} h_{lj} u_j, \qquad h_{lj} \in \mathcal{O}_\xi. \tag{2}$$

Since in our open set $u_j = u_i s_j$, we have

$$\begin{aligned} v_l &= u_i \sum_{j=1}^{n} \sigma^*(h_{lj})\, s_j = u_i g_l, \\ g_l &= \sum_{j=1}^{n} \sigma^*(h_{lj})\, s_j. \end{aligned} \tag{3}$$

We set $\varphi(x; t_1 : \dots : t_n) = (x; g_1 : \dots : g_n)$. Evidently our mapping coincides with (1) in their common domain of definition, because there $g_l = v_l/u_i$. Let us verify that φ is regular. For this purpose we have to show that $g_1, \dots, g_n$ do not vanish simultaneously at any point $\eta \in \sigma^{-1}(\xi)$. Suppose that all the $g_l(\eta) = 0$. Since not all the $s_j(\eta) = 0$ (because $s_j = 1$), it follows from (3) that $|h_{lj}(\xi)| = 0$. But

$$v_l \equiv \Sigma h_{lj}(\xi) \cdot u_j (\mathrm{mod}\ \mathfrak{m}_\xi^2),$$

and from this it would follow that the v_l are linearly dependent in $\mathfrak{m}_\xi/\mathfrak{m}_\xi^2$, whereas they form a system of local parameters at ξ. In this way we define a unique mapping $\varphi: Y \to Y'$, and similarly $\psi: Y' \to Y$. The fact that they are inverses of one another need only be verified on an open set, where the formulae (1) hold. But there it is obvious.

3. Behaviour of Subvarieties under a σ-Process. Let X be an irreducible quasiprojective subvariety of $\mathbb{P}^N$, and $\sigma:\Pi\to\mathbb{P}^N$ the σ-process defined in § 4.1. We investigate the inverse image $\sigma^{-1}(X)$ of X, which is, of course, a quasiprojective subvariety of Π.

Theorem 1. *If $X\subset\mathbb{P}^N$, X is non-singular at ξ and $X\neq\mathbb{P}^N$, then relative to the σ-process centered at ξ the inverse image $\sigma^{-1}(X)$ is reducible and consists of two components:*

$$\sigma^{-1}(X)=(\xi\times\mathbb{P}^{N-1})\cup Y. \tag{1}$$

On the component Y the mapping $\sigma: Y\to X$ determines a regular mapping. It is an isomorphism between some neighbourhood U of a point $x\in X$ and $\sigma^{-1}(U)$ provided that $x\neq\xi$, and it is the local σ-process $\sigma^{-1}(U)\to U$ when $x=\xi$.

Proof. We denote by Y the closure $\overline{\sigma^{-1}(X-\xi)}$ of the set $\sigma^{-1}(X-\xi)$. Since σ^{-1} is an isomorphism in $\mathbb{P}^N-\xi$, we see that $\sigma^{-1}(X-\xi)$ is isomorphic to $X-\xi$ and hence irreducible. Consequently so is Y. From the definition it is clear that (1) holds: if $x\in X-\xi$, then

$$\sigma^{-1}(x)\in Y,\quad \sigma^{-1}(\xi)=\xi\times\mathbb{P}^{N-1}.$$

The fact that $\sigma: Y\to X$ is an isomorphism in a neighbourhood of an arbitrary point $x\in X$, except $x=\xi$, has already been mentioned. It remains to investigate this mapping in a neighbourhood of ξ.

Here we can use the fact that in an affine space containing ξ, a σ-process can be described as a local σ-process and that a local σ-process does not depend on the choice of local coordinates. For by Theorem 5 of § 3 we can choose a system of local coordinates $u_1,\dots,u_N$ at a point $\xi\in\mathbb{P}^N$ such that in some neighbourhood of this point the variety X is given by the equations

$$u_{n+1}=\cdots=u_N=0, \tag{2}$$

and that the functions $u_1,\dots,u_n$ determine a local system of coordinates on X at ξ. We can choose a neighbourhood $U\subset\mathbb{P}^N$ of ξ so that $u_1,\dots,u_N$ satisfy the conditions a) and b) of the lemma in § 4.2, and therefore the proof of the theorem reduces to the special case when X is given by (2).

From the conditions a), b), and $u_it_j=u_jt_i$ we deduce that $t_{n+1}(x)=\cdots=t_N(x)=0$ for $x\neq\xi$. Therefore Y is contained in the subspace Y' defined in $X\times\mathbb{P}^{N-1}$ by the equations

$$t_{n+1}=\cdots=t_N=0, \tag{3}$$

$$u_it_j=u_jt_i,\quad 1\leqslant i,j\leqslant n. \tag{4}$$

If we denote by $\mathbb{P}^{n-1}$ the subspace of the projective space $\mathbb{P}^{N-1}$ defined by (3), then we see that $Y' \subset X \times \mathbb{P}^{n-1}$ is determined by (4). Thus, Y' coincides with the variety obtained by the local σ-process. We have now shown that $Y' = \overline{\sigma^{-1}(X-\xi)}$. Therefore $Y = Y'$, and this proves the theorem.

We can now give the most general definition of a σ-process. If X is a quasiprojective variety, $X \subset \mathbb{P}^n$, ξ a simple point of it, and Y the variety introduced in the statement of Theorem 1, then $\sigma: Y \to X$ is called the *σ-process centered at ξ*. From what we have proved about the local σ-process it follows that Y is irreducible if X is, that all points of $\sigma^{-1}(\xi)$ are simple on Y, and that $\sigma^{-1}(\xi) \simeq \xi \times \mathbb{P}^{n-1}$.

Note that the σ-process is an isomorphism if X is a curve. Thus, the presence of a non-trivial σ-process is a characteristic feature of many-dimensional algebraic geometry.

4. Exceptional Subvarieties. The example of the σ-process points to a fundamental difference between algebraic curves and varieties of dimension $n > 1$. Whereas a birational isomorphism for non-singular projective curves is an isomorphism, the σ-process gives an example to show that this need not be the case for higher dimensions.

We mention one peculiarity of the σ-process: it is a regular mapping and fails to be an isomorphism only because the rational mapping σ^{-1} is non-regular (at ξ).

We now investigate the mapping $f: X \to Y$, where f is a regular mapping and a birational isomorphism, that is, $f^{-1} = g$ is a rational, but non-regular mapping $Y \to X$. In the example of the σ-process we have seen that a subvariety of codimension 1 in Y is contracted to a point ξ. Let us show that the analogous property always holds in this situation.

Theorem 2. *Suppose that $f: X \to Y$ is a regular mapping and a birational isomorphism, that $y = f(x)$ is a simple point on Y for $x \in X$, and that the mapping $g = f^{-1}$ is non-regular at y. Then there exists a subvariety $Z \subset X$ with $x \in Z$, such that* $\operatorname{codim} Z = 1$, $\operatorname{codim} f(Z) \geqslant 2$.

Proof. We can replace X, if necessary, by an affine neighbourhood of x and may therefore assume that X is affine. Suppose that $X \subset \mathbb{A}^N$ and that $g = f^{-1}$ is given by the formulae $t_i = g_i (i = 1, \ldots, N)$, $g_i \in k(Y)$, where $t_1, \ldots, t_N$ are coordinates in $\mathbb{A}^N$.

Evidently $g_i = g^*(t_i)$, and since g is non-regular at y, at least one of the functions g_i is non-regular at y. Suppose that this is g_1, so that $g_1 \notin \mathcal{O}_y$. We can represent g_1 in the form $g_1 = u/v$, u, $v \in \mathcal{O}_y$, $v(y) = 0$, and since prime factorization in $\mathcal{O}_y$ is unique (by assumption y is a simple point), we can choose u and v relatively prime. Since $g = f^{-1}$, we have $t_1 = f^*(g_1) = f^*(u/v) = f^*(u)/f^*(v)$, therefore

$$f^*(v)t_1 = f^*(u). \tag{1}$$

Clearly $f^*(v)(x) = 0$, so that $x \in V(f^*(v))$. We set $Z = V(f^*(v))$. By the theorem on the dimension of an intersection $\operatorname{codim} Z = 1$, because $x \in Z$, and therefore Z is not empty. From (1) it follows that $f^*(u) = 0$ on Z, because t_1 is a regular function. Therefore $u = 0$ and $v = 0$ on $f(Z)$, and hence $f(Z) \subset V(u) \cap V(v)$.

It remains to verify that $\operatorname{codim}(V(u) \cap V(v)) \geqslant 2$. But if $V(u) \cap V(v)$ contained a component Y' with $y \in Y'$, $\operatorname{codim} Y' = 1$, then according to Theorem 1 of § 3, Y' would have a local equation h. This would mean that $u \in (h)$, $v \in (h)$, which contradicts the fact that u and v have no common factor in $\mathcal{O}_y$.

Definition. Let $f: X \to Y$ be a regular mapping and a birational isomorphism. A subvariety $Z \subset X$ is called *exceptional* if $\operatorname{codim} Z = 1$, $\operatorname{codim} f(Z) \geqslant 2$.

Corollary 1. *If a regular mapping of smooth varieties $f: X \to Y$ is a birational isomorphism, but not an isomorphism, then it has an exceptional subvariety.*

Corollary 2. *If $f: X \to Y$ is a regular mapping and a birational isomorphism, where X and Y are curves and Y is smooth, then $f(X)$ is open in Y and f determines an isomorphism between X and $f(X)$.*

The fact that $f(X)$ is open in Y follows from the existence of isomorphic open subsets U and V in X and Y. Since $f(U) = V$ is obtained from Y by removing finitely many points, a fortiori $f(X)$ is obtained in this way, hence is open in Y. If the mapping $f: X \to f(X)$ were not an isomorphism, then we would come to a contradiction to Theorem 2, because in our case only the empty set has codimension $\geqslant 2$.

5. Isomorphism and Birational Isomorphism. We consider the class of all birationally isomorphic algebraic quasiprojective varieties. All the representatives of this class are called its *models*.

In the next section we show that every class of birationally isomorphic curves contains a projective smooth model X_0. Corollary 2 to Theorem 3 of § 3 asserts that there is only one such model (to within isomorphism). Therefore, if we associate with every class the unique non-singular projective model contained in it, we reduce the classification problem for algebraic curves to within a birational isomorphism to the same problem for non-singular projective curves to within an isomorphism.

Fields of functions on algebraic curves are finitely generated extensions of transcendence degree 1 of a field k. We can therefore set up a one-to-one correspondence between such fields K and non-singular projective curves. Under this correspondence $K=k(X)$. We also call X a *model* of K.

One could try to find a model X directly, starting out from algebraic properties of K. We make this problem more precise by asking how the local rings of all points of a curve X can be characterized within K. It is easy to verify that every local ring $\mathcal{O}_x$ of a point $x\in X$ has the following properties:

1) $\mathcal{O}$ is a subring of K, $k\subsetneqq\mathcal{O}\subsetneqq K$;
2) $\mathcal{O}$ is a local ring and its maximal ideal $\mathfrak{m}$ is principal: $\mathfrak{m}=(u)$;
3) the field of fractions of $\mathcal{O}$ is K.

It can be shown (Exercises 7, 8, 9) that any subring $\mathcal{O}$ of K having the properties 1), 2), and 3), is the local ring $\mathcal{O}_x$ of a suitable point $x\in X$. Thus, X is a universal model: it contains all local rings of K satisfying the natural conditions 1), 2), and 3).

How can one solve these problems for varieties of dimension $n>1$? When a projective non-singular model exists, things are comparatively well-behaved: the existence was proved for $n=2$ and 3 (Walker and Zariski for fields of characteristic 0, and Abhyankar for any finite characteristic greater than 5), and for an arbitrary n and characteristic 0 (Hironaka). The existence is highly probable for an arbitrary field and arbitrary n. On the other hand, the uniqueness of a non-singular projective model is an exceptional feature of the case $n=1$. This is clear from the example of the projective plane $\mathbb{P}^2$ and a quadric, which are birationally isomorphic, but not isomorphic.

One could ask whether perhaps in every class of birationally isomorphic varieties there exists a model that is universal in the sense that the local rings of its points, as in the case $n=1$, exhaust all the local subrings of the field $K=k(X)$ satisfying the conditions 1), 2), and 3) [except that in 2) $\mathfrak{m}=(u_1,\dots,u_n)$ instead of $\mathfrak{m}=(u)$]. However, for the same reasons such models cannot exist. For if $\sigma: X'\to X$ is the σ-process centered at $\xi\in X$, then the local rings of points $y\in\sigma^{-1}(\xi)$ do not coincide with any of the local rings $\mathcal{O}_x$, $x\in X$. The reader can easily prove this as an exercise. True, by combining all the non-singular models of one class we can obtain a certain object having this property of universality; however, it is not a finite-dimensional algebraic variety. Some information on this "infinite model" can be found in [37], Vol. 2, Ch. VI, § 17.

Owing to the absence of a distinguished model the problem arises of studying the connections between various non-singular projective models of one class of birationally isomorphic varieties. Without proof we quote the relevant fundamental results. In what follows, all varieties are assumed to be irreducible, smooth, and projective.

We begin with two terms. A model X' *dominates* X if there exists a birational regular mapping $f: X' \to X$.

A variety is called a *relatively minimal model* if it does not dominate any variety not isomorphic to it. For example, a smooth projective curve is always a relatively minimal model. By Theorem 2 a variety is a relatively minimal model if it has no exceptional subvarieties.

It can be shown that every variety dominates at least one relatively minimal model. Thus, every class of birationally isomorphic varieties contains at least one relatively minimal model.

Now the important question of its uniqueness arises. If in every class there were such a unique model, then again it would reduce the birational classification to the classification to within isomorphism.

However, for $n > 1$ this is not the case. An example is the projective plane $\mathbb{P}^2$ and a quadric Q, which we know to be birationally isomorphic, so that they are models of one and the same class of birationally isomorphic surfaces. We show that $\mathbb{P}^2$ and Q are both relatively minimal models, in other words, do not have exceptional curves. Since $\mathbb{P}^2$ and Q are non-isomorphic (Remark 1 to Ch. I, § 6.2), this gives the required example.

In our case an irreducible exceptional curve $C \subset X$ must contract to a point $y \in Y: f(C) = y$ under a regular birational mapping $f: X \to Y$. Here X and Y are projective surfaces. Such curves have a number of very special properties (which explains the term "exceptional"). We mention only one of them.

According to Theorem 3 of § 3 the mapping f^{-1} fails to be regular at only finitely many points $y_i \in Y$. Let U be a sufficiently small affine neighbourhood of y such that f^{-1} is regular at all points of U other than y. We set $V = f^{-1}(U)$, $C = f^{-1}(y)$. Obviously, V is an open subset of X and $V \supset C$. We show that in V there is no irreducible curve C' closed in X and not contained in C. For C' is a projective curve and its image $f(C')$ is also projective. But $f(C') \subset U$, which is affine. By Corollary 2 to Theorem 3 of Ch. I, § 5, this is only possible when $f(C') = y'$ is a point. If $y' \neq y$, then C' must also be a point, because f^{-1} is an isomorphism in U, apart from y. But if $y' = y$, then $C' \subset f^{-1}(y) = C$.

Thus, C lies isolated in X: in some neighbourhood V of it there are no irreducible projective curves not contained in C. In other words, C cannot "move just a little". From this one can derive that many surfaces do not contain exceptional curves.

For example, let $X = \mathbb{P}^2$, $V = \mathbb{P}^2 - D \supset C$, where C is an exceptional curve. Then $\dim D = 0$, because otherwise C and D would intersect by the theorem on the dimension of an intersection. But if $\dim D = 0$, that is, D is a finite point set, then there exist arbitrarily many curves C that do not intersect D, for example straight lines.

Let $X=Q$. Here we use the existence of a group of projective transformations G carrying Q into itself. We recall that transformations of G are given by matrices A of order 4 satisfying the relation $A^*FA=F$, where F is the matrix of the equation of Q. Hence it follows that G is an algebraic subvariety in the space of all matrices of order 4. In what follows we may therefore take G to be an algebraic affine variety.

If C is a curve and $C\subset Q-D$, then we construct a transformation $\varphi\in G$ such that $\varphi(C)\not\subset C$, $\varphi(C)\subset Q-D$, but this contradicts the above property of exceptional curves. It is enough to show that the set of those $\varphi\in G$ for which $\varphi(C)\cap D\neq\emptyset$ is closed. Then we have at our disposal an entire neighbourhood of the identity transformation $e\in G$ consisting of elements with the required property. To describe the set S of those $\varphi\in G$ for which $\varphi(C)\cap D\neq\emptyset$ we consider in the direct product $G\times Q$ the set Γ of those pairs (φ, x) for which $x\in C$, $\varphi(x)\in D$. Clearly Γ is closed. If $f: G\times Q\to G$ is the natural projection, then $S=f(\Gamma)$, and $f(\Gamma)$ is closed according to Theorem 3 of Ch. I, § 5. This completes the proof that two distinct minimal models exist.

It is all the more surprising that nevertheless uniqueness of the minimal model holds for algebraic surfaces, provided that some special types are excluded. Namely, Enriques has shown that in a class of surfaces the minimal model is unique if the class does not contain a surface of the form $C\times\mathbb{P}^1$, where C is an algebraic curve. (Surfaces birationally isomorphic to $C\times\mathbb{P}^1$ are called *ruled*.)

For this result see [3], Ch. II, § 4, 42.

About minimal models for varieties of dimension $n\geqslant 3$ nothing is known.

Exercises

1. Suppose that $\dim X=2$, that ξ is a simple point of X, that C_1 and $C_2\subset X$ are two curves passing through ξ and not singular there; $\sigma: Y\to X$ the σ-process centered at ξ, $C_i'=\overline{\sigma^{-1}(C_i-\xi)}$, $Z=\sigma^{-1}(\xi)$. Show that $C_1'\cap Z=C_2'\cap Z$ if and only if C_1 and C_2 touch at ξ.

2. Suppose that $\dim X=2$, that ξ is a simple point of X, $C\subset X$ a curve with $\xi\in C$, and f the local equation of C in a neighbourhood of ξ. Let $f\equiv\prod_{i=1}^{r}(\alpha_i u+\beta_i v)^{l_i}\ (\mathfrak{m}_\xi^{l+1})$, $\Sigma l_i=l$, where u and v are local parameters at ξ and the forms $\alpha_i u+\beta_i v$ are not proportional to each other.

As in Exercise 1, $\sigma: Y\to X$; $C'=\overline{\sigma^{-1}(C-\xi)}$. Show that $C'\cap Z$ consists of r points.

3. The notation is that of Exercise 2, but firstly, $f\equiv(\alpha_1 u+\beta_1 v)(\alpha_2 u+\beta_2 v)\ (\mathfrak{m}_\xi^3)$, and secondly, the linear forms $\alpha_1 u+\beta_1 v$ and $\alpha_2 u+\beta_2 v$ are not propertional. Show that the two points of $C'\cap Z$ are simple on C'.

4. Consider the rational mapping $\varphi : \mathbb{P}^2 \to \mathbb{P}^4$ given by the formula

$$\varphi(x_0 : x_1 : x_2) = (x_0 x_1 : x_0 x_2 : x_1^2 : x_1 x_2 : x_2^2).$$

Show that φ is a birational isomorphism and that the inverse mapping $\overline{\varphi(\mathbb{P}^2)} \to \mathbb{P}^2$ is a σ-process.

5. As in Exercise 4, investigate the mapping $\mathbb{P}^2 \to \mathbb{P}^6$ given by all monomials of degree 3 other than x_0^3, x_1^3 and x_2^3.

6. Construct an example of a birational isomorphism $X \to Y$ under which an exceptional subvariety of codimension 1 is carried into a subvariety of codimension 2 ($\dim X = n$, where n is arbitrary).

7. Let $\mathcal{O}$ be a local ring of the field $k(X)$ satisfying the conditions 1) – 3) of § 4.5 (X is a projective algebraic curve). Show that for every $u \in k(X)$ either $u \in \mathcal{O}$ or $u^{-1} \in \mathcal{O}$. Let $X \subset \mathbb{P}^n$, and let $x_0, \ldots, x_n$ be the homogeneous coordinates in $\mathbb{P}^n$. Show that there exists an i such that $x_j/x_i \in \mathcal{O}\ (j = 0, \ldots, n)$.

8. The notation is the same as in Exercise 7. Let X' be an affine curve, $X' = X \cap \mathbb{A}_i^n$. Show that $k[X'] \subset \mathcal{O}$, that $k[X] \cap \mathfrak{m}$ is the ideal of some point $x \in X'$, and that $\mathcal{O}_x \subset \mathcal{O}$.

9. Show that if two rings $\mathcal{O}_1$ and $\mathcal{O}_2$ satisfy the conditions 1) – 3) of § 4.5 and $\mathcal{O}_1 \subset \mathcal{O}_2$, then $\mathcal{O}_1 = \mathcal{O}_2$. Hence and from Exercises 7 and 8 derive that (in the notation of Exercise 8) $\mathcal{O} = \mathcal{O}_x$.

10. Let V be the quadric cone given by the equation $xy = z^2$ in $\mathbb{A}^3$, and let $X' \to \mathbb{A}^3$ be the σ-process centered at the origin of coordinates, and V' the closure of the subvariety $\sigma^{-1}(V - O)$ in X'. Show that V' is a smooth variety and that the inverse image of the origin of coordinates under the mapping $\sigma : V' \to V$ is a smooth rational curve.

§ 5. Normal Varieties

1. Normality. We begin by recalling one algebraic concept. A ring A without divisors of zero is called integrally closed if every element of the field of fractions K of A that is integral over A belongs to A.

Definition. An irreducible affine variety X is called *normal* if the ring $k[X]$ is integrally closed. An irreducible quasiprojective variety X is called *normal* if each of its points has an affine normal neighbourhood.

We shall prove presently that smooth varieties are normal (Theorem 1). Here is an example of a non-normal variety. On the curve X with the equation

$$y^2 = x^2 + x^3$$

the function $t = y/x \in k(X)$ is integral over $k[X]$, because $t^2 = 1 + x$, however, $t \notin k[X]$ (Exercise 9 of Ch. I, § 3). This example shows that the point set of normality has some relation to the singular points of the variety.

Our next example is a variety that has a singular point but is normal. This is the cone X with the equation $x^2 + y^2 = z^2$ in $\mathbb{A}^3$ (we assume that the characteristic of the ground field is not 2).

Let us prove that the ring $k[X]$ is integrally closed in $k(X)$. In doing this we use the simplest properties of integral elements (see [37],

Vol. 1, Ch. V, § 1). The field $k(X)$ consists of the elements of the form $u+vz$, where u, $v \in k(x, y)$ and x and y are independent variables. Similarly $k[X]$ consists of those elements of $k(X)$ for which u, $v \in k[x, y]$, so that $k[X]$ is a finite module over $k[x, y]$ and hence all the elements of $k[X]$ are integral over $k[x, y]$. If $\alpha = u+vz \in k(X)$ is integral over $k[X]$, then it must also be integral over $k[x, y]$. Its minimal polynomial has the form $T^2 - 2uT + (u^2 - (x^2+y^2)v^2)$, hence $2u \in k[x, y]$ and $u \in k[x, y]$. Similarly, $u^2 - (x^2+y^2)v^2 \in k[x, y]$, and hence $(x^2+y^2)v^2 \in k[x, y]$.

Since $x^2+y^2 = (x+iy)(x-iy)$ is the product of two coprime elements, we see that $v \in k[x, y]$, and this means that $\alpha \in k[X]$.

We now establish some simple properties of normal varieties.

Lemma. *An irreducible variety X is normal if and only if the local rings $\mathcal{O}_x$ of all its points are integrally closed.*

Since the definition of normality is of local character, we may confine our attention to the case when X is affine. Let X be normal, $x \in X$. We prove that $\mathcal{O}_x$ is integrally closed. Let $\alpha \in k(X)$ be integral over $\mathcal{O}_x$, that is,

$$\alpha^n + a_1\alpha^{n-1} + \cdots + a_n = 0. \tag{1}$$

Here $a_i \in \mathcal{O}_x$, and therefore $a_i = b_i/c_i$, b_i, $c_i \in k[X]$, $c_i(x) \neq 0$. Setting $d_0 = c_1 \ldots c_n$ and multiplying (1) by d_0 we find that

$$d_0\alpha^n + d_1\alpha^{n-1} + \cdots + d_n = 0, \tag{2}$$

where $d_i \in k[X]$, $d_0(x) \neq 0$. Multiplying (2) by d_0^{n-1} and setting $d_0\alpha = \beta$ we find that β is integral over $k[X]$. By hypothesis, $k[X]$ is integrally closed, hence $d_0\alpha = \beta \in k[X]$. Then $\alpha = \beta/d_0 \in \mathcal{O}_x$, since $d_0(x) \neq 0$.

Suppose now that all the $\mathcal{O}_x$ are integrally closed. We show that then $k[X]$ is integrally closed. If $\alpha \in k(X)$ and α is integral over $k[X]$, then $\alpha^n + a_1\alpha^{n-1} + \cdots + a_n = 0$, $a_i \in k[X]$. But then all the more $a_i \in \mathcal{O}_x$ for every $x \in X$, and since $\mathcal{O}_x$ is integrally closed by hypothesis, we see that $\alpha \in \mathcal{O}_x$. Therefore $\alpha \in \bigcap_{x \in X} \mathcal{O}_x$. According to Theorem 4 of Ch. I, § 3, $\bigcap_{x \in X} \mathcal{O}_x = k[X]$ and hence $\alpha \in k[X]$.

Theorem 1. *Smooth varieties are normal.*

By virtue of the lemma it is sufficient to show that if x is a simple point, then the ring $\mathcal{O}_x$ is integrally closed. We know that factorization in $\mathcal{O}_x$ is unique (Theorem 2 of § 3). Every element $\alpha \in k(X)$ can be represented in the form $\alpha = u/v$, where u, $v \in \mathcal{O}_x$ have no common divisors. If α is integral over $\mathcal{O}_x$, then $\alpha^n + a_i\alpha^{n-1} + \cdots + a_n = 0$, $a_i \in \mathcal{O}_x$. Hence $u^n + a_1u^{n-1}v + \cdots + a_nv^n = 0$. So we see that $v | u^n$. Since u and v are coprime and factorization is unique, it follows that $\alpha \in \mathcal{O}_x$.

Theorem 1 shows that normality is a weaker form of smoothness. This becomes apparent also in properties of normal varieties. In particular, we show that the main property of smooth varieties (Theorem 1 of § 3) extends in a weaker version to normal varieties.

Theorem 2. *If X is a normal variety, $Y \subset X$, and codim $Y = 1$, then there exists an affine open set $X' \subset X$ such that $X' \cap Y \neq \emptyset$ and the ideal $Y' = X' \cap Y$ in $k[X']$ is principal.*

Of course, we may assume that X is affine. Let $f \in k[X]$, $f \neq 0$, $f \in \mathfrak{a}_Y$. Then $Y \subset V(f)$, and since codim $Y = 1$ and codim $V(f) = 1$ (by the theorem on the dimension of an intersection), Y consists of components of $V(f)$. Let $V(f) = Y \cup \bar{Y}$, $Y \not\subset \bar{Y}$. Setting $\bar{X} = X - \bar{Y}$ we find that $Y \cap \bar{X} \neq \emptyset$, $Y \cap \bar{X} = V(f) \cap \bar{X}$. Therefore we may assume at once that $Y = V(f)$.

To prove that $\mathfrak{a}_Y$ is a principal ideal means to find an element $u \in \mathfrak{a}_Y$ such that all the elements of $\mathfrak{a}_Y$ are divisible by u, that is, $\mathfrak{a}_Y u^{-1} \subset k[X]$. Such an element exists (possibly after replacing X by an open set) if there is an element $v \in k(X)$ with the properties

$$\mathfrak{a}_Y v \subset k[X], \tag{3}$$

$$\mathfrak{a}_Y v \not\subset \mathfrak{a}_Y. \tag{4}$$

For then there exists a $u \in \mathfrak{a}_Y$ such that $w = u \cdot v \notin \mathfrak{a}_Y$. Replacing X by $X - V(w)$ we achieve that w becomes invertible (in the ring $k[X - V(w)]$). Since $w \notin \mathfrak{a}_Y$, we have $Y \not\subset V(w)$ and $(X - V(w)) \cap Y \neq \emptyset$. The element we have found has the requisite two properties: $u \in \mathfrak{a}_Y$ by construction and $\mathfrak{a}_Y u^{-1} = \mathfrak{a}_Y v w^{-1} = \mathfrak{a}_Y v \subset k[X - V(w)]$, because w is invertible in $k[X - V(w)]$.

Finally, (4) holds if (3) does and $v \notin k[X]$. For a_Y has a finite basis over $k[X]$, and from the fact that $\mathfrak{a}_Y v \subset \mathfrak{a}_Y$ it follows that v is integral over $k[X]$: this is one of the simplest properties of integral elements. At this place we make use of the normality of X and conclude that then $v \in k[X]$.

Thus, it is sufficient to construct an element $v \in k(X)$ such that $v \notin k[X]$ and $\mathfrak{a}_Y v \subset k[X]$. We recall that $Y = V(f)$. By Hilbert's Nullstellensatz it follows that $a_Y^l \subset (f)$ for some $l > 0$, that is, the product of any l factors $\alpha_1, \dots, \alpha_l \in \mathfrak{a}_Y$ is divisible by f. We choose l as small as possible, subject to this property. Then there exist $\alpha_1, \dots, \alpha_{l-1} \in \mathfrak{a}_Y$ such that $g = \alpha_1 \dots \alpha_{l-1} \notin (f)$, and $g \cdot \alpha \in (f)$ for every $\alpha \in \mathfrak{a}_Y$, that is, $g\mathfrak{a}_Y \subset (f)$. So we see that we can set $v = gf^{-1}$.

Theorem 3. *The codimension of the set of singular points of a normal variety is not less than 2.*

Let X be normal, $\dim X = n$, S the set of singular points of X. We have seen that S is closed in X. Suppose that S contains an irreducible component Y of dimension $n-1$. Let X' be the open set whose existence was established in Theorem 2, and $Y'=Y\cap X'$. The variety Y' has at least one simple point (as a point of Y', but not necessarily as a point of X'). We denote it by y. Let $\mathcal{O}_{y,Y'}$ be its local ring on Y' and let $u_1,\dots,u_{n-1}$ be local parameters. By Theorem 2, $\mathfrak{a}_{Y'}=(u)$, hence $k[Y']=k[X']/(u)$. Similarly $\mathcal{O}_{y,Y'}=\mathcal{O}_{y,X'}/(u)$. Evidently $\mathfrak{m}_{y,X'}$ is the inverse image of $\mathfrak{m}_{y,Y'}$ under the natural homomorphism $\mathcal{O}_{y,X'}\to\mathcal{O}_{y,Y'}$. We denote by $v_1,\dots,v_{n-1}$ arbitrary inverse images of $u_1,\dots,u_{n-1}$. Then $\mathfrak{m}_{y,X'}=(v_1,\dots,v_{n-1},u)$. This shows that $\dim\mathfrak{m}_{y,X'}/\mathfrak{m}_{y,X'}^2\leqslant n$, hence y is a simple point on X, against the assumption that $y\in Y\subset S$. This proves the theorem.

Corollary. *For algebraic curves the concepts of smoothness and of normality are the same.*

Let us make a comparison between the properties of normal varieties we have deduced. First of all, observe that in the proof of Theorem 1 we have not used the smoothness of X to the full extent, but only the uniqueness of the decomposition into prime factors in the rings $\mathcal{O}_x$. In this context it is natural to single out the class of varieties in which the latter property holds. They are called *factorial*. Thus, a smooth variety is factorial, and a factorial variety is normal (this is shown essentially by Theorem 1). It can be shown that all these three classes of varieties are actually distinct. For example, it has been proved that if a hypersurface in $\mathbb{A}^n$, $n\geqslant 5$, has a unique singular point, then it is factorial ([14], XI, 3.14). A pretty example of a factorial surface that is not smooth is given by the equation $x^2+y^3+z^5=0$. An example of a normal, but not factorial, variety is the quadric cone we have already analysed: $z^2=(x+iy)\cdot(x-iy)$, which has two distinct decompositions of an element into prime factors.

Theorem 3 draws attention to a new property of varieties: the set of singular points is of codimension not less than 2. Varieties with this property are called *non-singular in codimension* 1. Theorem 3 asserts that such are, in particular, the normal varieties. These two classes of varieties are also distinct. We construct an example of a surface X that is not normal, but has only finitely many singular points. To do this it is enough to construct a regular finite mapping $f:\mathbb{A}^2\to\mathbb{A}^4$ such that $X=f(\mathbb{A}^2)$ is closed in $\mathbb{A}^4$, $f:\mathbb{A}^2\to X$ is a birational isomorphism, and that two points, say y_1, $y_2\in\mathbb{A}^2$, have the same image $z\in X$, and that $f:\mathbb{A}^2-\{y_1,y_2\}\to X-\{z\}$ is an isomorphism. Thus, f is very much like the parametrization (2) of the curve (1) in Ch. I, § 1.1. The existence of the mapping f contradicts the normality of X, and z is the only singular point on X. We specify f by the equation

$$f(x,y)=(x,xy,y(y-1),y^2(y-1)).$$

If coordinates in $\mathbb{A}^4$ are denoted by u, v, w, t, then it easy to verify that the equations of the variety X take the following form:

$$ut = vw, \quad w^3 = t(t-w), \quad u^2 w = v(v-u),$$

where $u = x$, $v = xy$, $w = y(y-1)$, $t = y^2(y-1)$. The relations $x = u$, $y^2 - y = w$ show that x and y are integral over $f^*k[X]$, hence that f is finite. The remaining properties of f we need are quite easy to verify.

2. Normalization of Affine Varieties. We consider the simplest example of a non-normal variety, the curve X defined by the equation $y^2 = x^2 + x^3$. Its parametrization by means of $t = y/x$ determines a mapping $f: \mathbb{A}^1 \to X$ or, what is the same, an embedding $k[X] \subset k[t]$. The mapping f is a birational isomorphism, therefore $k[X] \subset k[t] \subset k(X) = k(t)$. The line $\mathbb{A}^1$ is normal, and consequently the polynomial ring $k[t]$ is integrally closed. Furthermore, the ring $k[t]$ can be characterized as the collection of all elements $u \in k(X)$ that are integral relative to $k[X]$. For $t^2 = 1 + x$, hence t is integral over $k[X]$, therefore all the elements of $k[t]$ are integral over $k[X]$. If $u \in k(X)$ is integral over $k[X]$, then it is also integral over $k[t]$, and since $k[t]$ is integrally closed, $u \in k[t]$. Finally, the fact that $k[t]$ is integral over $k[X]$ means in geometrical terminology that the mapping f is finite. We show that for any irreducible affine variety X there exists a variety X' and a mapping $X' \to X$ with the same properties. We begin with a definition, which refers to arbitrary irreducible varieties.

Definition. *A normalization of an irreducible variety* X is an irreducible normal variety X^ν together with a regular mapping $\nu: X^\nu \to X$ that is finite and a birational isomorphism.

Theorem 4. *An affine irreducible variety has a normalization that is also affine.*

Proof. We denote by A the integral closure of $k[X]$ in $k(X)$, that is, the collection of all elements $u \in k(X)$ that are integral over $k[X]$. From the simplest properties of integral elements it follows that A is a ring and integrally closed. Suppose that we have found an affine variety X' such that $A = k[X']$. Then X' is normal, and the inclusion $k[X] \subset k[X']$ determines a regular mapping $f: X' \to X$. It is clear that X' is a normalization of X.

According to Theorem 5 of Ch. I, § 3, such a variety X' exists if A has no divisors of zero and if it is finitely generated. The first condition is satisfied because $A \subset k(X)$. The theorem will be proved if we can show that A is finitely generated. We show even more, namely, that A is finitely generated as a module over $k[X]$. If $A = k[X]\omega_1 + \cdots + k[X]\omega_m$, then $\omega_1, \ldots, \omega_m$, together with generators of the algebra $k[X]$ over k

form a system of generators of A as a k-algebra. We use Theorem 10 of Ch. I, § 5. According to this theorem there exists a ring $B\subset k[X]$ over which $k[X]$ is integral and which is isomorphic to a polynomial ring: $B\simeq k[T_1,\dots,T_r]$. Let us draw a picture of all these rings and fields:

$$\begin{array}{ccc} B\subset k[X]\subset A\subset k(X) \\ \cap \qquad\qquad \cup \\ k(T_1,\dots,T_r). \end{array}$$

From this diagram and the simplest properties of integral elements it is clear that A is the integral closure of B in $k(X)$. Next, the field $K=k(X)$ is a finite extension of $k(T_1,\dots,T_r)$, because $T_1,\dots,T_r$ is a transcendence basis of $k(X)$. Finally, B is integrally closed (the variety $\mathbb{A}^r$ is normal and even smooth). Therefore the final result we need, that A is finitely generated, is a consequence of the following proposition.

Proposition. *Let* $B=k[T_1,\dots,T_r]$, $L=k(T_1,\dots,T_r)$, K *a finite extension of* L, A *the integral closure of* B *in* K. *Then* A *is a* B*-module of finite type.*

The proof of the proposition differs depending on whether the extension K/L is separable or not. Let us show how to reduce everything to the case of a separable extension.

Let $K=L(\alpha_1,\dots,\alpha_s)$. If α_1 is not separable over L, then its minimal polynomial is of the form $\alpha_1^{p^s m}+a_1\alpha_1^{p^s(m-1)}+\dots+a_m=0$, where $a_i\in k(T_1,\dots,T_r)$ and $\alpha_1^{p^s}$ is separable over L. We set $a_i=b_i^{p^s}$, where $b_i\in k(T_1^{1/p^s},\dots,T_r^{1/p^s})$, $L'=k(T_1^{1/p^s},\dots,T_r^{1/p^s})$, $K'=K(T_1^{1/p^s},\dots,T_r^{1/p^s})$; $B'=k[T_1^{1/p^s},\dots,T_r^{1/p^s}]$, and let A' be the integral closure of B' in K'. Then $K'=L'(\alpha_1,\dots,\alpha_m)$ and $\alpha_1^m+b_1\alpha_1^{m-1}+\dots+b_m=0$, so that α_1 is separable over L'. On the other hand, $A\subset A'$, and if the proposition is proved for A', then A' is a module of finite type over B'. But B' is itself a module of finite type over B; a basis of it consists of the monomials $T_1^{i_1/p^s},\dots,T_r^{i_r/p^s}$, $0\leqslant i_1,\dots,i_r<p^s$. Therefore A', and hence A, is a module of finite type over B.

So we see that the proof of the proposition reduces to the case when α_1 is separable. By the theorem on the primitive element there exists an $\alpha_2'\in K$ such that $L(\alpha_1,\alpha_2)=L(\alpha_2')$. Then we have $L(\alpha_1,\dots,\alpha_s)=L(\alpha_2',\alpha_3,\dots,\alpha_s)$. Repeating the same arguments $s-1$ times we reduce the proof to the case of a separable extension.

For a proof in this case the reader is referred to [37], Vol. 1, Ch. V, § 4, Theorem 7. This proof does not depend on the remaining part of the book except for the rudiments of the theory of finite extensions.

Theorem 5. *If $g: Y \to X$ is a finite mapping and a birational isomorphism, then there exists a regular mapping $h: X^{\nu} \to Y$ such that the diagram*

$$\begin{array}{ccc} & X^{\nu} & \\ {}^{h}\swarrow & & \searrow^{\nu} \\ Y & \xrightarrow[g]{} & X \end{array}$$

commutes. If $g: Y \to X$ is a regular mapping, $g(Y)$ is dense in X, and Y is normal, then there exists a regular mapping $h: Y \to X^{\nu}$ such that the diagram

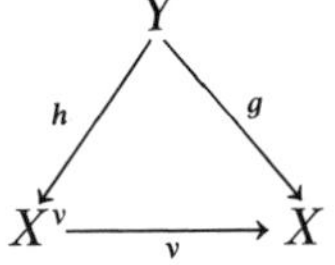

commutes.

Proof of the first part. By assumption we have the embeddings $k[X] \subset k[Y] \subset k(X)$, where $k[Y]$ is integral over $k[X]$. By the definition of the integral closure $k[Y] \subset k[X^{\nu}]$, and this gives the required regular mapping $h: X^{\nu} \to Y$.

Proof of the second part. An element u of $k[X^{\nu}]$ is integral over $k[X]$ and is contained in $k(X) \subset k(Y)$. Since $k[Y] \supset k[X]$, a fortiori u is integral over $k[Y]$, and because $k[Y]$ is integrally closed, we see that $u \in k[Y]$. Therefore $k[X^{\nu}] \subset k[Y]$, which gives a regular mapping $h: Y \to X^{\nu}$ with the required properties.

Corollary. *The normalization of an affine variety is unique. More accurately, if $\nu: X^{\nu} \to X$ and $\bar{\nu}: \bar{X}^{\nu} \to X$ are two normalizations, then there exists an isomorphism $g: X^{\nu} \to \bar{X}^{\nu}$ such that the diagram*

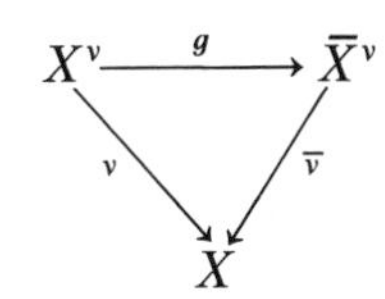

commutes.

This follows from either of the two parts of the theorem.

We do not prove the existence of a normalization for arbitrary quasiprojective varieties. We mention that for those varieties for which a normalization exists it has the property established in Theorem 5, as follows immediately by considering affine coverings.

3. Ramification. The concept of normality allows us to derive one important property of finite mappings. For a finite mapping $f: X \to Y$

the number of inverse images of a point $y \in Y$ is finite, as we have seen in Ch. I, § 5.3. Let us try to find this number. It is natural to expect, by analogy with the theorem on the dimension of the inverse image, that the number is one and the same for all points y in some open set and that a variation can only arise on some closed subset $Z \subset Y$.

This is so in the simplest example of the mapping

$$f: \mathbb{A}^1 \to \mathbb{A}^1, \qquad y = f(x) = x^2. \tag{1}$$

To state the peculiarity of this example in general form we introduce a concept.

Definition. If X and Y are irreducible varieties of equal dimension and if $f: X \to Y$ is a regular mapping for which $f(X)$ is dense in Y, then the degree of the extension $k(X)/f^*k(Y)$ (which under the present conditions is finite) is called the *degree of the mapping* f:

$$\deg f = [k(X) : f^*k(Y)].$$

In the case of the mapping (1) $\deg f = 2$, and if the characteristic of k is not 2, then any point $y \neq 0$ has two distinct inverse images, whereas $y = 0$ has only one. Is the number of inverse images always less than or equal to the degree of the mapping? This is not so in the example of Ch. I, § 1.1, of the parametrization (2) $f: \mathbb{A}^1 \to Y$ of the curve Y with a double point (1); however, the inverse image of the singular point consists of two points. It turns out that the reason for this is the fact that the curve Y is not normal.

Theorem 6. *If $f: X \to Y$ is a finite mapping of irreducible varieties and Y is normal, then the number of inverse images of any point $y \in Y$ does not exceed* $\deg f$.

By virtue of the definition of a finite mapping we may confine our attention to the case when X and Y are affine. We set

$$k[X] = A, \quad k[Y] = B, \quad k(X) = K, \quad k(Y) = L, \quad [K:L] = \deg f = n.$$

Since Y is normal, B is integrally closed, and since f is finite, A is a module of finite type over B. Therefore the coefficients of the minimal polynomial of any element $a \in A$ lie in B. This is a simple property of integrally closed rings, which the reader can find in [37], Vol. 1, Ch. V, § 3. Let $f^{-1}(y) = \{x_1, \ldots, x_m\}$. We consider an element $a \in A$ for which the values $a(x_i)$ are all distinct for $i = 1, \ldots, m$ (if

$X \subset \mathbb{A}^N$, then the matter comes to a construction of such a polynomial in N-dimensional space, and this is altogether elementary). Let $F \in B[T]$ be the minimal polynomial of a. Clearly $\deg F \leqslant n$. We replace in F all the coefficients by their values at y and we denote the polynomial so obtained by $\bar{F}(T)$. It has m distinct roots $a(x_i)$. Thus,

$$m \leqslant \deg \bar{F} = \deg F \leqslant n,$$

so that $m \leqslant n$, as required.

Henceforth we always consider finite mappings $f: X \to Y$ of irreducible varieties, where Y is assumed to be normal.

Definition. A mapping f is said to be *unramified* at a point $y \in Y$ if the number of inverse images of this point is equal to the degree of the mapping. Otherwise y is called a *ramification point*.

Theorem 7. *The set of points at which a mapping is unramified is open, and if the extension $k(X)/f^*k(Y)$ is separable, it is not empty.*

We keep to the notation introduced in the proof of Theorem 6. If f is unramified at y, then $\deg \bar{F} = \deg F = n$, and $\bar{F}$ has n distinct roots. We denote by $D(F)$ the discriminant of the polynomial F. As we have seen, the condition for being unramified at y can be written in the form

$$D(\bar{F}) = D(F)(y) \neq 0. \tag{2}$$

But then $D(F)(y') \neq 0$ for the points y' of some neighbourhood of y. This is what we had to show.

Thus, the set of ramification points is closed. It is called the subvariety of ramification of f.

The question remains whether it is a proper subvariety. If the extension $k(X)/f^*k(Y)$ is inseparable, then $D(F) = 0$ for the minimal polynomial F of any element of this extension. Therefore the condition (2) does not hold for even one point—all the points are ramification points.

Suppose that the extension $k(X)/f^*k(Y)$ is separable. Then f is also called separable. Again we may take X and Y to be affine, and we use the previous notation. If $a \in A$ is a primitive element of the extension $k(X)/f^*k(Y)$ and $F(T)$ its minimal polynomial, then $\deg F = n$, $D(F) \neq 0$. Therefore there exist points $y \in Y$ at which $D(F)(y) \neq 0$, and hence f is unramified. This proves Theorem 7.

We see that if a mapping $f: X \to Y$ is finite and separable, the varieties X and Y are irreducible, and Y is normal, then the picture is the same as in example (1): the points of some non-empty subset $U \subset Y$ have $\deg f$ distinct inverse images, and the points of the complement have fewer inverse images.

The following proposition expresses a fundamental property of unramified coverings:

Theorem 8.* *If $f: X \to Y$ is unramified at $y \in Y$, then for any point $x \in f^{-1}(y)$ and any $r > 0$ the homomorphism*

$$\varphi : \mathcal{O}_y/\mathfrak{m}_y^r \to \mathcal{O}_x/\mathfrak{m}_x^r , \tag{3}$$

induced by the homomorphism $f^*: \mathcal{O}_y \to \mathcal{O}_x$ is an isomorphism.

Intuitively this means that f is "an isomorphism to within infinitesimals of any order".

The assertion of the theorem is local, therefore we may assume that X and Y are affine and use the previous notation. For points $x \in X$ and $y \in Y$ we denote by $\bar{\mathfrak{m}}_x$ and $\bar{\mathfrak{m}}_y$ the maximal ideals of the points x and y in the rings A and B, where A and B are $k[X]$ and $k[Y]$ respectively.

Consider the ring $\tilde{\mathcal{O}} = A \circ \mathcal{O}_y$ and denote by $\tilde{\mathfrak{m}}$ and $\tilde{\mathfrak{m}}_i$ its ideals generated by the ideals $\mathfrak{m}_y \subset \mathcal{O}_y$ and $\bar{\mathfrak{m}}_{x_i} \subset A$. It is easy to see that the rings $\mathcal{O}_{x_i}$ are isomorphic to the local rings $\tilde{\mathcal{O}}_{\mathfrak{m}_i}$ of the maximal ideals $\tilde{\mathfrak{m}}_i$ of $\tilde{\mathcal{O}}$. Therefore

$$\mathcal{O}_{x_i}/\mathfrak{m}_{x_i}^r \simeq \tilde{\mathcal{O}}/\tilde{\mathfrak{m}}_i^r$$

and the assertion of the theorem is equivalent to the fact that the natural homomorphisms

$$\mathcal{O}_y/\mathfrak{m}_y^r \to \tilde{\mathcal{O}}/\tilde{\mathfrak{m}}_i^r \tag{4}$$

generated by the embedding $\mathcal{O}_y \subset \tilde{\mathcal{O}}$ are isomorphisms.

The proof of this fact is based on two lemmas. In both we are concerned with ideals $\mathfrak{a}_1, \ldots, \mathfrak{a}_n$, $\mathfrak{a}$ of some ring R, $\mathfrak{a}_i \supset \mathfrak{a}$, $i = 1, \ldots, n$. In this case the natural homomorphisms $R/\mathfrak{a} \to R/\mathfrak{a}_i$ determine a homomorphism $R/\mathfrak{a} \to \bigoplus R/\mathfrak{a}_i$.

Lemma 1. *The homomorphism*

$$\tilde{\mathcal{O}}/\tilde{\mathfrak{m}} \to \bigoplus \tilde{\mathcal{O}}/\tilde{\mathfrak{m}}_i \tag{5}$$

is an isomorphism.

The meaning of this homomorphism is that it associates with a function $f \in \tilde{\mathcal{O}}$ the collection of its values at the points x_i. Hence it is clear that it is an epimorphism. We need only prove that its kernel is trivial, that is,

$$\tilde{\mathfrak{m}} = \bigcap \tilde{\mathfrak{m}}_i . \tag{6}$$

Consider the ring $\bar{\mathcal{O}} = \tilde{\mathcal{O}}/\tilde{\mathfrak{m}}$ and its homomorphisms

$$\lambda_i : \bar{\mathcal{O}} \to k ,$$

whose kernels are the ideals $\tilde{\mathfrak{m}}_i/\tilde{\mathfrak{m}}$. We denote by $\bar{u}$ the image of an element $u \in \tilde{\mathcal{O}}$ in $\bar{\mathcal{O}}$.

As before, let $a \in A$ be an element for which all the $a(x_i)$ are distinct for $i = 1, \ldots, n$, and let $F(T) \in \mathcal{O}_y[T]$ be its minimal polynomial. Going over to the residue classes modulo $\tilde{\mathfrak{m}}$ we obtain an element $\bar{a} \in \bar{\mathcal{O}}$ and a polynomial $\bar{F} \in k[T]$, with $\bar{F}(\bar{a}) = 0$. Since the $\lambda_i(\bar{a}) = a(x_i)$ are all distinct and are roots of $\bar{F}$, and since $\deg \bar{F} = n$, we have

$$\bar{F}(T) = \prod_{i=1}^{n} (T - \lambda_i(\bar{a})) .$$

Suppose that $\bigcap \tilde{\mathfrak{m}}_i \neq \tilde{\mathfrak{m}}$. Then there exists an element $u \in \bigcap \tilde{\mathfrak{m}}_i$ such that $u \notin \tilde{\mathfrak{m}}$. By Hilbert's Nullstellensatz, $u^s \in \tilde{\mathfrak{m}}$ for some $s > 0$, and when we replace u by one of its

* Theorem 8, which is not as easy to prove as the remaining results of this chapter, is used in the whole book only once: in the last section. Therefore the proof can be omitted at a first reading.

powers, we may assume that $u \notin \tilde{\mathfrak{m}}$, $u^2 \in \tilde{\mathfrak{m}}$. We apply to the element $a+u$ the same arguments as to a. Since $\lambda_i(\bar{u})=0$, $i=1,\ldots,n$, we see that $\bar{a}+\bar{u}$ is a root of the same polynomial $\bar{F}(T)$. Since $\bar{u}^2=0$, we have

$$0=\bar{F}(\bar{a}+\bar{u})=\bar{F}(\bar{a})+\bar{F}'(\bar{a})\bar{u}=\bar{F}'(\bar{a})\bar{u}\,. \tag{7}$$

From the fact that the polynomial $\bar{F}(T)$ does not have multiple roots it follows that

$$\bar{F}'(T)\,U(T)+\bar{F}(T)\,V(T)=1$$

for suitable $U,\ V\in k[T]$. Therefore $\bar{F}'(\bar{a})\,U(\bar{a})=1$, and from (7) it follows that $\bar{u}=0$, $u\in\tilde{\mathfrak{m}}$. This proves (6) and hence also Lemma 1.

Lemma 2. *If* $\mathfrak{a}_i\supset\mathfrak{a}$ *for a system of ideals* $\mathfrak{a}_1,\ldots,\mathfrak{a}_n$, $\mathfrak{a}$ *of a ring* R *and if the homomorphism*

$$R/\mathfrak{a}\to\bigoplus R/\mathfrak{a}_i \tag{8}$$

is an isomorphism, then the homomorphisms

$$R/\mathfrak{a}^r\to\bigoplus R/\mathfrak{a}_i^r$$

are isomorphisms for all $r>0$.

Lemma 2 follows from elementary properties of ideals, which we state without proof. The reader can verify them or can consult [33], Vol. 2, § 89. The fact that the homomorphism (8) is an isomorphism is equivalent to

$$\mathfrak{a}=\bigcap\mathfrak{a}_i\,,\qquad \mathfrak{a}_i+\mathfrak{a}_j=R\,,\qquad i\neq j\,.$$

From the second relation it follows that

$$\mathfrak{a}_i^r+\mathfrak{a}_j^r=R\,,\qquad i\neq j\,,$$

and $\bigcap\mathfrak{a}_i^r=\mathfrak{a}_1^r\ldots\mathfrak{a}_n^r$. Hence

$$\bigcap\mathfrak{a}_i^r=\mathfrak{a}_1^r\ldots\mathfrak{a}_n^r=(\mathfrak{a}_1\ldots\mathfrak{a}_n)^r=(\cap\mathfrak{a}_i)^r=\mathfrak{a}^r\,,$$

which proves Lemma 2.

Proof of Theorem 8. From Lemmas 1 and 2 it follows that

$$\tilde{\mathcal{O}}/\tilde{\mathfrak{m}}^r=\bigoplus\tilde{\mathcal{O}}/\tilde{\mathfrak{m}}_i^r\cdot e_i \tag{9}$$

where e_i is the unit element of the ring $\tilde{\mathcal{O}}/\mathfrak{m}_i^r$.

On the other hand, $\tilde{\mathcal{O}}$ is a module over the local ring $\mathcal{O}_y$. We denote by u_i elements of $\tilde{\mathcal{O}}$ whose images are the e_i. From Nakayama's lemma and (9) for $r=1$ it follows that the u_i, $i=1,\ldots,n$, generate $\tilde{\mathcal{O}}$ over $\mathcal{O}_y$. We show that they are even free generators of $\tilde{\mathcal{O}}$ over $\mathcal{O}_y$. This follows from the fact that if K and L are the fields of fractions of A and B, respectively, [that is, $K=k(X)$, $L=k(Y)$], then

$$K=\tilde{\mathcal{O}}\cdot L=Lu_i+\cdots+Lu_n\,.$$

Since $[K:L]=n$, $u_1,\ldots,u_n$ are linearly independent not only over $\mathcal{O}_y$ but even over L. Thus, as a module over $\mathcal{O}_y$

$$\tilde{\mathcal{O}}=\mathcal{O}_yu_1\oplus\cdots\oplus\mathcal{O}_yu_n$$

and hence

$$\tilde{\mathcal{O}}/\tilde{\mathfrak{m}}^r=\mathcal{O}_y/\mathfrak{m}_y^r\,e_1\oplus\cdots\oplus\mathcal{O}_y/\mathfrak{m}_y^r\,e_n\,.$$

Comparing this with the decomposition (9) we find that the homomorphisms (4) are isomorphisms.

Corollary 1. *Under the conditions of Theorem 8 the homomorphism of local rings $f^*:\mathcal{O}_y\to\mathcal{O}_x$ determines an isomorphism of their completions $\hat{f}^*:\hat{\mathcal{O}}_y \xrightarrow{\sim} \hat{\mathcal{O}}_x$.*

The proof follows by a trivial verification from the definition of a completion. The assertion becomes perfectly obvious if we make use of the fact that $\hat{\mathcal{O}}_x$ is the projective limit of the rings $\mathcal{O}_x/\mathfrak{m}_x^r$.

Corollary 2. *Under the assumptions of Theorem 8 the tangent mapping*

$$d_x f:\Theta_{x,X}\to\Theta_{y,Y}$$

is an isomorphism.

This follows immediately from the theorem and the fact that

$$\Theta_{x,X}=(\mathfrak{m}_x/\mathfrak{m}_x^2)^*\,,\qquad \Theta_{y,X}=(\mathfrak{m}_y/\mathfrak{m}_y^2)^*\,.$$

Corollary 3. *Under the conditions of Theorem 8, X is smooth at x if and only if Y is smooth at y.*

This follows immediately from Corollary 2, the definition of a simple point, and the fact that $\dim X=\dim Y$.

In conclusion we explain the topological interpretation of the concept of being unramified when X and Y are defined over the field of complex numbers. We assume that the mapping $f:X\to Y$ is finite, that X and Y are smooth, and that f is unramified at all points of Y. In our case X and Y, as we have seen in § 2.3, are varieties of the same dimension $2\dim X=2\dim Y$, and f is a continuous mapping. We show that f defines X as an unramified covering of Y. This means that every point $y\in Y$ has a neighbourhood U with $y\in U$ such that $f^{-1}(U)$ splits into connected components V_i each of which is mapped by f homeomorphically onto U.

Let $f^{-1}(y)=\{x_1,\dots,x_m\}$, $u_1,\dots,u_n$ be local parameters in a neighbourhood of y, and $v_1^{(i)},\dots,v_n^{(i)}$ be local parameters at x_i. Corollary 2 shows that $|(\partial v_l/\partial u_j)(y)|\neq 0$ for all $i=1,\dots,m$. By the implicit function theorem it follows that there exist neighbourhoods V_i of x_i and U of y such that f determines a homeomorphism between V_i and U for all i. We can choose these neighbourhoods sufficiently small so that V_i and V_j are disjoint for $i\neq j$. We show that $f^{-1}(U)=\bigcup V_i$. If $y'\in U$, then $f^{-1}(y')$ consists of m points because f is unramified. Since y' has m inverse images already in $V_1,\dots,V_m$, we have $f^{-1}(U)=\bigcup V_i$.

4. Normalization of Curves

Theorem 9. *A quasiprojective irreducible curve X has a normalization X^ν (which is also quasiprojective).*

Proof. Let $X=\bigcup U_i$ be a covering of X by affine open sets. We denote by U_i^ν a normalization of U_i, which exists by Theorem 4, and by $f_i:U_i^\nu\to U_i$ the natural regular mapping, which is a birational isomorphism.

We embed the affine space containing U_i^ν in a projective space and denote by V_i the closure of U_i^ν in this projective space. Note that all the varieties occurring so far are birationally isomorphic to X: U_i is open in X, f_i is a birational isomorphism between U_i^ν and U_i, U_i^ν is open in V_i. Consequently U_i^ν and V_j are birationally isomorphic. Let $\varphi_{ij}:U_i^\nu\to V_j$ be the corresponding mapping. According to the corollary to Theorem 3, U_i^ν is a smooth curve, and since V_j is projective, φ_{ij} is regular by the corollary to Theorem 3 of § 3. We set $W=\prod_j V_j$, $\varphi_i=\prod_j \varphi_{ij}$, that is, $\varphi_i(u)=(\varphi_{i1}(u), \varphi_{i2}(u), \ldots, \ldots)$. We denote by X' the union of all the $\varphi_i(U_i^\nu)$ in W. We claim that $X'=X^\nu$. To substantiate this we have to show that: a) X' is quasiprojective, b) X' is irreducible, c) X' is normal, d) there exists a finite mapping $\nu: X'\to X$ that is a birational isomorphism.

To prove all this we set $U_0=\bigcap U_i$, which is an open subset of X. From the construction of φ_i it follows easily that $U_0^\nu\subset U_i^\nu$ and that all the φ_i coincide on U_0^ν. Let φ denote their restriction to U_0^ν. Then $\varphi(U_0^\nu)\subset\varphi_i(U_i^\nu)\subset\overline{\varphi(U_0^\nu)}$, where $\overline{\varphi(U_0^\nu)}$ is the closure of $\varphi(U_0^\nu)$ in W. Clearly $\varphi(U_0^\nu)$ is an irreducible quasiprojective curve, and $\overline{\varphi(U_0^\nu)}-\varphi(U_0^\nu)$ consists of finitely many points. By construction, $\varphi(U_0^\nu)\subset X'\subset\overline{\varphi(U_0^\nu)}$, therefore $\overline{\varphi(U_0^\nu)}-X'$ consists of finitely many points. This proves a) and b).

Let $x\in X'$ then $x\in\varphi_i(U_i^\nu)$ for some i, and $\varphi_i(U_i^\nu)$ is a neighbourhood of x. We show that φ_i is an isomorphism, and since U_i^ν is normal, it then follows that X' is normal, which proves c). For this purpose we observe that by construction φ_{ii} is an isomorphic embedding of U_i^ν in its closure V_i. Therefore the mapping $(u_1, u_2, \ldots)\to\varphi_{ii}^{-1}(u_i)$ is inverse to φ_i, which shows that it is an isomorphism.

Finally, to prove d) we construct a mapping $g_i:\varphi_i(U_i^\nu)\to X$; $g_i=f_i\varphi_i^{-1}$. By the preceding all the g_i are finite mappings. We show that all the g_i determine on X' a single finite mapping $f: X'\to X$. To see this we observe that all the g_i coincide on $\varphi_i(U_0^\nu)$: if $g: U_0^\nu\to U_0$ is a normalizing mapping, then $g_i=g$ on U_0^ν. Therefore the mappings g_i and g_j agree on the open subset $\varphi(U_0^\nu)$, which is contained in $\varphi_i(U_i^\nu)\cap\varphi_j(U_j^\nu)$. But two regular mappings that agree on a non-empty open subset agree everywhere; this follows from the corresponding property of functions. Thus, g_i and g_j coincide at all points where they are both defined, and this means that all the g_i determine a single regular mapping $\nu: X'\to X$. Clearly ν is a birational isomorphism, and the theorem is proved.

Theorem 10. *The normalization of a projective curve is projective.*

Let X be a projective curve, X^ν its normalization, and $\nu: X^\nu\to X$ the normalizing mapping. We assume that the curve X^ν is not projective and denote by $\bar{X}$ its closure in a projective space. Let $x\in\bar{X}-X^\nu$, let U be some affine neighbourhood of x on $\bar{X}$, U^ν the normalization of U, and $\nu': U^\nu\to U$ the normalizing mapping. Then we have the diagram

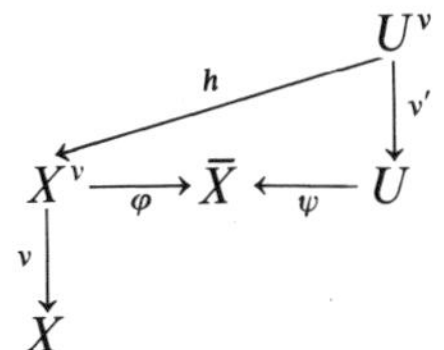

where φ and ψ are isomorphic embeddings. The mapping $v\varphi^{-1}\psi v'$ is a birational isomorphism, and by Corollary 1 to Theorem 3 of § 3 and the fact that U^v is a smooth curve, this mapping is regular. By Theorem 5 the regular mapping h drawn on the diagram exists, and it satisfies $\varphi h = \psi v'$. However, its existence leads to a contradiction: $\varphi h(U^v) \subset X^v$, and $x \in \psi v'(U^v)$, because the normalizing mapping is finite and hence epimorphic according to Theorem 4 of Ch. I, § 5. This proves the theorem.

Corollary. *An irreducible algebraic curve is birationally isomorphic to a smooth projective curve.*

This is a combination of the corollary to Theorem 3 and of Theorem 10.

The concept of normalization enables us to investigate properties of curves in more detail.

Theorem 11. *A regular mapping $\varphi: X \to Y$ is finite if X is an irreducible smooth projective curve,* $\dim Y > 0$, *and* $Y = \varphi(X)$.

Proof. Let V be an affine neighbourhood of a point $y \in Y$ and $B = k[V]$. We regard $k(Y)$ as a subfield of $k(X)$, the embedding being effected by a mapping φ^*. In particular, $B \subset k(X)$; let A be the integral closure of B in $k(X)$. In proving the existence of a normalization of an affine variety we have explained that A is a ring of finite type over B, hence $A = k[U]$, where U is an affine normal curve. Since it is birationally isomorphic to X, by Corollary 2 to Theorem 2 of § 4 we may assume that U is an open subset of X. We show that $U = \varphi^{-1}(V)$, which guarantees the mapping is finite.

Suppose that for some point $y_0 \in V$ there is a point $x_0 \notin U$, $\varphi(x_0) = y_0$. We consider a function $f \notin \mathcal{O}_{x_0}$, $f \in \mathcal{O}_{x_i}$ for all $x_i \in U$, $\varphi(x_i) = y_0$, $x_i \neq x_0$. Such a function is easy to construct by including the points x_0 and x_i each in an affine open set. If f has poles at points $x' \in U$, then $\varphi(x') = y' \neq y_0$, therefore we can find a function $h \in B$ such that $h(y_0) \neq 0$, $fh \in \mathcal{O}_{x'}$, that is, $fh \in A$. We have to take a function that vanishes at the points y' and raise it to a sufficiently high power. Now $f_1 = fh$ is integral over B, that is,

$$f_1^n + b_1 f_1^{n-1} + \cdots + b_n = 0\,, \quad b_i \in B\,, \quad f_1 = b_1 - b_2/f_1 - \cdots b_n/f_1^{n-1}\,.$$

Since $f_1 \notin \mathcal{O}_{x_0}$, we have $f_1^{-1} \in \mathfrak{m}_{x_0}$. Therefore the last equation leads to a contradiction: the right-hand side is regular at x_0, but the left-hand side is not.

Other applications are related to properties of singular points. In fact, the existence of a normalization enables us to introduce some useful characteristics of such points.

Let X be a curve and x a point of it, possibly singular; let $\nu: X^\nu \to X$ be a normalization of X and $\bar{x}_1, \ldots, \bar{x}_l$ inverse images of x on X^ν. The points $\bar{x}_i$ are called *branches* of X passing through x. This terminology is explained by the fact that if k is the field of complex (or of real) numbers and the U_i sufficiently small complex (or real) neighbourhoods of the $\bar{x}_i$, then some neighbourhood of x is the union of the "branches" $\nu(U_i)$.

We denote by Θ_i the tangent space to X^ν at $\bar{x}_i$. The mapping $d_{\bar{x}_i}\nu$ carries Θ_i into a linear subspace of the tangent space to X at x. Obviously $(d_{\bar{x}_i}\nu)(\Theta_i)$ is either the point x or a line. In the second case the branch $\bar{x}_i$ is called linear, and the line $(d_{\bar{x}_i}\nu)(\Theta_i)$ the tangent to this branch.

The branch $\bar{x}_i$ is linear if and only if the mapping ν^* carries $\mathfrak{m}_x/\mathfrak{m}_x^2$ into the whole space $\mathfrak{m}_{\bar{x}_i}/\mathfrak{m}_{\bar{x}_i}^2$. Suppose that x is the origin of coordinates in the space $\mathbb{A}^n$ with the coordinates $t_1, \ldots, t_n$. Then

$$\nu^*(t_1) + \mathfrak{m}_{\bar{x}_i}^2, \ldots, \nu^*(t_n) + \mathfrak{m}_{\bar{x}_i}^2$$

generate $\nu^*(\mathfrak{m}_x/\mathfrak{m}_x^2)$. Since $\bar{x}_i$ is simple, $\dim \mathfrak{m}_{\bar{x}_i}/\mathfrak{m}_{\bar{x}_i}^2 = 1$, and therefore the branch $\bar{x}_i$ is linear if and only if $\nu^*(t_s) \notin \mathfrak{m}_{\bar{x}_i}^2$ for at least one s between 1 and n. To put it differently, $\nu^*(t_s)$ must be a local parameter at $\mathfrak{m}_{\bar{x}_i}$. Since $\mathfrak{m}_x = (t_1, \ldots, t_n)$, this condition of linearity in invariant form becomes $\nu^*(\mathfrak{m}_x) \not\subset \mathfrak{m}_{\bar{x}_i}^2$. As a measure of the deviation of a branch $\bar{x}_i$ from linearity we can take the number l such that $\nu^*(\mathfrak{m}_x) \subset \mathfrak{m}_{\bar{x}_i}^l$, $\nu^*(\mathfrak{m}_x) \not\subset \mathfrak{m}_{\bar{x}_i}^{l+1}$. This number is called the multiplicity of the branch $\bar{x}_i$.

The point $(0,0)$ of the curve $y^2 = x^2 + x^3$ gives an example of two linear branches with the tangents $y = x$ and $y = -x$, and the point $(0,0)$ on the semicubical parabola $y^2 = x^3$ an example of a two-fold non-linear branch.

If x is the centre of a unique linear branch, then x is a simple point. This is a consequence of a lemma that will be proved in the next subsection. Thus, the simplest characterization of the "singularity" of a point is the number of branches corresponding to it and their multiplicities.

A singular point of a plane algebraic curve is called *ordinary* (or a *point with distinct tangents*) if only linear branches correspond to it and if tangents to distinct branches are distinct.

5. Projective Embeddings of Smooth Varieties. The smooth projective model of an algebraic curve constructed in the preceding subsection is situated in some projective space $\mathbb{P}^n$. The question arises how small this n can be chosen. We answer it by proving a general result on varieties of arbitrary dimension.

Theorem 12. *A smooth projective variety of dimension n is isomorphic to a subvariety of* $\mathbb{P}^{2n+1}$.

Let X be a smooth projective variety, $X \subset \mathbb{P}^N$. Theorem 12 will be proved if we can choose for $N > 2n+1$ a point $\xi \in \mathbb{P}^N - X$ such that the projection from ξ is an isomorphic embedding of X in $\mathbb{P}^{N-1}$. Therefore we begin by clarifying when a regular mapping is an isomorphic embedding.

Lemma. *A finite mapping f of a smooth variety is an isomorphic embedding if it is one-to-one and if $d_x f$ is an isomorphic embedding of the tangent space Θ_x for every $x \in X$.*

We set $f(X) = Y$, $\varphi = f^{-1}$. The lemma will be proved if we can show that φ is regular. This is an assertion of local character. Let $y \in Y$ and $f(x) = y$, $x \in X$. We denote by U and V affine neighbourhoods of x and y such that $f(U) = V$ and that $k[U]$ is integral over $k[V]$. The restriction of f to U is also denoted by f.

It is enough to show that f is an isomorphism for a suitable choice of U and V. Then $\varphi = f^{-1}$ is regular at y.

We recall that the space Θ_x is dual to $\mathfrak{m}_x/\mathfrak{m}_x^2$, where $\mathfrak{m}_x$ is the maximal ideal of the local ring $\mathcal{O}_x$. The second condition of the lemma means that the mapping $f^*: \mathfrak{m}_y/\mathfrak{m}_y^2 \to \mathfrak{m}_x/\mathfrak{m}_x^2$ is an epimorphism. In other words, if $\mathfrak{m}_y = (u_1, \dots, u_l)$, then $f^*(u_i) + \mathfrak{m}_x^2$ generate $\mathfrak{m}_x/\mathfrak{m}_x^2$. Applying Nakayama's lemma (§ 2.1) to $\mathfrak{m}_x$ as a module over $\mathcal{O}_x$ we see that then $\mathfrak{m}_x = (f^*(u_1), \dots, f^*(u_l))$ or

$$\mathfrak{m}_x = f^*(\mathfrak{m}_y)\, \mathcal{O}_x \,. \tag{1}$$

Now we apply Nakayama's lemma to $\mathcal{O}_x$ as a module over $f^*(\mathcal{O}_y)$. Let us show that $\mathcal{O}_x$ is of finite type over $f^*(\mathcal{O}_y)$. Let $\xi = \alpha/\beta \in \mathcal{O}_x$, that is, $\alpha, \beta \in k[U]$, $\beta(x) \neq 0$. We denote by Z the set $V(\beta)$, which is closed in U. From the Corollary to Theorem 4 in Ch. I, § 5, it follows that $T = f(Z)$ is closed. Since f is one-to-one, $f^{-1}(y) = x$, therefore $y \notin T$. Hence there exists a function $b \in k[V]$ such that $b = 0$ on T, $b(y) \neq 0$. We set $f^*(b) = a$. Then $a = 0$ on Z, $a(x) \neq 0$. By Hilbert's Nullstellensatz there exists a natural number l and a $\gamma \in k[U]$ such that $a^l = \beta\gamma$. Hence $\xi = \alpha\gamma/a^l \in k[U]\, f^*(\mathcal{O}_y)$ and our assertion follows from the fact that $k[U]$ is of finite type over $k[V]$. Now (1) shows that $\mathcal{O}_x/f^*(\mathfrak{m}_y)\,\mathcal{O}_x = \mathcal{O}_x/\mathfrak{m}_x = k$, hence is generated by the single element 1. From Nakayama's lemma it now follows that $\mathcal{O}_x = f^*(\mathcal{O}_y)$. Let $u_1, \dots, u_m$ be a basis of $k[U]$ over $k[V]$. By hypothesis, $u_i \in \mathcal{O}_x = f^*(\mathcal{O}_y)$. We denote by $V' = V - V(h)$ a smooth affine neighbourhood of y such that all the $(f^*)^{-1}(u_i)$ are regular in $U' = U - V(f^*(h))$. Then $k[U'] = \Sigma f^* k[V'] u_i$. By hypothesis, $u_i \in f^* k[V']$, from which it follows that $k[U'] = f^*(k[V'])$, and this means that f is an isomorphism between U' and V'. The lemma is now proved.

Corollary 1. *If every line passing through a point ξ intersects X in at most one point and if the tangent space to X at any of its points does not contain ξ, then the projection centered at ξ is an isomorphism.*

It is sufficient to use Theorem 7 of Ch. I, § 5.

Now we can turn to the proof of Theorem 12. We need only show that if X is a smooth variety, $\dim X = n$, $X \subset \mathbb{P}^N$, $N > 2n+1$, then there is a point ξ satisfying the conditions of the corollary.

We denote by U_1 and U_2 the sets of points $\xi \in \mathbb{P}^N$ relative to which ξ fails to satisfy the first or second condition of the corollary, respectively.

In $\mathbb{P}^N \times X \times X$ we consider the set Γ consisting of collinear points (a, b, c), $a \in \mathbb{P}^N$, $b, c \in X$. Clearly Γ is a closed subset of $\mathbb{P}^N \times X \times X$. The projections of $\mathbb{P}^N \times X \times X$ onto $\mathbb{P}^N$ and onto $X \times X$ determine regular mappings $\varphi: \Gamma \to \mathbb{P}^N$ and $\psi: \Gamma \to X \times X$. It is evident that if $y \in X \times X$, $y = (b, c)$, $b, c \in X$ and in addition $b \neq c$, then $\psi^{-1}(y)$ consists of points (a, b, c), where a is any point of the line passing through b and c. Therefore $\dim \psi^{-1}(y) = 1$, and from Theorem 7 of Ch. I, § 6 it follows that $\dim \Gamma = 2n+1$. By definition, $U_1 = \varphi(\Gamma)$, and from the same theorem it follows that $\dim U_1 \leqslant \dim \Gamma = 2n+1$.

Similarly, to investigate the set U_2 we consider in $\mathbb{P}^N \times X$ the set Γ consisting of those points (a, b) for which $a \in \Theta_b$. In exactly the same way we have mappings $\psi: \Gamma \to X$ and $\varphi: \Gamma \to \mathbb{P}^N$. For $x \in X$ we have $\dim \psi^{-1}(x) = n$, therefore $\dim \Gamma = 2n$, and since $U_2 = \varphi(\Gamma)$, $\dim U_2 \leqslant 2n$.

So we see that $\dim U_1 \leqslant 2n+1$, $\dim U_2 \leqslant 2n$, therefore, if $N > 2n+1$, then $U_1 \cup U_2 \neq \mathbb{P}^N$, which is what we had to prove.

Corollary 2. *Any quasiprojective smooth curve is isomorphic to a curve situated in three-dimensional projective space.*

We shall see later that not every curve is isomorphic to a curve contained in a projective plane. Therefore not every algebraic curve has a smooth plane projective model.

However, it can be shown that by extending the projection process which we had used in proving Theorem 12 one can obtain a plane curve whose singular points are all ordinary double points. According to Theorem 12 every smooth surface is isomorphic to a surface situated in a five-dimensional space. But generally speaking it cannot be projected into four-dimensional space. However, the projection can always be chosen so that apart from finitely many points it is an isomorphism. This leads naturally to examples of isolated points whose local rings are not integrally closed such as the one constructed in § 5.1.

Exercises

1. Let X be an affine variety, and K a finite extension of $k(X)$. Show that there exist an affine variety Y and a mapping $f: Y \to X$ having the following properties: 1) f is finite, 2) Y is normal, 3) $k(Y) = K$ and $f^*: k(X) \to k(Y)$ determines the given embedding of $k(X)$ in K. Show also that Y is uniquely determined by these properties. It is called the normalization of X in K.

2. Let X be the cone $z^2 = xy$. Show that the normalization of X in the field $k(X)(\sqrt{x})$ coincides with the affine plane and that the normalization mapping has the form $x = u^2$, $y = v^2$, $z = uv$.

3. Prove propositions similar to Exercise 1 for an arbitrary projective curve X. Show that if X is projective, then so is Y.

4. How is the normalization of $X \times Y$ connected with the normalizations of X and of Y?

5. Show that a point x is normal if the ring $k[[X]]$ (see Exercise 14 of § 3) has no divisors of zero and is normal. *Hint:* Use the result of Exercise 15 in § 3. Extend Exercise 7 of § 3 to singular points and apply it.

6. Show that the cone $X \subset \mathbb{A}^n$ given by the equation $x_1^2 + \cdots + x_n^2 = 0$ is normal for $n \geqslant 3$.

7. Show that on the hypersurface X of Exercise 13 in § 3 the origin of coordinates has an integrally closed local ring.

8. Is the Steiner surface normal? (Exercises 15 and 16 of § 1).

9. Show that every algebraic curve has a plane projective model in which the singular points have linear branches only.

Chapter III. Divisors and Differential Forms

§ 1. Divisors

1. Divisor of a Function. A polynomial in one variable is uniquely determined to within a constant factor by its roots and their multiplicities, that is, by a collection of points $x_1, \ldots, x_r \in \mathbb{A}^1$ with multiplicities $l_1, \ldots, l_r$. A rational function $\varphi(x) = f(x)/g(x)$, $f, g \in k[X]$ is determined by the zeros of the polynomials f and g, that is, by the points at which it vanishes or is non-regular. To distinguish the roots of g from those of f we take their multiplicities with a minus sign. Thus, φ is given by points $x_1, \ldots, x_r$ with arbitrary integral multiplicities $l_1, \ldots, l_r$. Now we set ourselves the task of specifying a rational function on an arbitrary algebraic variety in a similar way.

We start out from the fact that according to the theorem on the dimension of an intersection the set of points at which a regular function vanishes forms a subvariety of codimension 1. Therefore the object we associate with a function is a collection of irreducible subvarieties of codimension 1 with preassigned multiplicities. These multiplicities can have positive and negative integer values.

Definition. A collection of irreducible closed subvarieties $C_1, \ldots, C_r$ of codimension 1 in an irreducible variety X with preassigned integral multiplicities $l_1, \ldots, l_r$ is called a *divisor*.

A divisior D is written in the form

$$D = l_1 C_1 + \ldots + l_r C_r. \tag{1}$$

If all the $l_i = 0$, then we write $D = 0$, and if all the $l_i > 0$, then we write $D > 0$ and we say that D is *effective*. Irreducible subvarieties C_i of codimension 1 taken with the coefficient 1 are called *prime* divisors. If in (1) all the $l_i \neq 0$, then the variety $C_1 \cup \ldots \cup C_r$ is called the *support* of D and is denoted by Supp D. If in (1) the prime divisors C_i are different, we denote l_i by $v_{C_i}(D)$.

Now we define an operation of addition of divisors. We observe that if we allow the coefficients in (1) to take the value 0, any two divisors can be written in the form

$$D' = l'_1 C_1 + \cdots + l'_r C_r, \quad D'' = l''_1 C_1 + \cdots + l''_r C_r$$

with the same $C_1, \ldots, C_r$. Then, by definition,

$$D' + D'' = (l'_1 + l''_1)\, C_1 + \cdots + (l'_r + l''_r)\, C_r\,.$$

Thus, the divisors on a variety X form a group isomorphic to a free $\mathbb{Z}$-module whose generators are irreducible subvarieties of codimension 1 in X. This group is denoted by $\operatorname{Div}(X)$.

Now we proceed to assigning a divisor to a function $f \in k[X]$, $f \neq 0$. Let C be a prime divisor; first we associate with every function $f \in k(X)$, $f \neq 0$, an integer $v_C(f)$. If $X = \mathbb{A}^1$, this is the order of the zero or pole of f at a point.

This can be done only under a restriction on the variety X. Namely, we assume that X is smooth in codimension 1, that is, that the set of singular points of X is of codimension $\geqslant 2$. Let $C \subset X$ be an irreducible subvariety of codimension 1 and U an affine open set consisting of simple points that intersects C and is such that C is determined in U by a local equation. Such a set U exists under the restriction imposed on X, by virtue of Theorem 1 in Ch. II, § 3. Thus, $\mathfrak{a}_C = (\pi)$ in $k[U]$. We show that for every function $f \in k[U]$, $f \neq 0$, there exists an integer $l \geqslant 0$, such that $f \in (\pi^l)$, $f \notin (\pi^{l+1})$. If this were not the case that is, if $f \in (\pi^l)$ for all $l > 0$, then $f \in \bigcap (\pi^l)$, therefore $f = 0$ by Theorem 5 of Ch. II, § 2.

We denote the so defined number l by $v_C(f)$. It has the properties

$$\left.\begin{aligned} v_C(f_1 f_2) &= v_C(f_1) + v_C(f_2)\,, \\ v_C(f_1 + f_2) &\geqslant \min(v_C(f_1)\,; v_C(f_2)) \\ &\text{for} \quad f_1 + f_2 \neq 0 \end{aligned}\right\} \tag{2}$$

which follow easily from the definition and the irreducibility of C.

If X is irreducible, then any function $f \in k(X)$ can be represented in the form $f = g/h$, g, $h \in k[U]$. For $f \neq 0$ we set $v_C(f) = v_C(g) - v_C(h)$. From (2) it follows at once that $v_C(f)$ does not depend on the representation of f in the form g/h and that (2) holds for all $f \in k(X)$ other than zero.

Our definition of the number $v_C(f)$ so far depends on the choice of the open set U, and we should write $v_C^U(f)$ instead of $v_C(f)$. Let us show that, in fact, $v_C^U(f)$ does not depend on U.

To begin with we assume that V is an affine open set, $V \subset U$, and $V \cap C \neq \emptyset$. Then π is a local equation of C also in V, and evidently, $v_C^V(f) = v_C^U(f)$. But if V is any open set satisfying the same conditions as U, then $U \cap C$ and $V \cap C$ are open in C and non-empty, and since C is irreducible, their intersection is non-empty. Taking for W an affine neighbourhood in $U \cap V$ of some point $x \in U \cap V \cap C$, we see that by the preceding remark $v_C^U(f) = v_C^W(f)$, $v_C^V(f) = v_C^W(f)$, hence $v_C^U(f) = v_C^V(f)$. Thus, the notation $v_C(f)$ is justified. Observe that if $X = \mathbb{A}^1$, if $C = x$

is the point with the coordinate α, and $f \in k[\mathbb{A}^1] = k[T]$, then $v_x(f)$ is the multiplity of the root α of the polynomial $f(T)$, and the general definition in essence copies this special case.

If $v_C(f) = l > 0$, we say that f has a zero of order l on C. If $v_C(f) = -l < 0$, then f has a pole of order l on C. Observe that these concepts are defined for subvarieties of codimension 1 and not for points. For example, for the function x/y on $\mathbb{A}^2$ the point $(0, 0)$ belongs both to the subvariety of zeros $(x = 0)$ and the subvariety of poles $(y = 0)$ of the function.

Now we show that to a given function $f \in k(X)$ there correspond only finitely many irreducible subvarieties of codimension 1 for which $v_C(f) \neq 0$. First we consider the case when X is an affine variety and $f \in k[X]$. It follows from the definition that if C is not a component of the subvariety $V(f)$, then $v_C(f) = 0$. If X is affine, as before, but $f \in k(X)$, then $f = g/h$, $g, h \in k[X]$, and we see that $v_C(f) = 0$ if C is not a component of $V(g)$ or $V(h)$. Finally, in the general case, let $X = \bigcup U_i$ be a finite covering of X by affine open sets. Then every C intersects at least one U_i, so that $v_C(f) \neq 0$ only for those C that are closures of irreducible subvarieties $\tilde{C} \subset U_i$ such that $v_{\tilde{C}}(f) \neq 0$ in U_i. Since the numbers of U_i and of $\tilde{C}$ in any U_i are finite, so is the number of C with $v_C(f) \neq 0$. Thus, we can consider the divisor

$$\Sigma v_C(f)\, C, \tag{3}$$

where the sum extends over all irreducible subvarieties of codimension 1 for which $v_C(f) \neq 0$. This is called the *divisor of the function* f and is denoted by (f).

A divisor of the form $D = (f)$, $f \in k(X)$, is called *principal*. If $(f) = \Sigma l_i C_i$, then the divisors $(f)_0 = \sum_{i, l_i > 0} l_i C_i$ and $(f)_\infty = -\sum_{j, l_j < 0} l_j C_j$ are called *divisors of the zeros* and *of the poles* of f. Obviously, $(f)_0 \geqslant 0$, $(f)_\infty \geqslant 0$, $(f) = (f)_0 - (f)_\infty$. We draw attention to some simple properties:

$$(f_1 f_2) = (f_1) + (f_2), \quad (f) = 0 \quad \text{if} \quad f \in k; (f) \geqslant 0 \quad \text{if} \quad f \in k[X].$$

Let us show that for a smooth irreducible variety X the converse is also true: if $(f) \geqslant 0$, then f is a regular function on X. Let $x \in X$ be a point at which f is non-regular. Then $f = g/h$, $h, g \in \mathcal{O}_x$ while $f \notin \mathcal{O}_x$. From the unique prime factorization in $\mathcal{O}_x$ (Theorem 2 of Ch. II, § 3) it follows that h and g can be taken to be relatively prime in $\mathcal{O}_x$. Let π be a prime element of $\mathcal{O}_x$ that occurs in h but not in g. In some affine neighbourhood U of x the variety $V(\pi)$ is irreducible and of codimension 1. We denote its closure in X by C. Then clearly $v_C(f) < 0$. This result is also true when X is a normal variety, but we shall not prove this here.

Since on a projective irreducible variety X a function that is regular at all points is a constant (Corollary 1 to Theorem 2 of Ch. I, § 5), from the result we have just proved it follows that if $(f) \geqslant 0$, then $f = \alpha \in k$ on a smooth projective variety X. In particular, on a smooth projective irreducible variety a rational function is uniquely determined by its divisor to within a constant factor: if $(f) = (g)$, then $(f \cdot g^{-1}) = 0$ and $f = \alpha \cdot g$, $\alpha \in k$.

Example 1. $X = \mathbb{A}^n$. By Theorem 3 of Ch. I, § 6, any irreducible subvariety C of codimension 1 is given by a single equation: $\mathfrak{A}_C = (F)$, $F \in k[X]$. Hence $C = (F)$, that is, every prime divisor, and hence quite generally every divisor, is principal.

Example 2. $X = \mathbb{P}^n$. Every irreducible subvariety C of codimension 1 is given by a single homogeneous equation F, and $\mathfrak{a}_C = (T_i^{-l} F)$ in an affine open set U_i if the degree of F is l. This leads to the following method of constructing the divisor of a function $f \in k(\mathbb{P}^n)$: we represent f in the form $f = F/G$, where F and G are forms of the same degree, and we decompose these forms into products of irreducible forms: $F = \prod H_i^{l_i}$, $G = \prod L_j^{m_j}$; then

$$(f) = \sum l_i C_i - \sum m_j D_j, \tag{4}$$

where C_i and D_j are the irreducible divisors defined by the equations $H_i = 0$ and $L_j = 0$.

Let $\deg F$ denote the degree of the form F. Since $\deg F = \deg G$, we have $\Sigma l_i \deg H_i = \Sigma m_j \deg L_j$. We recall that in accordance with the definition of the degree of a projective variety in Ch. I, § 6.5, $\deg C_i$ is the same as $\deg H_i$. Therefore in (4) $\Sigma l_i \deg C_i = \Sigma m_j \deg D_j$. Defining the degree of the divisor $D = \Sigma l_i C_i$ as the number $\deg D = \Sigma l_i \deg C_i$, we have shown that if D is a principal divisor, then $\deg D = 0$.

It is easy to verify the converse: if $\Sigma l_i \deg C_i = 0$ and C_i is given by the equation H_i, where H_i is a form, then the function $f = \Pi H_i^{l_i}$ is homogeneous of degree 0 and $\Sigma l_i C_i = (f)$.

Example 3. The case $X = \mathbb{P}^{n_1} \times \ldots \times \mathbb{P}^{n_l}$ is analysed similarly. Again a subvariety C of codimension 1 is given by a single equation $H = 0$ (Theorem 3 of Ch. I, § 6); however, H is homogeneous in each group of coordinates of the spaces $\mathbb{P}^{n_i}$ and accordingly has l distinct degrees $\deg_i H$ $(i = 1, \ldots, l)$. Just as in Example 2, we introduce the degrees $\deg_i D$ of a divisor D on X, and D is principal if and only if $\deg_i D = 0$ $(i = 1, \ldots, l)$.

The principal divisors form a subgroup $P(X)$ of the group $\mathrm{Div}(X)$ of all divisors. The factor group $\mathrm{Div}(X)/P(X)$ is called *the group of divisor classes* and is denoted by $\mathrm{Cl}(X)$. Divisors belonging to one and the same coset in $\mathrm{Div}(X)/P(X)$ are called *equivalent*: $D_1 \sim D_2$ if $D_1 - D_2 = (f)$, $f \in k(X)$. The cosets in $\mathrm{Div}(X)/P(X)$ are called *divisor classes*. In the examples above we have

1. $\mathrm{Cl}(\mathbb{A}^n)=0$ 2. $\mathrm{Cl}(\mathbb{P}^n)=\mathbb{Z}$, 3. $\mathrm{Cl}(\mathbb{P}^{n_1} \times \ldots \times \mathbb{P}^{n_l})=\mathbb{Z}^l$.

2. Locally Principal Divisors. We assume the variety X to be smooth. In this case for every prime divisor $C \subset X$ and every point $x \in X$ there exists an open set U with $x \in U$ in which C is given by a local equation π. If D is any divisor, $D = \Sigma l_i C_i$, and if any of the C_i is given in U by the local equation π_i, then we have $D = (f)$, $f = \Pi \pi_i^{l_i}$. Thus, every point x has a neighbourhood in which D is a principal divisor. From all such neighbourhoods we can choose a finite covering $X = \bigcup U_i$, where in every U_i we have $D = (f_i)$.

Evidently the functions f_i cannot be chosen arbitrarily: the f_i are not identically zero, and in $U_i \cap U_j$ the divisors (f_i) and (f_j) coincide. As we have seen above, it follows that $f_i f_j^{-1}$ is a regular function in $U_i \cap U_j$ and does not vanish there. If a system of functions $\{f_i\}$ corresponding to the sets of the covering $\{U_i\}$ satisfies the conditions that $f_i f_j^{-1}$ is regular and does not vanish in $U_i \cap U_j$, then we call it compatible.

Conversely, every compatible system of functions determines a divisor on X. In fact, for every prime divisor C we set $l_C = v_C(f_i)$ if $U_i \cap C \neq \emptyset$, where f_i and C are regarded as a function and a prime divisor in the variety U_i. From the compatibility of the system of functions it follows that this number does not depend on the choice of U_i. Obviously there are only finitely many C such that $l_C \neq 0$, namely the closures of the irreducible components of the divisors (f_i). Therefore we can consider the divisor $D = \Sigma l_C C$. Obviously the given system of functions $\{f_i\}$ corresponds to it.

Finally, it is easy to clarify when two systems of functions $\{f_i\}$ and $\{g_j\}$ corresponding to coverings $\{U_i\}$ and $\{V_j\}$, respectively, give one and the same divisor. For this it is necessary and sufficient that in $U_i \cap V_j$ the functions $f_i g_j^{-1}$ are everywhere regular and do not vanish. The simple verification is left to the reader.

The specification of divisors by systems of functions enables us to investigate their behaviour under regular mappings. Let $\varphi: X \to Y$ be a regular mapping of smooth irreducible varieties, and let D be a divisor on Y. We assume that $\varphi(X) \not\subset \mathrm{Supp}\, D$. We show that under this restriction we can determine the inverse image $\varphi^*(D)$ of D by analogy with the determination of the inverse image of a regular function. First of all we clarify when the inverse image of a rational function f on Y can be constructed, and when it does not vanish identically on X. For this it is sufficient that there exists at least one point $y \in \varphi(X)$ at which f is regular and $f(y) \neq 0$. Then such points form a non-empty open set V. Now f is regular on V, and hence $\varphi^*(f)$ is a regular function on $\varphi^{-1}(V)$ that does not vanish identically (in fact, nowhere). Since $\varphi^{-1}(V)$ is open in X, we see that $\varphi^*(f)$ determines a rational function on X. In terms of divisors our condition on the mapping φ and the function f reduces to the fact that $\varphi(X) \not\subset \operatorname{Supp}(f)$.

Suppose now that the divisor D is given by a compatible system of functions $\{f_i\}$ and a covering $\{U_i\}$. We consider those U_i for which $\varphi(X) \cap U_i$ is not empty, and we show that $\varphi(X) \cap U_i \not\subset \operatorname{Supp}(f_i)$. For it follows from the irreducibility of X that $\overline{\varphi(X)}$ is irreducible in Y. If $\overline{\varphi(X)} \cap U_i \subset \operatorname{Supp}(f_i)$, then it follows from the irreducibility of $\overline{\varphi(X)}$ and the fact that $\varphi(X) \cap U_i$ is non-empty, that $\overline{\varphi(X)} \subset \operatorname{Supp}(f_i)$. Finally, the facts that $\operatorname{Supp}(f_i) \cap U_i = \operatorname{Supp} D \cap U_i$, that $\varphi(X)$ is irreducible, and that it intersects U_i imply that $\varphi(X) \subset \operatorname{Supp} D$, against the assumption.

Hence for all U_i that intersect $\varphi(X)$ the rational functions $\varphi^*(f_i)$ are defined in $\varphi^{-1}(U_i)$. The sets $\varphi^{-1}(U_i) = V_i$ for which $\varphi(X)$ intersects U_i are open and form a covering of X, and the functions $\varphi^*(f_i)$ form a compatible system, which determines some divisor on X. Obviously this divisor does not change when D is given by another system of functions. The divisor so obtained is called the inverse image of D and is denoted by $\varphi^*(D)$.

In particular, if $\varphi(X)$ is dense in Y, then the inverse image of any divisor $D \in \operatorname{Div}(Y)$ is defined.

If D and D' are two divisors on Y given by systems of functions of $\{f_i\}$ and $\{g_j\}$, corresponding to coverings $\{U_i\}$ and $\{V_j\}$, then the divisor $D + D'$ is given by the system of functions $\{f_i \cdot g_j\}$ and the covering $\{U_i \cap V_j\}$. From this it follows at once that $\varphi^*(D+D') = \varphi^*(D) + \varphi^*(D')$, so that if $\varphi(X)$ is dense in Y, then φ^* defines a homomorphism

$$\varphi^* : \operatorname{Div} Y \to \operatorname{Div} X .$$

The principal divisor (f) is given by the system of functions $f_i = f$, consequently $\varphi^*((f)) = (\varphi^*(f))$.

Therefore φ^* maps $P(Y)$ into $P(X)$ and defines a homomorphism $\varphi^* : \operatorname{Cl}(Y) \to \operatorname{Cl}(X)$.

As an application of the specification of divisors by compatible systems of functions we show how to associate a divisor not with a function, but with a form in the coordinates on a smooth projective variety. Let $X \subset \mathbb{P}^N$ and let F be a form in the coordinates in $\mathbb{P}^N$ that does not vanish identically on X. For every point $x \in X$ we consider a form G of the same degree as F such that $G(x) \neq 0$. Such forms exist: for example, if $x = (\alpha_0 : \ldots : \alpha_N)$ and $\alpha_i \neq 0$, we can take $G = T_i^{\deg F}$. Then $f = F/G$ is a rational function on X and is regular in the open set in which $G \neq 0$.

It is easy to see that there exist forms G_i such that the open sets $U_i = X - X_{G_i}$ form a covering of X. It is just as easy to verify that the functions $f_i = F/G_i$ and the open subsets U_i form a compatible system of functions and hence determine a divisor on X. Another choice of forms G_i does not change this divisor, which therefore depends only the form F. It is called the *divisor of* F and is denoted by (F). Since the f_i are regular in the sets U_i, we have $(F) \geqslant 0$. If F_1 is another form, $\deg F_1 = \deg F$, then $(F)-(F_1)$ is the divisor of the rational function F/F_1. Therefore $(F) \sim (F_1)$ if $\deg F = \deg F_1$.

In particular, all divisors (L), where L is a linear form, are equivalent to each other. Evidently $\operatorname{Supp}(L) = X_L$, the section of X by the hyperplane $L=0$. Therefore they are called *divisors of a hyperplane section.*

Taking above for F_1 the form $L^{\deg F}$ we obtain that $(F) \sim \deg F \cdot (L)$, where (L) is the divisor of a hyperplane section.

All the arguments connected with the specification of a divisor of a compatible system of functions can be generalized to arbitrary, not necessarily smooth, varieties. However, here the possibility of specification by a compatible system of functions must be taken as the definition of a divisor. The object at which we arrive in this way is called a *locally principal divisor.*

Strictly speaking, a locally principal divisor on an irreducible variety is a system of rational functions $\{f_i\}$ corresponding to the open sets of a covering $\{U_i\}$ and satisfying the conditions:

1) the f_i do not vanish identically and
2) $f_i f_j^{-1}$ and $f_j f_i^{-1}$ are regular on $U_i \cap U_j$.

Here two sets of functions $\{f_i\}$ and $\{g_j\}$ and coverings $\{U_i\}$ and $\{V_j\}$, respectively, determine the same divisor if $f_i g_j^{-1}$ and $f_i^{-1} g_j$ are regular in $U_i \cap V_j$.

Every function $f \in k(X)$ determines a locally principal divisor by setting $f_i = f$. Such divisors are called *principal.*

The product of two locally principal divisors given by functions $\{f_i\}$ and $\{g_j\}$ and coverings $\{U_i\}$ and $\{V_j\}$, respectively, is the divisor given by the functions $\{f_i g_j\}$ and the covering $\{U_i \cap V_j\}$. All locally principal divisors form a group, and the principal divisors a subgroup of it. The factor group is called the *Picard group* of the variety X and is denoted by $\operatorname{Pic}(X)$.

Every locally principal divisor has a *support*: this is the closed subvariety consisting of those points in U_i at which f_i is non-regular or zero. Just as for divisors on smooth varieties, so we can define the inverse image of a locally principal divisor D on Y under a regular mapping $\varphi: X \to Y$ if $\varphi(X)$ is not contained in $\operatorname{Supp} D$.

We mention one important special case. If X is a smooth variety and Y a subvariety, not necessarily smooth, then any divisor D on X for which $\operatorname{Supp} D \not\supset Y$ determines a locally principal divisor $\tilde{D}$ on Y. To see this we have to consider the embedding $\varphi: Y \to X$ and to set $\tilde{D} = \varphi^*(D)$. We call $\tilde{D}$ the *restriction* of D to Y and denote it by $\varrho_Y(D)$. From the definition it follows that for principal divisors $\varrho_Y((f)) = (\tilde{f})$, where $\tilde{f}$ is the restriction of f to Y.

Of course, the distinction between divisors and locally principal divisors appears only in the case of non-smooth varieties.

3. How to Shift the Support of a Divisor Away from Points

Theorem 1. *For every divisor D on a smooth variety X and finitely many points $x_1, \ldots, x_m \in X$ there exists a divisor D' such that $D' \sim D$, $x_i \notin \operatorname{Supp} D'$ $(i = 1, \ldots, m)$.*

We can take D to be a prime divisor, because otherwise it would be enough to apply the theorem to each of its components. In X we choose an affine open set containing the points $x_1, \ldots, x_m$. It is sufficient to prove the theorem for this set, so that we may assume X to be an affine variety. Using induction on m we may assume that $x_1, \ldots, x_i \notin \operatorname{Supp} D$, $x_{i+1} \in \operatorname{Supp} D$. It remains to construct a divisor D' such that $D' \sim D$, $x_1, \ldots, x_{i+1} \notin \operatorname{Supp} D'$. We consider some local equation π' of the prime divisor D in a neighbourhood of x_{i+1}. Let us show that π' can be chosen so that $\pi' \in k[X]$ (by assumption, X is affine). Indeed, π' is regular at x_{i+1}, and hence, if $(\pi')_\infty = \Sigma k_l F_l$, then $x_{i+1} \notin F_l$. Hence for every l there exists a function $f_l \in k[X]$ vanishing on F_l and such that $f_l(x_{i+1}) \neq 0$. Evidently the function $\pi = \pi' \prod f_l^{k_l}$ is regular on X and is a local equation of D in a neighbourhood of x_{i+1}. Since by hypothesis $x_j \notin \operatorname{Supp} D \cup x_1 \cup \ldots \cup x_{j-1} \cup x_{j+1} \cup \ldots \cup x_i$ $(j = 1, \ldots, i)$, for every $j = 1, \ldots, i$ there exists a function $g_j \in k[X]$ such that $g_j|_D = 0$, $g_j(x_l) = 0$ $(l = 1, \ldots, j-1, j+1, \ldots, i)$, $g_j(x_j) \neq 0$.

We consider the function

$$f = \pi + \sum_{j=1}^{i} \alpha_j g_j^2, \qquad \alpha_j \in k,$$

and choose the constants α_j so that

$$f(x_j) \neq 0 (j = 1, \ldots, i). \tag{1}$$

It is sufficient to take $\alpha_j \neq -\pi(x_j)/g_j(x_j)^2$. Since all the $g_{j|D} = 0$, in the local ring $\mathcal{O}_{x_{i+1}}$ we have $g_j \equiv 0(\pi)$ and $\Sigma \alpha_j g_j^2 = \pi^2 h$, $h \in \mathcal{O}_{x_{i+1}}$, $f = \pi(1 + \pi h)$. Since $(1 + \pi h)(x_{i+1}) = 1$, it follows that f is a local equation of D in a neighbourhood of x_{i+1}. Therefore $(f) = D + \Sigma r_s D_s$, and none of the prime divisors of D_s passes through x_{i+1}. This means that if we set

$D' = D - (f)$, then $x_{i+1} \notin \operatorname{Supp} D'$. Furthermore, (1) shows that $x_j \notin \operatorname{Supp}(f)$ ($j = 1, \ldots, i$), therefore the divisor D' satisfies the conditions of the theorem.

Here is a first application of Theorem 1. In § 1.2 we have defined the inverse image $f^*(D)$ of a divisor D of a variety X under a regular mapping $f\colon Y \to X$, provided that $f(Y) \not\subset \operatorname{Supp} D$. Theorem 1 enables us to replace D by an equivalent divisor D' for which $x \notin \operatorname{Supp} D'$, where x is an arbitrarily chosen point in $f(Y)$. Then automatically $f(Y) \not\subset \operatorname{Supp} D'$, and the inverse image $f^*(D')$ is defined. This shows that without any restrictions on the regular mapping f we can define the inverse image of a divisor class $C \in \operatorname{Cl}(X)$. For this purpose we have to choose in C a divisor D such that $f(Y) \not\subset \operatorname{Supp} D$ and consider the class on Y containing the divisor $f^*(D)$. It is easy to verify that in this way we obtain a homomorphism

$$f^* : \operatorname{Cl}(X) \to \operatorname{Cl}(Y).$$

In other words, $\operatorname{Cl}(X)$ is a functor from the category of irreducible smooth algebraic varieties into the category of Abelian groups.

4. Divisors and Rational Mappings. The correspondence between functions and divisors is useful for the investigation of rational mappings of varieties into a projective space. Let X be a smooth variety and $\varphi : X \to \mathbb{P}^n$ a rational mapping. We wish to find out at what points φ is non-regular.

A rational mapping is given by formulae

$$\varphi = (f_0 : \ldots : f_n), \qquad f_i \in k(X), \tag{1}$$

where we may assume that none of the functions f_i vanishes identically on X. Let

$$(f_i) = \sum_{j=1}^{m} k_{ij} C_j,$$

where the C_j are prime divisors. Here we allow some k_{ij} to be zero.

To clarify whether φ is regular at a point $x \in X$ we specify C_j by a local equation π_j at x. Then

$$f_i = \Bigl(\prod_j \pi_j^{k_{ij}}\Bigr) u_i, \qquad u_i \in \mathcal{O}_x, \qquad u_i(x) \neq 0.$$

By the unique prime factorization in $\mathcal{O}_x$ there exists a greatest common divisor d of the elements $f_0, \ldots, f_n$, that is, an element $d \in k(X)$ such that $f_i d^{-1} \in \mathcal{O}_x$ and if $d_1 \in k(X)$ is such that $f_i d_1^{-1} \in \mathcal{O}_x$, then $d_1 \mid d$, that is, $d d_1^{-1} \in \mathcal{O}_x$.

Since local equations of irreducible varieties are prime elements of $\mathcal{O}_x$, we have

$$d = \Pi \pi_j^{k_j}, \quad k_j = \min_{i=0, \ldots, n} k_{ij}.$$

The mapping φ is regular at x if there exists a function $g \in k(X)$, such that $f_i g^{-1} \in \mathcal{O}_x$ $(i=0, \ldots, n)$, and the $(f_i g^{-1})(x)$ are not all zero. By the definition of the greatest common divisor it follows that $g|d$. If $d = g \cdot h$, $h \in \mathcal{O}_x$, and $h(x) = 0$, then $h|(f_i g^{-1})$, hence all the $(f_i g^{-1})(x) = 0$. Thus, only a function g for which $d = g \cdot h$, $h(x) \neq 0$, can satisfy the necessary conditions. Then $f_i g^{-1} = (f_i d^{-1})\, h$, that is,

$$f_i g^{-1} = \left(\prod_j \pi_j^{k_{ij} - k_j}\right)(u_i h),$$

and φ is a regular mapping if and only if not all the functions $\prod\limits_j \pi_j^{k_{ij} - k_j}$ vanish at x.

To express this result in the language of divisors we define the g.c.d. of divisors $D_i = \Sigma k_{ij} C_j$ $(i = 1, \ldots, n)$ as the divisor

$$\text{g.c.d.}\,(D_1, \ldots, D_n) = \Sigma k_j C_j, \quad k_j = \min_{i=1, \ldots, n} k_{ij}.$$

Obviously $D_i' = D_i - \text{g.c.d.}(D_1, \ldots, D_n) \geqslant 0$, and the divisors D_i' have no common components. In particular, let us set

$$D = \text{g.c.d.}((f_0), \ldots, (f_n)), \quad D_i' = (f_i) - D.$$

Then in some neighbourhood of x

$$\left(\prod_j \pi_j^{k_{ij} - k_j}\right) = D_i',$$

and we can say that the mapping φ is regular at x if and only if not all the varieties $\operatorname{Supp} D_i'$ pass through this point.

So we have proved the following result.

Theorem 2. *The rational mapping* (1) *is non-regular precisely at the points of the set*

$$\bigcap \operatorname{Supp} D_i', \quad D_i' = (f_i) - \text{g.c.d.}((f_0), \ldots, (f_n)) \quad (i = 0, \ldots, n).$$

Since the divisors D_i' do not have common irreducible components, the set $\bigcap \operatorname{Supp} D_i'$ is of codimension $\geqslant 2$. Thus, Theorem 2 is a sharper form of Theorem 3 in Ch. II, § 3.

Remark. The divisors D_i' can be interpreted as the inverse images of the hyperplanes $x_i = 0$ under the mapping $\varphi: X \to \mathbb{P}^n$. For if $x \notin \bigcap \operatorname{Supp} D_i'$ and $D = (h)$ in a neighbourhood of x, then in the same neighbourhood a regular mapping is given by the formulae:

$$\varphi = (f_0/h : \ldots : f_n/h).$$

The inverse image of the hyperplane $x_i=0$ has the local equation f_i/h, hence coincides with D_i'.

More generally, if $\lambda=(\lambda_0:\ldots:\lambda_n)$ and $E_\lambda\subset\mathbb{P}^n$ is the hyperplane $\Sigma\,\lambda_i x_i=0$, then

$$\varphi^*(E_\lambda)=(\Sigma\,\lambda_i f_i)-D\,.$$

5. The Space Associated with a Divisor. The fact that all polynomials $f(t)$ of degree $\leqslant n$ form a finite-dimensional vector space can be interpreted in terms of divisors in the following way. We denote by x_∞ the point at infinity on the projective line $\mathbb{P}^1$ with the coordinate t. A polynomial in t of degree l has a pole of order l at x_∞ and has no other poles. Therefore the condition $\deg f\leqslant n$ can be expressed as follows: the divisor $(f)+nx_\infty$ is effective.

By analogy, for an arbitrary divisor D on a smooth variety X we can consider the set consisting of zero and of those functions $f\in k(X)$, $f\neq 0$, for which

$$(f)+D\geqslant 0\,. \tag{1}$$

This is a linear space over k under the usual operations on functions. For if $D=\Sigma\,n_i C_i$, then (1) is equivalent to the fact that

$$v_{C_i}(f)\geqslant -n_i,\quad v_C(f)\geqslant 0\quad\text{for}\quad C\neq C_i$$

and by virtue of this our assertion follows immediately from the formulae in § 1.1.

The space of functions satisfying the conditions (1) is called the *space associated with the divisor* D and is denoted by $\mathscr{L}(D)$.

Just as polynomials of degree $\leqslant n$ form a finite-dimensional space, so the space $\mathscr{L}(D)$ is finite-dimensional if D is an arbitrary divisor and X a projective variety.

In § 2 this theorem will be proved for the case of algebraic curves. By an induction on the dimension it can then be proved in the general case without any particular difficulty. However, the place of this theorem becomes more intelligible if it is obtained as a special case of a vastly more general proposition on coherent sheaves. In this form it will be proved in Ch. VI, § 3.

The dimension of the space $\mathscr{L}(D)$ is also called the *dimension of the divisor* D and is denoted by $l(D)$.

Theorem 3. *Equivalent divisors have equal dimensions.*

Let $D_1\sim D_2$; this means that $D_1-D_2=(g)$, $g\in k(X)$. If $f\in\mathscr{L}(D_1)$, then $(f)+D_1\geqslant 0$. From this it follows that $(f\cdot g)+D_2=f+D_1\geqslant 0$, that is, $f\cdot g\in\mathscr{L}(D_2)$, $g\cdot\mathscr{L}(D_1)=\mathscr{L}(D_2)$. Thus, multiplication of all functions $f\in\mathscr{L}(D_1)$ by a function g determines an isomorphism of the spaces $\mathscr{L}(D_1)$ and $\mathscr{L}(D_2)$, and the theorem follows.

So we see that we can talk of the dimension $l(C)$ of a divisor class C, understanding by this the common dimension of all the divisors in this class. This number has the following meaning. If $D \in C$, $f \in \mathscr{L}(D)$, then the divisor $D_f = (f) + D$ is effective. Clearly $D_f \sim D$, therefore $D_f \in C$. Conversely, every effective divisor $D' \in C$ is of the form D_f, where $f \in \mathscr{L}(D)$. Obviously, if X is projective, then the function f is determined by the divisor D_f uniquely to within a constant factor. Thus, we can set up a one-to-one correspondence between the effective divisors of the class C and the points of the $(l(C)-1)$-dimensional projective space $\mathbb{P}(\mathscr{L}(D))$ corresponding to D. (We recall that the projective space $\mathbb{P}(L)$ corresponding to a vector space L consists of all lines of L).

The space $\mathscr{L}(D)$ is useful in specifying rational mappings by divisors, as this was described in § 1.4. If

$$\varphi = (f_0 : \ldots : f_n) : X \to \mathbb{P}^n \tag{2}$$

is a rational mapping and, as in § 1.4,

$$D = \text{g.c.d.}((f_0), \ldots, (f_n)), \qquad D_i = (f_i) - D, \tag{3}$$

then $D_i \geqslant 0$, hence all the $f_i \in \mathscr{L}(-D)$.

The choice of the functions f_i depends on the chosen system of projective coordinates in $\mathbb{P}^n$. Therefore, to the mapping φ there corresponds, in an invariant fashion, the totality of all the functions $\sum_{i=0}^{n} \lambda_i f_i$ that are linear combinations of the f_i. These functions form a linear subspace $M \subset \mathscr{L}(-D)$. In what follows we assume that $\varphi(X)$ is not contained in any proper linear subspace of $\mathbb{P}^n$. Then $\Sigma \lambda_i f_i \neq 0$ on X if not all the $\lambda_i = 0$. The set of effective divisors corresponding to this set of functions, that is, the divisors $(g) - D$, $g \in M$, is called a *linear system* of divisors. If $M = \mathscr{L}(-D)$, then the linear system is called complete. The meaning of the divisors $(f) - D$, $f \in M$, is very simple: they are the inverse images of the divisors of the hyperplanes in $\mathbb{P}^n$ under φ. Thus, we can construct all rational mappings of a given smooth variety X into various projective spaces. For this purpose we must take an arbitrary divisor D, and in the space $\mathscr{L}(-D)$ a linear finite-dimensional subspace M. If $f_0, \ldots, f_n$ is a basis of it, then the formula (2) gives the required mapping. Observe that the divisors D_i for these $f_i \in \mathscr{L}(-D)$ have an additional property: they have no common components.

Since multiplication of all functions f_i by a common factor $g \in k(X)$ does not change the mapping φ, but a divisor D is changed into an equivalent divisor $(g) + D$, the class of the divisor D is an invariant of the rational mapping. Thus, we have the following method of constructing all those rational mappings φ of a variety X into a projective space $\mathbb{P}^m$ for which $\varphi(X)$ is not contained in any proper subspace of $\mathbb{P}^m$: we choose an arbitrary divisor class on X, and for every divisor D of this class we choose in $\mathscr{L}(-D)$ a linear finite-dimensional subspace M such that the effective divisors $(f)-D$ have no common components. If $f_0, \ldots, f_n$ is a basis of M, then our mapping is given by (2). Of course, it can happen that $\mathscr{L}(-D)=0$ or that all the divisors $(f)-D$, $f \in \mathscr{L}(-D)$ have a common component; then this divisor class does not lead to such a mapping.

We draw attention to one interesting property of this situation. Among all rational mappings corresponding to a given class C there exists a maximal one: this is obtained by taking for M the whole space $\mathscr{L}(-D)$, $D \in C$. (Here we rely on the unproved theorem that the space $\mathscr{L}(-D)$ is finite-dimensional.)

All other mappings corresponding to this class are obtained by constructing the compositum of this mapping with various projection mappings. For if $\varphi=(f_0:\ldots:f_N)$ and, say, $\psi=(f_0:\ldots:f_n)$, $n<N$, then $\psi=\pi\varphi$, where $\pi(x_0:\ldots:x_N)=(x_0:\ldots:x_n)$ is a projection, which we now regard as a rational mapping.

Let us see how this scheme works if we take for X the projective space $\mathbb{P}^m$. We know that $\mathrm{Cl}(\mathbb{P}^m)\simeq\mathbb{Z}$ and that the class C_l corresponding to an integer l consists of the divisors of degree l.

Clearly, if $l>0, D \in C_l$, then $\mathscr{L}(-D)=0$. If $l \leqslant 0$, then we can take for $-D$ the divisor $-lE$, where E is the divisor of the hyperplane at infinity $x_0=0$. In this case $\mathscr{L}(-lE)$ consists of polynomials of degree $\leqslant -l$ in inhomogeneous coordinates $x_1/x_0, \ldots, x_m/x_0$ (see Exercise 15). By multiplying the resulting formulae for the mapping by x_0^l, we obtain a Veronese mapping $v_l: \mathbb{P}^m \to \mathbb{P}^{\nu_{l,m}}$. So we see that every rational mapping of $\mathbb{P}^m$ can be obtained by combining a Veronese mapping with a projection.

Exercises

1. Determine the divisor of the function x/y on the quadric $xy-zt=0$ in $\mathbb{P}^3$.

2. Determine the divisor of the function $x-1$ on the circle $x_1^2+x_2^2=x_0^2$, $x=x_1/x_0$.

3. Determine the inverse image $f^*(D_a)$, where $f(x,y)=x$ is the projection of the circle of $x^2+y^2=1$ onto the x-axis, and D_a is the divisor on the line $\mathbb{A}^1$, $D_a=p$, $p \in \mathbb{A}^1$, with the coordinate a.

4. X is a smooth projective curve, $f \in k(X)$. Regarding f as a regular mapping $f: X \to \mathbb{P}^1$, prove that $(f)=f^*(D)$, where $D=0-\infty$ is a divisor on $\mathbb{P}^1$.

5. X is a smooth affine variety. Show that $\mathrm{Cl}(X)=0$ if and only if factorization in $k[X]$ is unique.

6. X is a smooth projective variety, $X \subset \mathbb{P}^N$, $k[S]$ is the ring of polynomials in inhomogeneous coordinates in $\mathbb{P}^N$, and $\mathfrak{A}_X \subset k[S]$ is an ideal of X. Prove that if in the ring

$k[S]/\mathfrak{A}_X$ factorization is unique, then $\mathrm{Cl}(X)=\mathbb{Z}$, and a generator is the class of hyperplane sections.

7. Find $\mathrm{Cl}(\mathbb{P}^n \times \mathbb{A}^n)$.

8. The projection $p: X \times \mathbb{A}^1 \to X$ determines a homomorphism $p^*: \mathrm{Cl}(X) \to \mathrm{Cl}(X \times \mathbb{A}^1)$. Show that p^* is an epimorphism. Hint: Use the mapping $q^*: \mathrm{Cl}(X \times \mathbb{A}^1) \to \mathrm{Cl}(X)$, where $q: X \to X \times \mathbb{A}^1$ is given by $q(x) = x \times 0$.

9. Show that for every divisor on $X \times \mathbb{A}^1$ there exists an open set $U \subset X$ such that on $U \times \mathbb{A}^1$ this divisor is principal. Hint: X can be regarded as affine, and the divisor as irreducible. Then it is given by a prime ideal in $k[X \times \mathbb{A}^1] = k[X][T]$. Use the fact that in $k(X)[T]$ all ideals are principal, then replace X by some principal affine open subset.

10. Show that $\mathrm{Cl}(X \times \mathbb{A}^1) \simeq \mathrm{Cl}(X)$. Use the results of Exercises 8 and 9.

11. Let X be the projective curve given by the equation $y^2 = x^2 + x^3$ in affine coordinates. Show that every locally principal divisor X is equivalent to a divisor whose support does not contain the point $(0, 0)$. Use this and the normalization mapping $\varphi: \mathbb{P}^1 \to X$ for which $\varphi^{-1}(0, 0)$ consists of two points x_1 and $x_2 \in \mathbb{P}^1$ to describe $\mathrm{Pic}(X)$ as D/P, where D is the group of all divisors on $\mathbb{P}^1$ whose supports do not contain x_1 and x_2, and P is the group of those principal divisors (f) for which f is regular at x_1 and x_2 and $f(x_1) = f(x_2) \neq 0$. Show that $\mathrm{Pic}(X)$ is isomorphic to the multiplicative group of non-zero elements of k.

12. Find $\mathrm{Pic}(X)$, where X is the curve with the equation $y^2 = x^3$.

13. Let X be a quadric cone. Use the mapping $\varphi: \mathbb{A}^2 \to X$ described in Exercise 2 to Ch. II, § 5, to determine the image $\varphi^*(\mathrm{Div}(X))$ in $\mathrm{Div}\,\mathbb{A}^2$. Show that $D = (F) \in \mathrm{Div}\,\mathbb{A}^2$ belongs to $\varphi^*(\mathrm{Div}(X))$ if and only if $F(-u, -v) = \pm F(u, v)$, that is, F is either an even or an odd function. Show that principal divisors on X correspond to even functions. Show also that $\mathrm{Cl}(X) \simeq \mathbb{Z}/2\mathbb{Z}$.

14. Using Theorem 2 determine the points at which the birational mapping $\varphi: X \to \mathbb{P}^2$ is non-regular, where X is a quadric in $\mathbb{P}^3$, and φ the projection from a point $x \in X$. The same for φ^{-1}.

15. Show that if E is the hyperplane $x_0 = 0$ in $\mathbb{P}^n$, then the space $\mathscr{L}(lE)$ consists of the polynomials in inhomogeneous coordinates $x_1/x_0, \ldots, x_n/x_0$ of degree $\leqslant l$. Hint: Use the fact that if $f \in \mathscr{L}(lE)$, then $f \in k[\mathbb{A}^n_0]$.

16. Show that every automorphism of $\mathbb{P}^n$ carries divisors of hyperplanes into each other. Hint: The class of hyperplanes is determined by invariant properties in $\mathrm{Cl}(\mathbb{P}^n)$, and the divisors of hyperplanes as effective divisors in it.

17. Show that every automorphism of the variety $\mathbb{P}^n$ is a projective transformation. Hint: Use the result of Exercise 16.

18. Let $\sigma: X \to Y$ be the σ-process centred at $y \in Y$, where Y is smooth. Show that $\mathrm{Cl}(X) \simeq \mathrm{Cl}(Y) \oplus \mathbb{Z}$.

§ 2. Divisors on Curves

1. The Degree of a Divisor on a Curve. We consider a projective smooth curve X. A divisor on X is a linear combination of points $D = \Sigma k_i x_i$, $k_i \in \mathbb{Z}$, $x_i \in X$. The *degree of the divisor* D is defined as the number $\deg D = \Sigma k_i$.

Example 2 of § 1.1 for $n = 1$ shows that on $X = \mathbb{P}^1$ a divisor D is principal if and only if $\deg D = 0$. We now show that $\deg D = 0$ for a principal divisor on any smooth projective curve. For this purpose we make use of the concept of the degree of a mapping f, $\deg f$, which was introduced in Ch. II, § 5.3.

Theorem 1. *If $f: X \to Y$ is a regular mapping of smooth projective curves and $f(X) = Y$, then* $\deg f = \deg f^*(y)$ *for every point $y \in Y$.*

In Theorem 1 $f^*(y)$ is the divisor on X that is the inverse image of the divisor on Y consisting of the point y with the coefficient 1. Thus, $\deg f$ is equal to the number of inverse images of any point $y \in Y$ (taken with appropriate multiplicities). This makes the intuitive meaning of the degree of f clearer—it shows how many times X covers Y under f.

Corollary. *The degree of a principal divisor on a smooth projective curve X is equal to zero.*

For every non-constant function $f \in k(X)$ determines a regular mapping $f: X \to \mathbb{P}^1$. Here $f^*(0) = (f)_0$ for $0 \in \mathbb{P}^1$—this follows at once from the definition of the two divisors. Similarly $f^*(\infty) = (f)_\infty$. By Theorem 1,

$$\deg(f) = \deg(f)_0 - \deg(f)_\infty = \deg f^*(0) - \deg f^*(\infty) = \deg f - \deg f = 0.$$

If X and Y are two varieties of the same dimension and if f is a regular mapping $f: X \to Y$ such that $f(X)$ is dense in Y, then it determines an embedding $f^*: k(Y) \to k(X)$; utilizing this we shall henceforth regard $k(Y)$ as a subfield of $k(X)$ (that is, for $u \subset k(Y)$ we write u instead of $f^*(u)$ when this cannot lead to misunderstandings).

Theorem 1 follows from two results. To state them we introduce the following notation. Let $x_1, \ldots, x_r$ be points on the curve X. We set

$$\tilde{\mathcal{O}} = \bigcap_{i=1,\ldots,r} \mathcal{O}_{x_i}. \tag{1}$$

Thus, $\tilde{\mathcal{O}}$ consists of the functions that are regular at all the points $x_1, \ldots, x_r$. If $\{x_1, \ldots, x_r\} = f^{-1}(y)$, $y \in Y$, then the ring $\mathcal{O}_y$, which we have agreed above to regard as a subring of $k(X)$, is contained in $\tilde{\mathcal{O}}$.

Theorem 2. *$\tilde{\mathcal{O}}$ is a principal ideal ring with finitely many prime ideals. There exist elements $t_i \in \tilde{\mathcal{O}}$ such that*

$$v_{x_i}(t_j) = \delta_{ij}, \qquad 1 \leqslant i, j \leqslant r. \tag{2}$$

If $u \in \tilde{\mathcal{O}}$, then

$$u = t_1^{l_1} \cdots t_r^{l_r} v, \tag{3}$$

where $l_i = v_{x_i}(u)$, and v is invertible in $\tilde{\mathcal{O}}$.

Theorem 3. *If $\{x_1, \ldots, x_r\} = f^{-1}(y)$, then $\tilde{\mathcal{O}}$ is a free module over $\mathcal{O}_y$ and $\tilde{\mathcal{O}} \simeq \mathcal{O}_y^n$, where $n = \deg f$.*

Let us first show how Theorem 1 follows from Theorems 2 and 3. Let t be a local parameter at y, and $\{x_1, \ldots, x_r\} = f^{-1}(y)$. According to Theorem 2, $t = t_1^{l_1} \cdots t_r^{l_r} v$, where $l_i = v_{x_i}(t)$. Recalling the definition of the inverse image of a divisor we see that

$$f^*(y)=\Sigma l_i x_i \quad \text{and} \quad \deg f^*(y)=\sum_{i=1}^{r} l_i .$$

Since the elements $t_1, \ldots, t_r$ are pairwise coprime in $\tilde{\mathcal{O}}$, we have

$$\tilde{\mathcal{O}}/(t) \simeq \bigoplus_{i=1}^{r} \tilde{\mathcal{O}}/(t_i^{l_i}) .$$

It is easy to see that every element $w \in \tilde{\mathcal{O}}$ has a unique representation in the form

$$w \equiv \alpha_0 + \alpha_1 t_i + \cdots + \alpha_{l_i-1} t_i^{l_i-1} \pmod{t_i^{l_i}}, \qquad \alpha_i \in k . \tag{4}$$

For if we have already got the representation

$$w \equiv \alpha_0 + \alpha_1 t_i + \cdots + \alpha_{s-1} t_i^{s-1} \pmod{t_i^s},$$

then

$$v = t_i^{-s}(w - \alpha_0 - \cdots - \alpha_{s-1} t_i^{s-1}) \in \tilde{\mathcal{O}} \subset \mathcal{O}_{x_i} .$$

We set $v(x_i)=\alpha_s$. Then $v_{x_i}(v-\alpha_s)>0$, and from Theorem 2 it follows that $v \equiv \alpha_s \pmod{t_i}$, that is,

$$w \equiv \alpha_0 + \alpha_1 t_i + \cdots + \alpha_{s-1} t^{s-1} + \alpha_s t_i^s \pmod{t_i^{s+1}} .$$

This proves (4) by induction.

From the representation (4) it follows that $\dim \tilde{\mathcal{O}}/(t_i^{l_i}) = l_i$. Therefore

$$\dim \tilde{\mathcal{O}}/(t) = \sum_{i=1}^{r} l_i . \tag{5}$$

When we now apply Theorem 3, it follows that $\tilde{\mathcal{O}}/(t) \simeq (\mathcal{O}_y/(t))^n$. But t is a local parameter at y, therefore

$$\mathcal{O}_y/(t) \simeq k , \qquad \dim \tilde{\mathcal{O}}/(t) = n = \deg f . \tag{6}$$

Now (5) and (6) prove Theorem 1.

Proof of Theorem 2. We denote by u_i a local parameter at x_i. Then x_i occurs in the divisor (u_i) with the coefficient 1, that is, $(u_i)=x_i+D$, where x_i does not occur in D. By Theorem 1 of § 1 we can shift the support of D away from $x_1, \ldots, x_r$, that is, we can find a function f_i such that these points do not occur in $D+(f_i)$. This means that for $t_i = u_i f_i$ the relations (2) hold. Let $u \in \tilde{\mathcal{O}}$. We set $v_{x_i}(u)=l_i$. By hypothesis, $l_i \geqslant 0$. For the element $v = u t_1^{-l_1} \cdots t_r^{-l_r}$ we have $v_{x_i}(v)=0$ for all $i=1, \ldots, r$, from which it follows that $v \in \tilde{\mathcal{O}}$ and $v^{-1} \in \tilde{\mathcal{O}}$. So we obtain a representation (3) for u.

It remains to verify that $\tilde{\mathcal{O}}$ is a principal ideal ring. Let $\mathfrak{a}$ be an ideal of $\tilde{\mathcal{O}}$. We set $l_i = \inf_{u \in \mathfrak{a}} v_{x_i}(u)$ and $a = t_1^{l_1} \ldots t_r^{l_r}$. Then $ua^{-1} \in \tilde{\mathcal{O}}$, that is, $\mathfrak{a} \subset (a)$.

Let us show that $\mathfrak{a}=(a)$. To do this we denote by $\mathfrak{a}'$ the set of functions ua^{-1}, $u\in\mathfrak{a}$. Evidently $\mathfrak{a}'$ is an ideal of $\tilde{\mathcal{O}}$ and $\inf_{u\in\mathfrak{a}'} v_{x_i}(u)=0$. Hence for every $i=1,\dots,r$ there exists a $u_i\in\mathfrak{a}'$ for which $v_{x_i}(u_i)=0$, that is, $u_i(x_i)\neq 0$. An obvious verification shows that $v_{x_i}(c)=0$ $(i=1,\dots,r)$ for the element $c=\sum_{j=1}^{r} u_j t_1\dots\hat{t}_j\dots t_r\in\mathfrak{a}'$ (the symbol $\hat{t}_j$ indicates that the corresponding factor is absent). This means that $c^{-1}\in\tilde{\mathcal{O}}$, therefore $\mathfrak{a}'=\tilde{\mathcal{O}}$, $\mathfrak{a}=(a)$. This proves the theorem.

Now we turn to the proof of Theorem 3. First of all, we show that $\tilde{\mathcal{O}}$ is a module of finite type over $\mathcal{O}_y$. For this purpose we recall that according to Theorem 11 of Ch. II, § 5, the mapping f is finite. Therefore the point y has an affine neighbourhood V such that the curve $U=f^{-1}(V)$ is also affine and that the ring $A=k[U]$ is a module of finite type over $B=k[V]$. As always, the embedding $B\subset A$ is effected by the mapping f^*.

Lemma. *In the previous notation $\tilde{\mathcal{O}}=A\mathcal{O}_y$, even if Y is not a normal curve.*

For if $\varphi\in\tilde{\mathcal{O}}$ and z_i are the poles of φ on U, then $f(z_i)=y_i\neq y$. There exists a function $h\in B$ such that $h(y)\neq 0$, $h(y_i)=0$, and $\varphi h\in\mathcal{O}_{z_i}$, hence $\varphi h\in A$. Since $h^{-1}\in\mathcal{O}_y$, we have $\varphi\in A\mathcal{O}_y$. So we have shown that $\tilde{\mathcal{O}}\subset A\mathcal{O}_y$. The reverse inclusion is obvious, and the lemma is proved.

Now we can complete the proof of Theorem 3. Clearly, generators of the module A over $k[V]$ are at the same time generators of $A\mathcal{O}_y$ over $\mathcal{O}_y$. Therefore $\tilde{\mathcal{O}}$ is a module of finite type. By the main theorem on modules over a principal ideal ring, $\tilde{\mathcal{O}}$ is a direct sum of a free module and a torsion module. However, $\mathcal{O}_y$ and $\tilde{\mathcal{O}}$ are contained in the field $k(X)$, from which it follows that this torsion module is zero and that $\tilde{\mathcal{O}}\simeq\mathcal{O}_y^m$ for some m.

It remains to determine m, that is, the rank of $\tilde{\mathcal{O}}$. It is equal to the maximum number of linearly independent elements over $\mathcal{O}_y$ contained in $\tilde{\mathcal{O}}$. Since linear independence over a ring and over its field of fractions is one and the same thing, and since the field of fractions of $\mathcal{O}_y$ is $k(Y)$, we see that m is equal to the maximal number of linearly independent elements of $\tilde{\mathcal{O}}$ over $k(Y)$.

By hypothesis, $[k(X):k(Y)]=n$, so that necessarily $m\leqslant n$. It remains to show that $\tilde{\mathcal{O}}$ contains n linearly independent elements relative to $k(Y)$. Let $\alpha_1,\dots,\alpha_n$ be a basis of the extension $k(X)/k(Y)$. We denote by l the maximum order of the poles of the functions α_i at the points x_j, and by t a local parameter of y. Evidently the functions $\alpha_i t^l$ are regular at these points, hence are contained in $\tilde{\mathcal{O}}$. Consequently, they are linearly independent over $k[Y]$. This completes the proof of the theorem.

2. Bezout's Theorem on Curves. Here we give the simplest applications of the theorem on the degree of a principal divisor. They are very special cases of more general theorems, which we shall prove in connection with the theory of intersection indices. However, it is convenient to give an account of these simple cases now, because they will be useful for us in the next subsection.

Let X be a smooth projective curve, $X \subset \mathbb{P}^n$, F a form in the point coordinates of $\mathbb{P}^n$ that is not identically zero on X and x a point on X.

In § 1.2 we have introduced the divisor (F) of F on X. The degree $\deg(F)$ of this divisor is also denoted by (X, F) and is called the intersection index of X with the hypersurface $\mathbb{P}^n_F$.

Theorem 1 leads at once to an important consequence: this number is one and the same for all forms of the same degree.

For if $\deg F = \deg F_1$, then $f = F/F_1 \in k(X)$. From the definition of the divisor (F) it follows at once that $(F) = (F_1) + (f)$, hence $(F) \sim (F_1)$. By the corollary to Theorem 1, $\deg(F) = \deg(F_1)$.

To find out how the number (X, F) depends on the degree of the form F it is sufficient to take for F any form of degree $m = \deg F$. In particular, we may set $F = L^m$, where L is a linear form. Then

$$(X, F) = m(X, L) = (\deg F)(X, L). \tag{1}$$

Finally, we explain the meaning of the number (X, L). In Ch. I we have introduced the concept of the degree $\deg X$ of a curve X as the maximum number of points of intersection of X with a hyperplane not containing X.

Since $(X, L) = \sum_{L(x)=0} v_x(L)$, we have $\deg X \leqq (X, L)$.

Let us find out when $v_x(F) = 1$ for the case of an arbitrary form F. By virtue of the additivity of the function $v_x(F)$ it is sufficient to consider the case of an irreducible form.

Lemma. *Let $X \subset \mathbb{P}^n$, F an irreducible form, and $Y = \mathbb{P}^n_F$. The equality $v_x((F)) = 1$ is equivalent to the fact that $F(x) = 0$, and $\Theta_{x,Y} \not\supset \Theta_{x,X}$. Both these spaces are regarded as subspaces of $\Theta_{x,\mathbb{P}^n}$.*

The proof comes from a comparison of some definitions in Ch. II. Let G be a form for which $G(x) \neq 0$, $\deg G = \deg F$. By definition, $v_x(F) = v_x(f)$, where $f = (F/G)|_X$. We know that $v_x(f) > 1$ is equivalent to the fact that $f \in \mathfrak{m}_x^2$, or, what is the same, $d_x f = 0$. But $d_x f \in \Theta^*_{x,X}$ is also the restriction to $\Theta_{x,X}$ of the differential $d_x(F/G)$ of the function F/G, which is rational on $\mathbb{P}^n$ and regular at x. Thus, $v_x(F) > 1$ is equivalent to $d_x(F/G) = 0$ on $\Theta_{x,X}$. Furthermore, F/G is a local equation of Y in a neighbourhood of x in which $G \neq 0$. Therefore $d_x(F/G) = 0$ is the equation of $\Theta_{x,Y}$ and $d_x(F/G) = 0$ on $\Theta_{x,X}$ if and only if $\Theta_{x,Y} \supset \Theta_{x,X}$.

We apply this to calculate the intersection index (X, L).

Since the number (X, L) is one and the same for all linear forms L, the number of points $x \in X$ for which $L(x)=0$ assumes its maximum when all the $v_x(L)=1$. By the lemma this is equivalent to the fact that the hyperplane L does not touch X at any point. Taking for L such a linear form we find that

$$\deg X = (X, L). \tag{2}$$

It only remains to verify that linear forms with the required property actually exist. This is easily done by means of arguments we have used many times: in the product $X \times \tilde{\mathbb{P}}^n$ (where $\tilde{\mathbb{P}}^n$ is the space of hyperplanes in $\mathbb{P}^n$) we consider the set Γ of pairs (x, ξ) such that ξ touches X at x. A standard application of the theorem on the dimension of fibres of mappings then shows that the image of Γ under the projection $X \times \tilde{\mathbb{P}}^n \to \tilde{\mathbb{P}}^n$ is of codimension $\geqslant 1$.

Comparing (1) and (2) we obtain the relation

$$(X, F) = \deg F \cdot \deg X, \tag{3}$$

which is called *Bezout's theorem*. This theorem has many applications in elementary geometry, which one can find, for example, in [34], Ch. III.

3. Cubic Curves. From the corollary to Theorem 1 it follows that all equivalent divisors on a smooth projective curve have the same degree. Hence we can speak of the degree of a divisor class. We have therefore the homomorphism

$$\deg : \mathrm{Cl}(X) \to \mathbb{Z}$$

whose image is the whole group $\mathbb{Z}$ and whose kernel consists of the classes of degree zero and is denoted by $\mathrm{Cl}^\circ(X)$. The role of this group will already be clear from the following result.

Theorem 4. *A smooth projective curve X is rational if and only if* $\mathrm{Cl}^\circ(X)=0$.

For if $X \approx \mathbb{P}^1$, we are concerned with Example 2 of § 1.1 (for $n=1$). There we have seen that $\mathrm{Cl}(\mathbb{P}^1)=\mathbb{Z}$ and hence $\mathrm{Cl}^\circ(\mathbb{P}^1)=0$. Conversely, let $\mathrm{Cl}^\circ(X)=0$. This means that every divisor of degree zero is principal. In particular, if x, $y \in X$, $x \neq y$, then there exists a function $f \in k(X)$ such that $x-y=(f)$. Regarding f as a mapping $X \to \mathbb{P}^1$ we deduce from Theorem 1 that $k(X)=k(f)$, that is, f is a birational isomorphism. Since X and $\mathbb{P}^1$ are smooth projective curves, f is an isomorphism.

Now we analyse the simplest case when $\mathrm{Cl}^\circ(X) \neq 0$. These are plane smooth projective curves of degree 3. In Ch. I, § 1 we have seen such curves need not be rational; for example, the curve with the equation

$x^3+y^3=1$ is non-rational. In § 5.4 we show that all plane smooth projective curves of degree 3 are non-rational. We shall now make use of this fact.

Theorem 5. *We choose an arbitrary point x_0 on a smooth projective plane curve X of degree 3 and associate with any point $x \in X$ the class C_x containing the divisor $x-x_0$. The mapping $x \to C_x$ determines a one-to-one correspondence between points $x \in X$ and classes $C \in \mathrm{Cl}^\circ(X)$.*

If $C_x = C_y$, $x-x_0 \sim y-x_0$ and $x \sim y$. From the proof of Theorem 4 it follows that for every $x \neq y$ this would lead to the curve X being rational, whereas we know that it is not.

It remains to show that in every class C of degree zero there is a divisor of the form $x-x_0$. To begin with, let D be any effective divisor. We show that there exists a point $x \in X$ such that

$$D \sim x + l x_0 . \tag{1}$$

If $\deg D = 1$, then (1) is true with $l=0$. If $\deg D > 1$, then $D = D' + y$, $\deg D' = \deg D - 1$, $D' > 0$. Applying induction we may assume that (1) is proved for $D' : D' \sim z + m x_0$. Then $D \sim y + z + m x_0$. If we can find a point x such that

$$y + z \sim x + x_0 , \tag{2}$$

then (1) follows.

First let $y \neq z$. We draw the line through these points with the equation $L=0$. By Bezout's theorem $(L, X) = 3$, and hence

$$(L) = y + z + u , \qquad u \in X . \tag{3}$$

Next we suppose that $u \neq x_0$, and we draw the line through u and x_0 with the equation $L_1 = 0$. As in (3) we find that $(L_1) = u + x_0 + x$. Since $(L) \sim (L_1)$, we have $y + z + u \sim u + x + x_0$, hence (2) follows.

We still have to analyse the cases when $y = z$ or $u = x_0$. If $y = z$, then we draw the tangent to X at y. Let $L=0$ be its equation. By the Lemma in § 2, $v_y((L)) \geqslant 2$, and therefore $(L) = 2y + u$. Thus, (2) also holds in this case. The case $u = x_0$ is treated similarly.

Now let $\deg D = 0$. Then $D = D_1 - D_2$, $D_1 \geqslant 0$, $D_2 \geqslant 0$, $\deg D_1 = \deg D_2$. Applying (1) to D_1 and D_2 we see that $D_1 \sim y + l x_0$, $D_2 \sim z + l x_0$ with one and the same l, because $\deg D_1 = \deg D_2$. Therefore

$$D = D_1 - D_2 \sim y - z ,$$

and it is sufficient to find a point x for which $y - z \sim x - x_0$. This is equivalent to $y + x_0 \sim z + x$ and is the same as (2) apart from the notation.

4. The Dimension of a Divisor. In § 1.5 we have associated with a divisor D on a smooth variety a vector space $\mathscr{L}(D)$.

Theorem 6. *The space $\mathscr{L}(D)$ is finite-dimensional for every divisor D on a smooth projective algebraic curve.*

First of all it is easy to reduce the assertion of the theorem to the case $D \geqslant 0$. For let $D = D_1 - D_2$, $D_1 \geqslant 0$, $D_2 \geqslant 0$. Then $\mathscr{L}(D) \subset \mathscr{L}(D_1)$: if $f \in \mathscr{L}(D)$, then $(f) + D_1 - D_2 = D' \geqslant 0$, hence $(f) + D_1 = D' + D_2 \geqslant 0$, that is $f \in \mathscr{L}(D_1)$. This gives the required reduction. Now let $D \geqslant 0$, $D = \sum_{i=1}^{r} n_i x_i$, $n_i \geqslant 0$. At the points x_i we choose local parameters t_i. The condition $f \in \mathscr{L}(D)$ is equivalent to $v_{x_i}(f) \geqslant -n_i (i = 1, \ldots, r)$, $v_x(f) \geqslant 0$ for $x \neq x_i$, that is $f \in t_i^{-n_i} \mathcal{O}_{x_i}$ $(i = 1, \ldots, r)$, $f \in \mathcal{O}_x$ for $x \neq x_i$.

In view of all this we can consider the linear mapping

$$\varphi : \mathscr{L}(D) \to \bigoplus_{i=1}^{r} t_i^{-n_i} \mathcal{O}_{x_i} / \mathcal{O}_{x_i}$$

that associates with a function $f \in \mathscr{L}(D)$ all its residue classes in the spaces $t_i^{-n_i} \mathcal{O}_{x_i} / \mathcal{O}_{x_i}$. If $\varphi(f) = 0$, then $f \in \mathcal{O}_{x_i}$ $(i = 1, \ldots, r)$, and since $f \in \mathscr{L}(D)$, we have $f \in \mathcal{O}_x$ for $x \neq x_i$. Therefore f is regular at all the points $x \in X$. Since X is a projective curve, such a function must be a constant. Thus, the kernel of φ is k, hence one-dimensional. To show that $\mathscr{L}(D)$ is finite-dimensional it remains to verify that the space $\bigoplus_{i=1}^{r} t_i^{-n_i} \mathcal{O}_{x_i} / \mathcal{O}_{x_i}$ is finite-dimensional. Obviously multiplication by $t_i^{n_i}$ determines an isomorphism $t_i^{-n_i} \mathcal{O}_{x_i} / \mathcal{O}_{x_i} \simeq \mathcal{O}_{x_i} / t_i^{n_i} \mathcal{O}_{x_i}$, and in the proof of Theorem 2 we have seen that the space $\mathcal{O}_{x_i} / t_i^{n_i} \mathcal{O}_{x_i}$ is of finite dimension n_i. Thus, $\bigoplus_{i=1}^{r} t_i^{-n_i} \mathcal{O}_{x_i} / \mathcal{O}_{x_i}$ is a direct sum of finite-dimensional spaces, hence itself finite-dimensional.

Together with the proof of the theorem we have obtained the estimate $\dim \mathscr{L}(D) \leqslant \deg D + 1$ for $D \geqslant 0$.

Exercises

1. Let X be a smooth affine curve, and $x_1, \ldots, x_m \in X$. Show that the functions t_i in Theorem 2 can be taken to be the left-hand sides of the equations of those hyperplanes E_i for which $x_i \in E_i$, $x_j \notin E_i$ for $i \neq j$ and $E_i \not\supset \Theta_{x_i, X}$ (so that it does not touch X at x_i).

2. Show that if a curve X is non-rational, then the estimate $l(D) \leqslant \deg D + 1$ in Theorem 6 can be improved to $l(D) \leqslant \deg D$ for every $D \geqslant 0$.

3. Let X be the projective closure of the affine curve $y^2 = x^3 + Ax + B$, where the polynomial $x^3 + Ax + B$ does not have multiple roots and the characteristic of k is different from 2. Show that X is a smooth curve and that its intersection with the line at infinity consists of a single point x_0. Find a local parameter at x_0 and the numbers $v_{x_0}(x)$, $v_{x_0}(y)$.

4. Under the conditions of Exercise 3, find the general form of a function in $\mathscr{L}(mx_0)$. In particular, show that $l(mx_0) = m$ for $m > 0$. Utilize the fact that every function $f \in k(X)$ can be written in the form $P(x) + Q(x)\,y$, $P, Q \in k(X)$, and find out when $f \in \mathscr{L}(mx_0)$.

5. Under the conditions of Exercise 3 find out how to express addition of classes in Cl^0 in terms of the points corresponding to them, according to Theorem 5. More

accurately, if $x_1, x_2 \in X$, $C_{x_3} = C_{x_1} + C_{x_2}$, find out how to express the coordinates of x_3 in terms of coordinates of x_1 and x_2. Here x_0 can be taken to be the point at infinity on X.

6. In the notation of Exercise 3 show that $C_{x_1} + C_{x_2} + C_{x_3} = 0$ if and only if the points x_1, x_2, and x_3 are collinear.

7. In the notation of Exercise 3 show that if $x = (\alpha, \beta)$, $-C_x = C_y$, then $y = (\alpha, -\beta)$. Show that the group Cl^0 has exactly four elements of order two. Find the points on X corresponding to them.

8. A simple point $x \in X$, where X is a plane curve, is called a point of inflexion if $v_x((\Theta_{x,x})) \geqslant 3$. Show that under the conditions of Exercises 3, 4, and 5 x is a point of inflexion if and only if $3C_x = 0$.

9. Show that the line passing through two points of inflexion of the curve X of Exercises 3–7 intersects it in a third point of inflexion.

§ 3. Algebraic Groups

The results of the preceding sections lead to an interesting branch of algebraic geometry: the theory of algebraic groups. We do not go deeply into this topic, but to give the reader at least some idea of it, we give in this section an account of some of its main results, omitting most of the proofs.

1. Addition of Points on a Plane Cubic Curve. Theorem 5 of §2 establishes a one-to-one correspondence between the points of a smooth projective plane cubic curve X and the elements of the group $\mathrm{Cl}^\circ(X)$. To a point $x \in X$ there corresponds the class C_x containing the divisor $x - x_0$, where x_0 is a fixed point, which serves to specify the correspondence.

Making use of this we can transfer the group law from $\mathrm{Cl}^\circ(X)$ to the set X itself. The resulting operation on points of X is called *addition* and is denoted by $\oplus$. According to the definition, $x \oplus y = z$ if $C_x + C_y = C_z$, that is,

$$x + y \sim z + x_0 . \tag{1}$$

Evidently the point x_0 is the null element. We denote it henceforth by o, so that (1) can be rewritten in the form

$$x + y \sim (x \oplus y) + o . \tag{2}$$

The proof of Theorem 5 of §2 makes it possible for us to describe the operation $\oplus$ and the operation $\ominus$ of taking the opposite element in elementary geometric terms. Namely, if the tangent to X at o intersects X at p, and if the line passing through p and x intersects X at x', then

$$2o + p \sim p + x + x' , \qquad x + x' \sim 2o \tag{3}$$

which means that $x' = \ominus x$ (Fig. 7). If $x = p$, then the line through x we have drawn must be replaced by the tangent at p.

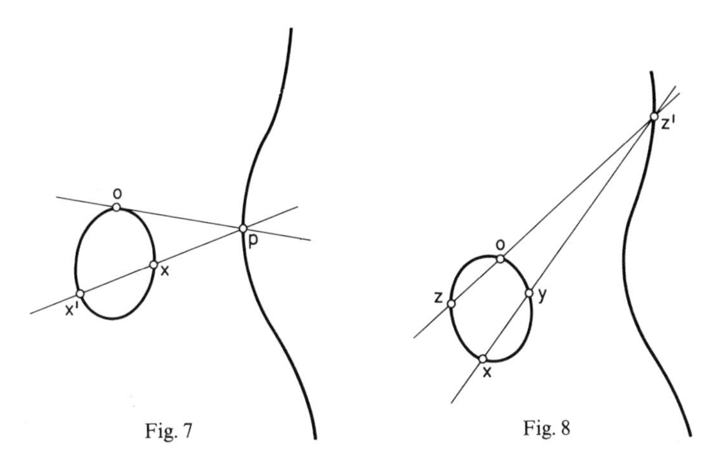

Fig. 7 Fig. 8

Similarly, to describe the operation $\oplus$ we draw the line through x and y. Let z' be its third point of intersection with X and z the third point of intersection with X of the line passing through z' and o. Then (Fig. 8).

$$\begin{gathered} x+y+z' \sim z'+z+o\,, \\ x+y \sim z+o\,, \\ z = x \oplus y\,. \end{gathered} \tag{4}$$

If $x=y$ (or $z'=o$), the secant through x and y must be replaced by the tangent at x (or at z').

Now we prove the important property of "algebraicity" of the group law on X. It will be the basis of the definition of an algebraic group in the next subsection.

Theorem 1. *The mappings $\varphi: X \to X$, $\varphi(x) = \ominus x$, and $\psi: X \times X \to X$, $\psi(x, y) = x \oplus y$, are regular.*

Lemma. *Let $a \in X$ and let $s_a: X \to X$ be the mapping that associates with a point x the third point of intersection of L with X, where L is the line passing through x and a if $x \neq a$, and the tangent at a if $x = a$. Then the mapping s_a is regular.*

We begin by showing that s_a is regular at all points $x \neq a$. For this purpose we choose a coordinate system so that a is the origin of the coordinates and that the third point of intersection with X of the line L joining x to a lies in the finite part of the plane. The latter condition is easily seen to mean that if $x = (\xi, \eta)$ and $f(u, v) = f_3(u, v) + f_2(u, v) + f_1(u, v)$ is the equation of X, then $f_3(\xi, \eta) \neq 0$. The equation of L is of the form $u = t\xi$, $v = t\eta$. Substituting this in the equation of X we obtain the equation $t^3 f_3(\xi, \eta) + t^2 f_2(\xi, \eta) + t f_1(\xi, \eta) = 0$. We know two of its roots $t=0$ and $t=1$ corresponding to the points of intersection a and x. Therefore the value t corresponding to the third point of intersection is obtained from the relation $1+t = -f_2/f_3$. So we see that

$$s_a(x) = \left(-\frac{f_3(\xi, \eta) + f_2(\xi, \eta)}{f_3(\xi, \eta)} \xi, \ -\frac{f_3(\xi, \eta) + f_2(\xi, \eta)}{f_3(\xi, \eta)} \eta \right).$$

Since $f_3(\xi, \eta) \neq 0$, this shows that s_a is regular for $x \neq a$.

To prove the same for $x = a$ we observe that a rational mapping of a smooth projective curve into itself is regular. Hence there exists a regular mapping $\tilde{s}_a$ that coincides with s_a for $x \neq a$. Since $s_a^2 = 1$, s_a and $\tilde{s}_a$ are one-to-one. But as they agree at all points except possibly one, there must also agree at this point. Hence $s_a = \tilde{s}_a$, and this means that s_a is regular.

Proof of Theorem 1. The assertion about φ follows immediately from the lemma, because according to (3), $\varphi = s_p$. The mapping $\psi(x, y)$ for $x \neq y$ is defined by means of the secant through x and y. Arguing as in the proof of the lemma, we can easily verify that ψ is a rational mapping. For any point $a \in X$ the mapping t_a, $t_a(x) = a \oplus x$, is regular, because according to (4) $t_a = s_0 s_a$. Obviously the relation

$$\psi(x, y) = t_{a \oplus b}^{-1} \psi(t_a(x), t_b(y))$$

holds for arbitrary points $a, b \in X$. Therefore, if ψ is regular at (x_0, y_0), then it is also regular at $(t_a(x_0), t_b(y_0))$. But it must be regular at some points, because it is rational. Hence it follows that it is regular everywhere.

2. Algebraic Groups. Plane cubic curves are one of the most important examples of a general concept, which we now introduce.

An *algebraic group* is an algebraic variety G which at the same time is a group for which the following conditions hold: the mappings $\varphi: G \to G$, $\varphi(g) = g^{-1}$, and $\psi: G \times G \to G$, $\psi(g_1, g_2) = g_1 g_2$ are regular (here g^{-1} and $g_1 g_2$ are the inverse elements and the product in G).

Examples of Algebraic Groups

Example 1. A plane cubic curve with the group law $\oplus$. The fact that the conditions in the definition of an algebraic group are satisfied is the contents of Theorem 1.

Example 2. The affine line $\mathbb{A}^1$ on which the group law is given by addition of point coordinates. This group is called *additive*.

Example 3. The variety $\mathbb{A}^1 - o$, where o is the origin; the group law is given by multiplication of point coordinates. This group is called *multiplicative*.

Example 4. In the space $\mathbb{A}^{n^2}$ of square matrices of order n the open set of non-singular matrices with the usual law of matrix multiplication. This is called the *general linear group.*

Example 5. In the space $\mathbb{A}^{n^2}$ the closed subset consisting of the orthogonal matrices. Naturally, the group law is the same as in Example 4.

Let us show by a very simple example how the fact that G is an algebraic group can influence the geometry of the variety G.

Theorem 2. *The variety of an algebraic group is smooth.*

From the definition of an algebraic group it follows that for any $h \in G$ the mapping

$$t_h : G \to G, \ t_h(g) = hg$$

is an automorphism of G.

Since $t_h(g_1) = g_2$ for any $g_1, g_2 \in G$ where $h = g_2 g_1^{-1}$, and since the property of a point of being singular is invariant under automorphisms, we see that if at least one point of G is singular, then so are all the points. But this contradicts the fact that in any algebraic variety the singular points form a closed proper subvariety. Therefore G cannot have singular points.

3. Factor Groups. Chevalley's Theorem. This subsection contains the statement of some basic theorems on algebraic groups. Proofs of these theorems are not provided.

A *subgroup* of an algebraic group G is a subgroup of G that is a closed subset of G.

A subgroup $H \subset G$ is called *normal,* as in the abstract theory of groups, if $g^{-1}Hg = H$ for all $g \in H$. Finally, a *homomorphism* of algebraic groups $\varphi : G_1 \to G_2$ is a regular mapping that is a homomorphism of abstract groups.

The problem of constructing the factor group of a given normal subgroup N is very delicate. The difficult question is, of course, how to turn the set G/N into an algebraic variety.

Theorem A.* *The abstract group G/N can be made into an algebraic group in such a way that the following conditions hold:*

1. The natural mapping $\varphi : G \to G/N$ is a homomorphism of algebraic groups.

2. For any homomorphism of algebraic groups $\psi : G \to G_1$ whose kernel contains N there exists a homomorphism $f : G/N \to G_1$ for which $\psi = f \cdot \varphi$.

---

* Letters denote theorems that are stated without proof.

Obviously the algebraic group G/N is uniquely determined by the conditions 1 and 2. It is called the *factor group* of N in G.

An algebraic group G is called *affine* if the algebraic variety G is affine, and it is called an *Abelian variety* if the algebraic variety G is projective and irreducible.

Theorem B. *An affine algebraic group is isomorphic to a subgroup of a general linear group* (Example 4 above).

Evidently the general linear group, and hence each of its subgroups, is affine.

Theorem C (Chevalley's Theorem). *Every algebraic group G has a normal subgroup N such that N is affine and G/N is an Abelian variety. N is uniquely determined by the these properties.*

4. Abelian Varieties. The condition of projectivity of the variety of an algebraic group G, which defines Abelian varieties, contains a surprising amount of information. Many unexpected properties of algebraic varieties follow from it. We derive the simplest of these here, because they only require appliations of simple theorems that were proved in Ch. I.

We need a property of arbitrary projective varieties. We define a family of mappings of the variety X into Z as a regular mapping $f: X \times Y \to Z$, where Y is some algebraic variety, the so-called base of the family.

Evidently, for every $y \in Y$ we have the mapping $f_y(x) = f(x, y)$, which justifies our terminology.

Lemma. *If X and Y are irreducible varieties, X is projective, and, for a family f of mappings of X into Z with base Y and some point $y_0 \in Y$, $f(X \times y_0)$ is a single point $z_0 \in Z$, then $f(X \times y)$ is a single point for every $y \in Y$.*

Proof. Consider the graph Γ of f. Obviously $\Gamma \subset X \times Y \times Z$ and Γ is isomorphic to $X \times Y$. We denote by p the projection $X \times Y \times Z \to Y \times Z$, and by $\bar{\Gamma}$ the set $p(\Gamma)$. Since X is projective, $\bar{\Gamma}$ is closed by Theorem 3 of Ch. I, § 5. We denote by $q: \bar{\Gamma} \to Y$ the mapping defined by the projection $Y \times Z \to Y$. The fibre of q over y obviously has the form $(y, f(x, y))$, hence is not empty, so that $q(\bar{\Gamma}) = Y$. On the other hand, by hypothesis, for $y = y_0$ the fibre consists of the single point (y_0, z_0). Applying Theorem 7 of Ch. I, § 6, we see that the fibres over an open set are zero-dimensional and that $\dim \bar{\Gamma} = \dim Y$.

We take an arbitrary point $x_0 \in X$; clearly $\bar{\Gamma} \supset \{(y, f(x_0, y)), y \in Y\}$. Since both varieties are irreducible and have the same dimension, they are identical, and this means that $f(X \times y) = f(x_0, y)$.

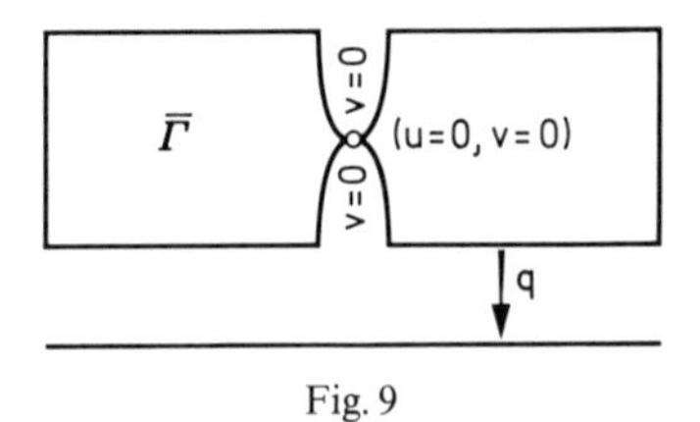

Fig. 9

Note. Without the assumption that X is projective the lemma is false, as is shown by the example of the family of mappings $f: \mathbb{A}^1 \times \mathbb{A}^1 \to \mathbb{A}^1$, $f(x, y) = x\,y$. The reason for this is that the set $\bar{\Gamma}$ is not closed and that Theorem 7 of Ch. I, § 6 is not applicable to it. In our example $\bar{\Gamma} \subset \mathbb{A}^1 \times \mathbb{A}^1 = \mathbb{A}^2$ consists of all the points (u, v) except those with $u = 0$, $v \neq 0$. This is a plane from which the line $u = 0$ has been removed but the point $u = 0$, $v = 0$ retained. Actually, Theorem 7 of Ch. I, § 6 is not true for the projection $q: (u, v) \to u$: the dimension of the fibre over the point $u = 0$ is 0, but the dimension of the image is 1, and the dimension of the variety to be mapped is 2 (Fig. 9).

Theorem 3. *An Abelian variety is commutative.*

Consider the family of mappings of G into G with basis G:

$$f: G \times G \to G, f(g, g') = g^{-1} g' g .$$

Evidently $f(g, e) = e$, hence by the lemma $f(G, g')$ consists of a single point. Therefore $f(G, g') = f(e, g') = g'$, but this means that the group G is commutative.

Theorem 4. *If $\psi: G \to H$ is a regular mapping of an Abelian variety G into an algebraic group H, then $\psi(g) = \psi(e)\, \varphi(g)$, where $e \in G$ is the unit element and $\varphi: G \to H$ is a homomorphism.*

Proof. We set $\varphi(g) = \psi(e)^{-1}\, \psi(g)$ and show that φ is a homomorphism. For this purpose we consider the family of mappings of the variety G into H whose base coincides with G:

$$f: G \times G \to H, f(g', g) = \varphi(g')\, \varphi(g)\, \varphi(g' g)^{-1} .$$

Since $\varphi(e) = e'$ is the unit element of H, we have $f(G, e) = e'$. By the lemma, the image $f(G, g)$ for every element $g \in G$ consists of a single point, so that $f(g', g)$ does not depend on g'. Setting $g' = e$ we see that $f(g', g) = f(e, g) = e'$, which means that φ is a homomorphism.

Corollary. *If two Abelian varieties are isomorphic as algebraic varieties, then they are also isomorphic as groups – "the geometry determines the algebra".*

5. Picard Varieties. The only examples of Abelian varieties which we have encountered so far are plane cubic curves. We have defined a group law on them, starting out from their group of divisor classes. This example is typical for a far more general situation. Starting out from arbitrary smooth projective variety X we can construct an Abelian variety whose group of points is isomorphic to a subgroup of $\mathrm{Cl}(X)$ (or $\mathrm{Cl}^\circ(X)$ in the case of a cubic curve). We give this definition, but omit the proofs of all assertions except the very simplest. Our aim is to study divisors on smooth varieties, but in the course of the arguments we come across divisors on arbitrary varieties. In that case we understand by divisors only locally principal divisors.

We now define a new relationship of equivalence for divisors: algebraic equivalence. It is coarser than (that is, follows from) the equivalence we have considered before.

Let X and T be any two irreducible varieties. For every $t \in T$ the mapping $j_t: x \to (x, t)$ defines an embedding of X in $X \times T$. Every divisor C on $X \times T$ for which $\mathrm{Supp}\, C \not\supset X \times t$ determines a divisor $j_t^*(C)$ on X. In that case we say that the divisor $j_t^*(C)$ is defined.

A *family of divisors* on X with base T is a mapping $f: T \to \mathrm{Div}(X)$. A family f is called *algebraic* if there exists a divisor $C \in \mathrm{Div}(X \times T)$ such that the divisor $j_t^*(C)$ is defined for all $t \in T$ and $j_t^*(C) = f(t)$.

Two divisors D_1 and D_2 on X are called *algebraically equivalent* if there exists an algebraic family of divisors f on X with base T and two points $t_1, t_2 \in T$ such that $f(t_1) = D_1$, $f(t_2) = D_2$. This relation is written as $D_1 \equiv D_2$. Thus, algebraic equivalence of two divisors indicates that it is possible to "deform them algebraically" into each other. Clearly algebraic equivalence is reflexive and symmetric. It is easy to show that it is also transitive. If the algebraic equivalence of two divisors D_1 and D_2 is realized by a divisor C on $X \times T$ and the equivalence of D_2 and D_3 by a divisor C' on $X \times T'$, then to show that D_1 and D_3 are equivalent we have to consider the divisor

$$(C \times T') + (C' \times T) - D_2 \times T \times T'$$

on $X \times T \times T'$. The detailed verification is left to the reader.

Finally, it is easy to see that algebraic equivalence is compatible with addition in the group $\mathrm{Div}(X)$: the divisors D with $D \equiv 0$ form a subgroup, which we denote by $\mathrm{Div}^a(X)$.

Equivalence of divisors implies their algebraic equivalence. It is enough to verify this for equivalence of a divisor to zero. Let $D \in \mathrm{Div}(X)$, $D \sim 0$, that is, $D = (g)$, $g \in k(X)$. Consider the variety $T = \mathbb{A}^2 - (0, 0)$ and denote by u and v coordinates on $\mathbb{A}^2$. We regard g, u and v as functions on $X \times T$, understanding by this, as usual, $p^*(g)$, $q^*(u)$, and $q^*(v)$, where $p: X \times T \to X$ and $q: X \times T \to T$ are the projections. We set $C = (u + vg)$ and consider the algebraic family determined by the divisors C on $X \times T$. It is easy to verify that $f(1,0)=0$ (the divisor zero), $f(0,1)=D$, hence $D \equiv 0$.

Finally, we consider the notion of algebraic equivalence on the example of a smooth projective curve X. For any two points $x, y \in X$ we have $x \equiv y$. To see this it is sufficient to consider the family of divisors f parameterized by X itself and defined by the diagonal on $X \times X$. It is easy to verify that $f(x)=x$ for all $x \in X$. Therefore $D \equiv (\Sigma n_i) x_0$ for every divisor $D = \Sigma n_i x_i$ and every point $x_0 \in X$, that is, two divisors of the same degree are algebraically equivalent.

The converse is somewhat more difficult to prove: algebraically equivalent divisors on a smooth projective curve have the same degree. We do not prove this here. Thus, divisors on a smooth projective curve X are algebraically equivalent if and only if they have the same degree. Therefore

$$\mathrm{Div}(X)/\mathrm{Div}^a(X) = \mathrm{Cl}(X)/\mathrm{Cl}^\circ(X) = \mathbb{Z}\,.$$

A generalization of this is the following theorem, which was proved by Severi (for fields of characteristic zero) and Néron (in general).

Theorem D. *For a smooth projective variety X the group* $\mathrm{Div}(X)/\mathrm{Div}^a(X)$ *is finitely generated.*

It can be shown that for $X = \Pi\, \mathbb{P}^{n_i}$ algebraic equivalence of divisors is the same thing as equivalence. This example shows that the group $\mathrm{Div}(X)/\mathrm{Div}^a(X)$ can be more complicated than $\mathbb{Z}$.

In the case of a plane cubic curve X the group $\mathrm{Cl}^\circ(X) = \mathrm{Div}^a(X)/P(X)$, where $P(X)$ is the group of principal divisors, is a one-dimensional Abelian variety. Similarly, for every projective smooth variety there exists an Abelian variety G whose group of points is isomorphic to $\mathrm{Div}^a(X)/P(X)$, and which has the following property. For every algebraic family of divisors f on X with basis T there exists a regular mapping $\varphi: T \to G$ such that $f(t) - f(t_0) \in \varphi(t)$, where t_0 is some fixed point of T [and G is identified with $\mathrm{Div}^a(X)/P(X)$, hence $\varphi(t)$ is regarded as a class of divisors].

The Abelian variety G is uniquely determined by this property. It is called the *Picard variety* of X.

The Picard variety of a smooth projective algebraic curve X is also called its *Jacobian variety*.

Exercises

1. Let G be an algebraic group, $\psi: G \times G \to G$ a regular mapping defining a group law, Θ_e the tangent space to G at the unit element, Θ'_e the tangent space of $G \times G$ at the unit element. Show that $\Theta'_e = \Theta_e \oplus \Theta_e$, and that $d_e\psi: \Theta_e \oplus \Theta_e \to \Theta_e$ is given by vector addition.

2. In the notation of Exercise 1, let G be a commutative group and $\varphi_n: G \to G$ be given by $\varphi_n(g) = g^n$. Assuming that the characteristic of the ground field is zero, show that $d_e\varphi_n(x)$ is a non-singular linear transformation. Deduce that in a commutative algebraic group the number of elements of order n is finite and that an n-th root can be extracted from any element.

§ 4. Differential Forms

1. One-Dimensional Regular Differential Forms. In Ch. II we have introduced the concept of the differential $d_x f$ of a function f that is regular at a point x of an algebraic variety X. By definition $d_x f$ is a linear form on the tangent space Θ_x of x, so that $d_x f \in \Theta_x^*$. We now investigate how this notion depends on the point x.

If the function f is fixed and regular on the whole of X, then $d_x f$ in its dependence on x is an object of a new type we have not met so far: it associates with every point $x \in X$ a vector of the space Θ_x^* dual to the tangent space at this point. Later we shall all the time come across objects of a similar kind. The following explanation may help. In linear algebra we are concerned with constants, but also with other quantities: vectors, linear forms, and tensor products. In geometry the analogue to constants are functions (whose values are constants). The analogues of vectors, linear forms, etc. are "functions" associating with every point x of an algebraic (or differentiable) variety X a vector, a linear form, etc. in the tangent space Θ_x at this point.

We consider the set $\Phi[X]$ of all mappings φ that associate with every point $x \in X$ a vector $\varphi(x)$ of the space Θ_x^*. Of course, this is far too large a set, just as the set of all functions on X with values in k is far too large to be interesting. Similarly to the way in which among all functions we have selected the regular ones, so we select in the set $\Phi[X]$ a part that is more closely connected with the structure of the variety X. For this purpose we observe that $\Phi[X]$ is an Abelian group if we set $(\varphi + \psi)(x) = \varphi(x) + \psi(x)$. Furthermore, $\Phi[X]$ becomes a module over the ring of all functions on X with values in k if we set $(f \cdot \varphi)(x) = f(x) \cdot \varphi(x)$ for a function f on X and for $\varphi \in \Phi[X]$. In particular, we can regard $\Phi[X]$ as a module over the ring $k[X]$ of all regular functions on X.

As we have seen, every function that is regular on X determines a differential $d_x f \in \Phi[X]$.

Therefore every function $f \in k[X]$ determines a function $\varphi \in \Phi[X]: \varphi(x) = d_x f$, which we denote by df.

Definition. An element $\varphi \in \Phi[X]$ is called a *regular differential form on* X if every point $x \in X$ has a neighbourhood U such that the restriction of φ to U belongs to the submodule of $\Phi[U]$ that is generated over $k[U]$ by the df, $f \in k[U]$.

Obviously, all regular differential forms on X form a module over $k[X]$, which we denote by $\Omega[X]$. Thus, $\varphi \in \Omega[X]$ if in a neighbourhood of every point $x \in X$ there is a representation

$$\varphi = \sum_{i=1}^{m} f_i dg_i \tag{1}$$

where $f_1, \dots, f_m, g_1, \dots, g_m$ are regular in a neighbourhood of x.

Taking the differential of a function determines a mapping $d: k[X] \to \Omega[X]$. The properties (1) of Ch. II, § 1.3 now assume the form

$$d(f+g) = df + dg, \quad d(f \cdot g) = f \cdot dg + g \cdot df. \tag{2}$$

From these formulae it is easy to derive an identity, which is true for any polynomial $F \in k[T_1, \dots, T_m]$ and any functions $f_1, \dots, f_m \in k[X]$:

$$d(F(f_1, \dots, f_m)) = \sum_{i=1}^{m} \frac{\partial F}{\partial T_i}(f_1, \dots, f_m) df_i. \tag{3}$$

For this purpose we have to reduce the proof to the case of a monomial, using (2), and then prove it by induction on the degree of the monomial, again using (2). The details of this verification are left to the reader.

Once (3) has been proved for polynomials, it generalizes immediately to the case of rational functions F. Here we have to keep in mind that if a rational function F is regular at x, then so are all the functions $\partial F/\partial T_i$ at this point. For then $F = P/Q$, where P and Q are polynomials and $Q(x) \neq 0$. Therefore

$$\frac{\partial F}{\partial T_i} = Q^{-2}\left(Q \frac{\partial P}{\partial T_i} - P \frac{\partial Q}{\partial T_i}\right):$$

from which its regularity follows.

Example 1. $X = \mathbb{A}^n$. Since at every point $x \in \mathbb{A}^n$ the differentials of the coordinates $d_x t_1, \dots, d_x t_n$ form a basis of the space Θ_x^*, every element $\varphi \in \Phi[\mathbb{A}^n]$ has a unique representation in the form $\varphi = \sum_{i=1}^{n} \psi_i dt_i$, where ψ_i are functions on $\mathbb{A}^n$ with values in k.

If $\varphi \in \Omega[\mathbb{A}^n]$, then the decomposition (1) holds in a neighbourhood of every point. Applying (3) to the g_i we obtain the decomposition $\varphi = \Sigma\, h_i\, dt_i$, in which the h_i are regular at x. Since such a representation is unique, the ψ_i must be regular at every point $x \in \mathbb{A}^n$, so that $\psi_i \in k[\mathbb{A}^n]$. Therefore $\Omega[\mathbb{A}^n] = \bigoplus k[\mathbb{A}^n]\, dt_i$.

Example 2. Let $X = \mathbb{P}^1$ and let t be a coordinate on X.

Then $X = \mathbb{A}_0^1 \cup \mathbb{A}_1^1$, with $\mathbb{A}_0^1 \simeq \mathbb{A}_1^1 \simeq \mathbb{A}^1$. By the result of Example 1 every element $\varphi \in \Omega[P^1]$ can be represented in the form $\varphi = P(t)\, dt$ on

$\mathbb{A}_0^1$, $\varphi = Q(u)\,du$ on $\mathbb{A}_1^1$, where $ut=1$. From the last relation it follows that $du = -dt/t^2$, and in $\mathbb{A}_0^1 \cap \mathbb{A}_1^1$ we have

$$P(t)\,dt = -\frac{Q(t^{-1})}{t^2}\,dt\,, \quad \text{that is,} \quad P(t) = -\frac{Q^*(t)}{t^{n+2}}\,,$$

if $\deg Q = n$. Here $Q^*(t) = t^n Q(1/t)$ and $Q^*(0) \neq 0$.

Such a relation among polynomials is possible only when $P = Q = 0$. Therefore $\Omega[\mathbb{P}^1] = 0$.

Example 3. Let X be given by the equation $x_0^3 + x_1^3 + x_2^3 = 0$ in $\mathbb{P}^2$ and k be of characteristic different from 3.

We denote by U_{ij} the open set in which $x_i \neq 0$, $x_j \neq 0$. Then $X = U_{01} \cup U_{12} \cup U_{20}$. We set

$$\text{in } U_{01}: x = \frac{x_1}{x_0}, \qquad y = \frac{x_2}{x_0}, \qquad \varphi = \frac{dy}{x^2},$$

$$\text{in } U_{12}: u = \frac{x_2}{x_1}, \qquad v = \frac{x_0}{x_1}, \qquad \psi = \frac{dv}{u^2},$$

$$\text{in } U_{20}: s = \frac{x_0}{x_2}, \qquad t = \frac{x_1}{x_2}, \qquad \chi = \frac{dt}{s^2}.$$

Evidently $\varphi \in \Omega[U_{01}]$, $\psi \in \Omega[U_{12}]$, $\chi \in \Omega[U_{20}]$. It is easy to verify that $\varphi = \psi$ in $U_{01} \cap U_{12}$, $\varphi = \chi$ in $U_{01} \cap U_{20}$. Therefore these formulae determine a single form $\omega \in \Omega[X]$. This example is interesting in that $\Omega[X] \neq 0$, whereas X is a projective variety, and there are no non-constant regular functions on it.

In the general case we can prove a fact that is analogous to, but weaker than, that in Example 1.

Theorem 1. *Every simple point x of an algebraic variety X has an affine neighbourhood U such that the module $\Omega[U]$ is free over $k[U]$. Its rank is equal to* $\dim_x X$.

Proof. Let $X \subset \mathbb{A}^N$ and suppose that $F_1, \ldots, F_m$ form a basis of the ideal of X. Then $F_i = 0$ on X, and therefore by (3),

$$\sum_{j=1}^{N} \frac{\partial F_i}{\partial T_j}\,dt_j = 0\,. \tag{4}$$

If x is a simple point and $\dim_x X = n$, then the rank of the matrix $((\partial F_i/\partial T_j)(x))$ is equal to $N-n$. Suppose, for example, that $t_1, \ldots, t_n$ are local parameters at x. Then it follows from (4) that all the dt_j can be expressed in terms of $dt_1, \ldots, dt_n$ with coefficients that are rational functions and regular at x.

We consider a neighbourhood U of x in which all these functions are regular. Then $d_y t_1, \ldots, d_y t_n$ form in it a basis of Θ_y^* for every point

$y \in U$. Let $\varphi \in \Omega[U]$. By what we have said above, in U there is a unique representation

$$\varphi = \sum_{i=1}^{n} \psi_i \, dt_i, \tag{5}$$

where the ψ_i are functions on U with values in k. From the representation (1) and the formula (3) it follows that φ can be expressed in a neighbourhood of every point $y \in U$ as a linear combination of $dt_1, \ldots, dt_N$ whose coefficients are functions regular in U. As we have seen, $dt_1, \ldots, dt_N$ can similarly be expressed in terms of $dt_1, \ldots, dt_n$. Therefore $\varphi = \sum_{i=1}^{n} g_i \, dt_i$, where the g_i are regular in a neighbourhood of y. From the uniqueness of the representation (5) it follows that $\psi_i = g_i$ in a neighbourhood of y, and hence that $\psi_i \in k[U]$. So we see that $\Omega[U] = \sum_{i=1}^{n} k[U] \, dt_i$.

Let us assume that among the $dt_1, \ldots, dt_n$ there is a relation $\sum_{i=1}^{n} g_i \, dt_i = 0$ and that $g_n \neq 0$, say. Then $dt_1, \ldots, dt_n$ are linearly dependent in the open set where $g_n \neq 0$, but this contradicts the fact that the $d_y t_i$ are independent in Θ_y^* for all $y \in U$. This proves the theorem.

Corollary. *If $u_1, \ldots, u_n$ is any system of local parameters at x, then in some neighbourhood U of x the differentials $du_1, \ldots, du_n$ generate the module $\Omega[U]$.*

Let $dt_1, \ldots, dt_n$ be a basis of the free module $\Omega[U]$ in a neighbourhood U of x, which exists according to Theorem 1. Then $du_i = \sum_{j=1}^{n} g_{ij} \, dt_j$, and since the u_i are local parameters, $|g_{ij}(x)| \neq 0$. Therefore in a neighbourhood U' in which $|g_{ij}| \neq 0$ the $du_1, \ldots, du_n$ generate the module $\Omega[U']$.

2. Algebraic Description of the Module of Differentials. We have seen in Ch. I that the category of affine varieties is equivalent to the category of rings of a special type. Therefore we can view the whole theory of affine varieties from a purely algebraic angle, and in particular, we can try to grasp the algebraic meaning of the module of differential forms.

Consider an affine variety X and denote by A the ring $k[X]$ and by Ω the module $\Omega[X]$. Taking a differential determines a homomorphism of k-modules $d: A \to \Omega$.

Proposition 1. *The module Ω is generated over A by the elements df, $f \in A$.*

This is an analogue to Theorem 4 of Ch. I, § 3 and is proved in the same way. If $\omega \in \Omega$, then by definition for every point $x \in X$ there exists a representation $\omega = \Sigma f_{i,x} dg_{i,x}$, $f_{i,x}, g_{i,x} \in \mathcal{O}_x$. For every function $u \in \mathcal{O}_x$ there exists a representation $u = v/w$, $v, w \in A$, $w(x) \neq 0$. Utilizing such a representation for $f_{i,x}$ and $g_{i,x}$ and taking the least common denominator of all the fractions, we obtain a function p_x such that $p_x(x) \neq 0$,

$$p_x \cdot \omega = \Sigma\, r_{i,x}\, dh_{i,x}\,, \; r_{i,x}, h_{i,x} \in A\,.$$

Since $p_x(x) \neq 0$, there exist functions $q_x \in A$ such that $\Sigma\, p_x q_x = 1$, hence $\omega = \Sigma\, q_x r_{i,x} dh_{i,x}$. This proves Proposition 1.

Proposition 1 suggests the idea of describing the module Ω in terms of its generators df, $f \in A$. Clearly the following relations hold:

$$\begin{gathered} d(f+g) = df + dg\,, \qquad dfg = f\,dg + g\,df \\ d\alpha = 0 \quad \text{for} \quad \alpha \in k\,. \end{gathered} \tag{1}$$

Proposition 2. *If X is a smooth affine variety, and $A = k[X]$, then the A-module Ω is determined by the relations* (1).

Proof. We denote by R the module defined over A by generators df in one-to-one correspondence with the elements of A, and by the relations (1). There is an obvious homomorphism $\xi: R \to \Omega$, and Proposition 1 shows that ξ is an epimorphism.

It remains to show that the kernel of ξ is trivial. Let $\varphi \in R$ and $\xi(\varphi) = 0$. Observe that the arguments in the proof of Theorem 1 use only the relations (1). Therefore they are applicable to the module R and show that for every point $x \in X$ there exists a function $D \in A$ such that $D(x) \neq 0$ and $D \cdot \varphi = \Sigma g_i dt_i$, $g_i \in A$, where now the local parameters t_i are chosen to be elements of A. If $\xi(\varphi) = 0$, then $\Sigma\, g_i dt_i = 0$ in the module Ω, and from Theorem 1 it follows that all the $g_i = 0$. Thus, $D \cdot \varphi = 0$. So we see for every point x there exists a function $D \in A$, such that $D(x) \neq 0$, $D \cdot \varphi = 0$. Arguing as in the proof of Proposition 1 we find that $\varphi = 0$, and the proposition is proved.

Thus, in this case the module $\Omega[X]$ can be described purely algebraically, starting out from the ring $k[X]$. This suggests the idea of considering a similar module for every ring A that is an algebra over a subring A_0. The module R determined by the generators da and the relations (1) (of course, $\alpha \in A_0$ in the latter), is called the *module of differentials of the ring A* over A_0.

If the variety X is not smooth, then this module of differentials R, which is defined purely algebraically, does not, in general, coincide with $\Omega[X]$. (See Exercise 9). Proposition 1 which is also true for non-smooth varieties shows that R contains more information on X than the module $\Omega[X]$. However, later we shall be concerned essentially with smooth varieties, and this difference will not be important for us.

3. Differential Forms of Higher Degrees. The differential forms we have considered in § 4.1 associate with every point $x \in X$ an element of the space Θ_x^*. Now we consider more general differential forms, which associate with a point $x \in X$ a linear skew-symmetric form on the space Θ_x, that is, an element of the r-th exterior power $\Lambda^r \Theta_x^*$ of Θ_x^*.

The definition is entirely analogous to that in § 4.1. We denote by $\Phi^r[X]$ the set of all correspondences between points $x \in X$ and elements of $\Lambda^r \Theta_x^*$. Thus, if $\omega \in \Phi^r[X]$, $x \in X$, then $\omega(x) \in \Lambda^r \Theta_x^*$. In particular, $\Phi^0[X]$ is the ring of arbitrary mappings $X \to k$; $\Phi^1[X]$ is the $\Phi[X]$ considered in the previous subsection. Therefore $df \in \Phi^1[X]$ for $f \in k[X]$.

We recall that the operation of exterior multiplication $\wedge$ is defined for every vector space L: if $\varphi \in \Lambda^r L$, $\psi \in \Lambda^s L$, then $\varphi \wedge \psi \in \Lambda^{r+s} L$, furthermore, $\varphi \wedge \psi$ is distributive, associative, and $\psi \wedge \varphi = (-1)^{rs} \varphi \wedge \psi$. If $e_1, \ldots, e_n$ is a basis of L, then a basis of $\Lambda^r L$ consists of all products $e_{i_1} \wedge \ldots \wedge e_{i_r}$, $i_1 < i_2 < \ldots < i_r$. Therefore $\dim \Lambda^r L = \binom{n}{r}$, in particular, $\dim \Lambda^n L = 1$, $\Lambda^r L = 0$ for $r > n$.

Let us define an operation of exterior multiplication in the sets $\Phi^r[X]$: for $\omega_r \in \Phi^r[X]$, $\omega_s \in \Phi^s[X]$ we define $\omega = \omega_r \wedge \omega_s$ by the equation $\omega(x) = \omega_r(x) \wedge \omega_s(x)$ for all $x \in X$. Evidently $\omega \in \Phi^{r+s}[X]$. For $r = 1$, $s = 0$ we arrive at the multiplication of elements of $\Phi^1[X] = \Phi[X]$ by functions. Setting $s = 0$, r arbitrary, we see that a multiplication of $\Phi^r[X]$ by functions on X is defined. In particular, all the $\Phi^r[X]$ are modules over the ring $k[X]$.

Definition. An element $\varphi \in \Phi^r[X]$ is said to be an *r-dimensional regular differential form on X* if every point $x \in X$ has a neighbourhood U such that on U the element $\varphi|_u$ belongs to the submodule of $\Phi^r[U]$ generated over $k[U]$ by the elements $df_1 \wedge \ldots \wedge df_r$, $f_1, \ldots, f_r \in k[U]$.

All the r-dimensional regular differential forms on X form a module over $k[X]$, which we denote by $\Omega^r[X]$.

Thus, an element $\omega \in \Omega^r[X]$ can be expressed in a neighbourhood of an arbitrary point $x \in X$ in the form

$$\omega = \Sigma\, g_{i_1 \ldots i_r}\, df_{i_1} \wedge \ldots \wedge df_{i_r} \tag{1}$$

where the $g_{i_1 \ldots i_r}$, $f_{i_1}, \ldots, f_{i_r}$ are regular at x.

The operation of exterior multiplication is defined for regular forms, and obviously for $\omega_r \in \Omega^r[X]$, $\omega_s \in \Omega^s[X]$ we have $\omega_r \wedge \omega_s \in \Omega^{r+s}[X]$. In particular, every $\Omega^r[X]$ is a module over $k[X]$.

The differential forms we have considered in the preceding subsection are, from the point of view of the new definition, one-dimensional.

Theorem 1 has an analogue for forms in $\Omega^r[X]$ for every r.

Theorem 2. *Every simple point of an n-dimensional variety has a neighbourhood U such that the module $\Omega^r[U]$ is free over $k[U]$ and of rank* $\binom{n}{r}$.

Proof. In the proof of Theorem 1 we have seen that there exist a neighbourhood U of a simple point x and n functions $u_1, \ldots, u_n$, regular in U, such that $d_y u_1, \ldots, d_y u_n$ form a basis of Θ_y^* for every $y \in U$. Hence it follows that every element $\varphi \in \Phi^r[U]$ can be represented in the form

$$\varphi = \Sigma\, \psi_{i_1 \ldots i_r} du_{i_1} \wedge \ldots \wedge du_{i_r}$$

where $\psi_{i_1 \ldots i_r}$ are functions on U with values in k.

If $\varphi \in \Omega^r[U]$, then φ can be represented in the form (1) for every point $y \in U$. Applying Theorem 1 to the forms du_i we see that the functions $\psi_{i_1 \ldots i_r}$ are regular at y. Since y is an arbitrary point on U, they are regular in U. Thus, the forms $du_{i_1} \wedge \ldots \wedge du_{i_r}$, $i_1 < i_2 < \ldots < i_r$, generate the module $\Omega^r[U]$. It remains to show that these forms are linearly independent over $k[U]$. But every dependence

$$\Sigma\, g_{i_1 \ldots i_r} du_{i_1} \wedge \ldots \wedge du_{i_r} = 0$$

gives at $x \in U$ the relation

$$\Sigma\, g_{i_1 \ldots i_r}(x)\, d_x u_{i_1} \wedge \ldots \wedge d_x u_{i_r} = 0\,. \tag{2}$$

Since $d_x u_1, \ldots, d_x u_n$ is a basis of the space Θ_x^*, we see that the $d_x u_{i_1} \wedge \ldots \wedge d_x u_{i_r}$ form a basis in $\Lambda^r \Theta_x^*$. Therefore it follows from (2) that $g_{i_1} \ldots {}_{i_1}(x) = 0$ for all $x \in U$, that is, $g_{i_1 \ldots i_r} = 0$.

Of particular importance is the module $\Omega^n[U]$, which under the assumptions of Theorem 2 is of rank 1 over $k[U]$. Thus, if $\omega \in \Omega^n[U]$, then

$$\omega = g\, du_1 \wedge \ldots \wedge du_n,\ g \in k[U]\,. \tag{3}$$

The expression for ω in this form depends essentially on the choice of the local parameters $u_1, \ldots, u_n$. Let us clarify what this dependence is. Let $v_1, \ldots, v_n$ be another n regular functions on X such that $v_1 - v_1(x), \ldots, v_n - v_n(x)$ are local parameters at any point $x \in U$. Then

$$\Omega^1[U] = k[U]\, dv_1 + \cdots + k[U]\, dv_n$$

and, in particular, all the du_i are representable in the form

$$du_i = \sum_{j=0}^{n} h_{ij} dv_j\, (i = 1, \ldots, n)\,. \tag{4}$$

Since $d_x u_1, \ldots, d_x u_n$ form a basis of Θ_x^* for all $x \in U$, it follows from (4) that $\det(h_{ij}(x)) \neq 0$. By analogy with analysis $\det(h_{ij})$ is called the *Jacobian function* of $u_1, \ldots, u_n$ with respect to $v_1, \ldots, v_n$. We denote it by $J\left(\frac{u_1, \ldots, u_n}{v_1, \ldots, v_n}\right)$ As we have seen, $J\left(\frac{u_1, \ldots, u_n}{v_1, \ldots, v_n}\right) \in k[U]$ and for all $x \in U$

$$J\left(\frac{u_1, \ldots, u_n}{v_1, \ldots, v_n}\right)(x) \neq 0 . \tag{5}$$

Substitution of (4) in the expression for ω and a simple calculation in the exterior algebra show that

$$\omega = g \cdot J\left(\frac{u_1, \ldots, u_n}{v_1, \ldots, v_n}\right) dv_1 \wedge \ldots \wedge dv_n . \tag{6}$$

Thus, although the form $\omega \in \Omega^n[U]$ is given by a function $g \in k[X]$, such a specification is possible only when a choice is made of local parameters and depends essentially on this choice.

We recall that, as a rule, a representation (3) is possible only locally (see Theorems 1 and 2). If $X = \bigcup U_i$ and if in each U_i such a representation is possible, it may happen that we cannot associate with ω a single function g on the whole of X: the functions g_i obtained in the various U_i need not agree. We have seen an example of this in § 4.1 (Example 3).

4. Rational Differential Forms. Example 2 in § 4.1 shows that on an algebraic variety X there may be very few regular differential forms ($\Omega^1[\mathbb{P}^1] = 0$), whereas there are plenty of open subsets on which there are many such forms ($\Omega^1[U] = k[U]du$). We have come across a similar phenomenon in connection with the notion of a regular function, and starting out from just these arguments we have introduced the concept of a rational function as a function that is regular on some open subset. We now introduce a similar notion for differential forms.

We consider a smooth irreducible quasiprojective variety X. Let ω be an r-dimensional differential form on X. We recall that it makes sense to speak of ω vanishing at a point $x \in X : \omega(x) \in \Lambda^r \Theta_x^*$, and in particular, it may be zero.

Lemma. *The set of points at which a regular differential form ω vanishes is closed.*

Let Y be the set of zeros of the form ω. Since closure is a local property, we may restrict our attention to a sufficiently small neighbourhood U of an arbitrary point $x \in X$. In particular, we may choose U so that Theorems 1 and 2 hold in it. Then there exists functions $u_1, \ldots, u_n \in k[U]$ such that $\Omega^r[U]$ is a free module with the generators

$du_{i_1} \wedge \ldots \wedge du_{i_r}$, $i_1 < \ldots < i_r$. Therefore ω has a unique representation in the form $\omega = \sum g_{i_1 \ldots i_r} du_{i_1} \wedge \ldots \wedge du_{i_r}$ and $\omega(x) = 0$ is equivalent to the equalities $g_{i_1 \ldots i_r}(x) = 0$, which determine a closed set.

From the lemma it follows, in particular, that if $\omega(x) = 0$ for all points x of an open set U, then $\omega = 0$ on the whole of X.

We now introduce a new object, which consists of an open set $U \subset X$ and a differential form $\omega \in \Omega^r[U]$. For such pairs (ω, U) we define an equivalence relation $(\omega, U) \sim (\omega', U')$ if $\omega = \omega'$ on $U \cap U'$. By the remark made above it is sufficient to require that ω and ω' agree on some open set contained in U and U', and the transitivity of this equivalence relation follows from this. The class defined by it is called a *rational differential form* on X. The set of all r-dimensional rational differential forms on X is denoted by $\Omega^r(X)$. Clearly $\Omega^0(X) = k(X)$.

Operations on representatives carry over to classes and define a multiplication: if $\omega_r \in \Omega^r(X)$, $\omega_s \in \Omega^s(X)$, then $\omega_r \wedge \omega_s \in \Omega^{r+s}(X)$. For $s = 0$ we see that $\Omega^r(X)$ is a module over $k(X)$.

If a rational differentional form ω (which is an equivalence class of pairs) contains a pair $(\bar{\omega}, U)$, then ω is called *regular* in U. The union of all open sets in which ω is regular is an open set U_ω, the so-called *domain of regularity* of ω. Evidently ω determines a certain regular form belonging to $\Omega^r[U_\omega]$. If $x \in U_\omega$, then we say that ω is *regular* at x. Obviously $\Omega^r(X)$ does not change when X is replaced by an open subset, in other words, it is a birational invariant.

Let us clarify the structure of the module $\Omega^r(X)$ over $k(X)$.

Theorem 3. *$\Omega^r(X)$ is a vector space of dimension $\binom{n}{r}$ over $k(X)$.*

We consider any open set $U \subset X$ for which the module $\Omega^r[U]$ is free over $k[U]$ (Theorems 1 and 2). Then there exist n functions $u_1, \ldots, u_n \in k[U]$ such that the products

$$du_{i_1} \wedge \cdots \wedge du_{i_r}, \quad 1 \leqslant i_1 < \cdots < i_r \leqslant n \tag{1}$$

form a basis of $\Omega^r[U]$ over $k[U]$. Every form $\omega' \in \Omega^r(X)$ is regular in some open set $U' \subset U$ for which, as before, the forms (1) give a basis in $\Omega^r[U']$ over $k[U']$. Therefore ω' is uniquely representable in the form

$$\sum_{1 \leq i_1 < \cdots < i_r \leq n} g_{i_1 \cdots i_r} du_{i_1} \wedge \cdots \wedge du_{i_r},$$

where the $g_{i_1 \cdots i_r}$ are regular in some open set $U' \subset U$, that is, rational on X. But this means that the forms (1) are a basis of $\Omega^r(X)$ over $k(X)$.

When do n functions $u_1, \ldots, u_n \in k(X)$ have the property that $du_{i_1} \wedge \cdots \wedge di_{i_r} (1 \leqslant i_1 < \cdots < i_r \leqslant n)$ is a basis of $\Omega^r(X)$ over $k(X)$? We derive a sufficient condition for this. The condition is also necessary, but we do not need this.

Theorem 4. *If $u_1, \ldots, u_n$ is a separable transcendence basis of $k(X)$, then the forms $du_{i_1} \wedge \cdots \wedge du_{i_r}$, $1 \leqslant i_1 < \cdots < i_r \leqslant n$, are a basis of $\Omega^r(X)$ over $k(X)$.*

Since $\Omega^r(X)$ and $k(X)$ are birational invariants, we may take X to be affine: $X \subset \mathbb{A}^N$.

Let $u_1, \ldots, u_n$ be a separable transcendence basis of $k(X)$. Then every element $v \in k(X)$ satisfies a relation $F(v, u_1, \ldots, u_n) = 0$ that is separable with respect to v.

In particular, the relations $F_i(t_i, u_1, \ldots, u_n) = 0 (i = 1, \ldots, N)$ hold for all coordinates t_i in $\mathbb{A}^N$. It follows from them that

$$F'_{i,t_i} dt_i + \sum_{j=1}^{n} F'_{i,u_j} du_j = 0 \, (i = 1, \ldots, N) \quad \text{on} \quad X.$$

From the separability of the polynomials F_i with respect to t_i it follows that $F'_{i,t_i} \neq 0$ on X. Therefore

$$dt_i = \sum_{j=1}^{n} \left(- \frac{F'_{i,u_j}}{F'_{i,t_i}} \right) du_j. \tag{2}$$

On some open set $U \subset X$ all the functions $-F'_{i,u_j}/F'_{i,t_i}$ and u_i are regular, and then (2) shows that at every point $y \in U$ the differentials $d_y u_j$ generate Θ^*_y. Since the number of these differentials is equal to the dimension of the space, they form a basis of it. Therefore the du_i form a basis of the module $\Omega^1[U]$ over $k[U]$, the products (1) form a basis of $\Omega^r[U]$ over $k[U]$, and a fortiori a basis of $\Omega^r(X)$ over $k(X)$.

Exercises

1. Show that on the affine circle with the equation $x^2 + y^2 = 1$ the rational differential form dx/y is regular. It is assumed that the characteristic of the ground field is not 2.

2. In the notation of Exercise 1 show that $\Omega^1[X] = k[X] \dfrac{dx}{y}$. Hint: Write an arbitrary form $\omega \in \Omega^1[X]$ in the form $\omega = f \cdot \dfrac{dx}{y}$ and use the fact that $\dfrac{dx}{y} = -\dfrac{dy}{x}$.

3. Show that in Example 3 of § 4.1 $\dim \Omega^1[X] = 1$.

4. Show that $\Omega^n[\mathbb{P}^n] = 0$.

5. Show that $\Omega^1[\mathbb{P}^n] = 0$.

6. Show that $\Omega^r[\mathbb{P}^n] = 0$ for $r > 0$.

7. Let $\omega = \dfrac{P(t)}{Q(t)} dt$, where P and Q are polynomials $\deg P = m$, $\deg Q = n$, be a rational form on $\mathbb{P}^1$ (t is the coordinate on $\mathbb{P}^1$). At what points $x \in \mathbb{P}^1$ is the form ω not regular?

8. Show that the tangent fibre of the smooth variety X introduced in Ch. II, § 1.4, is birationally isomorphic to the direct product $X \times \mathbb{A}^n$. Hint: For the open set U in Theorem 1 construct an isomorphism of the tangent fibre to U onto $U \times \mathbb{A}^n$: $(x, \xi) \to x \times ((d_x u_1)(\xi), \ldots, (d_x u_n)(\xi))$, $\xi \in \Theta_x$.

9. Compute the module R constructed in the proof of Proposition 1 in § 4.2 for the curve $y^2 = x^3$, and show that $\xi(3y\,dx - 2x\,dy) = 0$. Hint: Use the fact that

$$k[X] = k[x] + k[x]y.$$

10. Let K be an extension of k. A derivation of K over k is a k-linear mapping $D: K \to K$ satisfying the condition $D(xy) = D(x)\,y + xD(y)$, $x, y \in K$. Show that if $u \in K$ and if D is a derivation, then so is the mapping $D_1(x) = uD(x)$, so that all the derivations of K over k form a vector space over K, which is denoted by $D_k(K)$.

11. Let D be a derivation of the field $K = k(X)$ over k, $\omega \in \Omega^1(X)$, $\omega = \Sigma f_i\, dg_i$. Show that the function $(D, \omega) = \Sigma f_i D(g_i)$ does not depend on the representation of ω in the form $\Sigma f_i dg_i$. Show also that this scalar product establishes an isomorphism

$$D_k(K) \simeq (\Omega^1(X))^* = \mathrm{Hom}_{k(X)}(\Omega^1(X), k(X)) .$$

§ 5. Examples and Applications of Differential Forms

1. Behaviour under Mappings. We begin by investigating the behaviour of differential forms under regular mappings. If $\varphi: X \to Y$ is such a mapping, $x \in X$, then $d_x\varphi$ is a mapping $\Theta_{x,X} \to \Theta_{\varphi(x),Y}$, and the adjoint transformation $(d_x\varphi)^*$ maps $\Theta^*_{\varphi(x),Y}$ into $\Theta^*_{x,X}$. Hence for $\omega \in \Phi[Y]$ we have $\varphi^*(\omega) \in \Phi[X]$, where $\varphi^*(\omega)\,(x) = (d_x\varphi)^*(\omega\,(\varphi(x)))$.

From the definition it follows easily that the mapping $(d_x\varphi)^*$ is compatible with taking the differential, that is, $\varphi^*(d_{\varphi(x)}f) = d_x(\varphi^*(f))$ for $f \in k[Y]$. Hence it follows that if $\omega \in \Omega^1[Y]$, then $\varphi^*(\omega) \in \Omega^1[X]$, and φ^* determines a homomorphism $\varphi^*: \Omega^1[Y] \to \Omega^1[X]$, which is compatible with taking the differential for $f \in k[Y]$.

Finally, from linear algebra it is standard knowledge that a linear transformation of linear spaces $\varphi: L \to M$ determines a linear transformation $\Lambda^r\varphi: \Lambda^r L \to \Lambda^r M$. Applying this to the mapping $(d_x\varphi)^*$ we obtain a mapping $\Lambda^r(d_x\varphi)^*: \Lambda^r\Theta^*_{\varphi(x),Y} \to \Lambda^r\Theta^*_{x,X}$ and mappings $\Phi^r[Y] \to \Phi^r[X]$ and $\Omega^r[Y] \to \Omega^r[X]$. We denote the latter again by φ^*.

From all we have said above it follows that the effective computation of the action of the operator φ^* on a differential form is very simple: if

$$\omega = \Sigma\, g_{i_1 \dots i_r} du_{i_1} \wedge \dots \wedge du_{i_r}$$

then

$$\varphi^*(\omega) = \Sigma\, \varphi^*(g_{i_1 \dots i_r})\, d(\varphi^*(u_{i_1})) \wedge \dots \wedge d(\varphi^*(u_{i_r})) . \tag{1}$$

Now let X be irreducible, $\varphi: X \to Y$ a rational mapping, and $\varphi(X)$ dense in Y. Since φ is a regular mapping of an open set $U \subset X$ into Y and every open set $V \subset Y$ intersects $\varphi(U)$, the preceding arguments determine a mapping $\varphi^*: \Omega^r(Y) \to \Omega^r(X)$. This mapping is again given by formula (1).

We know that for $r = 0$, that is, for functions, the mapping φ^* is an embedding. For differential forms this is not always the case. For example, let $X = Y = \mathbb{P}^1$, $k(X) = k(t)$, $k(Y) = k(u)$, let k be of finite characteristic p, and let φ be given by the formula $u = t^p$. Then $\varphi^*(f(u)) = f(t^p)$ and $\varphi^*(df) = d(f(t^p)) = 0$, $(f \in k(u))$, so that $\varphi^*(\Omega^1(Y)) = 0$. The situation is clarified by the following result.

Theorem 1. *If the field $k(X)$ has a separable transcendence basis over $k(Y)$, then the mapping $\varphi^*: \Omega^r(Y) \to \Omega^r(X)$ is an embedding.*

Here we identify the field $k(Y)$ with the subfield $\varphi^* k(Y)$ of $k(X)$.

Suppose that $k(X)/k(Y)$ has a separable transcendence basis $v_1, \ldots, v_s$. This means that $v_1, \ldots, v_s$ are algebraically independent over $k(Y)$, and that $k(X)$ is a finite separable extension of $k(Y)(v_1, \ldots, v_s)$. The field $k(Y)$ has a separable transcendence basis over k (see Note 1 to Theorem 6 of Ch. I, § 3), which we denote by $u_1, \ldots, u_r$. Then $u_1, \ldots, u_r, v_1, \ldots, v_s$ is a separable transcendence basis of $k(X)$ over k.

When we write an arbitrary differential form $\omega \in \Omega^r(Y)$ as

$$\omega = \Sigma g_{i_1 \ldots i_r} du_{i_1} \wedge \ldots \wedge du_{i_r} \tag{2}$$

and apply (1) to it, we obtain an expression for $\varphi^*(\omega)$ in terms of products $d\varphi^*(u_{i_1}) \wedge \ldots \wedge d\varphi^*(u_{i_r})$, which form part of a basis of $\Omega^r(X)$ over $k(X)$, because $\varphi^*(u_i)$ is part of the separable transcendence basis $u_1, \ldots, u_r, v_1, \ldots, v_s$ (Theorem 4 of § 4). Therefore $\varphi^*(\omega) = 0$ only if all the $\varphi^*(g_{i_1 \ldots i_r}) = 0$, but this is possible only when $g_{i_1 \ldots i_r} = 0$, that is, $\omega = 0$.

All the preceding results were more or less obvious. Now we come to an unexpected fact.

Theorem 2. *If X and Y are smooth varieties, Y projective, and $\varphi: X \to Y$ a rational mapping such that $\varphi(X)$ is dense in Y, then $\varphi^* \Omega^r[Y] \subset \Omega^r[X]$.*

In other words, φ^* carries regular differential forms into regular ones. Since φ is only rational, this seems altogether improbable even for functions, that is, for $r = 0$. In this case the situation is saved by the fact that owing to Y being projective, regular functions on Y are constants, and the theorem is trivial.

In the general case the theorem is less obvious. We utilize the fact that by Theorem 3 of Ch. II, § 3, the mapping φ is regular on $X - Z$, where Z is closed in X and $\operatorname{codim}_X Z \geqslant 2$. If $\omega \in \Omega^r[Y]$, then $\varphi^*(\omega)$ is regular on $X - Z$. Let us show that this implies its regularity on the whole of X. For this purpose we write $\varphi^*(\omega)$ in an open set U in the standard form (2) [with ω replaced by $\varphi^*(\omega)$], where now $u_1, \ldots, u_n$ are regular functions on U such that $du_{i_1} \wedge \ldots \wedge du_{i_r}$ is a basis of $\Omega^r[U]$ over $k[U]$. Then the regularity of the forms $\varphi^*(\omega)$ on $X - Z$ implies the regularity of all the functions $g_{i_1 \ldots i_r}$ in $U - (Z \cap U)$. But $\operatorname{codim}_U (Z \cap U) \geqslant 2$, and this means that the set of points where the $g_{i_1 \ldots i_r}$ are not regular is of codimension $\geqslant 2$. On the other hand, this set is a divisor $(g_{i_1 \ldots i_r})_\infty$. This is possible only when $(g_{i_1 \ldots i_r})_\infty = 0$ and hence the function $g_{i_1 \ldots i_r}$ is regular.

Corollary. *If two smooth projective varieties X and Y are birationally isomorphic, then the vector spaces $\Omega^r[X]$ and $\Omega^r[Y]$ over k are isomorphic.*

The significance of Theorem 2 and its corollary is enhanced by the fact that for a projective variety X the space $\Omega^r[X]$ is finite-dimensional over k. This result is a consequence of a general theorem on coherent sheaves, which will be proved in Ch. VI. For the case of curves we prove it in Subsection 3. We set $h^r = \dim \Omega^r[X]$. The corollary to Theorem 2 indicates that the numbers $h^r (r = 0, 1, \ldots, n)$ are birational invariants of a smooth projective variety X.

2. Invariant Differential Forms on a Group. Let X be an algebraic variety, ω a differential form of it, and g an automorphism of X. The form ω is said to be invariant under g if

$$g^*(\omega) = \omega .$$

In particular, let G be an algebraic group. From the definition given in § 3.2 it follows at once that for every element $g \in G$ the mapping

$$t_g(x) = g \cdot x$$

is regular and is an automorphism of G quâ algebraic variety. A differential form on G is called *invariant* if it is invariant under all the transformations t_g.

An invariant differential form is regular. For if a form ω is regular at a point $x_0 \in G$, then $t_g^* \omega$ is regular at $g^{-1} x_0$. But $t_g^* \omega = \omega$, hence ω is regular at all points $g x_0, g \in G$, and these are all the points of G.

We show how to find all the invariant differential forms on an algebraic group. For this purpose we observe that a mapping $f: X \to Y$ determines a mapping f^* not only on differential forms, but also on the vector spaces Φ^r:

$$f^*: \Phi^r[Y] \to \Phi^r[X] .$$

In particular, the t_g^* are automorphisms of the vector spaces $\Phi^r[G]$. We begin by determining the set of elements $\varphi \in \Phi^r[G]$ that are invariant under all t_g^*, $g \in G$. This set contains, in particular, the invariant differential forms.

The condition

$$t_g^*(\varphi) = \varphi$$

means that for every point $x \in G$

$$\varphi(x) = (\Lambda^r \, dt_g^*)(\varphi(gx)) . \tag{1}$$

In particular, for $g = x^{-1}$,

$$(\Lambda^r \, dt_{x^{-1}}^*)(\varphi(e)) = \varphi(x) . \tag{2}$$

This formula shows that φ is uniquely determined by the element $\varphi(e)$ of the finite-dimensional vector space $\Lambda^r\Theta_e^*$. Conversely, by specifying an arbitrary $\eta\in\Lambda^r\Theta_e^*$, we can construct by (2) an element $\varphi\in\Phi^r[G]$:

$$\varphi(x)=(\Lambda^r\, dt_{x^{-1}}^*)(\eta)\,.$$

A simple substitution shows that it also satisfies (1), in other words, is invariant under t_g^*. Thus, the subspace of elements $\varphi\in\Phi^r[G]$ that are invariant under the automorphisms t_g^* is isomorphic to $\Lambda^r\Phi_e^*$, and the isomorphism is given by the correspondence

$$\varphi\to\varphi(e)\,.$$

Let us now show that all the elements φ we have constructed are regular differential forms, that is, contained in $\Omega^r[G]$. Owing to the invariance, regularity of a form φ need only be verified at an arbitrary single point, for example, at the unit point e. Furthermore, we may restrict ourselves to the case $r=1$. For if

$$\eta=\Sigma\,\alpha_{i_1}\wedge\ldots\wedge\alpha_{i_r},\ \alpha_j\in\Lambda^1\Theta_e^*\,,$$

and if the forms φ_j corresponding to the α_j by (2) are regular, then the form $\varphi=\Sigma\,\varphi_{i_1}\wedge\ldots\wedge\varphi_{i_r}$ is regular and corresponds to η.

We take an affine neighbourhood V of e such that the module $\Omega^1[V]$ is free, and let $du_1,\ldots,du_n$ be a basis of it. There exists an affine neighbourhood U of e such that $\mu(U\times U)\subset V$, where μ is the mapping that defines multiplication in G. Like every function in $k[U\times U]$, $\mu^*(u_l)$ can be written in the form

$$\mu^*(u_l)(g_1,g_2)=\Sigma\, v_{lj}(g_1)\,w_{lj}(g_2),\ v_{lj},w_{lj}\in k[U]\,,$$
$$(g_1,g_2)\in U\times U\subset G\times G\,.$$

By definition $t_h=\mu s_h$, where s_h is the embedding $G\to G\times G$, $s_h(g)=(h,g)$. Therefore

$$(t_g^*\,du_l)(g)=\Sigma\, v_{lj}(g)\,d_e w_{lj}\,.$$

When we express dw_{l_j} in terms of the du_k, we obtain the relations

$$t_g^*du_l=\Sigma\, c_{ml}(g)\,du_m,\ c_{ml}\in k[U]\,, \tag{3}$$

where

$$c_{ml}(g)=\sum_j v_{lj}(g)\frac{\partial w_{lj}}{\partial u_m}(e)\,. \tag{4}$$

Now we write the invariant form φ as $\varphi=\Sigma\psi_m du_m$ and consider the relation $t_g^*\varphi=\varphi$ at e. Substituting the expressions (3) and equating coefficients of du_m, we obtain

$$\Sigma\, c_{ml}\psi_l=\psi_m(e)\,. \tag{5}$$

Since $(c_{ml}(e))$ is the unit matrix, we have $\det(c_{ml})(e) \neq 0$, and from the system of Eqs. (5) it follows that $\psi_m \in \mathcal{O}_e$.

Let us state the result we have proved:

Proposition. *The mapping $\omega \to \omega(e)$ establishes an isomorphism between the space of r-dimensional invariant regular differential forms on G and the space $\Lambda^r \Theta_e^*$.*

3. The Canonical Class. Now we make a special analysis of n-dimensional rational differential forms on an n-dimensional smooth variety X. In some neighbourhood of a point $x \in X$ such a form can be represented as $\omega = g\, du_1 \wedge \ldots \wedge du_n$. We cover the whole of X with affine sets U_i such that in each of them this representation $\omega = g^{(i)} du_1^{(i)} \wedge \ldots \wedge du_n^{(i)}$ holds. In the intersection $U_i \cap U_j$ we find, according to (6) of § 4.3, that

$$g^{(j)} = g^{(i)} J\left(\frac{u_1^{(i)}, \ldots, u_n^{(i)}}{u_1^{(j)}, \ldots, u_n^{(j)}}\right).$$

Since the Jacobian J is regular and non-zero in $U_i \cap U_j$ [see (5) in § 4.3], the system of functions $g^{(i)}$ in U_i is compatible in the sense of § 1.2 and therefore determines a divisor on X. This is called the *divisor of the form* ω and is denoted by (ω).

The following properties of a divisor of an n-dimensional differential form on an n-dimensional variety follow easily from the definition:

a) $(f \cdot \omega) = (f) + (\omega)$ if $f \in k(X)$;

b) $(\omega) \geqslant 0$ if and only if $\omega \in \Omega^n[X]$.

According to Theorem 3 of § 4 (for $r = n$) the space $\Omega^n(X)$ is one-dimensional over $k(X)$. Therefore, if $\omega_1 \in \Omega^n(X)$, $\omega_1 \neq 0$, then every form $\omega \in \Omega^n(X)$ can be represented as $\omega = f\omega_1$. Hence property a) shows that the divisors of all the forms $\omega \in \Omega^n(X)$ are equivalent to each other and form a single divisor class on X.

This divisor class is called the *canonical class* of X and is denoted by K or K_X.

Let ω_1 be a fixed form in $\Omega^n(X)$ in terms of which every form can be expressed as $\omega = f\omega_1$. Property b) shows that ω is regular on X if and only if $(f) + (\omega_1) \geqslant 0$. In order words, $\Omega^n[X] \simeq \mathscr{L}((\omega_1))$, where we make use of the concept of the space associated with a divisor that was introduced in § 1.5.

Thus, $h^n = \dim_k \Omega^n[X] = l((\omega_1)) = l(K)$. So we see that the invariant h^n introduced in § 5.1 coincides with the dimension of the canonical class.

Example. Let us assume that X is the variety of an algebraic group. In § 5.2 we have shown that the space of r-dimensional invariant differential forms on X is isomorphic to $\Lambda^r \Theta_e^*$, where Θ_e is the tangent space to X at the unit point e. In particular, the space of n-dimensional invariant differential forms is one-dimensional because $\Lambda^n\Theta_e^* \simeq k$. If ω is a non-zero invariant form, then $\omega \in \Omega^n[X]$, that is, $(\omega) \geqslant 0$. But if $\omega(x)=0$ for some point $x \in X$, then by the invariance also $\omega(y)=0$ for every point $y \in X$. Therefore $\omega(x) \neq 0$ for all $x \in X$, that is, ω is regular and does not vanish on X. This means that $(\omega)=0$ or that $K_X=0$.

In § 2 we have shown that the number $l(D)$ is finite for any divisor D on a smooth projective algebraic curve. Hence it follows, in particular, that the number $h^1=\dim_k \Omega^1[X]$ is finite for any smooth projective algebraic curve X. This number is called the genus of the curve and is denoted by $g(X)$ or g; here $h^1=g$ if $\dim X=1$.

When $\dim X=1$, we know that all the divisors of one class have one and the same degree, so that we can speak of the degree $\deg C$ of a class C. In particular, the degree $\deg K_X$ of the canonical class is a birational invariant of the curve X.

The invariants we have introduced: the genus $g(X)$ and $\deg K_X$ are not independent. It can be shown that they are linked by the relation $\deg K_X=2g(X)-2$. (See § 5.6.) In particular, if a smooth projective curve X is an algebraic group, then $K_X=0$, as we have just seen. Therefore $g_X=1$, that is, among all projective curves only on those of genus 1 can the law of an algebraic group be defined. We shall see in § 5.6 that curves of genus 1 are precisely the smooth cubic curves.

4. Hypersurfaces. Next we compute the canonical class and determine h^n for the case when X is a smooth hypersurface in $\mathbb{P}^N$, $n=\dim X=N-1$. Let X be given by an equation $F(x_0:\ldots:x_N)=0$, $\deg F=\deg X=m$. We consider an affine open set U in which $x_0 \neq 0$. In it X is given by an equation $G(y_1,\ldots,y_N)=0$, $G(y_1,\ldots,y_N)=F(1,y_1,\ldots,y_N)$, where $y_i=x_i/x_0$.

In the open subset $U_i \subset U$ in which $G'_{y_i} \neq 0$ local parameters are $y_1,\ldots,\hat{y}_i,\ldots,y_N$, and the form $dy_1\wedge\cdots\wedge\widehat{dy_i}\wedge\cdots\wedge dy_N$ is a basis of $\Omega^n[U_i]$ over $k[U_i]$. However, it is convenient to take as a basis the form

$$\omega_i=\frac{1}{G'_{y_i}}\,dy_1\wedge\cdots\wedge\widehat{dy_i}\wedge\cdots\wedge dy_N$$

(which is possible, because $G'_{y_i} \neq 0$ in U_i). The fact is that the forms $\omega_1,\ldots,\omega_N$ are very simply connected with each other: multiplying the relation

$$\sum_{i=1}^{N} G'_{y_i}\,dy_i=0$$

by $dy_1\wedge\cdots\wedge\widehat{dy_i}\wedge\cdots\wedge\widehat{dy_j}\wedge\cdots\wedge dy_N$ we see that

$$\omega_j=(-1)^{i+j}\,\omega_i. \tag{1}$$

Since X is smooth, $U=\bigcup U_i$, and it follows from (1) that all the forms ω_j are regular in the whole of U and that the divisor of these forms in U is equal to 0.

It remains to investigate the points that do not belong to U. Let us consider, for example, an open subset V in which $x_1 \neq 0$. Coordinates in this affine variety are $z_1, \ldots, z_N : z_1 = 1/y_1,\ z_i = y_i/y_1\ (i=2, \ldots, N)$. Evidently

$$y_1 = \frac{1}{z_1}, \quad y_i = \frac{z_i}{z_1} \quad (i=2, \ldots, N). \tag{2}$$

Therefore

$$dy_1 = -\frac{dz_1}{z_1^2}, \quad dy_i = \frac{z_1\, dz_i - z_i\, dz_1}{z_1^2} \quad (i=2, \ldots, N).$$

We substitute these expression in ω_N. Using the fact that $dz_1 \wedge dz_1 = 0$, we obtain

$$\omega_N = -\frac{1}{z_1^N G'_{y_N}} dz_1 \wedge \cdots \wedge dz_{N-1}.$$

The equation of X in V is of the form

$$H(z_1, \ldots, z_N) = 0, \quad \text{where} \quad H = z_1^m G\left(\frac{1}{z_1}, \frac{z_2}{z_1}, \ldots, \frac{z_N}{z_1}\right).$$

From the relation

$$H'_{z_N} = z_1^{m-1} G'_{y_N}\left(\frac{1}{z_1}, \frac{z_2}{z_1}, \ldots, \frac{z_N}{z_1}\right) = z_1^{m-1} G'_{y_N}(y_1, \ldots, y_N)$$

it follows that

$$\omega_N = -\frac{1}{z_1^{N-m+1} H'_{z_N}} dz_1 \wedge \cdots \wedge dz_{N-1}. \tag{3}$$

All the arguments referring to U also apply to V and show that

$$\Omega^n[V] = k[V] \frac{1}{H'_{z_N}} dz_1 \wedge \cdots \wedge dz_{N-1}. \tag{4}$$

Therefore in V we have $(\omega_N) = -(N-m+1) \cdot (z_1)$. Evidently (z_1) in V is a divisor of the form x_0 on X, as it was defined in § 1.2. Ultimately we find that the relation $(\omega_N) = (m-N-1) \cdot (x_0) = (m-n-2) \cdot (x_0)$ holds on X. Thus, K_X is the divisor class containing $(m-n-2)L$, where L is a section of X by a hyperplane.

Now let us find $\Omega^n[X]$. We know that $\Omega^n[U] = k[U]\, \omega_N$. Let $\omega = P(y_1, \ldots, y_N)\, \omega_N$, $P \in k[y_1, \ldots, y_N]$.

Substituting (2) and using (3) we find that in V

$$\omega = -\frac{\tilde{P}(z_1, \ldots, z_N)}{z_1^{l+N-m+1}} \frac{1}{H'_{z_N}} dz_1 \wedge \cdots \wedge dz_{N-1}, \quad l = \deg P,$$

$$\tilde{P}(z_1 \ldots z_N) = z_1^l P\left(\frac{1}{z_1}, \frac{z_2}{z_1}, \ldots, \frac{z_N}{z_1}\right).$$

From (4) it now follows at once that $(\omega) \geqslant 0$ in V if and only if $\tilde{P}/z_1^{l+N-m+1} \in k[V]$, that is, when $l+N-m+1 \leqslant 0$ or $l \leqslant m-N-1$. Thus, $\omega \in \Omega^n[X]$ if and only if $\omega = P \cdot \omega_N$,

$$\deg P \leqslant m - N - 1 = m - n - 2. \tag{5}$$

Hence it is easy to calculate the dimension of $\Omega^n[X]$. Namely, two distinct polynomials $P, Q \in k[y_1, \ldots, y_N]$ satisfying the condition (5) determine distinct elements of the ring $k[X]$, in other words, $P - Q \equiv 0(G)$, and this contradicts (5). Thus, the dimension of $\Omega^n[X]$ is the same as that of the space of the polynomials P satisfying (5). This dimension is $(m-1)\ldots(m-N)/N! = \binom{m-1}{N}$. Thus,

$$h^n(X) = \binom{m-1}{n+1}. \tag{6}$$

Here is the simplest case of this formula: for $N=2$, $n=1$

$$g(X) = \frac{(m-1)(m-2)}{2}$$

is the formula for the genus of a smooth plane curve of degree m.

From (6) we can draw at once an important conclusion. Interpreting $\binom{m-1}{n+1}$ as the number of combinations we see that for $m > m' > n+1$

$$\binom{m-1}{n+1} > \binom{m'-1}{n+1}.$$

Therefore (6) shows that hypersurfaces of distinct degrees $m, m' > n+1$ are birationally non-isomorphic. So we see that there exist infinitely many birationally non-isomorphic algebraic varieties of a given dimension.

In particular, for $N=2, m=3$ we obtain $g(X)=1$, and since $g(\mathbb{P}^1)=0$, we see that a smooth projective plane curve of degree 3 is non-rational.

From (6) it follows that $h^n(X)=0$ if $m \leqslant N$. In particular, $h^n(\mathbb{P}^n)=0$. For $n=1$ we have verified this directly in § 5.2.

Let us consider in more detail the case $m \leqslant N$. If $N=2$, this means that $m=1$ or 2. For $m=1$ we have $X=\mathbb{P}^1$ and we know already that $h^1(\mathbb{P}^1)=0$. For $m=2$ we are concerned with a smooth conic, which is isomorphic to $\mathbb{P}^1$, so that in this case the equality $h^1(X)=0$ does not tell us anything new.

Let $N=3$. For $m=1$ we are concerned with $\mathbb{P}^2$, and the equality $h^2=0$ is already known to us. For $m=2$, X is a quadric, which is birationally isomorphic to $\mathbb{P}^2$, so that the equality $h^2(X)=0$ is a consequence of $h^2(\mathbb{P}^2)=0$ and Theorem 2. For $m=3$, X is a cubic surface. If on such a surface there are two skew lines, then it is birationally isomorphic to $\mathbb{P}^2$ (see Example 2 of Ch. I, § 3.3). It can be shown that any smooth cubic surface contains two skew lines*, so that again the equality $h^2(X)=0$ is a consequence of Theorem 2 and the fact that $h^2(\mathbb{P}^2)=0$.

These examples lead to interesting questions on smooth hypersurfaces of small degree: $X \subset \mathbb{P}^N$, $m=\deg X \leqslant N$. We see that for $N=2$ or 3 the hypersurface X is birationally isomorphic to the projective space $\mathbb{P}^{N-1}$, which gives an "explanation" of the equality $h^n(X)=0$, $n=N-1$.

For $N=4$ we come across a new phenomenon. When $m=3$, for example, even for the hypersurface

$$x_0^3+x_1^3+x_2^3+x_3^3+x_4^3=0 \tag{7}$$

the question whether it is birationally isomorphic to $\mathbb{P}^3$ is very delicate. However, it can be shown that there exists a rational mapping $\varphi:\mathbb{P}^3 \to X$ such that $\varphi(\mathbb{P}^3)$ is dense in X and $k(\mathbb{P}^3)$ separable over $k(X)$ (see Exercise 18). Together with the equality $h^3(\mathbb{P}^3)=0$ and Theorem 2 this gives $h^3(X)=0$. In this context we introduce the following terminology: a variety X is called *rational* if it is birationally isomorphic to $\mathbb{P}^n$, $n=\dim X$, and *unirational* if there exists a rational mapping $\varphi:\mathbb{P}^n \to X$ such that $\varphi(P^n)$ is dense in X and $k(\mathbb{P}^n)/k(X)$ is separable. From Theorem 2 and Exercise 6 in § 4 it follows that for a unirational variety X all the $h^i(X)=0$.

Typical for a number of difficulties that occur in algebraic geometry is the question whether the concepts of a rational and a unirational variety are one and the same. This is the so-called Lüroth problem. Clearly, it can be restated as a question in the theory of fields: let K be a subfield of the field of rational functions $k(T_1, \ldots, T_n)$ such that $k(T_1, \ldots, T_n)/K$ is finite and separable; is K isomorphic to a field of rational functions?

For $n=1$ the answer is in the affirmative, even without the assumption that k is algebraically closed and that $k(T)/K$ is separable.

For $n=2$ without these restrictions the answer is in the negative, with them it is in the affirmative, but the proof is very subtle. For fields of characteristic zero there is an account, for example, in [3], Ch. III.

* See cy. Manin, Y. Cubic Forms, North-Holland, 1974, p. 118.

For $n \geqslant 3$ the answer is in the negative even when k is the field of complex numbers. The simplest example of a unirational but not rational variety is a 3-dimensional smooth hypersurface of degree 3 in $\mathbb{P}^4$, in particular, the hypersurface (7). (See Ex. 18 and footnote to p. 425.)

5. Hyperelliptic Curves. As a second example we consider one type of curves. We denote by Y an affine plane curve with the equation $y^2 = F(x)$, where $F(x)$ is a polynomial without multiple roots of odd degree $n = 2m+1$ (in Ch. I, § 1, it was proved that the case of even degree reduces to that of odd degree). We assume that the characteristic of k is not 2. A smooth projective model X of the curve Y is called a *hyperelliptic curve*. We compute the canonical class and the genus of X.

The rational mapping $(x, y) \to x$ of the curve Y in $\mathbb{A}^1$ determines a regular mapping $f: X \to \mathbb{P}^1$. Clearly $\deg f = 2$, so that by Theorem 1 of § 2 for $\alpha \in \mathbb{P}^1$ either $f^{-1}(\alpha)$ consists of two points z' and z'' in each of which $v_{z'}(u) = v_{z''}(u) = 1$ for a local parameter u at α, or else $f^{-1}(\alpha) = z$ and $v_z(u) = 2$.

The affine curve Y is easily seen to be smooth. If $\bar{Y}$ is its projective closure, then X is a normalization of $\bar{Y}$ and we have the mapping $\varphi: X \to \bar{Y}$, which is an isomorphism between Y and $\varphi^{-1}(Y)$. Hence it follows that if a point $\xi \in \mathbb{A}^1$ has the coordinate α and $F(\alpha) \neq 0$, then $f^{-1}(\xi) = (z', z'')$, and if $F(\alpha) = 0$, then $f^{-1}(\xi) = z$.

Let us consider the point at infinity $\alpha_\infty \in \mathbb{P}^1$. If the coordinate on $\mathbb{A}^1$ is denoted by x, then $u = x^{-1}$ is a local parameter at α_∞. If $f^{-1}(\alpha_\infty)$ were to consist of two points z' and z'', then at z', say, the function u would be a local parameter. Hence it would follow that $v_{z'}(u) = 1$, $v_{z'}(F(x)) = -n$. But since n is odd, this contradicts the fact that $v_{z'}(F(x)) = 2v_{z'}(y)$. Thus, $f^{-1}(\alpha_\infty)$ consists of a single point, which we denote by z_∞, and $v_{z_\infty}(x) = -2$, $v_{z_\infty}(y) = -n$. It follows that $X = \varphi^{-1}(Y) \cup z_\infty$.

Let us now turn to differential forms on X. Consider, for example, the form $\omega = dx/y$. If $y(\xi) \neq 0$ at a point $\xi \in Y$, then x is a local parameter and $v_\xi(\omega) = 0$. But if $y(\xi) = 0$, then y is a local parameter and $v_\xi(x) = 2$, from which it follows again that $v_\xi(\omega) = 0$. Thus, $(\omega) = l \cdot z_\infty$, and it remains for us to determine l. For this purpose we recall that if t is a local parameter at z_∞, then $x = t^{-2}u$, $y = t^{-n}v$, where u and v are units in $\mathcal{O}_{z_\infty}$. Therefore $\omega = t^{n-3} w\, dt$, with $w, w^{-1} \in \mathcal{O}_{z_\infty}$, hence $(\omega) = (n-3) z_\infty$.

Now let us find $\Omega^1[X]$. As we have seen, ω forms a basis of the module $\Omega^1[Y]: \Omega^1[Y] = k[Y]\omega$, so that every form in $\Omega^1[X]$ can be written as $u\omega$, where $u \in k[Y]$, hence can be represented as $P(x) + Q(x)y$, $P, Q \in k[X]$.

It remains to clarify which of these forms are regular at z_∞. This is so if and only if

$$v_{z_\infty}(u) \geqslant -(n-3). \tag{1}$$

Let us find such $u \in k[Y]$. Since $v_{z_\infty}(x) = -2$, $v_{z_\infty}(P(x))$ is always even, and since $v_{z_\infty}(y) = -n$, $v_{z_\infty}(Q(x)y)$ is odd. Therefore

$$v_{z_\infty}(u) = v_{z_\infty}(P(x) + Q(x)y) \leqslant \min\left(v_{z_\infty}(P(x)), v_{z_\infty}(Q(x)y)\right),$$

hence, if $Q \neq 0$, then $v_{z_\infty}(u) \leqslant -n$. Thus, $u = P(x)$, and (1) shows that $2\deg P \leqslant n-3$.

We have found that $\Omega^1[X]$ consists of forms $P(x)\,dx/y$, where the degree of the polynomial $P(x)$ does not exceed $(n-3)/2$. Hence $g = h^1 = \dim \Omega^1[X] = (n-1)/2$.

It is intersting to compare the results of § 5.4 and § 5.5 for $N=2$. In the second case we have seen that there exist algebraic curves of any preassigned genus and in the first that the genus of a plane smooth curve is of the form $(n-1)(n-2)/2$, in other words, is by no means an arbitrary integer. Thus, not every smooth projective curve is isomorphic to a plane smooth curve. For example, this is not true for hyperelliptic curves with $n=9$.

6. The Riemann-Roch Theorem for Curves. One of the central results of the theory of algebraic curves is the *Riemann-Roch theorem.* It is expressed in the equation

$$l(D) - l(K-D) = \deg D - g + 1\,, \tag{1}$$

where D is an arbitrary divisor on a smooth projective curve, K its canonical class, and g its genus.

The proof of this theorem goes deep into the details of algebraic curves and will therefore not be given here. However, we can indicate some of its consequences which make its value for theory of curves quite manifest.

Corollary 1. *Setting $D=K$ we find that, since $l(K-K)=l(0)=1$, and $l(K)=g$, we have* $\deg K = 2g-2$.

Of this equation we have talked in § 5.3.

Corollary 2. *If* $\deg D > 2g-2$, *then* $l(K-D)=0$.

For otherwise there would exist a divisor D' such that $K-D \sim D' \geqslant 0$, but this is impossible because $\deg D' < 0$. Thus, the Riemann-Roch theorem shows that $l(D) = \deg D - g + 1$ for $\deg D > 2g-2$.

Corollary 3. *If $g=0$ and $D=x$ is a point on X, then by* (1), $l(D) \geqslant 2$.

This means that the space $\mathscr{L}(D)$ contains, apart from constants, also a nonconstant function f. For such a function $(f)_\infty = x$, that is, if we interpret f as a mapping $f: X \to \mathbb{P}^1$, then $\deg f = 1$ by Theorem 1 of § 2. Hence it follows that $X \simeq \mathbb{P}^1$, that is, the equality $g=0$ is not only necessary, but also sufficient for a curve X to be rational.

Corollary 4. *We consider a basis $f_0, \dots, f_n$ of the space $\mathscr{L}(D)$, $D \geqslant 0$, and the corresponding rational mapping $\varphi = (f_0 : \dots : f_n)$, $X \to \mathbb{P}^n$. Let us clarify when φ is an embedding. We show that this is so under the following conditions:*

$$\begin{aligned} l(D-x) &= l(D) - 1\,, \\ l(D-x-y) &= l(D) - 2\,, \end{aligned} \tag{2}$$

for arbitrary points $x, y \in X$.

From Corollary 2 it follows that the Eq. (2) are true if $\deg D \geqslant 2g+1$, so that in this case φ is an embedding.

We note first of all, that the first conditions in (2) guarantees that $-D = \text{g.c.d.}\ (f_i)$. For by definition g.c.d. $(f_i) \geqslant -D$. Now if we did not have equality, then there would exist a point x such that $(f_i) \geqslant -D + x$, that is, $\mathscr{L}(D) \subset \mathscr{L}(D-x)$, $l(D) \leqslant l(D-x)$, which contradicts (2). Thus, according to the remark at the end of § 1.4, the divisors $D_\lambda = (\Sigma \lambda_i f_i) + D$ are inverse images of hyperplanes under the mapping φ.

To prove that φ is an isomorphism we use the lemma in Ch. II, § 5.5, whose conditions we can verify by means of the remark made above. If $\varphi(x) = \varphi(y)$, then every hyperplane E passing through the point $\varphi(x)$ also passes through $\varphi(y)$. This means that if $D_\lambda - x \geqslant 0$, then $D_\lambda - x - y \geqslant 0$, that is, $l(D-x) \leqslant l(D-x-y)$, which contradicts the second condition in (2).

Let us show that the tangent spaces are mapped isomorphically. This is equivalent to the fact that

$$\varphi^* : \mathfrak{m}_{\varphi(x)}/\mathfrak{m}^2_{\varphi(x)} \to \mathfrak{m}_x/\mathfrak{m}_x^2$$

is an epimorphism. If this is not so, then $\varphi^*(\mathfrak{m}_{\varphi(x)}) \subset \mathfrak{m}_x^2$, because in our case $\dim \mathfrak{m}_x/\mathfrak{m}_x^2 = 1$. In other words, for any function $u \in \mathfrak{m}_{\varphi(x)}$ we have $v_x(\varphi^*(u)) \geqslant 2$. Applied to linear functions this shows that if $D_\lambda - x \geqslant 0$, then $D_\lambda - 2x \geqslant 0$. Again we find that $l(D-x) \leqslant l(D-2x)$, which contradicts the second condition in (2) and completes the proof.

Obviously, when a different basis is chosen in $\mathscr{L}(D)$, the mapping φ is multiplied by a projective transformation of the space $\mathbb{P}^n$. On the other hand, replacing D by another divisor $D + (f)$ leads to an isomorphism $u \to uf$ of the space $\mathscr{L}(D)$ and therefore does not change φ. Thus, it makes sense to talk of the mapping φ corresponding to a class of divisors.

For example, let X be a curve of genus 1, $x_0 \in X$. The conditions of Corollary 4 hold for the divisor $3x_0$. Therefore the mapping φ corresponding to this divisor maps X isomorphically onto a curve $X' \subset \mathbb{P}^2$ [because $l(3x_0) = 3$ by (1)]. As we have seen, $3x_0$ is the inverse image of a section of X' by a line, and since $\deg 3x_0 = 3$, also $\deg X' = 3$. Thus, every curve of genus 1 is isomorphic to a plane cubic curve.

Of the greatest interest are mappings φ that correspond to classes intrinsically connected with the curve X. Such are, for example, the multiples nK of the canonical class. The Eq. (1) shows that $\deg nK \geqslant 2g+1$ for $n \geqslant 2$ if $g>2$, and $n \geqslant 3$ if $g=2$. Thus, for $g>1$ the class $3K$ always satisfies the conditions of Corollary 4. The corresponding mapping φ_{3K} maps the curve X into $\mathbb{P}^m$, where $m=l(3K)-1=5g-6$ (by Corollary 2). Here two curves X and X' are isomorphic if and only if their images $\varphi_{3K}(X)$ and $\varphi_{3K}(X')$ are obtained from each other by projective transformations of the space. In this way the problem of a birational classification reduces to a projective classification.

The mapping φ corresponding to the canonical class is not always an embedding. However, all the cases when this is not so can be enumerated (see Exercises 16 and 17).

As a simple application of these arguments we consider plane curves of degree 4. According to § 5.4 their canonical class coincides with the class of the intersection with a line in $\mathbb{P}^2$. Therefore the mapping φ_K corresponding to the canonical class coincides with their natural embedding in a plane. From what we have said above it follows that two such curves are isomorphic if and only if they are projectively equivalent. This leads us to a very important conclusion. The set of plane curves of degree 4 can be identified with the space $\mathbb{P}^{14}$ (Ch. I, § 4.4). On the other hand, the group of all projective transformations of a plane is of dimension 8 (matrices of order 3 to within a constant factor). Using the theorem on the dimension of fibres it is easy to deduce that in $\mathbb{P}^{14}$ there exists an open set U and a mapping $f: U \to M$ onto some variety M such that two points u_1 and u_2 in U correspond to projectively equivalent curves only if they lie in one fibre of the mapping f. Therefore the dimension of the fibre is 8, and $\dim M = 14-8=6$.

Thus, it is by no means true that any two curves of degree 4 are isomorphic: they must also correspond to one and the same point on the 6-dimensional variety M. This shows that the genus is not a complete system of birational invariants of curves. Apart from the integral-valued invariant, the genus, curves also have "continuous" invariants, the so-called *moduli*. It can be shown that all curves of given genus $g>1$ form (in a sense we do not make precise here) a single continuous variety of dimension $3g-3$. In the case of curves of degree 4 we have $g=3$ and $3g-3=6=\dim M$. A similar also holds for curves of genus 1 (see Exercises 12 and 13). Only for $g=0$ are all the curves of this genus isomorphic.

7. Projective Immersions of Surfaces. Here we give an account of how the facts proved in the preceding subsection for algebraic curves can be generalized to surfaces. No proofs are given. The reader can find them in the book [3]. Furthermore, we restrict ourselves to the case of a field of characteristic 0.

An analogue to curves of genus greater than 1 are surfaces for which a multiple of the canonical class determines a birational isomorphism. They are called surfaces of general type, and for them a birational classification reduces in a certain sense to projective classification. The main result on surfaces of general type consists in the fact that for them already the five-fold canonical class $5K$ determines a regular mapping and a birational isomorphism.

It remains to list the surfaces that are not of general type. They play the role of curves of genus 0 and 1 and are given by similar constructions.

Analogues to rational curves are, firstly, *rational surfaces*, that is, surfaces birationally isomorphic to $\mathbb{P}^2$, and, secondly, *ruled surfaces*. These are surfaces X that can be mapped onto a curve C in such a way that all fibres of this mapping are isomorphic to a projective line $\mathbb{P}^1$. Thus, they are an algebraic family of lines.

Analogues to curves of genus 1 are three types of surfaces.

The first type are two-dimensional Abelian varieties. Surfaces of the second type (which are called *$K3$ surfaces*) have the property, in common with Abelian varieties, that their canonical class is 0. However, in contrast to Abelian varieties there are no regular one-dimensional differential forms on them (according to the results in § 5.2 on Abelian varieties there exist invariant, hence regular, one-dimensional differential forms). The third type are *elliptic surfaces*, that is, families of elliptic curves. These surfaces possess a mapping $f: X \to C$ onto a curve C such that for all $y \in C$ for which $f^{-1}(y)$ is a smooth curve (and such are all the y apart from finitely many) this curve has genus 1.

The main theorem states that all surfaces not of general type to within a birational isomorphism are exhausted by the five listed: rational, ruled, Abelian, $K3$ and elliptic.

To throw more light on these classes of surfaces it is convenient to classify them by an invariant $\varkappa$, the maximal dimension of the image of a surface X under the rational mappings given by the divisor classes nK, $n = 1, 2, \ldots$. If $l(nK) = 0$ for all n, then they are no such mappings and we set $\varkappa = -1$. Here is the result of the classification. Surfaces of general type are those for which $\varkappa = 2$. Surfaces with $\varkappa = 1$ are all elliptic. More accurately, they are the elliptic surfaces for which $nK \neq 0$ for $n \neq 0$. The order of the canonical class of an elliptic surface X in the group $\mathrm{Cl}(X)$ is infinite or is a divisor of 12. Surfaces with $\varkappa = 0$ are characterized by the condition $12K = 0$. Thus, they are elliptic surfaces for which $12K = 0$, surfaces of type $K3$, and two-dimensional Abelian varieties. Surfaces with $\varkappa = -1$ are rational or ruled.

For each of these types there is a characterization in terms of invariants similar to the way in which $g=0$ characterizes rational curves. We quote the characterization only for the first two types. For this purpose we use the result of Exercise 7, according to which the numbers $l(mK)$ for $m \geqslant 0$ are birational invariants of smooth projective varieties. They are called multiple genera and are denoted by P_m. In particular $P_1=h^n = \dim \Omega^n[X]$ for $n=\dim X$.

Criterion for Rationality. *A surface X is rational if and only if $\Omega^1[X] = 0$ and $P_1=P_2=0$.*

A solution of Lüroth's problem for surfaces follows easily from this criterion.

Criterion for Being Ruled. *A surface X is ruled if and only if $P_3 = P_4 = 0$.*

Generalizations of the result reported in this subsection to varieties of dimension >2 are not known.*

Exercises

1. Show that $df=0$ for an element $f \in k(X)$ if and only if $f \in k$ (when k is of characteristic 0) or $f=g^p$ (when k is of characteristic $p>0$). *Hint:* Use Theorem 1 and the following lemma: if K/L is a finite separable extension of characteristic p, $x \in K$, and its minimal polynomial is of the form $\Sigma a_i^p x^i, a_i \in L$, then $x=y^p$, $y \in K$.

2. Let X and Y be smooth projective curves, $\varphi: X \to Y$ a regular mapping such that $\varphi(X)=Y$, let $x \in X$, $y \in Y$, $\varphi(x)=y$, and let t be a local parameter at y. Show that the number $e_x = v_x(\varphi^* dt)$ does not depend on the choice of the local parameter t and that $e_x>0$ if and only if x is a ramification point of φ. The number e_x is called the *ramification multiplicity* of x.

3. In the notation of Exercise 2, let $\varphi^*(y)=\Sigma l_i x_i$, where y is the divisor consisting of the single point y. Suppose that the characteristic of k is either 0 or $p>l_i$. Show that $e_{x_i} = l_i - 1$.

4. In the notation of Exercises 2 and 3, let $Y=\mathbb{P}^1$. Show that $g(X)=\frac{1}{2}\sum_{x \in X} e_x - \deg\varphi + 1$. Generalize this relation to the case of an arbitrary curve Y.

5. Suppose that $\varphi: X \to Y$ satisfies the conditions of Exercise 2. Show that a differential $\omega \in \Omega^1(Y)$ is regular when the differential $\varphi^*\omega \in \Omega^1(X)$ is regular.

6. Denote by Ψ_m the set of all functions ψ in mn vectors x_{ij}, $i=1, \dots, m$, $j=1, \dots, n$, of an n-dimensional space L satisfying the following conditions:

a) ψ is linear in every argument,

b) ψ is skew-symmetric as a function of $x_{i_0 j}$, $j=1, \dots, n$ for any fixed i_0,

c) ψ is symmetric as a function of $x_{i j_0}$, $i=1, \dots, m$, for any given j_0. Suppose that the characteristic of k is greater than m. Show that every function $\psi \in \Psi_m$ is given by its values $\psi_{y_1 \dots y_n}$ on the vectors $x_{ij}=y_j$ and that $\psi_{y_1 \dots y_n} = d^m \psi_{e_1 \dots e_n}$, where d is the determinant of the coordinates of the vectors $y_1, \dots, y_n$ in the basis $e_1 \dots e_n$. Let $\xi_1, \dots, \xi_n \in L^*$. The function ψ for which $\psi_{y_1 \dots y_n} = (\det(\xi_i(y_j)))^m$ is denoted by $(\xi_1 \wedge \dots \wedge \xi_n)^m$. Show that the space Ψ_m is 1-dimensional and that $(\xi_1 \wedge \dots \wedge \xi_n)^m$ is a basis of it.

7. Generalize the construction of regular and rational n-dimensional differential forms on n-dimensional varieties, replacing everywhere the space $\Lambda^n \Theta_x^*$ by the corresponding

* Generalizations of results described here, to surfaces over fields of positive characteristic, have since been proved by Bombieri and Mumford. See E. Bombieri and D. Husemoller: "Classification and embeddings of surfaces". Algebraic Geometry. Azcata 1974. Proceedings of Symposia in Pure Math. XXIX. (Footnote to new printing).

space Ψ_m. The corresponding object is called a *differential form of weight m*. Show that in the analogue to formula (6) of § 4.3 J must be replaced by J^m. Show that a differential form of weight m has a divisor, and that all these divisors form a single class, which coincides with mK_X. Generalize Theorem 2.

8. Determine the space of regular differential forms of weight 2 on a hyperelliptic curve. *Hint:* Write them in the form $f(dx)^2/y^2$.

9. Verify the Riemann-Roch theorem for the case $X=\mathbb{P}^1$.

10. Let X be a smooth projective curve, $g(X)=1$, and $x\in X$. Show that for every $n>1$ there exists a rational function u_n on X for which $v_x(u_n)=-n$, $v_y(u_n)\geqslant 0$ for $y\neq x$. Use Corollary 2 in § 5.6.

11. Under the conditions of Exercise 10 show that $k(X)=k(u_2,u_3)$, where u_2 and u_3 are connected by a relationship of degree 3. *Hint:* To derive this relationship apply the Riemann-Roch theorem to $\mathscr{L}(6x)$. Prove that $[k(X):k(u_2,u_3)]=1$, using Theorem 1 of § 2.

Deduce that if the characteristic of the field is not 2 or 3, then any curve of genus 1 is isomorphic to a curve given by the equation

$$v^2=u^3+au+b\,,$$

where $\mathscr{L}(2x)=\{1,u\}$, $\mathscr{L}(3x)=\{1,u,v\}$.

12. Let X_1 and X_2 be two curves of genus 1 given by the equations $v_i^2=u_i^3+a_iu_i+b_i$, $i=1,2$. Show that any isomorphism between them carrying the pole u_1 into u_2 is determined by a linear transformation. Show that if the curves are isomorphic, then there exists an isomorphism with this property. *Hint:* Use the structure of the algebraic group on X_1 and X_2 and the existence of shifts carrying a point into any other given point.

13. Under the conditions of Exercise 12 and assuming that $b_1\neq 0$, $b_2\neq 0$, show that two curves X_1 and X_2 are isomorphic if and only if $a_1^3/b_1^2=a_2^3/b_2^2$.

14. Verify the relation $\deg K=2g-2$ for hyperelliptic curves and for smooth plane curves.

15. Show that for a hyperelliptic curve the ratios of regular differential forms generate a subfield of $k(X)$ isomorphic to a field of rational functions. Starting out from this prove that a smooth plane projective curve of degree $m>3$ is not hyperelliptic.

16. Show that for a hyperelliptic curve the rational mapping corresponding to the canonical class is not an isomorphism.

17. Show that if the mapping corresponding to the canonical class of a curve X is not an isomorphism, then X is rational or hyperelliptic. *Hint:* If one of the conditions (2) in § 5.6 is not satisfied, then the Riemann-Roch theorem gives $l(x)\geqslant 2$ or $l(x+y)\geqslant 2$.

18. Show that a smooth hypersurface X of degree 3 in $\mathbb{P}^4$ is unirational. *Hint:* Using Theorem 10 of Ch. I, § 6, show that X contains a line l. Using Exercise 8 in § 4 show that there exists an open set $U\subset X$, $U\cap l\neq\emptyset$ such that the tangent fibre to U is isomorphic to $U\times\mathbb{A}^3$. Denote by $\mathbb{P}^2$ the projective space consisting of the lines that pass through the origin of coordinates in $\mathbb{A}^3$. For a point $\xi=(u,\alpha)$, $u\in l\cap U$, $\alpha\in\mathbb{P}^2$, denote by $\varphi(\xi)$ the point of intersection of the line α lying in $\theta_{u,X}$ with X. Show that φ determines the rational mapping $\mathbb{P}^1\times\mathbb{P}^2\to X$.

19. Let o be a point of an algebraic curve X of genus g. Show (using the Riemann-Roch theorem) that any divisor D with $\deg D=0$ is equivalent to a divisor of the form D_0-go, where $D_0>0$, $\deg D_0=g$ (generalization of Theorem 5 in § 2).

20. Let $X\subset\mathbb{P}^2$ be a smooth plane irreducible curve with the equation $F=0$, let $a=(\alpha_0:\alpha_1:\alpha_2)\notin X$, $x\in X$. The multiplicity c_x with which x occurs in a divisor of the form $\sum_{i=0}^{2}\alpha_i\frac{\partial F}{\partial x_i}$ is called the contact multiplicity at x. Show that $c_x=e_x$ is the ramification multiplicity of the projection maping $\varphi:X\to\mathbb{P}^1$ from α. Deduce that the number $c=\sum_{x\in X}c_x$ – the number of tangents taken with the relevant multiplicities passing through α – does not depend on α. It is called the *class* of the curve. Show that $c=n(n-1)$, $n=\deg X$.

Chapter IV. Intersection Indices

§ 1. Definition and Basic Properties

1. Definition of an Intersection Index. The theorems on the dimension of an intersection of varieties, which we have proved in Ch. I, often allow us to assert that certain systems of equations have solutions. However, they say nothing about the number of solutions, assuming it to be finite. The difference is like that between the theorem on the existence of roots of a polynomial and the theorem that the total number of roots of a polynomial is equal to its degree. The latter theorem is true only if we count each root with its multiplicity. Similarly, to formulate general theorems on the number of points of intersection of subvarieties, we ought to assign certain multiplicities to these points. This will be done in the present subsection.

We consider the intersection of subvarieties of codimension 1 on a smooth variety X. We are interested in the case when the number of points of intersection is finite. If $\dim X = n$, and if $C_1, \ldots, C_l$ are subvarieties of codimension 1 with a non-empty intersection, then by the theorem on the dimension of an intersection $\dim(C_1 \cap \cdots \cap C_l) > 0$ if $l < n$. Therefore we naturally consider the case $l = n$. The theory to be applied later becomes simpler if instead of subvarieties of codimension 1 we consider arbitrary divisors. Thus, we consider n divisors $D_1, \ldots, D_n$ on a n-dimensional variety X. If $x \in X$, $x \in \bigcap \operatorname{Supp} D_i$ and $\dim_x \bigcap \operatorname{Supp} D_i = 0$, then we say that $D_1, \ldots, D_n$ *are in general position* at x. This means that $\bigcap \operatorname{Supp} D_i$ in some neighbourhood of x consists of x only. If $D_1, \ldots, D_n$ are in general position at all points of the subvariety $\bigcap \operatorname{Supp} D_i$, this subvariety consists of finitely many points or is empty. We then say that $D_1, \ldots, D_n$ are *in general position.*

We begin by defining the intersection index for effective divisors in general position. Let $D_1, \ldots, D_n$ be effective divisors in general position at a point x, having local equations $f_1, \ldots, f_n$ in some neighbourhood of this point. Then there exists a neighbourhood U of x in which $f_1, \ldots, f_n$ are regular and do not vanish anywhere except at x. From Hilbert's Nullstellensatz it follows that the ideal generated by the functions $f_1, \ldots, f_n$

in the local ring $\mathcal{O}_x$ of x contains some power of the maximal ideal $\mathfrak{m}_x$ of this ring. Let

$$(f_1, \ldots, f_n) \supset \mathfrak{m}_x^l. \tag{1}$$

We consider the factor space $\mathcal{O}_x/(f_1, \ldots, f_n)$ (over k). Its dimension over k is finite. To see this it is sufficient, by (1), to show that $\dim_k \mathcal{O}_x/\mathfrak{m}_x^l < \infty$. The latter follows at once from the theorem on expansion in power series: $\dim_k \mathcal{O}_x/\mathfrak{m}_x^l$ coincides with the dimension of the space of polynomials of degree less than l in n variables.

Henceforth we denote the dimension of a vector space E over k by the symbol $l(E)$.

Definition 1. If $D_1, \ldots, D_n$ are effective divisors on an n-dimensional variety X in general position at a point $x \in X$, having local equations $f_1, \ldots, f_n$ in some neighbourhood of this point, then the number

$$l(\mathcal{O}_x/(f_1, \ldots, f_n)) \tag{2}$$

is called the intersection *index* (or *multiplicity*) of $D_1, \ldots, D_n$ at this point and is denoted by $(D_1, \ldots, D_n)_x$.

Actually, the number (2) depends only on the divisors $D_1, \ldots, D_n$ and not on the choice of their local equations $f_1, \ldots, f_n$: if $f_1', \ldots, f_n'$ are other local equations, then $f_i' = f_i g_i$, where $g_i, g_i^{-1} \in \mathcal{O}_x$, therefore $(f_1, \ldots, f_n) = (f_1', \ldots, f_n')$.

Suppose now that $D_1, \ldots, D_n$ are divisors, but not necessarily effective. We represent them in the form $D_i = D_i' - D_i''$, $D_i' \geqslant 0$, $D_i'' \geqslant 0$, where the divisors D_i' and D_i'' do not have common components. This representation is unique. Assuming that $D_1, \ldots, D_n$ are in general position at x, then for any permutation $i_1, \ldots, i_n$ and any l the divisors $D_{i_1}', \ldots, D_{i_l}', D_{i_{l+1}}'', \ldots, D_{i_n}''$ are in general position at x, because $\operatorname{Supp} D_i = \operatorname{Supp} D_i' \cup \operatorname{Supp} D_i''$.

We now define the intersection index of $D_1, \ldots, D_n$ at x by additivity, that is, we set

$$(D_1, \ldots, D_n)_x = \sum_{i_1, \ldots, i_n} \sum_{0 \leqslant l \leqslant n} (-1)^{n-l} (D_{i_1}', \ldots D_{i_l}', D_{i_{l+1}}'', \ldots, D_{i_n}'')_x. \tag{3}$$

Definition 2. If $D_1, \ldots, D_n$ are divisors in general position on an n-dimensional variety X, then the number

$$\sum_{x \in \bigcap \operatorname{Supp} D_i} (D_1, \ldots, D_n)_x$$

is called their *intersection index* and is denoted by $(D_1, \ldots, D_n)$.

One could extend the sum formally to all points $x \in X$, but only the non-zero terms are written down above.

Note. The intersection index could also be defined without the condition that X should be a smooth variety; however, then we have to confine ourselves to locally principal divisors. All our definitions remain valid.

Now we look at some examples, with the object of showing that our definition of intersection multiplicity agrees with the geometric intuition.

Example 1. Let $\dim X = 1$, t a local parameter at x, f the local equation of the divisor D, $v_x(f) = v_x(D) = m$. Then $(D)_x = l(\mathcal{O}_x/(f)) = l(\mathcal{O}_x/(t^m)) = m$. Thus, in this case the index $(D)_x$ is equal to the multiplicity with which x occurs in D.

In the next example we assume that the D_i are prime divisors, that is, irreducible subvarieties of codimension 1.

Example 2. If $x \in D_1 \cap \cdots \cap D_n$, then according to the definition $(D_1, \ldots, D_n)_x \geqslant 1$. Let us clarify when $(D_1, \ldots, D_n)_x = 1$.

Since $f_i \in \mathfrak{m}_x$, hence $(f_1, \ldots, f_n) \subset \mathfrak{m}_x$, and $l(\mathcal{O}_x/\mathfrak{m}_x) = 1$, the condition $(D_1, \ldots, D_n)_x = 1$ is equivalent to $(f_1, \ldots, f_n) = \mathfrak{m}_x$. In other words, $f_1, \ldots, f_n$ must form a system of local parameters. In Ch. II, § 2.1, we have seen that this holds if and only if the subvarieties $D_1, \ldots, D_n$ intersect at x transversally, so that x is a simple point on all the D_i and $\bigcap \Theta_{x, D_i} = x$.

Example 3. Let $\dim X = 2$, let x be a simple point on two curves D_1 and D_2. According to Example 2, $(D_1, D_2)_x > 1$ if and only if the lines Θ_{x, D_1} and Θ_{x, D_2} are identical. Let $f_i (i = 1, 2)$ be local equations of the curves D_i, and let u and v be local parameters at x and $f_i \equiv (\alpha_i u + \beta_i v)(\mathfrak{m}_x^2)$ $(i = 1, 2)$. Then the equations of the lines Θ_{x, D_i} are of the form $\alpha_i \xi + \beta_i \eta = 0$ $(i = 1, 2)$ where $\xi = d_x u, \eta = d_x v$ are coordinates on $\Theta_{x, X}$. Therefore $\Theta_{x, D_1} = \Theta_{x, D_2}$ if and only if $\alpha_2 u + \beta_2 v = \gamma(\alpha_1 u + \beta_1 v)$ for some $\gamma \in k$, $\gamma \neq 0$, in other words, $f_2 \equiv \gamma f_1 (\mathfrak{m}_x^2)$. It is therefore natural to define the order of contact of the curves D_1 and D_2 at x as the number l such that there exists an invertible element g, $g^{-1} \in \mathcal{O}_x$, for which $f_2 \equiv g f_1 (\mathfrak{m}_x^{l+1})$, and that such a g does not exist for larger values of the exponent l. We show that the intersection index $(D_1, D_2)_x$ exceeds by 1 the order of contact of the curves D_1 and D_2 at x.

For this purpose we note that since x is a simple point on D_1, we may assume that f_1 is one of the elements of the system of local parameters at x. On the other hand, $g^{-1} f_2$ is a local equation of D_2. Therefore we may assume that u and v are local parameters, that a local equation of D_1 is u, and one of D_2 is f, with $f \equiv u (\mathfrak{m}_x^{l+1})$. Then $f \equiv u + \varphi(u, v) (\mathfrak{m}_x^{l+2})$, where φ is a form of degree $l + 1$. Here φ is not divisible by u, otherwise D_1 and D_2 would have contact of order $> l$. Therefore

$$\varphi(0, v) = C v^{l+1}, \quad C \neq 0. \tag{4}$$

According to the definition of the intersection index

$$(D_1, D_2)_x = l(\mathcal{O}_x/(u, f)) = l(\mathcal{O}_x/(u)/(u, f)/(u)).$$

Evidently $\mathcal{O}_x/(u) = \bar{\mathcal{O}}$ is the local ring of x on D_1, and the homomorphism $\mathcal{O}_x \to \bar{\mathcal{O}}$ is the restriction of a function on X to the curve D_1. Furthermore, $(u,f)/(u) = (\bar{f})$, where $\bar{f}$ is the image of f in $\bar{\mathcal{O}}$. Since $\bar{f} \in \bar{\mathfrak{m}}_x^{l+1}$, $\bar{f} \equiv \bar{\varphi}(\bar{\mathfrak{m}}_x^{l+2})$, and since by (4) $\bar{\varphi} \notin \bar{\mathfrak{m}}_x^{l+2}$ in $\bar{\mathcal{O}}$, we have $v_x(\bar{f}) = l+1$ and $l(\bar{\mathcal{O}}/(\bar{f})) = l+1$. Thus, $(D_1, D_2)_x = l+1$.

Example 4. Again, let $\dim X = 2$ and let x be a singular point on D. This means that $f \in \mathfrak{m}_x^2$, where f is a local equation of D. Therefore it is natural to define the *multiplicity of the singular point* as the largest m for which $f \in \mathfrak{m}_x^m$. We show that for every curve D' in general position at x together with D,

$$(D, D')_x \geqslant m \tag{5}$$

and that there exist curves for which $(D, D')_x = m$.

Let f' be a local equation of D'. We denote the ring $\mathcal{O}_x/\mathfrak{m}_x^m$ by $\bar{\mathcal{O}}$ and the image of f' in $\bar{\mathcal{O}}$ by $\bar{f}$. Since $f \in \mathfrak{m}_x^m$, we have

$$(D, D')_x = l(\mathcal{O}_x/(f, f')) \geqslant l(\bar{\mathcal{O}}/(\bar{f})).$$

By the theorem on power series expansions the ring $\bar{\mathcal{O}}$ is isomorphic to $k[u,v]/(u,v)^m$. Therefore, as a vector space it is isomorphic to the space of polynomials in u and v of degree $<m$ and is of dimension $1+2+\dots+m = \dfrac{m(m+1)}{2}$. If $f' \in \mathfrak{m}_x^l$, $f' \notin \mathfrak{m}_x^{l+1}$, then to the elements of the ideal $(\bar{f})$ there correspond polynomials of the form $\bar{f} \cdot g$, where g ranges over all polynomials of degree $<m-l$. Therefore

$$l((\bar{f})) \leqslant 1 + \dots + (m-l) = \frac{(m+1-l)(m-l)}{2}.$$

Since $f' \in \mathfrak{m}$, we have $l \geqslant 1$ and therefore $l(\bar{\mathcal{O}}/(\bar{f})) = l(\bar{\mathcal{O}}) - l((\bar{f})) \geqslant m$.

Now we show that equality in (5) can be attained.

Let $f \equiv \varphi(u,v)\,(\mathfrak{m}_x^{m+1})$, where φ is a form of degree m. We consider a linear form in u and v that does not divide φ. By a linear transformation of u and v we can achieve that it is u, that is, $\varphi(0,v) \neq 0$. For D' we take the curve with the local equation u. Then $(D, D')_x = l(\mathcal{O}_x/(u,f))$, and as we have seen in the discussion of Example 3, this number is equal to m.

2. Additivity of the Intersection Index

Theorem 1. *If $D_1, \dots, D_{n-1}, D'_n$ and $D_1, \dots, D_{n-1}, D''_n$ are divisors in general position at x, then*

$$\begin{aligned}(D_1, \dots, D_{n-1}, D'_n + D''_n)_x \\ = (D_1, \dots, D_{n-1}, D'_n)_x + (D_1, \dots, D_{n-1}, D''_n)_x.\end{aligned} \tag{1}$$

Proof. First of all, it is obvious that Theorem 1 need only be proved for effective divisors $D_1, \dots, D_{n-1}, D'_n, D''_n$. We assume from now on that the divisors are effective.

We denote local equations of the divisors $D_1, \ldots, D_{n-1}, D_n', D_n''$ by $f_1, \ldots, f_{n-1}, f_n', f_n''$, the ring $\mathcal{O}_x/(f_1, \ldots, f_{n-1})$ by $\bar{\mathcal{O}}$, and the images of f_n' and f_n'' in $\bar{\mathcal{O}}$ by f and g. Then

$$(D_1, \ldots, D_{n-1}, D_n' + D_n'')_x = l(\bar{\mathcal{O}}/(f \cdot g)),$$
$$(D_1, \ldots, D_{n-1}, D_n')_x = l(\bar{\mathcal{O}}/(f)),$$
$$(D_1, \ldots, D_{n-1}, D_n'')_x = l(\bar{\mathcal{O}}/(g)).$$

Since the sequence

$$0 \to (g)/(f\,g) \to \bar{\mathcal{O}}/(f\,g) \to \bar{\mathcal{O}}/(g) \to 0$$

is exact, we have

$$l(\bar{\mathcal{O}}/(f \cdot g)) = l(\bar{\mathcal{O}}/(g)) + l((g)/(f \cdot g)). \tag{2}$$

If g is not a divisor of zero in $\bar{\mathcal{O}}$, then multiplication by g determines an isomorphism $\bar{\mathcal{O}}/(f) \simeq (g)/(f\,g)$ and

$$l((g)/(f \cdot g)) = l(\bar{\mathcal{O}}/(f)). \tag{3}$$

Therefore, if we can show that g is not a divisor of zero in $\bar{\mathcal{O}}$, then (1) follows from (2) and (3).

A sequence of n elements $f_1, \ldots, f_n$ of the local ring $\mathcal{O}_x$ of a simple point of an n-dimensional variety is called a *regular sequence* if the f_i are not divisors of zero in $\mathcal{O}_x/(f_1, \ldots, f_{i-1})$ for $i = 1, \ldots, n$.

The arguments above show that Theorem 1 is a consequence of the following proposition.

Lemma 1. *If $D_1, \ldots, D_n$ are divisors in general position at a simple point x, then their local equations $f_1, \ldots, f_n$ form a regular sequence.*

The proof of Lemma 1 in its turn requires a simple auxiliary proposition.

Lemma 2. *Over a local ring, the property of being a regular sequence is preserved under a permutation of the elements of the sequence.*

Proof of Lemma 2. It is enough to show that under a permutation of two adjacent terms f_i, f_{i+1} in a regular sequence we again obtain a regular sequence. We set $(f_1, \ldots, f_{i-1}) = \mathfrak{a}$, $\mathcal{O}_x/\mathfrak{a} = A$, and we denote by a and b the images of f_i and f_{i+1} in A. Everything reduces to a proof of Lemma 2 for the regular sequence a, b in A. We have to prove that 1) b is not a divisor of zero in A and 2) that a is not a divisor of zero $\bmod b$.

1) Let $cb = 0$. We show that then

$$c \in (a)^l \tag{4}$$

for all l. From the fact that A is Noetherian and from Theorem 5 in Ch. II, § 2, it then follows that $c = 0$.

The relation (4) can be verified by induction. If $c = c_1 a^l$, then $c_1 a^l b = 0$. Since a, b is a regular sequence, a is not a divisor of zero, hence $c_1 b = 0$. Again from the simplicity of the sequence a, b it follows that $c_1 \in (a)$, that is, $c \in (a)^{l+1}$.

2) Let $xa = yb$. From the regularity of the sequence a, b it follows that $y = az$, $z \in A$ hence $x = zb$.

This proves Lemma 2.

Proof of Lemma 1. The proof proceeds by induction on the dimension n of the variety X. From the condition of the lemma and the theorem on the dimension of an intersection it follows that $\dim_x(\mathrm{Supp}(f_1) \cap \cdots \cap \mathrm{Supp}(f_{n-1})) = 1$. Therefore we can find a function u such that $u(x) = 0$, that x is a simple point on the subvariety $V(u)$, and that $(f_1), \ldots, (f_{n-1}), (u)$ are divisors in general position at x. It is sufficient to take for u the equation of the hyperplane passing through x and not containing $\Theta_{x,X}$ nor any component of the curve $\mathrm{Supp}(f_1) \cap \cdots \cap \mathrm{Supp}(f_{n-1})$. We consider the restrictions of the functions $f_1, \ldots, f_{n-1}$ to $V(u)$. Evidently they satisfy all the conditions of Lemma 1, therefore by the inductive hypothesis form a regular sequence on $V(u)$. Since the local ring of x on $V(u)$ is of the form $\mathcal{O}_x/(u)$, we see that $u, f_1, \ldots, f_{n-1}$ is a regular sequence. It follows from Lemma 2 that then the sequence $f_1, \ldots, f_{n-1}, u$ is also regular.

To prove that the sequence $f_1, \ldots, f_{n-1}, f_n$ is regular it remains for us to verify that f_n is not a divisor of zero in $\mathcal{O}_x/(f_1, \ldots, f_{n-1})$. From the condition on the functions $f_1, \ldots, f_n$ it follows that in some neighbourhood of x the equations $f_1 = \cdots = f_n = 0$ have no solution other than x. Hilbert's Nullstellensatz therefore shows that $(f_1, \ldots, f_n) \supset \mathfrak{m}_x^l$ for some l. In particular, $u^l \in (f_1, \ldots, f_n)$, that is, $u^l \equiv a f_n ((f_1, \ldots, f_{n-1}))$ for some $a \in \mathcal{O}_x$.

If f_n were a divisor of zero in $\mathcal{O}_x/(f_1, \ldots, f_{n-1})$, then it would follow that also u^l, hence u, is a divisor of zero in this ring. But this contradicts the fact, which we have proved, that $f_1, \ldots, f_{n-1}, u$ is a regular sequence.

Lemma 1, and with it Theorem 1, are now proved.

3. Invariance under Equivalence. We proceed to the proof of the main property of intersection indices, which lies at the basis of all their applications.

Theorem 2. *If X is a smooth projective variety and if both $D_1, \ldots, D_{n-1}, D_n$ and $D_1, \ldots, D_{n-1}, D_n'$ are divisors in general position and D_n and D_n' are equivalent, then*

$$(D_1, \ldots, D_{n-1}, D_n) = (D_1, \ldots, D_{n-1}, D_n') . \tag{1}$$

By the condition of the theorem $D_n - D_n' = (f)$, and (1) is equivalent to the fact that

$$(D_1, \ldots, D_{n-1}, (f)) = 0 \tag{2}$$

when $D_1, \ldots, D_{n-1}, (f)$ are in general position.

Representing D_i, $1 \leqslant i \leqslant n-1$, as differences of effective divisors we see that it is sufficient to prove (2) for $D_i > 0$, $1 \leqslant i \leqslant n-1$. From now on we assume this to be the case. The proof of Theorem 2 makes use of a more general concept of intersection index than we have used so far. Let $D_1, \ldots, D_l$, $l \leqslant n$, be effective divisors on an n-dimensional smooth variety X. We say that they are in general position if $\dim \bigcap_{i=1,\ldots,l} \operatorname{Supp} D_i = n-l$ or $\bigcap \operatorname{Supp} D_i$ is empty. Suppose that this property holds and that

$$\bigcap_{i=1,\ldots,l} \operatorname{Supp} D_i = \bigcup C_j \tag{3}$$

where C_j are irreducible varieties of dimension $n-l$.

Under these conditions we can assign to the components C_j the so-called intersection multiplicities, which coincide with the intersection indices when $l=n$, and consequently C_j consists of a single point.

The definition of intersection multiplicities uses certain more general concepts, which we now introduce.

Definition 1. We consider an irreducible subvariety C of an irreducible variety X and functions $f \in k(X)$ that are regular for at least one point $c \in C$ (of course, then they are regular in a whole open subset of C). Such functions form a ring $\mathcal{O}_C$, the so-called *local ring of the irreducible subvariety C.*

It is easy to see that the ring $\mathcal{O}_C$ does not change when X is replaced by an open subset of it whose intersection with C is not empty. Therefore we may regard X as affine. In that case the ring $\mathcal{O}_C$ is obtained by the construction described in Ch. II, § 1.1: $\mathcal{O}_C = k[X]_{\mathfrak{p}}$, where $\mathfrak{p}$ is the ideal of the subvariety C. In particular, $\mathcal{O}_C$ is a Noetherian local ring. We denote its maximal ideal by $\mathfrak{m}_C$. Every function $f \in \mathcal{O}_C$ determines by restriction a rational function on C. From this it is easy to deduce that $\mathcal{O}_C/\mathfrak{m}_C = k(C)$.

Next we give some properties of local rings at prime ideals whose verification is obvious. We recall that in Ch. II, § 1.1, we have defined a homomorphism $\varphi: A \to A_{\mathfrak{p}}$. To every ideal $\mathfrak{a} \subset A$ there corresponds an ideal $\varphi(\mathfrak{a}) \subset A_{\mathfrak{p}}$, generated by the elements $\varphi(x)$, $x \in \mathfrak{a}$. It consists of the pairs (a, b) for which there exists a $b' \notin \mathfrak{p}$ such that $ab' \in \mathfrak{a}$.

An immediate verification shows that $\varphi(\mathfrak{a}) = A_{\mathfrak{p}}$ when $\mathfrak{a} \not\subset \mathfrak{p}$, and if $\mathfrak{a} \subset \mathfrak{p}$, then

$$A_{\mathfrak{p}}/\varphi(\mathfrak{a}) = (A/\mathfrak{a})_{\overline{\mathfrak{p}}} \tag{4}$$

where $\overline{\mathfrak{p}}$ is the image of $\mathfrak{p}$ in $A/\mathfrak{a}$.

The latter property is equally easy to verify: let $\mathfrak{a}$ and $\mathfrak{b}$ be ideals of A, $\mathfrak{a} \supset \mathfrak{b}$, and suppose that there is an isomorphism of A-modules $\mathfrak{a}/\mathfrak{b} \simeq A/\mathfrak{q}$, where $\mathfrak{q}$ is a prime ideal in A. Then under localization with respect to a prime ideal $\mathfrak{p} \subset A$ we have $\varphi(\mathfrak{a}) \supset \varphi(\mathfrak{b})$ and

$$\begin{aligned} \varphi(\mathfrak{a}) &= \varphi(\mathfrak{b}), && \text{when} \quad \mathfrak{q} \not\subset \mathfrak{p}, \\ \varphi(\mathfrak{a})/\varphi(\mathfrak{b}) &\simeq A_{\mathfrak{p}}/\varphi(\mathfrak{p}), && \text{when} \quad \mathfrak{q} = \mathfrak{p}. \end{aligned} \tag{5}$$

The other concept required to define intersection multiplicities is the length of a module.

Definition 2. A module M over a ring A is said to be a *module of finite length* if it has a finite sequence of submodules

$$M = M_0 \supset M_1 \supset \cdots \supset M_n = 0, \quad M_i \neq M_{i+1} \tag{6}$$

for which the factor modules M_i/M_{i+1} are all simple, that is, do not contain submodules other than zero and the whole module. From the Jordan-Hölder theorem it follows that all such chains consist of the same number n of modules, which is called the *length of the module* and is denoted by $l(M)$.

If A is a field, then the concept of length becomes the dimension of a vector space.

Later we shall use two almost obvious properties of this concept.

If a module has finite length, then the same is true for any of its submodules and factor modules.

If a module M has a chain of submodules (6) in which the lengths of the modules M_i/M_{i+1} are finite, then the length of M is also finite and $l(M) = \Sigma\, l(M_i/M_{i+1})$.

Finally we are in a position to proceed to the definition of intersection multiplicities. It is an exact copy of the definition of an intersection index. Let C be one of the components C_j in (3). We choose a point $x \in C$ and local equations f_i of the divisors D_i in a neighbourhood of this point. Then $f_i \in \mathcal{O}_C$, and the ideal $\mathfrak{a} = (f_1, \ldots, f_l) \subset \mathcal{O}_C$ does not depend on the choice of local equations nor of the point x. For if $g_1, \ldots, g_l$ are other local equations in a neighbourhood of another point, then f_i and g_i are local equations of the divisor D_i on an entire open set intersecting C. Hence it follows that $f_i g_i^{-1} \in \mathcal{O}_C$ and $g_i f_i^{-1} \in \mathcal{O}_C$, therefore

$$(f_1, \ldots, f_l) = (g_1, \ldots, g_l).$$

Lemma. *The module $\mathcal{O}_C/\mathfrak{a}$ is of finite length.*

For since C is an irreducible component of the subvariety determined by the equation $f_1 = \cdots = f_l = 0$, there exists an open affine subset $U \subset X$ intersecting C in which these equations determine C. Then by Hilbert's Nullstellensatz $(f_1, \ldots, f_l) \supset \mathfrak{a}_C^r$ for some $r > 0$. Consider now the local ring $A_{\mathfrak{p}}$, where $A = k[U]$, $\mathfrak{p} = \mathfrak{a}_C$. As we have already said, $A_{\mathfrak{p}} = \mathcal{O}_C$, $\varphi((f_1, \ldots, f_l)) = \mathfrak{a}$, and $\varphi(\mathfrak{a}_C) = \mathfrak{m}_C$. Therefore in $\mathcal{O}_C$ we have

$\mathfrak{a} \supset \mathfrak{m}_C^r$. To check that the module $\mathcal{O}_C/\mathfrak{a}$ is of finite length it is sufficient to verify this for the module $M = \mathcal{O}_C/\mathfrak{m}_C^r$.

Considering the sequence of submodules $M_i = \mathfrak{m}_C^i/\mathfrak{m}_C^r$ we see that it is sufficient to check that the modules $\mathfrak{m}_C^i/\mathfrak{m}_C^{i+1}$ are of finite length. But under the action of A on this module the ideal $\mathfrak{m}_C$ annihilates all the elements. Therefore $A/\mathfrak{m}_C = k(C)$ acts on the module, so that $\mathfrak{m}_C^i/\mathfrak{m}_C^{i+1}$ is a vector space over the field $k(C)$, and its length is the same as its dimension over this field. Since A is a Noetherian ring, this module is finitely generated, hence is a finite-dimensional vector space, which proves the lemma.

Definition 3. The number $l(\mathcal{O}_C/\mathfrak{a})$ is called the *intersection multiplicity of the divisors* $D_1, \ldots, D_l$ in the component C. It is denoted by $(D_1, \ldots, D_l)_C$.

Theorem 2 is a simple consequence of two propositions, which we now state.

Proposition 1. *If* $D_1, \ldots, D_n$ *are divisors in general position at* x *and* $D_1 \geqslant 0, \ldots, D_{n-1} \geqslant 0$, *then*

$$(D_1, \ldots, D_n)_x = \sum_{j=1}^{r} (D_1, \ldots, D_{n-1})_{C_j} (\varrho_{C_j}(D_n))_x, \tag{7}$$

where $C_1, \ldots C_r$ *are all the irreducible components of the variety* $\operatorname{Supp} D_1 \cap \cdots \cap \operatorname{Supp} D_{n-1}$, *and* $\varrho_{C_j}(D_n)$ *is the restriction of* D_n *to* C_j (see Ch. III, § 1.2).

Note that since $D_1, \ldots, D_n$ are divisors in general position at x, we have $\dim C_j = 1$, $x \in \operatorname{Supp} \varrho_{C_j}(D_n)$, and the intersection index $(\varrho_{C_j}(D_n))_x$ is defined (on the curve C_j).

Proposition 2. *For a curve* C *and a locally smooth divisor* D *on it*

$$(D)_x = \sum_{\nu(y)=x} (\nu^*(D))_y, \tag{8}$$

where $\nu: C^\nu \to C$ *is the normalization of* C.

We derive at once Theorem 2 from these propositions and postpone their proof to the next subsection.

We write the intersection index in the form

$$(D_1, \ldots, D_n) = \sum_{x \in X} (D_1, \ldots, D_n)_x$$

According to Proposition 1

$$(D_1, \ldots, D_n) = \sum_{j=1}^{r} (D_1, \ldots, D_{n-1})_{C_j} \sum_{x \in C_j} (\varrho_{C_j}(D_n))_x,$$

and according to Proposition 2

$$\sum_x (\varrho_{C_j}(D_n))_x = \sum_{y \in C_j^\vee} ((v^* \varrho_{C_j})(D_n))_y .$$

If D_n is a principal divisor, $D_n = (f)$, then so is the divisor

$$(v^* \varrho_{C_j})(D_n) : (v^* \varrho_{C_j})(D_n) = (g) \quad \text{and} \quad ((g))_y = v_y(g) .$$

Since X is a projective variety, the curves C_j are projective, and by Theorem 10 of Ch. II, § 5, so are the $C_j^\vee$. According to the corollary to Theorem 1 of Ch. III, § 2, $\sum_{y \in C_j^\vee} v_y(g) = \deg((g)) = 0$, from which it follows that $(D_1, \ldots, D_{n-1}, (f)) = 0$.

4. End of the Proof of Invariance. Now we prove Proposition 1.

Let $f_1, \ldots, f_{n-1}$ be local equations of the divisors $D_1, \ldots, D_{n-1}$, $\mathfrak{a} = (f_1, \ldots, f_{n-1}) \subset \mathcal{O}_x$, $\mathcal{O}_x/\mathfrak{a} = \overline{\mathcal{O}}_x$, $\bar{f}$ the image of f (the local equation of D_n) in $\overline{\mathcal{O}}_x$. The definition of the intersection index shows that

$$(D_1, \ldots, D_n)_x = l(\overline{\mathcal{O}}_x/(\bar{f})) \tag{1}$$

and Lemma 1 of § 1.2 asserts that $\bar{f}$ is not a divisor of zero in $\overline{\mathcal{O}}_x$.

First of all we have to clarify what are the prime ideals of $\overline{\mathcal{O}}_x$. We denote by $\mathfrak{p}_i$ the collection of functions in $\mathcal{O}_x$ that vanish identically on C_i, and by $\bar{\mathfrak{p}}_i$ the image of $\mathfrak{p}_i$ in $\overline{\mathcal{O}}_x$. Evidently,

$$\overline{\mathcal{O}}_x/\bar{\mathfrak{p}}_i = \mathcal{O}_x/\mathfrak{p}_i = \mathcal{O}_{x, C_i} \tag{2}$$

is the local ring of x on C_i. We denote by $\bar{\mathfrak{m}}$ the maximal ideal of $\overline{\mathcal{O}}_x$, the image of the ideal $\mathfrak{m}_x \subset \mathcal{O}_x$.

Lemma 1. *The ideals $\bar{\mathfrak{p}}_1, \ldots, \bar{\mathfrak{p}}_r$ and $\bar{\mathfrak{m}}$ are the only prime ideals of $\overline{\mathcal{O}}_x$.*

The assertion of the lemma is equivalent to the fact that $\mathfrak{p}_1, \ldots, \mathfrak{p}_r$ and $\mathfrak{m}_x$ are the only prime ideals of $\mathcal{O}_x$ containing $\mathfrak{a}$. Let $\mathfrak{a} \subset \mathfrak{p} \subset \mathcal{O}_x$, $\mathfrak{p}$ being a prime ideal. We consider an affine neighbourhood U of x in which $f_1, \ldots, f_{n-1}$ are regular, and we set $A = k[U]$, $\mathfrak{P} = A \cap \mathfrak{p}$. Clearly $\mathfrak{P}$ is a prime ideal. We denote by V the subvariety which it defines in U. Since $\mathfrak{p} \supset \mathfrak{a}$, we have $V \subset C_1 \cup \cdots \cup C_r$, and since $\mathfrak{P}$ is prime, V is irreducible. Therefore V either coincides with one of the C_i, and then $\mathfrak{P} = A \cap \mathfrak{p}_i$, or else V is a point $y \in U$ (we recall that the C_i are one-dimensional). In the latter case, if $y \neq x$, then $\mathfrak{P}$ and hence also $\mathfrak{p}$ contains a function that does not vanish at x. Since $\mathcal{O}_x$ is a local ring, we would then have $\mathfrak{p} = \mathcal{O}_x$, (whereas the ring itself is not considered as one of its prime ideals.) Thus, the only remaining possibility is that $\mathfrak{P} = A \cap \mathfrak{m}_x$. Since $\mathfrak{p} = \mathfrak{P} \cdot \mathcal{O}_x$, it follows easily that $\mathfrak{p} = \mathfrak{p}_i$ for some i, $i = i, \ldots, r$, or $\mathfrak{p} = \mathfrak{m}_x$, as the lemma claims.

Lemma 2. *Every Noetherian ring A has a sequence of ideals $A=\mathfrak{q}_1\supset\mathfrak{q}_2\supset\cdots\supset\mathfrak{q}_s=0$ such that*

$$\mathfrak{q}_i/\mathfrak{q}_{i+1}\simeq A/\bar{\mathfrak{p}} \tag{3}$$

where $\bar{\mathfrak{p}}$ is some prime ideal of A ($\bar{\mathfrak{p}}$ depending on i).

Proof. We consider an arbitrary element $a\in A$, $a\neq 0$, and denote by $\operatorname{Ann}(a)$ the annihilator of a, that is, the set of all $x\in A$ for which $xa=0$. Since A is Noetherian, any sequence $\operatorname{Ann}(a)\subset\operatorname{Ann}(a_1)\subset\operatorname{Ann}(a_2)\subset\cdots$ breaks off, therefore we may assume that already a has the following property: from $\operatorname{Ann}(a)\subset\operatorname{Ann}(a')$, $a'\neq 0$, it follows that $\operatorname{Ann}(a)=\operatorname{Ann}(a')$. Let us show that then the ideal $\operatorname{Ann}(a)$ is prime. For if $bc\in\operatorname{Ann}(a)$, $b\notin\operatorname{Ann}(a)$, then $abc=0$, $ab\neq 0$ and therefore $\operatorname{Ann}(a)\subset\operatorname{Ann}(ab)$, hence $\operatorname{Ann}(a)=\operatorname{Ann}(ab)$, by the property of a. But $c\in\operatorname{Ann}(ab)$, hence $c\in\operatorname{Ann}(a)$. This shows that the ideal $\operatorname{Ann}(a)$ is prime. We set $\operatorname{Ann}(a)=\bar{\mathfrak{p}}$. The homomorphism $x\to ax$ determine an isomorphism of modules $(a)\simeq A/\bar{\mathfrak{p}}$.

It now remains to go over to the ring $A_1=A/(a)$ and to apply the same argument to it. So we obtain an ascending sequence of ideals $\mathfrak{q}_s\subset\mathfrak{q}_{s+1}\subset\cdots$, and (3) holds for every pair of adjacent ideals. Since A is Noetherian, the sequence must break off.

According to Lemma 2, $\bar{\mathcal{O}}_x$ contains a chain of ideals $\mathfrak{q}_i$ having the property (3).

Lemma 3. *If k_j is the number of times $\bar{\mathfrak{p}}=\bar{\mathfrak{p}}_j$ occurs in (3), then $(D_1,\ldots,D_{n-1})_{C_j}=k_j$.*

Proof. From the definition of a local ring and a prime ideal it follows at once that $(\mathcal{O}_x)_{\mathfrak{p}_j}=\mathcal{O}_{C_j}$. Applying the relation (4) of § 1.3 we find that

$$\mathcal{O}_{C_j}/(f_1,\ldots,f_{n-1})=(\mathcal{O}_x/(f_1,\ldots,f_{n-1}))_{\bar{\mathfrak{p}}_j}=(\bar{\mathcal{O}}_x)_{\bar{\mathfrak{p}}_j}.$$

To the chain of ideals $\mathfrak{q}_i$ we apply the mapping φ corresponding to the ideal $\bar{\mathfrak{p}}_j$. So we obtain the chain

$$\mathcal{O}_{C_j}/(f_1,\ldots,f_{n-1})=\varphi(\mathfrak{q}_1)\supseteq\varphi(\mathfrak{q}_2)\supseteq\cdots\supseteq\varphi(\mathfrak{q}_s)=0.$$

The relation (5) of § 1.3 shows what the factors in this chain are: if $\bar{\mathfrak{p}}\neq\bar{\mathfrak{p}}_j$, then $\varphi(\mathfrak{q}_i)=\varphi(\mathfrak{q}_{i+1})$, and if $\bar{\mathfrak{p}}=\bar{\mathfrak{p}}_j$, then $\varphi(\mathfrak{q}_i)/\varphi(\mathfrak{q}_{i+1})=(\bar{\mathcal{O}}_x/\bar{\mathfrak{p}}_j)_{\bar{\mathfrak{p}}_j}=k(C_j)$. Thus, the length of the module $\mathcal{O}_{C_j}/(f_1,\ldots,f_{n-1})$ is equal to k_j, as Lemma 3 claims.

Now we can prove Proposition 1. We have the exact sequence $0\to\mathfrak{q}_2\to\bar{\mathcal{O}}_x\to\bar{\mathcal{O}}_x/\mathfrak{q}_2\to 0$.

Here two cases are possible: 1) the prime ideal $\bar{\mathfrak{p}}$ corresponding to $\mathfrak{q}_2$ by virtue of (3) is equal to $\bar{\mathfrak{m}}$, 2) $\bar{\mathfrak{p}}=\bar{\mathfrak{p}}_j$ for some $j=1,\ldots,r$.

In case 1) $\bar{\mathcal{O}}_x/\mathfrak{q}_2\simeq\bar{\mathcal{O}}_x/\bar{\mathfrak{m}}=k$. Since $\bar{f}$ is not a divisor of zero in $\bar{\mathcal{O}}_x$, the mapping $a\to\bar{f}\cdot a$ carries any ideal in $\bar{\mathcal{O}}_x$ into an ideal that is isomorphic to it as an $\bar{\mathcal{O}}_x$-module. In particular,

$$\bar{\mathcal{O}}_x/\mathfrak{q}_2 \simeq \bar{f}\bar{\mathcal{O}}_x/f\mathfrak{q}_2\,. \tag{4}$$

From the diagram

$$\begin{array}{ccccc} & & \bar{\mathcal{O}}_x & & \\ & \cup & & \supset & \\ \bar{f}\bar{\mathcal{O}}_x & & & & \mathfrak{q}_2 \\ & \supset & & \cup & \\ & & \bar{f}\mathfrak{q}_2 & & \end{array}$$

we find that

$$l(\bar{\mathcal{O}}_x/\bar{f}\mathfrak{q}_2) = l(\bar{\mathcal{O}}_x/\bar{f}\bar{\mathcal{O}}_x) + l(\bar{f}\bar{\mathcal{O}}_x/\bar{f}\mathfrak{p}_2) = l(\bar{\mathcal{O}}_x/\mathfrak{q}_2) + l(\mathfrak{q}_2/\bar{f}\mathfrak{q}_2)$$

and all these numbers are finite. By (4) it follows that $l(\bar{\mathcal{O}}_x/\bar{f}\bar{\mathcal{O}}_x) = l(\mathfrak{q}_2/\bar{f}\mathfrak{q}_2)$.

In case 2) $\bar{\mathcal{O}}_x/\mathfrak{q}_2 \simeq \mathcal{O}_{x,C_j}$, and we have the sequence

$$0 \to \mathfrak{q}_2/\bar{f}\mathfrak{q}_2 \to \bar{\mathcal{O}}_x/\bar{f}\bar{\mathcal{O}}_x \to \mathcal{O}_{x,C_j}/\bar{f}\mathcal{O}_{x,C_j} \to 0\,. \tag{5}$$

This sequence is exact. The verification is completely obvious except at one place: namely, the fact that the homomorphism $\mathfrak{q}_2/\bar{f}\mathfrak{q}_2 \to \bar{\mathcal{O}}_x/\bar{f}\bar{\mathcal{O}}_x$ is an embedding. This follows at once from the fact that the image $\bar{f}$ is not a divisor of zero in the ring $\bar{\mathcal{O}}_x/\mathfrak{q}_2 \simeq \mathcal{O}_{x,C_j}$. For this ring has no divisors of zero at all, and $\bar{f}$ is not equal to zero in it, because $f \neq 0$ on C_j. From (5) we have

$$l(\bar{\mathcal{O}}_x/\bar{f}\bar{\mathcal{O}}_x) = l(\mathfrak{q}_2/\bar{f}\mathfrak{q}_2) + l(\mathcal{O}_{x,C_j}/\bar{f}\mathcal{O}_{x,C_j}) = l(\mathfrak{q}_2/\bar{f}\mathfrak{q}_2) + l(\mathcal{O}_{x,C_j}/(\bar{f}))\,.$$

Repeating the same argument s times we obtain the formula

$$(D_1, \ldots, D_n)_x = \Sigma k_j l(\mathcal{O}_{x,C_j}/(\bar{f})) = \Sigma k_j(\varrho_{C_j}((\bar{f})))_x\,.$$

Here k_j is the number of indices $t \leqslant s$ for which in the sequence $\bar{\mathcal{O}}_x = \mathfrak{q}_1 \supset \cdots \supset \mathfrak{q}_s = 0$ we have $\mathfrak{q}_t/\mathfrak{q}_{t+1} \simeq \bar{\mathcal{O}}_x/\bar{\mathfrak{p}}_j$. Lemma 3 guarantees that this number is equal to $(D_1, \ldots, D_{n-1})_{C_j}$.

Proof of Proposition 2. We denote by $y_1, \ldots, y_l$ inverse images of x under the normalization mapping $\nu: C^\nu \to C$, by $\mathcal{O}_{y_i}$ their local rings on C^ν, and we set $\tilde{\mathcal{O}} = \bigcap \mathcal{O}_{y_i}$. Obviously $\tilde{\mathcal{O}} \supset \mathcal{O}_x$ [if we regard $k(C)$ and $\mathcal{O}_x$ as embedded in the field $k(C^\nu)$ by means of the mapping ν^*]. We need the following property of these rings:

Lemma 4. *There exists an element $d \in \mathcal{O}_x$, $d \neq 0$, such that*

$$d \cdot \tilde{\mathcal{O}} \subset \mathcal{O}_x\,. \tag{6}$$

We consider an affine neighbourhood U of x and its normalization V, and we set $A = k[U]$, $B = k[V]$. One of the main steps in the proof of the theorem on the existence of a normalization of an affine variety consists in establishing that B is a module of finite type over A. Hence and from the fact that B is contained in the field of fractions of A it follows that

$$d \cdot B \subset A \tag{7}$$

for some $d \in A$, $d \neq 0$. It is easy to see that

$$\tilde{\mathcal{O}} = B \cdot \mathcal{O}_x \,. \tag{8}$$

For evidently $B \subset \tilde{\mathcal{O}}$, therefore $B\mathcal{O}_x \subset \tilde{\mathcal{O}}$. On the other hand, the lemma in Ch. III, § 2.1, shows that $\tilde{\mathcal{O}} \subset B\mathcal{O}_x$, and (6) follows at once from (7) and (8).

Now it is not difficult to conclude the proof of Proposition 2. By Lemma 4, $l(\tilde{\mathcal{O}}/\mathcal{O}_x) \leqslant l(\tilde{\mathcal{O}}/d\tilde{\mathcal{O}})$ and by Theorem 3 of Ch. III, § 2, $l(\tilde{\mathcal{O}}/d\tilde{\mathcal{O}}) = \sum_{y_i} l(\mathcal{O}_{y_i}/d\mathcal{O}_{y_i}) = \Sigma v_{y_i}(d)$, therefore $l(\tilde{\mathcal{O}}/\mathcal{O}_x) < \infty$. From the diagram

$$\begin{array}{ccccc} & & \tilde{\mathcal{O}} & & \\ & \cup & & \supset & \\ f\tilde{\mathcal{O}} & & & & \mathcal{O}_x \\ & \supset & & \cup & \\ & & f\mathcal{O}_x & & \end{array}$$

it follows that

$$l(\tilde{\mathcal{O}}/f\,\mathcal{O}_x) = l(\tilde{\mathcal{O}}/\mathcal{O}_x) + l(\mathcal{O}_x/f\,\mathcal{O}_x) = l(\tilde{\mathcal{O}}/f\,\tilde{\mathcal{O}}) + l(f\,\tilde{\mathcal{O}}/f\,\mathcal{O}_x)\,,$$

and all these number are finite. Since f is not a divisor of zero in $\tilde{\mathcal{O}}$, we have $l(\tilde{\mathcal{O}}/\mathcal{O}_x) = l(f\,\tilde{\mathcal{O}}/f\,\mathcal{O}_x)$, from which we obtain that $l(\mathcal{O}_x/f\,\mathcal{O}_x) = l(\tilde{\mathcal{O}}/f\tilde{\mathcal{O}}_x)$. Finally, Theorem 2 of Ch. III, § 1, yields $l(\tilde{\mathcal{O}}/f\,\tilde{\mathcal{O}}) = \sum_{y_i} v_{y_i}(f) = \sum_{y_i} (v^*(f))_{y_i}$. Since $l(\mathcal{O}_x/f\,\mathcal{O}_x) = (D)_x$, this proves Proposition 2.

5. General Definition of the Intersection Index. Theorem 2 and the theorem on shifting the support of a divisor away from a point (Theorem 1 of Ch. III, § 1) allow us to define the intersection index of any n divisors on an n-dimensional smooth projective variety, without any restrictions of the type of general position.

For this purpose we need two lemmas.

Lemma 1. *For any n divisors $D_1, \ldots, D_n$ on an n-dimensional variety X there are n divisors $D'_1, \ldots, D'_n$ such that $D_i \sim D'_i$ $(i = 1, \ldots, n)$ and $D'_1, \ldots, D'_n$ are in general position.*

Suppose that we have already found such divisors $D'_1, \ldots, D'_l$ such that $D_i \sim D'_i$ $(i = 1, \ldots, l)$ and that $\dim(\operatorname{Supp} D'_i \cap \cdots \cap \operatorname{Supp} D'_l) = n - l$, or that this intersection is empty. Let

$$\operatorname{Supp} D'_1 \cap \cdots \cap \operatorname{Supp} D'_l = C_1 \cup \cdots \cup C_r$$

be a decomposition into irreducible components. On each of the components C_j we chose one point x_j, and by using the theorem on shifting the support of a divisor we find a divisor D'_{l+1} such that $D'_{l+1} \sim D_{l+1}$ and $x_j \in \operatorname{Supp} D'_{l+1}$ $(j=1, \ldots, r)$. Then $\operatorname{Supp} D'_{l+1}$ a fortiori does not contain any of the components C_j, and by the theorem of the dimension of an intersection

$$\dim(\operatorname{Supp} D'_1 \cap \cdots \cap \operatorname{Supp} D'_{l+1}) = n - l - 1,$$

if this intersection is not empty. When we arrive at $l=n$ in this manner, we have obtained the required system of divisors.

Lemma 2. *If the divisors $D_1, \ldots, D_n$ and $D'_1, \ldots, D'_n$ are in general position and if $D_i \sim D'_i$ $(i=1, \ldots, n)$, then*

$$(D_1, \ldots, D_n) = (D'_1, \ldots, D'_n). \tag{1}$$

If $D_1 = D'_1, \ldots, D_{n-1} = D'_{n-1}$, this is the assertion of Theorem 2. Let us show that (1) is true if $D_1 = D'_1, \ldots, D_{n-l} = D'_{n-l}$. For $l=n$ we then obtain our proposition.

We use induction on l. Suppose that the assertion is true for values smaller than l. Since both system of divisors $D_1, \ldots, D_n$ and $D'_1, \ldots, D'_n$ are in general position, we have $\dim Y = \dim Y' = 1$, where $Y = \bigcap_{i \neq n-l+1} \operatorname{Supp} D_i$, $Y' = \bigcap_{i \neq n-l+1} \operatorname{Supp} D'_i$. On each component of each of the varieties Y and Y' we choose one point, and according to the theorem on the shift of the support of a divisor we find a divisor D''_{n-l+1} such that $\operatorname{Supp} D''_{n-l+1}$ does not pass through any of these points and that $D''_{n-l+1} \sim D_{n-l+1}$. Then both systems $D_1, \ldots, D_{n-l}, D''_{n-l+1}, \ldots, D_n$ and $D'_1, \ldots, D'_{n-l}, D''_{n-l+1}, \ldots, D'_n$ are in general position.

By Theorem 2

$$\begin{aligned} (D_1, \ldots, D_n) &= (D_1, \ldots, D_{n-l}, D''_{n-l+1}, \ldots, D_n), \\ (D'_1, \ldots, D'_n) &= (D'_1, \ldots, D'_{n-l}, D''_{n-l+1}, \ldots, D'_n). \end{aligned} \tag{2}$$

The right-hand sides of (2) are equal to each other by the inductive hypothesis (they have already $n-l+1$ equal divisors), and this proves Lemma 2.

Using Lemmas 1 and 2 we can define the *intersection index* $(D_1, \ldots, D_n)$ for any n divisors on a smooth n-dimensional variety, without requiring that they are in general position. For this purpose we find arbitrary divisors $D'_1, \ldots, D'_n$ satisfying the conditions of Lemma 1, so that the index $(D'_1, \ldots, D'_n)$ is defined, and we define $(D_1, \ldots, D_n)$ by the equality $(D_1, \ldots, D_n) = (D'_1, \ldots, D'_n)$.

We have to show that this definition does not depend on the choice of the auxiliary divisors $D_1', \ldots, D_n'$, but precisely this is guaranteed by Lemma 2.

We can now talk, for instance, of the sef-intersection (C, C) for a curve C on a surface X. This number is also denoted by (C^2). We give some examples how it can be computed.

Example 1. $X = \mathbb{P}^2$, C a straight line. By definition, $(C^2) = (C', C'')$, where $C' \sim C'' \sim C$ and C' and C'' are in general position. For C' and C''' we may take, for example, two distinct lines. They intersect in a unique point x, and $(C', C'') = (C', C'')_x = 1$, because they are transversal at this point. Therefore $(C^2) = 1$.

Example 2. X is a smooth surface in $\mathbb{P}^3$. Let X be given by the equation $F(x_0 : x_1 : x_2 : x_3) = 0$, $\deg F = m$.

We calculate (E^2), where E is the divisor of a plane section (in Ch. III, § 1.2, we have shown that the divisors of all plane sections are equivalent). By definition $(E^2) = (E', E'')$, where $E' \sim E'' \sim E$ and E' and E'' are in general position. For E' and E'' we choose distinct plane sections. Then $E' = \Sigma l_i C_i$, where C_i are plane curves and $\Sigma l_i \deg C_i = m$ (we simply have to substitute in F the equation of E' and decompose the resulting form in three variables into irreducible factors). We choose E'' so that it is transversal to all the C_i; that such a choice is possible follows easily from a count of the dimension of the set of non-transversal planes. Then, by Bezout's theorem in Ch. III, § 2.2

$$(E', E'') = \Sigma\, l_i (C_i, E'') = \Sigma\, l_i \deg C_i = m = \deg X\,.$$

Therefore $(E^2) = \deg X$.

Example 3. Suppose that on the surface X of Example 2 there lies a line L. We wish to compute (L^2).

We draw a plane through L and denote by E the corresponding plane section. Then L is contained as a component in E:

$$E = L + C, C = \Sigma l_i C_i,\ \Sigma l_i\ \deg C_i = m - 1\,.$$

First we compute (C^2). To do this we observe that at the points of intersection of L and C the curve E has a singular point, and this means that the plane containing it coincides with the tangent plane to X at this point. We consider another plane passing through L, but distinct from the tangent planes to X at the points of $L \cap C$. This plane determines a divisor $E' = L + C'$, and the points $L \cap C$ and $L \cap C'$ are all distinct. This means that $C \cap C' \doteq \emptyset$ and $(C^2) = (C, C') = 0$. So we have obtained the equations

$$m=(E^2)=(E,L+C)=(E,L)+(E,C)=1+(E,C)\,,$$
$$(C,E)=m-1\,,$$
$$m-1=(E,C)=(L,C)+(C^2)=(L,C)\,,$$
$$1=(E,L)=(L^2)+(L,C)=(L^2)+m-1\,,$$
$$(L^2)=2-m\,.$$

Observe that $(L^2)<0$ for $m>2$. In fact, lines can lie on surfaces of any degree, for example, the line $x_0=x_1, x_2=x_3$ on the surface $x_0^m-x_1^m+x_2^m-x_3^m=0$.

Exercises

1. Let X be a surface, x a simple point on it, u and v local parameters at x, and f the local equation of a curve C in a neighbourhood of x. If $f=(au+bv)(cu+dv)+g, g\in\mathfrak{m}_x^3$, and if the linear forms $au+bv$ and $cu+dv$ are not proportional, then x is called a *double point* of C with *distinct tangents*, and the lines of Θ_x with the equations $au+bv=0$ and $cu+dv=0$ are called the *tangents* at x. Under these assumptions let C' be a smooth curve on X passing through x. Show that $(C,C')_x>2$ if and only if $\Theta_{x,C'}$ is one of the tangents to C at x.

2. Let $C=V(F)$, $D=V(G)$ be two plane curves in $\mathbb{A}^2$, and x a simple point on each of them. Let f be the restriction of the polynomial F to the curve D, and $v_x(f)$ the order of the zero of this function at x on D. Show that this number does not change when F and G interchange places.

3. Let Y be a smooth irreducible subvariety of codimension 1 of an n-dimensional smooth variety X. Show that for divisors $D_1,\ldots,D_{n-1}$ in general position with Y at x, $(D_1,\ldots,D_{n-1},Y)_x=(\varrho_Y(D_1),\ldots,\varrho_Y(D_{n-1}))_x$, the second intersection index being computed on Y.

4. Find the degree of the surface $v_m(\mathbb{P}^2)$ (v_m is the Veronese mapping).

5. Let X be a smooth projective surface contained in the space $\mathbb{P}^n$, and L a projective subspace of $\mathbb{P}^n$ of dimension $n-2$. Suppose that L and X intersect in finitely many points, and that at l of these points the tangent plane to X intersects L in a line. Show that the number of points of intersection of X and L does not exceed $\deg X-l$.

6. The same as in Exercise 5, but the dimension of L is $n-m$, $m\geqslant 2$. Show that the number of points of intersection of X and L does not exceed $\deg X-l-m+2$. *Hint:* Draw through L a suitable linear subspace satisfying the conditions of Exercise 5.

7. Show that $(H^n)=\deg X$ for an n-dimensional smooth projective variety $X\subset\mathbb{P}^N$, where H is the divisor of the intersection of X with a hyperplane in $\mathbb{P}^N$.

8. Show that if $D_1,\ldots,D_{n-1}$ are effective divisors on an n-dimensional variety in general position and if C is an irreducible component of the intersection of their supports, then $(D_1,\ldots,D_{n-1})_C=\min(D_1,\ldots,D_{n-1},D)_x$, where the minimum is taken over all points $x\in C$ and all effective divisors D for which $x\in\operatorname{Supp}D$.

9. Calculate $(D_1,D_2)_C$, where D_1 and D_2 are given in $\mathbb{A}^3$ by the equations $x=0$ and $x^2+y^2+xz=0$, and C is the line $x=0$, $y=0$.

§ 2. Applications and Generalizations of Intersection Indices

1. Bezout's Theorem in a Projective Space and Products of Projective Spaces. Theorems 1 and 2 of § 1 put us in a position to calculate intersection indices of any divisors on a variety X, if only the group $\mathrm{Cl}(X)$ is sufficiently well known to us. We show in this two examples.

Example 1. $X = \mathbb{P}^n$. We know that $\mathrm{Cl}(X) \simeq \mathbb{Z}$, and for a generator of this group we can take the divisor E of a hyperplane. Every effective divisor D is a divisor of a form F, and if $\deg F = m$, then $D \sim mE$. Hence it follows that if $D_i \sim m_i E$ $(i = 1, \ldots, n)$, then

$$(D_1, \ldots, D_n) = m_1 \ldots m_n (E^n) = m_1 \ldots m_n \,, \tag{1}$$

because evidently $(E^n) = 1$.

If D_i are effective divisors, that is, correspond to forms F_i of degree m_i and are in general position, then the points of the set $\bigcap \mathrm{Supp}\, D_i$ coincide with the non-zero solutions of the system of equations

$$F_1(x_0 \ldots x_n) = 0 \,,$$
$$\ldots\ldots\ldots\ldots\ldots$$
$$F_n(x_0 \ldots x_n) = 0 \,.$$

For such a point (or solution) x the index $(D_1, \ldots, D_m)_x$ is naturally called the multiplicity of the solution. Then the Eq. (1) shows that the number of solutions of a system of n homogeneous equations in $n+1$ unknowns is either infinite or equal to the product of the degrees, provided that their solutions are counted with their multiplicities. Here only non-zero solutions are considered, and proportional solutions are counted as one. This result is called *Bezout's theorem* in the projective space $\mathbb{P}^n$.

Example 2. $X = \mathbb{P}^n \times \mathbb{P}^m$. In this case $\mathrm{Cl}(X) = \mathbb{Z} \oplus \mathbb{Z}$. Every effective divisor D is determined by a polynomial G homogeneous with respect to the variables $x_0, \ldots, x_n$ (coordinates in $\mathbb{P}^n$) and $y_0, \ldots, y_m$ (coordinates in $\mathbb{P}^m$). If G is of degree of homogeneity k and l, then $D \to (k, l)$ determines an isomorphism $\mathrm{Cl}(X) \simeq \mathbb{Z} \oplus \mathbb{Z}$. In particular, as generators of $\mathrm{Cl}(X)$ we can take a divisor E determined by linear forms in the x_i, and a divisor F determined by linear forms in the y_i. Then $D \sim kE + lF$.

Let $D_i \sim k_i E + l_i F$ $(i = 1, \ldots, n+m)$. Then

$$(D_1, \ldots, D_{n+m}) = \Sigma\, k_{i_1} \ldots k_{i_r} l_{j_1} \ldots l_{j_s} (\underbrace{E, \ldots, E}_{r}, \underbrace{F, \ldots, F}_{s}) \,,$$

where the summation extends over all permutations $(i_1 \ldots i_r\, j_1 \ldots j_s)$ of the numbers $1, 2, \ldots, n+m$ for which $i_1 < i_2 < \cdots < i_r; j_1 < j_2 < \cdots < j_s$.

Let us calculate the intersection index

$$(\underbrace{E, \ldots, E}_{r}, \underbrace{F, \ldots, F}_{s}). \tag{2}$$

If $r > n$, then we can find r linear forms $E_1, \ldots, E_r$ without common zeros, and therefore

$$(\underbrace{E, \ldots, E}_{r}, F, \ldots, F) = (E_1, \ldots, E_r, F, \ldots, F) = 0.$$

The matter is similar if $s > m$. Since $r + s = n + m$, the index (2) can be different from zero only for $r = n$, $s = m$. In this case we can take for $E_1, \ldots, E_n$, $F_1, \ldots, F_m$ the divisors determined by the forms $x_1, \ldots, x_n, y_1, \ldots, y_m$. These divisors have a unique point in common $(1:0:\ldots:0; 1:0\ldots:0)$. They intersect transversally at it, as is easy to verify by going over to the open set $x_0 \neq 0$, $y_0 \neq 0$, which is isomorphic to the affine space $\mathbb{A}^{n+m}$. Thus,

$$(k_1 E + l_1 F, \ldots, k_{n+m} E + l_{n+m} F) = \Sigma k_{i_1} \ldots k_{i_n} l_{j_1} \ldots l_{j_m}, \tag{3}$$

where the sum extends over all permutations $(i_1 \ldots i_n j_1 \ldots j_m)$ of the numbers $1, 2, \ldots, n+m$ in which $i_1 < i_2 < \cdots < i_n$; $j_1 < j_2 < \cdots < j_m$. This proposition is called *Bezout's theorem* in the variety $\mathbb{P}^n \times \mathbb{P}^m$.

A common feature of the examples we have analysed is the fact their groups $\mathrm{Cl}(X)$ are finitely generated. It is natural to ask whether this is true for any smooth projective variety X. This is not so, and a counter-example is given by a plane cubic curve for which $\mathrm{Cl}(X) \supset \mathrm{Cl}^0(X)$, $\mathrm{Cl}(X)/\mathrm{Cl}^0(X) \simeq \mathbb{Z}$, and the elements of the groups $\mathrm{Cl}^0(X)$ are in one-to-one correspondence with the points of X. Therefore, if k is the field of complex numbers, then the group $\mathrm{Cl}^0(X)$ is even uncountable.

However, this "bad" subgroup $\mathrm{Cl}^0(X)$ has no influence at all on the intersection index $(D) = \deg D$ – it consists of divisors of degree 0. A similar situation arises in the case of arbitrary smooth projective variety X. Namely, it can be shown that if a divisor D is algebraically equivalent to zero (for the definition see Ch. III, § 3.5), then $(D_1, \ldots, D_{n-1}, D) = 0$ for arbitrary divisors $D_1, \ldots, D_{n-1}$. Thus, the intersection indices depend only on the elements of the group $\mathrm{Div}(X)/\mathrm{Div}^a(X)$. For this group Theorem D of Ch. III, § 3.5, asserts that it is always finitely generated. Obviously, if $E_1, \ldots, E_r$ are generators of this group, then to know arbitrary intersection indices of divisors on X it is sufficient to know finitely many numbers $(E_{i_1}, \ldots, E_{i_n})$, just as we have seen in Example 1 and 2. In other words, an analogue to Bezout's theorem holds on X.

2. Varieties over the Field of Real Numbers. The various versions of Bezout's theorem proved above have some pretty applications to algebraic geometry over the field of real numbers.

We return to Example 1 of § 2.1 and assume that the equations $F_i = 0\,(i = 1, \ldots, n)$ have real coefficients; we are interested in real solutions. If $\deg F_i = m_i$ and the divisors D_i are in general position, then $(D_1, \ldots, D_n) = m_1 \ldots m_n$, as was shown in § 2.1. According to the definition, $(D_1, \ldots, D_n) = \Sigma(D_1, \ldots, D_n)_x$, where the sum extends over the solutions x of the system $F_1 = 0, \ldots, F_n = 0$. Of course, here we must consider both real and complex solutions. However, since the polynomials F_i have real coefficients, together with any complex solution x, the system also has the conjugate complex solution $\bar{x}$. From the definition of the intersection index it follows at once that $(D_1, \ldots, D_n)_x = (D_1, \ldots, D_n)_{\bar{x}}$, therefore $(D_1, \ldots, D_n) \equiv \Sigma(D_1, \ldots, D_n)_y \pmod 2$, where now the sum extends only over the real solutions. In particular, if $(D_1, \ldots, D_n)$ is odd (and this is equivalent to the fact that all the $m_i = \deg F_i$ are odd), we see that there exists at least one real solution. This proposition has been proved under the assumption that the divisors D_i are in general position. However, the following simple argument allows us to get rid of this restriction.

The fact is that the theorem on shifting the support of a divisor has in our case a perfectly simple proof and in a more explicit form. Namely, we can take a linear form l that is different from zero at all the points $x_1, \ldots, x_r$ from which we wish to shift the support of the divisor. If a divisor D is determined by a form F of degree m, then the divisor D' determined by the form $F_\varepsilon = F + \varepsilon l^m$ satisfies all the conditions of the theorem if only $F(x_j) + \varepsilon l(x_j)^m \neq 0$ $(j = 1, \ldots, r)$. These conditions can be satisfied for sufficiently small values of ε.

Now we show how to get rid of the restriction concerning general position in the proposition we have proved above about the existence of a real solution of a system of equation of odd degree. Let

$$F_1 = \cdots = F_n = 0 \tag{1}$$

be any such system. By what we have said above, we can find sufficiently small values of ε for which the divisors defined by the forms $F_{i,\varepsilon} = F_i + \varepsilon l_i^{m_i}$ are in general position.

By what we have already proved, the system $F_{1,\varepsilon} = 0, \ldots, F_{n,\varepsilon} = 0$ has a real solution x_ε. Since a projective space is compact, we can find a sequence of numbers $\varepsilon_m \to 0$ such that the points x_{ε_m} converge to a point $x \in \mathbb{P}^n$. Since then $F_{j,\varepsilon_m} \to F_j$, we see that x is a solution of the system (1).

Let us summarize what we have proved.

Theorem 1. *A system of n homogeneous real equations in $n+1$ unknowns has a non-zero real solution if the degrees of all the equations are odd.*

Very similar arguments apply to the variety $\mathbb{P}^n \times \mathbb{P}^m$ (see Example 2 of § 2.1). We obtain the following result.

Theorem 2. *A system of real equations*

$$F_i(x_0:\ldots:x_n;\, y_0:\ldots:y_m)=0\,(i=1,\ldots,n+m)$$

has a non-zero real solution if the number $\Sigma k_{i_1}\ldots k_{i_n} l_{j_1}\ldots l_{j_m}$ *is odd.*

Here l_i' and l_j are the degrees of homogeneity of a polynomial F_i in the first and second system of variables, respectively, and a zero solution is one for which $x_0=\cdots=x_n=0$ or $y_0=\cdots=y_m=0$.

Theorem 2 has interesting applications in algebra. One of these refers to the problem of division algebras over the field of real numbers $\mathbb{R}$. If the rank of such an algebra is n, then it has a basis $e_1,\ldots,e_n$ and is given by a multiplication table

$$e_i e_j=\sum_{l=1}^{n} c_{ij}^{l} e_l \quad (i,j=1,\ldots,n)\,. \tag{2}$$

We do not assume the algebra to be associative, therefore the c_{ij}^l can be arbitrary. A *division algebra* is one for which an equation

$$ax=b \tag{3}$$

is soluble for every $a\neq 0$ and every b. It is easy to see that this is equivalent to the absence of divisors of zero in the algebra. To see this it is sufficient to consider the linear transformation $\varphi:\varphi(x)=ax$ in the vector space formed by the elements of the algebra. Condition (3) indicates that the image of φ is the whole space. This is equivalent to the fact that the kernel of φ is zero. The latter condition means that there are no divisors of zero in the algebra, that is, from $xy=0$ it follows that $x=0$ or $y=0$. If

$$x=\sum_{i=1}^{n} x_i e_i\,,\quad y=\sum_{j=1}^{n} y_j e_j\,,$$

then it follows from (2) that

$$xy=\sum_{l=1}^{n} z_l e_l\,,\quad z_l=\sum_{i,j=1}^{n} c_{ij}^{l} x_i y_j \quad (l=1,\ldots,n)\,.$$

Thus, division is possible in the algebra if the system of equations

$$F_l(x,y)=\sum_{i,j=1}^{n} c_{ij}^{l} x_i y_j=0 \quad (l=1,\ldots,n) \tag{4}$$

has no real solutions in which $(x_1,\ldots,x_n)\neq(0,\ldots,0)$ and $(y_1,\ldots,y_n)\neq(0,\ldots,0)$. These equations almost fall under the conditions of Theorem 2. The only difference is that the polynomials F_l determine equations in $\mathbb{P}^{n-1}\times\mathbb{P}^{n-1}$, but their number n is not equal to the dimension $2n-2$ of this space. Therefore we choose an arbitrary integer $1\leqslant r\leqslant n-1$ and set $x_{r+2}=\cdots=x_n=0$, $y_{n-r+1}=\cdots=y_n=0$. The equations $F_l(x_1, \ldots, x_{r+1}, 0, \ldots, 0; y_1, \ldots, y_{n-r+1}, 0, \ldots, 0) = 0$ $(l = 1, \ldots, n)$ are now given in $\mathbb{P}^r \times \mathbb{P}^{n-r}$ and a fortiori have no non-zero real solutions. By Theorem 2 this is possible only if the sum

$$\Sigma k_{i_1} \ldots k_{i_r} l_{j_1} \ldots l_{j_{n-r}} \tag{5}$$

is even, and this must hold for all $r = 1, \ldots, n-1$. In our case the forms F_l are bilinear, so that $k_i = l_i = 1$ and the sum (5) is equal to the number of its terms, namely $\binom{n}{r}$. So we see that the system (4) has no non-zero real solutions only if all the numbers $\binom{n}{r}$ are even for $r = 1, \ldots, n-1$. This is possible only if $n = 2^l$. For our condition on $\binom{n}{r}$ can be expressed in the following way: in the field $\mathbb{F}_2$ of two elements $(T+1)^n = T^n + 1$. If $n = 2^l \cdot m$, where m is odd and $m > 1$, then in $\mathbb{F}_2$

$$(T+1)^{2^l m} = (T^{2^l} + 1)^m = T^{2^l m} + m T^{2^l(m-1)} + \cdots + 1 \neq T^n + 1 .$$

We have proved the following results:

Theorem 3. *The rank of a division algebra over the field of real numbers is a power of two.*

It can be shown that a division algebra exists only for $n = 1, 2, 4, 8$. The proof of this fact uses rather delicate topological reasoning.

Applying similar arguments we can investigate for what values of m and n the system of equations

$$\sum_{i,j=1}^{m} c_{ij}^l x_i y_j = 0 \quad (l = 1, \ldots, n)$$

does not have non-zero real solutions. This question is interesting in that it is equivalent to the problem of ellipticity of the system of differential equations

$$\sum_{l=1}^{n} \sum_{j=1}^{m} c_{ij}^l \frac{\partial u_j}{\partial x_l} = 0 \qquad (i = 1, \ldots, m) .$$

3. The Genus of a Smooth Curve on a Surface. In the geometry on a smooth projective surface X an important role is played by the following formula, which expresses the genus of a smooth curve $C \subset X$ in terms of certain intersections indices:

$$g_C = \frac{(C, C+K)}{2} + 1 ; \tag{1}$$

here g_C is the genus of C, and K is the canonical class of X.

This formula could be proved by using the tools already known to us. However, a clearer and geometrically more lucid proof follows from the simplest properties of vector bundles. This will be given in Ch. VI, § 1.4. Here we only quote some of its applications.

1. If $X=\mathbb{P}^2$, then $\mathrm{Cl}(X)=\mathbb{Z}$, and a generator is the class L containing all straight lines. If $\deg C=n$, then $C\sim nL$. Since $K=-3L$ and $(L^2)=1$, the formula (1) in this case gives $g=\frac{n(n-3)}{2}+1=\frac{(n-1)(n-2)}{2}$. This result was obtained by other means in Ch. III, § 5.4.

2. Let X be a smooth quadric in $\mathbb{P}^3$. Let us clarify how to classify smooth curves on X by their geometric properties.

An algebraic classification is perfectly clear. Since $X\cong\mathbb{P}^1\times\mathbb{P}^1$, any curve on X is given by an equation $F(x_0:x_1;y_0:y_1)=0$, where F is a polynomial homogeneous in x_0 and x_1 as well as y_0 and y_1. We denote the degrees of homogeneity by m and n, respectively. The number of coefficients of such a polynomial is $(m+1)(n+1)$, hence all curves given by equations of degree of homogeneity m and n correspond to points of the projective space $\mathbb{P}^{mn+m+n}$. Since for arbitrary positive m and n there exist smooth irreducible curves, for example, the curve with the equation

$$2x_0^m y_0^n + x_0^m y_1^n + x_1^m y_1^n = 0\,,$$

to smooth irreducible curves there correspond points of a non-empty open subset of $\mathbb{P}^{mn+m+n}$.

We have seen in § 2.1 that $\mathrm{Cl}(X)=\mathbb{Z}\oplus\mathbb{Z}$, and if the curve C is given by an equation with the degrees of homogeneity m and n, then

$$C\sim mE+nF\,, \tag{2}$$

where E and F are as in Ex. 2. Thus, curves corresponding to given numbers m and n are effective divisors of the class $mE+nF$.

The classes E and F correspond to two families of rectilinear generators on X. It is easy to find the intersection indices of curves given in the form (2): if

$$C\sim mE+nF\,,\quad C'\sim m'E+n'F\,, \tag{3}$$

then

$$(C,C')=mn'+nm'\,. \tag{4}$$

In particular,

$$m=(C,F)\,,\quad n=(C,E)\,. \tag{5}$$

This points to the geometric meaning of m and n: just as the degree of a plane curve is equal to the number of points of its intersection with a line, so m and n are the two "degrees" of the curve C with respect to the two systems of rectilinear generators E and F on X.

If we take the embedding $X \subset \mathbb{P}^3$ into account, a curve acquires a new geometric invariant: the degree. We know that a family of curves on X can be simply classified by the invariants m and n. Our object now is to obtain this classification in terms of the invariants $\deg C$ and g_C.

First of all we observe that

$$\deg C = (C, H) \tag{6}$$

where H is a plane section of X. For this purpose we find a plane intersecting C transversally. Then $\deg C$ is equal to the number of points of intersection of C and H. Next, if $x \in C \cap H$, then $H \not\supset \Theta_{x,C}$, therefore $(H, C)_x = 1$, from which it is clear that (C, H) is also equal to the number of points of intersection of C and H. This proves (6).

Now we mention that

$$H \sim E + F, \tag{7}$$

as is immediately clear from (5) and the fact that H and E, and also H and F, are transversal at their point of intersection. Substituting this expression in (6) and applying (4) we find that

$$\deg C = m + n. \tag{8}$$

Observe that m and n are positive for any irreducible curve C except when C is a line. For if C does not belong to, say, the first family of rectilinear generators, then by taking any point $x \in C$ and a line E of the first family passing through x we see that C and E are in general position and that $(C, E) = n \geqslant (C, E)_x > 0$.

We proceed to the computation of g_C. To apply the formula (1) we have to know the canonical class of the surface X. Let us turn to this now. We use the fact that $X \simeq \mathbb{P}^1 \times \mathbb{P}^1$. It is easy to solve an even more general problem: to find the canonical class of a surface $X = Y_1 \times Y_2$, where Y_1 and Y_2 are smooth projective curves. Denoting by π_1 and π_2 the projections $\pi_1 : X \to Y_1$, $\pi_2 : X \to Y_2$ we consider arbitrarily one-dimensional differential forms $\omega_1 \in \Omega^1(Y_1)$, $\omega_2 \in \Omega^1(Y_2)$ and associate with them forms $\pi_1^*(\omega_1)$ and $\pi_2^*(\omega_2)$ on X. The form $\omega = \pi_1^*(\omega_1) \wedge \pi_2^*(\omega_2)$ is two-dimensional and its divisor (ω) belongs to the canonical class. It is this divisor we wish to compute.

Let $x \in X$, $x = (y_1, y_2)$, $y_1 \in Y_1, y_2 \in Y_2$, and let t_1 and t_2 be local parameters on Y_1 and Y_2 in a neighbourhood of y_1 and y_2. Then an obvious verification shows that $\pi_1^*(t_1)$ and $\pi_2^*(t_2)$ form a system of local parameters for the point x on X. We represent ω_1 and ω_2 in the form $\omega_1 = u_1\, dt_1$, $\omega_2 = u_2\, dt_2$. Then $(\omega_1) = (u_1)$ and $(\omega_2) = (u_2)$ in a neighbourhood of y_1 and y_2. Clearly $\omega = \pi_1^*(u_1) \cdot \pi_2^*(u_2)\, d\pi_1^*(t_1) \wedge d\pi_2^*(t_2)$, from which it follows that in some neighbourhood of x

$$(\omega) = (\pi_1^*(u_1)) + (\pi_2^*(u_2)) = \pi_1^*((\omega_1)) + \pi_2^*((\omega_2)).$$

Since this is true for any point $x \in X$, we see that $(\omega) = \pi_1^*((\omega_1)) + \pi_2^*((\omega_2))$, or in other words,

$$K_X = \pi_1^*(K_{Y_1}) + \pi_2^*(K_{Y_2}) . \tag{9}$$

Now we turn to the case $X = \mathbb{P}^1 \times \mathbb{P}^1$. We know that $-2y \in K_{\mathbb{P}^1}$, $y \in \mathbb{P}^1$. Therefore the formula (9) in our case gives

$$-2(\pi_1^*(y_1) + \pi_2^*(y_2)) \in K_X .$$

Since $\pi_1^*(y_1) = E$, $\pi_2^*(y_2) = F$, we now obtain the final formula

$$-2E - 2F \in K_X . \tag{10}$$

To find the genus of the curve $C \sim mE + nF$ we have to substitute this formula in (1) and use (4). So we obtain

$$g_C = (m-1)(n-1) . \tag{11}$$

Thus, the numbers m and n are uniquely determined, to within a permutation, by the degree and the genus of C. We see that for a given degree d there exist $d+1$ families of curves on $X: M_0, M_1, \ldots, M_d$. The genus of the family M_j is equal to $j(d-j) - d + 1$, and the families M_k and M_l have one and the same genus only if $j + l = d$, that is, if they are obtained from each other by the automorphism of $\mathbb{P}^1 \times \mathbb{P}^1$ that interchanges the factors. The dimension of the family M_j is $(j+1)(d-j+1) - 1$ or, expressed in terms of degree and genus: $g + 2d$.

In his "Lectures on the development of mathematics in the nineteenth century" Felix Klein gives a classification of curves of degree 3 and 4

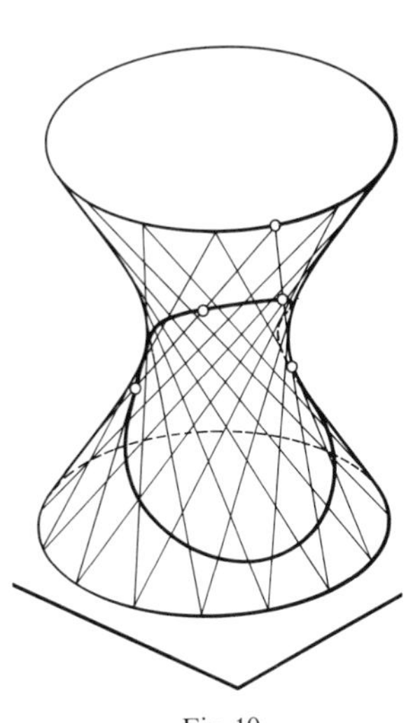

Fig. 10

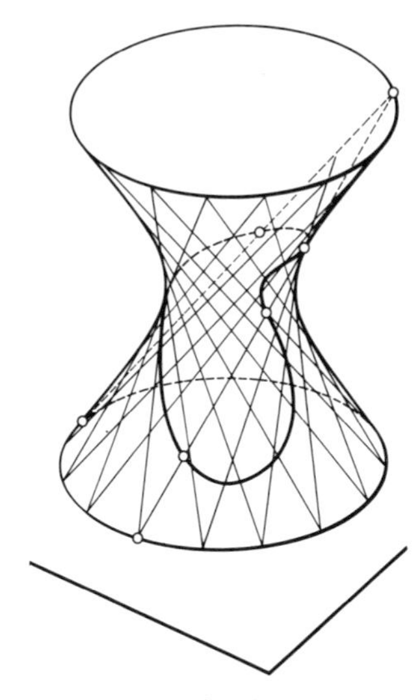

Fig. 11

on a hyperboloid as an example of the application of the idea of birational geometry. Our figures, which illustrate curves with $d=4$, are taken from the same source: In Fig. 10 $m=n=2$, and in Fig. 11 $m=1$, $n=3$.

3. As another application of formula (1) let us find out what negative values the index of self-intersection of a smooth curve C on a surface of degree 3 in $\mathbb{P}^3$ can take. By a result in Ch. III, § 5.4, in this case $K=-E$, where E is a hyperplane section. Therefore (1) takes the form

$$(C^2)-\deg C=2g-2.$$

Evidently, $(C^2)<0$ only if $g=0$ and $\deg C=1$, that is, C is a line on the surface. In that case $(C^2)=-1$.

4. The Ring of Classes of Cycles. Our theory of intersection indices of divisors is a special case of a general theory concerning subvarieties of arbitrary dimension. The concept of a divisor is replaced here by that of a *l-dimensional cycle*. This is the name for the elements of the free Abelian group generated by the irreducible subvarieties of dimension l. Two irreducible subvarieties Y_1 and Y_2, by definition, are in general position if all the irreducible components Z_i of the intersection $Y_1 \cap Y_2$ are of one and the same dimension and if

$$\operatorname{codim} Z_i=\operatorname{codim} Y_1+\operatorname{codim} Y_2 \text{ or } Y_1 \cap Y_2=\emptyset.$$

The basis of the theory in this case is that components Z_i are provided with positive integral multiplicities $n_i(Y_1, Y_2)$. These multiplicities are not, in general, lengths of certain rings, as in our theory. They are defined as sums in which only the first terms are of this form. The whole theory turns out to be very complicated and requires a considerably larger apparatus of commutative and homological algebra. The reader can find an account of it in [29].

The cycle $\Sigma n_i(Y_1, Y_2)Z_i$ is called the *product of the subvarieties* Y_1 and Y_2. By additivity this concept extends to any two cycles in general position. (Two cycles are, by definition, in general position if each component of one is in general position with every component of the other.)

The main property of this multiplication is its invariance under a notion of equivalence, which we shall explain presently. It generalizes the algebraic equivalence of divisors introduced in Ch. III, § 3.5, and is defined in a completely analogous manner. Let T be an arbitrary irreducible smooth variety and $Z \subset X \times T$ a cycle such that Z and $X \times t$ for every point $t \in T$ are in general position. The set of cycles $C_t=Z\cdot(X \times t)$ is called an algebraic family. Two cycles C_1 and C_2 are called *algebraically equivalent* if there exists a family of cycles C_t, $t \in T$, such that

$C_{t_1}=C_1$, $C_{t_2}=C_2$ for two points $t_1, t_2 \in T$. The set of classes of cycles under algebraic equivalence forms a group.

Multiplication of cycles on a projective variety is invariant under algebraic equivalence. There is a theorem on reduction to general position according to which for two cycles C_1 and C_2 there exist C_1' and C_2' such that C_1' is equivalent to C_1, C_2' equivalent to C_2, and C_1' and C_2 are in general position. These two results make it possible to define a product of any two classes of cycles.

We denote by $\mathfrak{A}_r$ the group of classes (under algebraic equivalence) of cycles of codimension r on a smooth projective variety X. The group

$$\mathfrak{A} = \bigoplus_{r=0}^{n} \mathfrak{A}_r, \qquad n = \dim X,$$

is a ring if we define multiplication for individual components as we have done above, and for arbitrary elements by additivity. This ring is commutative and associative. By the formula for the dimension of an intersection [formula (4) of Ch. I, § 6.2],

$$\mathfrak{A}_r \cdot \mathfrak{A}_s \subset \mathfrak{A}_{r+s},$$

that is, $\mathfrak{A}$ is a graded ring. It is easy to show that all points of X, regarded as zero-dimensional cycles, are equivalent, and that the cycle of x, $x \in X$, is not equivalent to zero. Therefore the group $\mathfrak{A}_n = \mathbb{Z} \cdot u$ has the standard generator u: the class of cycles of $x \in X$. The classes of divisors under algebraic equivalence form the group $\mathfrak{A}_1$. For n elements $\alpha_1, \ldots, \alpha_n \in \mathfrak{A}_1$ the product $\alpha_1 \ldots \alpha_n$ belongs $\mathfrak{A}_n = \mathbb{Z}u$:

$$\alpha_1 \ldots \alpha_n = m \cdot u, \quad m \in \mathbb{Z}.$$

The number m coincides with the intersection index $(\alpha_1, \ldots, \alpha_n)$, which we have defined in § 1.

The ring $\mathfrak{A}$ is a very interesting, but not very well explored invariant of X. The group $\mathfrak{A}_0$ is isomorphic to $\mathbb{Z}$ – a generator of it is X itself. We have already stated that $\mathfrak{A}_n \simeq \mathbb{Z}$. The group $\mathfrak{A}_1$ is finitely generated: this is the assertion of Theorem D in Ch. III, § 3.5. Whether the remaining groups $\mathfrak{A}_r$ are finitely generated is not known. This is called the basis problem.

Exercises

1. Determine $\deg v_m(\mathbb{P}^n)$, where v_m is the Veronese mapping.
2. Let C be a smooth plane curve of degree r on a smooth surface of degree m in $\mathbb{P}^3$. Determine (C^2) (generalization of Example 3 in § 1.5).

3. Suppose that on a smooth projective surface of degree m in $\mathbb{P}^3$ the divisor of a form of degree l consists of a single component with multiplicity 1, which is a smooth curve. Find its genus.

4. Show that the number of solutions of the system of equations

$$f_i(x_0^{(1)}, \ldots, x_{n_1}^{(1)}; \ldots; x_0^{(l)}, \ldots, x_{n_l}^{(l)}) = 0,$$

linear in each of the system of variables $x_0^{(j)}, \ldots, x_{n_j}^{(j)}$, is equal to $(\Sigma n_i)!/\Pi(n_i!)$ if the number of equations is equal to $\sum_{i=1}^{l} n_i$. As always, the number of solutions is understood in the sense of the corresponding intersection index.

5. Show that the system of n real equations

$$f_i(x_1, \ldots, x_n; y_1, \ldots, y_n) = 0 \quad (i = 1, \ldots, n),$$

homogeneous and of odd degree in each system of variables $x_1, \ldots, x_n$ and $y_1, \ldots, y_n$, has a real non-zero solution provided that $n \neq 2^l$.

6. Let X be a smooth curve, D the diagonal in $X \times X$ [the set of points of the form (x, x)]. Show that $(D^2) = -\deg K_X$. *Hint:* Use the fact that D and X are isomorphic.

7. Show that if a smooth curve C lies on a smooth surface X of degree 4 in $\mathbb{P}^3$ and if $(C^2) < 0$, then $(C^2) = -2$.

8. Show that the self-intersection indices of smooth curves on a smooth surface of even degree in $\mathbb{P}^3$ are always even.

§3. Birational Isomorphisms of Surfaces

In this section we explain how intersection indices can be applied to establish some basic properties of birational isomorphisms of surfaces. We begin by deriving some of the simplest properties of a σ-process of an algebraic surface.

1. σ-Processes of Surfaces. Let X be an algebraic surface, $\xi \in X$ a simple point, x and y local parameters at ξ, and $\sigma: Y \to X$ a σ-process centred at this point. According to Theorem 1 of Ch. II, §4, there exists a neighbourhood U of ξ such that $V = \sigma^{-1}(U)$ can be described by equations $t_0 y = t_1 x$ in $U \times \mathbb{P}^1$, where $(t_0 : t_1)$ are coordinates in $\mathbb{P}^1$. Furthermore, in the open set $t_0 \neq 0$ the σ-process is given by the simple equations

$$x = u, \qquad y = uv \tag{1}$$

where $v = t_1/t_0$. At any point $\eta \in \sigma^{-1}(\xi)$ the functions u and $v - v(\eta)$ form a system of local parameters. We set $L = \sigma^{-1}(\xi)$. Evidently, the local equation of the curve L is $u = 0$.

Let C be an irreducible curve on X passing through ξ. Just as in Theorem 1 of Ch. II, §4, in our case the inverse image $\sigma^{-1}(C)$ of C consists of two components: L and a curve C', which can be defined as the closure of $\sigma^{-1}(C - \xi)$ in Y. The curve C' is called the *characteristic inverse image* of C. We denote it by $\sigma'(C)$. Now we regard C as an irreducible divisor on X. Then

$$\sigma^*(C) = \sigma'(C) + mL, \tag{2}$$

where $\sigma'(C)$ occurs with the coefficient 1, because σ is an isomorphism of $Y - L$ onto $X - \xi$. Let us find the coefficient m in (2). For this purpose we assume that C has ξ as an r-fold point. This means that if f is a local equation of C in a neighbourhood of ξ, then $f \in \mathfrak{m}_\xi^r$, $f \notin \mathfrak{m}_\xi^{r+1}$. Now $\sigma^*(C)$ has a local equation $\sigma^*(f)$ in a neighbourhood of any point $\eta \in \sigma^{-1}(\xi)$. We set

$$f = \varphi(x, y) + \psi, \qquad \psi \in \mathfrak{m}_\xi^{r+1} \tag{3}$$

where φ is a form of degree r.

Substituting the transformation formulae (1) in (3) we obtain $(\sigma^* f)(u, v) = \varphi(u, uv) + \sigma^*\psi$. Since $\psi \in \mathfrak{m}_\xi^{r+1}$, we have $\psi = F(x, y)$, where F is a form of degree r, with coefficients in $\mathfrak{m}_\xi$. Therefore $\sigma^*(\psi) = (\sigma^* F)(u, vu)$, and finally,

$$(\sigma^* f)(u, v) = u^r(\varphi(1, v) + u(\sigma^* F)(1, v)); \tag{4}$$

since $\varphi(1, v)$ is not divisible by u, it follows that in (2) $m = r$ is the multiplicity of the singular point ξ on C.

Theorem 1. *The inverse image of a prime divisor C on X containing the centre ξ of a σ-process is given by the formula $\sigma^*(C) = \sigma'(C) + mL$, where $\sigma'(C)$ is a prime divisor, $L = \sigma^{-1}(\xi)$, and m is the multiplicity of ξ on C.*

2. Some Intersection Indices. We begin with a general property of birational regular mappings $f: Y \to X$ of smooth projective surfaces.

Theorem 2. *If D_1 and D_2 are divisiors on X, then*

$$(f^*(D_1), f^*(D_2)) = (D_1, D_2). \tag{1}$$

If $\bar{D}$ is a divisor on Y whose components are all exceptional curves, then

$$(f^*(D), \bar{D}) = 0 \tag{2}$$

for any divisor D on X.

We denote by $S \subset X$ the finite set of points at which the mapping f^{-1} is non-regular, and we set $T = f^{-1}(S)$ (set-theoretically). Then f determines an isomorphism

$$Y - T \to X - S. \tag{3}$$

If neither $\operatorname{Supp} D_1$ nor $\operatorname{Supp} D_2$ intersects S and if D_1 and D_2 are in general position, then (1) is obvious by virtue of the isomorphism (3). Otherwise we use the theorem on shifting the support of a divisor away from points, (Theorem 1 of Ch. III, § 1.3). Let $D_1' \sim D_1$ and $D_2' \sim D_2$ be divisors such that $(\operatorname{Supp} D'_1) \cap S = (\operatorname{Supp} D'_2) \cap S = \emptyset$ and D'_1 and D'_2 are in general position. Then $(D_1, D_2) = (D'_1, D'_2)$, and by what we have said above, $(D'_1, D'_2) = (f^*(D'_1), f^*(D'_2))$. Since $f^*(D'_i) \sim f^*(D_i)$, the equation (1) now follows.

(2) is equally obvious if $(\operatorname{Supp} D) \cap S = \emptyset$. The general case reduces to this one by very similar arguments.

Now we derive corollaries which refer directly to a σ-process. We use the notation of §3.1.

Corollary 1.

$$(L^2) = -1. \tag{4}$$

We consider a curve $C \subset X$ with the local equation y. According to Theorem 1, $\sigma^*(C) = \sigma'(C) + L$, and by (1) in §3.1 it is clear that a local equation of $\sigma'(C)$ is v. Since a local equation of L is u, we see that $(\sigma'(C), L) = 1$, therefore (4) follows from (2).

Corollary 2. *$(\sigma'(C), L) = m$, where m is the multiplicity of a singular point ξ on C.*

This follows at once from (2), (4) and by (2) in §3.1.

Corollary 3.

$$(\sigma'(C_1), \sigma'(C_2)) = (C_1, C_2) - m_1 m_2 \tag{5}$$

where m_1 and m_2 are the multiplicities of ξ on C_1 and C_2, respectively.

By Theorem 1 and 2

$$\begin{aligned}(C_1, C_2) &= (\sigma^*(C_1), \sigma^*(C_2)) = (\sigma'(C_1) + m_1 L, \sigma^*(C_2)) \\ &= (\sigma'(C_1), \sigma^*(C_2)) = (\sigma'(C_1), \sigma'(C_2) + m_2 L) \\ &= (\sigma'(C_1), \sigma'(C_2)) + m_1 m_2,\end{aligned}$$

from which (5) follows.

3. Elimination of Points of Indeterminacy. We can now derive an important property of rational mappings of algebraic surfaces.

Theorem 3. *If $\varphi: X \to \mathbb{P}^n$ is a rational mapping of a smooth projective surface, then there exists a sequence of σ-processes such that the mapping $\psi = \varphi\sigma_1 \dots \sigma_m$ is regular.*

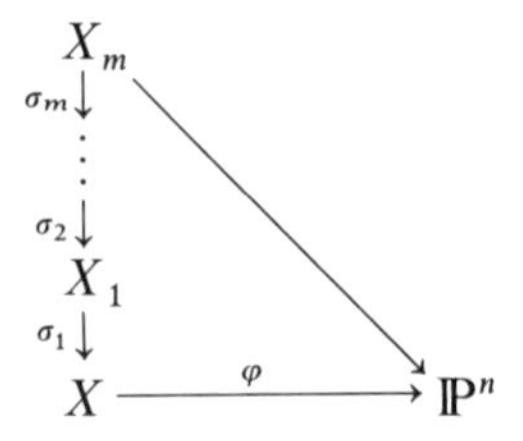

Proof. We know that φ is non-regular only at finitely many points, (Theorem 3 of Ch. II, § 3.1), and Theorem 2 of Ch. III, § 1.4, gives a more detailed description of this set, which we now recall. Let $\varphi = (f_0 : \ldots : f_n)$, $\bar{D} = \text{g.c.d.}\,((f_0), \ldots, (f_n))$, and $D_i = (f_i) - \bar{D}$. Then the set of points of non-regularity of φ is $\bigcap_{i=0}^{m} \operatorname{Supp} D_i$.

We introduce the following invariant of the rational mapping φ. Evidently, all the divisors D_i are equivalent to each other. Therefore we can set

$$d(\varphi) = (D_i^2)\,.$$

We show that $d(\varphi) \geqslant 0$. For this purpose we set $\lambda = (\lambda_0, \ldots, \lambda_n)$, $D_\lambda = \left(\sum_{i=0}^{n} \lambda_i f_i\right) - \bar{D}$. Obviously $D_\lambda \geqslant 0, D_\lambda \sim D_i$. We have to find a λ such that D_0 and D_λ have no common components; then $d(\varphi) = (D_0, D_\lambda) \geqslant 0$. By hypothesis, all the D_i do not have common components. Hence for each irreducible component $C \subset D_0$ there exists an $i \geqslant 1$ such that $v_C(D_i) = 0$. The condition $v_C(D_\lambda) > 0$ means that $\Sigma \lambda_i \cdot g_i|_C = 0$, where the g_i are local equations of the divisor D_i in a neighbourhood of some point $c \in C$. In view of this, $v_C(D_\lambda) = 0$ for all λ in some non-empty set in $\mathbb{A}^{n+1}$. Therefore, there exists a λ belonging to all the open sets corresponding to all the irreducible curves $C \subset G_0$, and for it D_0 and D_λ have no common components.

If $x_0 \in \bigcap \operatorname{Supp} D_i$, then all the $\operatorname{Supp} D_\lambda$ contain x_0. Therefore $d(\varphi) > 0$ if $\bigcap \operatorname{Supp} D_i$ is not empty, that is, if the mapping φ is non-regular. In that case we denote by $\sigma : X' \to X$ the σ-process centred at the point $x_0 \in \bigcap \operatorname{Supp} D_i$, and we set $\varphi' = \varphi\sigma$. We show that $d(\varphi') < d(\varphi)$, and of course, Theorem 3 follows from this.

For the divisor $D = \Sigma l_i C_i$ we define the multiplicity of a point ξ on D as the number $m = \Sigma m_i l_i$, where m_i are the multiplicities of ξ on the curve C_i. Evidently, Theorem 1 then becomes true for every effective divisor, and if $D \geqslant 0$, then $m \geqslant 0$, where $m = 0$ indicates that $\xi \notin \operatorname{Supp} D$.

Similarly we set $\sigma'(D) = \Sigma l_i \sigma'(C_i)$. Then $\sigma^* D = \sigma' D + mL$.

We denote by v_i the multiplicities of x_0 on the divisors D_i, and we set $v = \min v_i$. The mapping φ' is given by the functions $f_i' = \sigma^* f_i$ and

$$(f_i') = (\sigma^* f_i) = \sigma'(D_i) + (v_i - v)\,L + v\,L + \sigma^* \bar{D}\,,$$

where the divisors $D_i' = \sigma'(D_i) + (v_i - v)\,L,\ i = 0, \ldots, n$, have no common components.

We choose an i such that $v_i = v$, and then, by definition,

$$d(\varphi') = (D_i'^2) = ((\sigma' D_i)^2)\,.$$

From the relation $\sigma^* D_i = \sigma' D_i + \nu L$ and from Theorem 2 it follows that $((\sigma' D_i)^2) = ((\sigma^* D_i - \nu L)^2) = ((\sigma^* D_i)^2) - \nu^2 = (D_i^2) - \nu^2$, and therefore $d(\varphi') = d(\varphi) - \nu^2$. This proves Theorem 3.

Note. In Theorem 3 there is no need to assume that X is a projective surface. In the proof this property was used only in referring to the fact that $(C, (f)) = 0$ for every curve $C \subset X$. However, this proposition was applied only to curves of the form $\sigma^{-1}(\xi)$, which are projective even when X is not. It is easy to see that the required property holds for such curves C.

The simplest example of Theorem 3 is the mapping $f: \mathbb{A}^2 \to \mathbb{P}^1$ in the definition of a projective line: $f(x, y) = (x : y)$. The mapping f is non-regular at the point $\xi = (0, 0)$. Substituting (1) of §3.1 we see that $f(x, y) = (1 : v)$ at those points belonging to $\sigma^{-1}(\xi)$ and to the set $t_0 \neq 0$, hence $f\sigma$ is regular there.

4. Decomposition into σ-Processes. Now we have everything at our disposal to prove the main result on birational isomorphisms of surfaces.

Theorem 4. *Let $\varphi: X \to Y$ be a birational isomorphism of smooth projective surfaces. Then there exists a surface Z and surfaces and mappings.*

$$\sigma_i : X_i \to X_{i-1} \quad (i = 1, \ldots, l)$$

$$\tau_j : Y_j \to Y_{j-1} \quad (j = 1, \ldots, m)$$

such that $X_0 = X$, $Y_0 = Y$, $X_l = Y_m = Z$, σ_i and τ_j are σ-processes, and $\varphi\sigma_1 \ldots \sigma_l = \tau_1 \ldots \tau_m$. *In other words, the diagram*

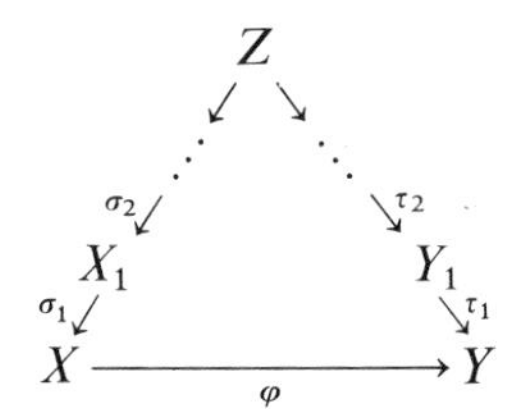

commutes.

Theorem 4 is an obvious consequence of Theorem 3 and the following proposition.

Theorem 5. *Let $\varphi: X \to Y$ be a regular mapping of smooth projective surfaces and a birational isomorphism. Then there exists a sequence of surfaces and mappings $\sigma_i : Y_i \to Y_{i-1}$ $(i = 1, \ldots, l)$, such that σ_i is a σ-process, $Y_0 = Y$, $Y_l = X$ and*

$$\varphi = \sigma_1 \ldots \sigma_l .$$

We precede the proof of Theorem 5 with some general remarks on birational isomorphisms of surfaces.

First of all, for an arbitrary rational mapping $\varphi: X \to Y$, where X is a smooth surface and Y a projective variety, we can talk of the image $\varphi(C)$ of a curve $C \subset X$. For φ is regular at all points of C except possibly a finite set S of points. By $\varphi(C)$ we mean the closure of $\varphi(C-S)$ in Y.

Then the theorem on the existence of exceptional subvarieties for regular mappings (Theorem 2 of Ch. II, §4) remains true.

Lemma. *If $\varphi: X \to Y$ is a birational isomorphism of smooth projective surfaces and if φ^{-1} is non-regular at a point $y \in Y$, then there exists a curve $C \subset X$ such that $\varphi(C) = y$.*

Proof. We consider open sets $U \subset X$ and $V \subset Y$ on which φ establishes an isomorphism, and we denote by Z the closure of the graph of the isomorphism $\varphi: U \to V$ in $X \times Y$. The projections onto X and Y determine regular birational isomorphisms $p: Z \to X$ and $q: Z \to Y$. Evidently $\varphi^{-1} = p \circ q^{-1}$ and since φ^{-1}, by hypothesis, is non-regular at y, neither is q^{-1}.

We may now apply the theorem on the existence of exceptional subvarieties (Theorem 2 of Ch. II, §4) to the regular mapping $q: Z \to Y$. This theorem shows that there exists a curve $D \subset Z$ such that $q(D) = y$. We set $p(D) = C$ and verify that C satisfies the condition of the lemma. Actually, we need only verify that $\dim C = 1$, that is, $\dim C = \dim D$. If this were not the case, then $p(C)$ would be a point $x \in X$, and for all points $z \in D$ we would obtain $p(z) = x$, $q(z) = y$, that is, $z = (x, y)$, but this contradicts the fact that $D \subset X \times Y$ is a curve.

Now we proceed to the proof of Theorem 5. We assume that φ is not an isomorphism, that is, φ^{-1} is non-regular at a point $y \in Y$. We consider the σ-process $\sigma: Y' \to Y$ centred at y and we define $\varphi': X \to Y'$ so that the diagram

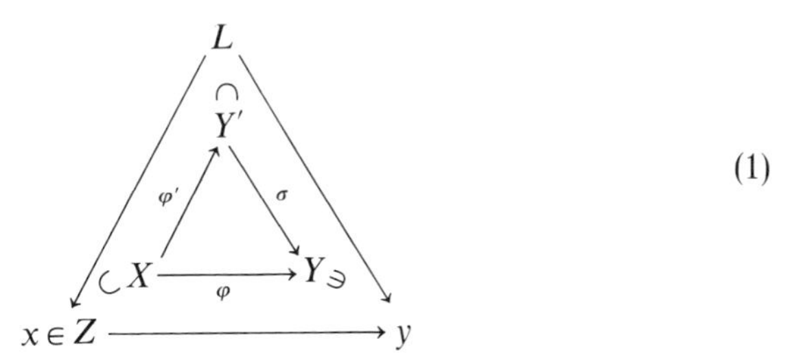

(1)

commutes.

The theorem will be proved if we can show that φ' is a regular mapping. For from the commutativity of the diagram (1) it then follows that the subvariety $\varphi^{-1}(y)$ is mapped by φ' into $\sigma^{-1}(y) = L = \mathbb{P}^1$. From the fact that φ' maps X onto the whole of Y' it follows that $\varphi^{-1}(y)$ is mapped onto the whole of L. Therefore, not all the components of $\varphi^{-1}(y)$ are mapped into a single point. Hence for $y' \in L$ the number of components of $(\varphi')^{-1}(y')$ is less than the number of components of $\varphi^{-1}(y)$.

Consequently, by performing finitely many σ-processes we can achieve that there are no exceptional subvarieties on X, so that our mapping becomes an isomorphism.

It remains to prove that φ' is regular. Suppose that this is not so. Then, by the lemma, $\psi = (\varphi')^{-1}$ maps some curve lying on Y' into a point $x \in X$. From the commutativity of the diagram (1) it follows that this curve can only be L, so that $\psi(L) = x$.

According to Theorem 3 of Ch. II, § 3, there exists a finite set $E \subset L$ such that ψ is regular at all points $y' \in L - E$. Since $\sigma(y') = y$, it follows from the commutativity of the diagram (1) that also $\varphi(x) = y$.

Let us show that the mapping

$$d_x\varphi : \Theta_{x,X} \to \Theta_{y,Y} \tag{2}$$

is an isomorphism. It is sufficient to show that it is an epimorphism. Let $d_x\varphi\Theta_{x,X} \subseteq l \subset \Theta_{x,Y}$, where l is a line in the plane $\Theta_{y,Y}$. Then it follows from the commutativity of the diagram (1) that

$$(d_{y'}\sigma)(\Theta_{y',Y'}) \subset l \tag{3}$$

for all points $y' \in L - E$. However, this contradicts the simplest properties of a σ-process. For let C be a smooth curve on Y, $y \in C$, and $\Theta_{y,C} \neq l$, for example, $C = V(\alpha u + \beta v)$, where u and v are local parameters at y. Then according to (2) of §3.1 $\sigma(\sigma'(C)) = C$, and $\sigma'(C)$ intersects L in a single point y' which on L has the coordinates $(-\beta : \alpha)$, $\sigma'(C)$ is smooth at this point, and $\sigma : \sigma'(C) \to C$ is an isomorphism. We can choose α and β so that $y' \notin E$, and then already $(d_{y'}\sigma)(\Theta_{y',\sigma'(C)}) \not\subset l$.

The facts that (2) is an isomorphism and that φ^{-1} is non-regular at y contradict each other. For by applying the theorem on exceptional subvarieties (Ch. II, §4.2) we can find a curve $Z \subset X$, with $x \in Z$, such that $\varphi(Z) = y$. Then $\Theta_{x,Z} \subset \Theta_{x,X}$ (we recall that the tangent space is also defined when x is a singular point on Z). Since $\varphi(Z) = y$, we see that $(d_x\varphi)\Theta_{x,Z} = 0$, hence that the mapping (2) has a kernel. This contradiction proves Theorem 5.

5. Notes and Examples. We consider a birational isomorphism $f: X \to Y$ of smooth projective surfaces which is also a regular mapping. We assume that f^{-1} is non-regular only at a single point $\eta \in Y$ and that the curve $C = f^{-1}(\eta)$ is irreducible. According to Theorem 5 f is a product of σ-processes: $f = \sigma_1 \dots \sigma_m$, and since in every σ-process there arises a curve that is contractible to a point, C is irreducible only if $m = 1$ and f

itself is a σ-process. Then C coincides with the curve L for which we proved in §3.1 and §3.2 that

$$L \simeq \mathbb{P}^1, \quad (L^2) = -1. \tag{1}$$

The converse is also true: if C is a curve on a smooth projective surface X satisfying the conditions (1), then there exists a regular mapping $f: X \to Y$ which is a birational isomorphism, such that Y is smooth, $f(C) = \eta \in Y$, and that f coincides with a σ-process. Thus, the conditions (1) are necessary and sufficient for a curve C to be contractible to a point in the sense indicated above. This result, which is due to Castelnuovo, will not be proved here; the reader can find a proof in the book [3], Ch. II.

In conclusion we construct, in accordance with Theorem 4, a decomposition into σ-processes for a simple birational isomorphism. This is the birational automorphism f of the projective plane $\mathbb{P}^2$, the so-called *quadratic transformation*, which is given by the formulae:

$$\begin{gathered} f(x_0 : x_1 : x_2) = (y_0 : y_1 : y_2)\,, \\ y_0 = x_1 x_2,\, y_1 = x_0 x_2,\, y_2 = x_0 x_1\,, \end{gathered} \tag{2}$$

We regard f as a birational isomorphism of two copies: $\mathbb{P}^2$ and $\overline{\mathbb{P}}^2$ of the projective plane $\mathbb{P}^2$, and we denote in one of them the coordinates by $(x_0 : x_1 : x_2)$, in the other by $(y_0 : y_1 : y_2)$. Clearly f is non-regular at the three points $\xi_0 = (1:0,0)$, $\xi_1 = (0:1:0)$, $\xi_2 = (0:0:1)$. According to Theorem 3 we have to start by carrying out the σ-processes σ_0, σ_1 and σ_2 at these points. We arrive at a surface X and a regular mapping $\varphi: X \to \mathbb{P}^2$, $\varphi = \sigma_2 \sigma_1 \sigma_0$. We prove that the mapping $\psi = f\varphi: X \to \overline{\mathbb{P}}^2$ is regular. For ψ is regular at a point z if $\varphi(z) \neq \xi_i$. At the points $\zeta \in \sigma_0^{-1}\xi_0$ the mapping $f\sigma_0$ is also regular. To verify this it is sufficient to set $x = x_1/x_0$, $y = x_2/x_0$, and to substitute (1) of §3.1 in (2). So we see that

$$f(x,y) = (xy : y : x), \quad f_{\sigma_0}(u,v) = (uv : v : 1). \tag{3}$$

Since σ_1 and σ_2 induce isomorphisms in neighbourhoods of the points ζ, we see that ψ is regular at the points z for which $\varphi(z) = \xi_0$. The situation is similar for ξ_1 and ξ_2.

According to Theorem 4 ψ is a product of σ-processes: $\psi = \tau_1 \ldots \tau_m$. Let us clarify what curves $C \subset X$ can be mapped by ψ into points. Evidently this can only be either curves $M'_i = \sigma_i^{-1}(\xi_i)$ $(i = 0, 1, 2)$ or characteristic inverse images of such curves $L \subset \mathbb{P}^2$ that are mapped by f into points. It is easy to see that f determines an isomorphism between $\mathbb{P}^2 - L_0 - L_1 - L_2$ and $\overline{\mathbb{P}}^2 - M_0 - M_1 - M_2$, where L_i is the line in $\mathbb{P}^2$ defined by the equation $x_i = 0$, and M_i the line in $\overline{\mathbb{P}}^2$ with the equation $y_i = 0$. Therefore ψ can contract to a point only the curves M'_0, M'_1, M'_2, L'_0,

L'_1, L'_2 where L'_i are the characteristic inverses of the curves L_i in X. But from (3) we see that, for example, M'_0 (given by the local equation $u=0$) is mapped to the whole curve $y_0=0$. Similarly M'_i is mapped onto M_i for $i=1,2$. Thus, ψ can shrink into points only the curves L'_i. Furthermore, ψ^{-1} is non-regular at the points $\eta_0=(1:0:0)$, $\eta_1=(0:1:0)$, $\eta_2=(0:0:1)$, otherwise f^{-1} would be regular at one of these points, but f^{-1} is given by the same formulae as f, as is clear from (2). Thus, on the one hand, in the decomposition $\psi=\tau_1\ldots\tau_m$ there cannot occur more than three σ-processes, and on the other hand, there must occur the σ-processes at the point η_0, η_1, η_2. So we see that

$$f=\tau_2\tau_1\tau_0\sigma_0^{-1}\sigma_1^{-1}\sigma_2^{-1}.$$

It is easy to give an idea of the disposition of the curves $M'_0, M'_1, M'_2, L'_0, L'_1, L'_2$, on X. The arrows in Fig. 12 indicate into what points the curves contract.

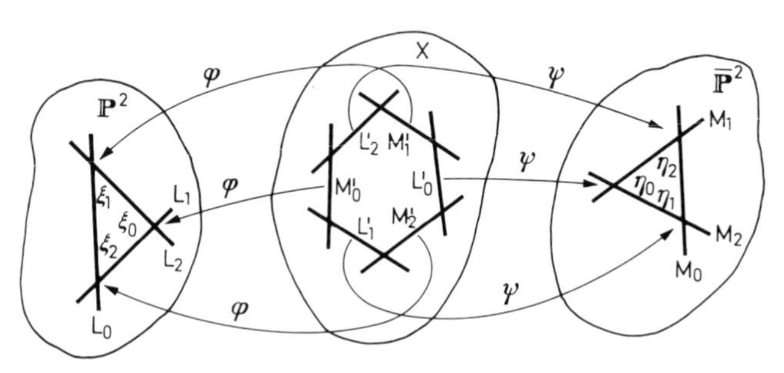

Fig. 12

Of course, the quadratic transformation depends on the choice of the system of coordinates in $\mathbb{P}^2$ or, what is the same, on the choice of the points ξ_0, ξ_1, ξ_2. By multiplying out several such transformations we obtain new birational automorphisms of the plane. Noether has proved a theorem to the effect that every birational automorphism of the plane can be represented as a product of quadratic transformations and a projective transformation. We do not give the very subtle proof of this theorem, which can be found in [3], Ch. V. The interesting question to what extent the representation of a birational automorphism in terms of quadratic ones is unique has not been investigated so far.

Exercises

1. For every integer l (positive, negative or zero) construct a smooth projective surface X and on it an irreducible curve C such that $(C^2)=l$. *Hint:* Obtain X by blowing up some points on $\mathbb{P}^2$.

2. Let X be a smooth projective surface, C_1 and C_2 two curves on it. Let x be a non-singular point on both C_1 and C_2. Let $\sigma: Y\to X$ be a σ-process at x, C'_1 and C'_2 char-

acteristic inverse images of C_1 and C_2. Show that C'_1 and C'_2 intersect at points $y \in \sigma^{-1}(x)$ if and only if C_1 and C_2 touch at x. Here $\sigma^{-1}(x) \cap C'_1 \cap C'_2$ consists of the single point y and the order of contract of C'_1 and C'_2 at y is one less than the order of contact of C_1 and C_2 at x.

3. Suppose that a mapping $f: \mathbb{P}^2 \to \mathbb{P}^1$ is given by the formula

$$f(x_0 : x_1 : x_2) = (P(x_0, x_1, x_2) : Q(x_0, x_1, x_2)),$$

where P and Q are forms of degree n. How many σ-processes have to be performed so as to obtain a surface $\varphi: X \to \mathbb{P}^2$ for which $f\varphi$ is regular?

4. Let $X \subset \mathbb{P}^3$ be a smooth quadric and $f: X \to \mathbb{P}^2$ the birational isomorphism consisting in the projection of X from a point $x \in X$. Decompose f into a product of σ-processes.

5. Let f be the birational automorphism of $\mathbb{P}^2$ given in inhomogeneous coordinates by the formulae $x' = x$, $y' = y + x^2$. Decompose f into a product of σ-processes.

6. Let $L \subset \mathbb{P}^2$ be a line, x and y two points of it, $X \to \mathbb{P}^2$ the product of σ-processes at x and y, and L' the characteristic inverse image of L. Show that $(L')^2 = -1$. By Castelnuovo's theorem stated in § 3.5 there exists a regular mapping $f: X \to Y$ that is a birational isomorphism and contracts L' to a point. Construct it in the given case. *Hint:* Search for it among the preceding exercises.

7. Let $f: X \to Y$ be a regular mapping of smooth projective varieties and a birational isomorphism. Show that for $D_1, \ldots, D_n \in \operatorname{Div}(Y)$ we have $(f^*(D_1), \ldots, f^*(D_n)) = (D_1, \ldots, D_n)$.

8. Let $\sigma: X \to Y$ be a σ-process centred at a point $y \in Y$, $\Gamma = \sigma^{-1}(y)$, $D_1, \ldots, D_{n-1} \in \operatorname{Div}(Y)$. Show that $(\Gamma, \sigma^*(D_1), \ldots, \sigma^*(D_{n-1})) = 0$.

9. In the notation of Exercise 8 find Γ^n for every $n > 1$.

10. Let X be a smooth projective surface, and $\sigma: X' \to X$ a σ-process. Show that the canonical classes K_X and $K_{X'}$ of surfaces X and X' are connected by the relation $K_{X'} = \sigma^* K_X + L$, where $L \subset X'$ is an exceptional curve.

11. Let X be a smooth projective surface, $C \subset X$ an irreducible curve of which it is known that its characteristic inverse image relative to finitely many σ-processes is a smooth curve (it can be shown that such a sequence of σ-processes exists for every curve). Show that

$$g(C^{\nu}) = \frac{(C, C+K)}{2} + 1 - \Sigma \frac{l_i(l_i - 1)}{2}$$

where C^{ν} is the normalization of C, and l_i are the multiplicities of all singular points at which the σ-processes are performed.

12. Show that if $C \subset \mathbb{P}^2$ is a curve of degree n with d simple singular points, then

$$g(C^{\nu}) = \frac{(n-1)(n-2)}{2} - d.$$

13. Generalize the concept of class (Exercise 20 to Ch. III, § 5) to curves with singular points. Show that under the conditions of Exercise 12 the class of C is $n(n-1) - d$.

Part II. Schemes and Varieties

Chapter V. Schemes

In this chapter we return to the starting point of our entire investigation – the concept of an algebraic variety, and we attempt to throw light on it from a more general and invariant point of view. This leads us, on the one hand, to new concepts and methods, which are exceptionally fruitful even in the study of quasiprojective varieties with which we have been concerned previously. On the other hand, we arrive in this way at a generalization of this concept, which makes the field of applicability of algebraic geometry far more extensive.

What causes us to look again at the definition of an algebraic variety? If we recall how affine, projective, and quasiprojective varieties are defined, we see that in the last analysis they are all defined by systems of equations. Of course, one and the same variety can be given by various systems of equations, and it is precisely the wish to abstract from the accidental choice of the system of equations and of the embedding in an enveloping space that leads to the concept of isomorphism of varieties. In this form the system of basic concepts of algebraic geometry reminds us of the theory of finite extensions of fields at the time when it was formulated in terms of polynomials: the basic object was the equation and the idea of independence of the accidentally chosen equation was formulated by means of the "Tschirnhaus transformation". In the theory of fields an invariant formulation of the basic concept was connected with the analysis of a finite extension K/k which, although (for separable extensions) it can be represented in the form $K=k(\theta)$, $f(\theta)=0$, nevertheless reflects the properties of the equation $f=0$ that are invariant under the Tschirnhaus transformation. As another parallel we can point to the concept of a topological manifold, which in the works of Poincaré was still defined as a subset of a Euclidean space, until it was invariantly defined as a special case of the more general concept of a topological space.

The central problem of this and the following chapter are the formulation and investigation of the "abstract" concept of an algebraic variety, independent of its concrete specification. This concept therefore has in algebraic geometry the same function as that of a finite extension in the theory of fields or of a topological space in topology.

The path on which we are led to this definition is based on two observations on the definition of quasiprojective varieties. Firstly, the basic concepts (for example, that of a regular mapping) are defined for quasiprojective varieties, starting from their covering by affine open sets. Secondly, all the properties of an affine variety X are reflected in the ring $k[X]$, which is invariantly connected with it. These arguments indicate that the general concept of an algebraic variety must in some sense reduce to that of an affine variety. In the definition of affine varieties we have to start out from rings of a certain special type and define a variety as a geometric object connected with this ring.

The programme we have outlined is not hard to put into effect: in Ch. I we have investigated in detail how properties of an affine variety X are reflected in the ring $k[X]$, and this enables us to come to a definition of the variety X, starting out from some ring which *post factum* turns out to be $k[X]$. However, in this way we may obtain considerably more than an invariant definition of an algebraic variety. The fact of the matter is that the coordinate rings of affine varieties are of a very special kind: they are finitely generated algebras over some field and have no nilpotent elements (according to Exercises 1 and 2 in Ch. I, §2, all such rings arise from affine varieties). But since the just developed definition of an affine variety, starting out from some ring A, satisfies these three conditions, it makes sense at once to replace in this definition A by a perfectly arbitrary commutative ring. So we arrive at a wide generalization of affine varieties. Since the general definition of an algebraic variety reduces to that of an affine variety, the concept of an algebraic variety is also generalized to the same extent. The general concept at which we arrive in this way is called a *scheme*.

The concept of a scheme enables us to include an incomparably wider range of objects than algebraic varieties. We can point to two reasons why this generalization has turned out to be so exceptionally useful, both for the "classical" algebraic geometry and for other domains. Firstly, the rings occurring in the definition of a scheme (analogues to the rings $K[U]$, where U is an affine open subset of an algebraic variety) now need not by any means be algebras over a field. For example, they may be of the type of the ring of integers $\mathbb{Z}$, the rings of integers in algebraic number fields, or the polynomial rings $\mathbb{Z}[T]$. The introduction of these objects makes it possible to apply the theory of schemes in number theory and provide the best of all the known methods of using geometrical intuition in number-theoretical problems. Secondly, the rings occurring in the definition of a scheme may contain nilpotent elements. For example, the use of these schemes makes it possible to apply in algebraic geometry the concepts of differential geometry connected with infinitely small changes of points or subvarieties $Y \subset X$ on an algebraic variety, even when X and Y are quasiprojective varieties. One should not forget the fact that as a special case of schemes we obtain an invariant definition of an algebraic variety, which as we shall see is vastly more convenient in applications, even when it does not lead to a more general concept.

In the hope that the reader has already in his possession sufficient factual material, we abandon the usual style of our book "from the special to the general". In Ch. V we introduce the general concept of a scheme and establish its simplest properties. In Ch. VI we define "abstract" algebraic varieties, which we simply call varieties. By several examples we show how the concepts and ideas introduced in this chapter enable us to solve some specific problems which we had encountered earlier on several occasions in the theory of quasiprojective varieties.

§ 1. Spectra of Rings

1. Definition of a Spectrum. Let us proceed to the execution of the programme outlined. We consider a ring A, which we always assume to be commutative, with an identity; otherwise it is completely arbitrary. We try to connect with it a certain geometric object, which in case A is the coordinate ring of an affine variety X must bring us back to X. To begin with, this object is defined only as a set, but later we endow it with a number of structures (for example, a topology), which must justify its claim to be geometrical.

The very first definition requires as a preliminary some explanations. If we wish to recuperate the affine variety X starting out from the ring $k[X]$, then it is most natural to utilize the connection between subvarieties $Y \subset X$ and their ideals $\mathfrak{a}_Y \subset K[X]$. In particular, to a point $x \in X$ there corresponds a maximal ideal M_x, and it is easy to verify that the mapping $x \to M_x \subset k[X]$ establishes a one-to-one correspondence between points $x \in X$ and maximal ideals of the ring $k[X]$. Therefore it appears natural to associate with every ring A as a "geometrical object" the set of its maximal ideals. This set is called the *maximal spectrum* of A and is denoted by Specm A. However, in the generality in which we have considered the problem, the mapping $A \to \operatorname{Specm} A$ has certain defects, one of which we wish to examine.

Clearly it is to be expected that the association of a ring A with some set should have the basic properties that link the coordinate ring of an affine variety with the variety itself. The most important of these properties is that homomorphisms of rings correspond to regular mappings of a variety. Is there a natural way of associating with a homomorphism of rings $f: A \to B$ a mapping of Specm B into Specm A? More generally, how can we associate with an ideal $\mathfrak{b} \subset B$ an ideal $\mathfrak{a} \subset A$? Clearly there is only one reasonable method, namely to consider the inverse image $f^{-1}(\mathfrak{b})$ of this ideal. But alas – the inverse image of a maximal ideal is not always maximal. For example, if A is a ring without divisors of zero, but not a field, and if f is an embedding of it in a field K, then the inverse image of the zero ideal, which is maximal in K, is the zero ideal in A, which is not maximal.

This complication does not arise if instead of maximal ideals we consider prime ideals; an elementary verification shows that the inverse image of a prime ideal under any homomorphism is prime. In the case when $A = k[X]$ and X is an affine variety, the set of prime ideals of A has a clear geometrical meaning – it is the set of all irreducible closed subvarieties of X: points, irreducible curves, surfaces, etc. Finally, for a very wide class of rings the set of prime ideals is determined by that of maximal ideals (Exercise 8). All this motivates the following definition:

The set of prime ideals of a ring A is called its *spectrum* and is denoted by Spec A. The prime ideals are called the *points of the spectrum*.

Since we consider only rings with an identity, the ring itself will not be counted among its prime ideals (so that the factor ring always exists).

Each ring has at least one maximal ideal. This follows easily from Zorn's lemma (see, for example [37], Vol. I, Ch. III, §8). Therefore the spectrum of any ring is non-empty.

We have already talked of the geometrical significance of Spec A, when A is the coordinate ring of an affine variety. Let us look at some other examples.

Example 1. Spec $\mathbb{Z}$ consists of the prime ideals (2), (3), (5), (7), (11)... and the zero ideal.

Example 2. Let $\mathcal{O}_x$ be the local ring of a point x on an irreducible algebraic curve. Then Spec $\mathcal{O}_x$ consists of two points: the maximal ideal and the zero ideal.

Consider a ring homomorphism $\varphi: A \to B$. Henceforth we shall always be concerned only with such homomorphisms that carry the identity of one ring into that of the other. As we have remarked above, for any prime ideal of B its inverse image is a prime ideal of A. By associating with every prime ideal its inverse image we determine a mapping

$$^a\varphi : \operatorname{Spec} B \to \operatorname{Spec} A,$$

which is said to be *associated* with the homomorphism φ.

Example 3. Consider the ring $\mathbb{Z}[i]$, $i^2 = -1$, and let us try to find its spectrum, by using the embedding $\varphi: \mathbb{Z} \to \mathbb{Z}[i]$. This determines the mapping

$$^a\varphi : \operatorname{Spec} \mathbb{Z}[i] \to \operatorname{Spec} \mathbb{Z} .$$

We denote by ω and ω' the points of $\operatorname{Spec} \mathbb{Z}$ and $\operatorname{Spec} \mathbb{Z}[i]$ that correspond to the zero ideal. Clearly $({}^a\varphi)^{-1}(\omega) = \omega'$. Other points in $\operatorname{Spec} \mathbb{Z}$ correspond to prime ideals. By definition $({}^a\varphi)^{-1}(p)$ consists of the prime ideals of $\mathbb{Z}[i]$ that divide p. It is standard knowledge that these ideals are principal and that there are two of them if $p \equiv 1 \pmod 4$ and one if $p = 2$ or $p \equiv 3 \pmod 4$. All this can be illustrated as follows (see Fig. 13):

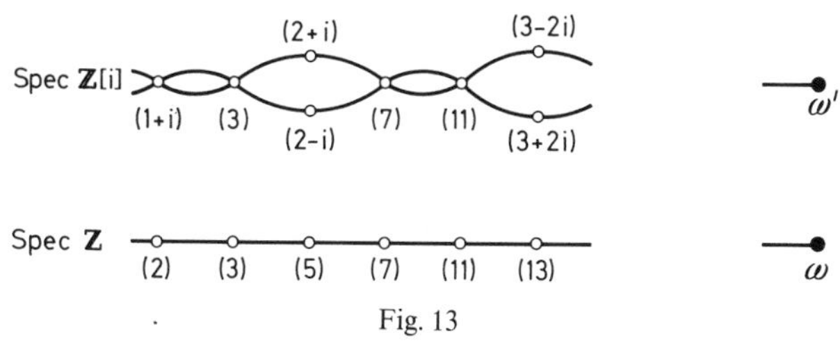

Fig. 13

The reader is recommended to work out the more complicated example of the spectrum of the ring $\mathbb{Z}[T]$, using the embedding $\mathbb{Z} \to \mathbb{Z}[T]$.

Example 4. We recall that a set $S \subset A$ that does not contain 0 and is closed under multiplication is called multiplicative. For every multiplicative set we can construct the ring of fractions A_S consisting of the pairs (a, s), $a \in A$, $s \in S$, with an identification according to the rule:

$$(a, s) = (a', s')$$

if there exists an element $s'' \in S$ such that

$$s''(as' - a's) = 0 .$$

The operations are defined by the natural rules

$$(a, s) + (a', s') = (as' + sa', ss') ,$$
$$(a, s) \cdot (a', s') = (aa', ss') .$$

The reader can find a more detailed description of this construction in [37], Vol. I, Ch. IV, §9. From now on we write a pair (a, s) in the form a/s. In particular, if S is the set $A - \mathfrak{p}$, where $\mathfrak{p}$ is a prime ideal, then the ring A_S is the same as the local ring $A_{\mathfrak{p}}$ of $\mathfrak{p}$.

Suppose S is a nonempty multiplicative set and $s \in S$. The correspondence $a \to (as, s)$ determines a homomorphism

$$\varphi : A \to A_S ,$$

hence a mapping

$$^a\varphi : \operatorname{Spec} A_S \to \operatorname{Spec} A .$$

The reader can easily verify that $^a\varphi$ is an embedding and that $^a\varphi\,(\operatorname{Spec} A_S) = U_S$ is the set of all prime ideals of A that do not contain any element of S. The inverse mapping $\psi : U_S \to \operatorname{Spec} A_S$ has the form $\psi(\mathfrak{p}) = \{x/s;\, x \in \mathfrak{p}, s \in S\}$. In particular, if $f \in A$ is a non-nilpotent element, then the system $S = \{f^n, n = 1, 2, \ldots\}$ does not contain 0. In this case the ring A_S is denoted by A_f.

2. Properties of the Points of a Spectrum. With every point $x \in \operatorname{Spec} A$ we can associate the field of fractions of the quotient ring by the corresponding prime ideal. This field is denoted by $k(x)$. Thus, there is a homomorphism

$$A \to k(x),$$

whose kernel is the prime ideal we have denoted by x. The image of an element $f \in A$ under this homomorphism is denoted by $f(x)$. If $A = k[X]$, where X is an affine variety and $x \in X$ determines a maximal ideal of A, then $k(x)$ coincides with k, and for $f \in A$ the element $f(x)$ defined above coincides with the value of the function f at the point x. In the general case every element $f \in A$ also determines a "function" on $\operatorname{Spec} A$:

$$x \to f(x),$$

however, with the proviso that its values at distinct points belong, speaking generally, to distinct sets. For example, for $A = \mathbb{Z}$ we can regard every integer as a "function" whose value at the point (p) belongs to the field $\mathbb{Z}/(p)$, and at the point (0) to the field of rational numbers $\mathbb{Q}$.

Here we come across one of the most serious difficulties, where the "classical" geometrical intuition turns out to be inapplicable in our more general situation. The fact of the matter is that the element $f \in A$ is not always uniquely determined by the corresponding function on $\operatorname{Spec} A$. For example, the elements to which there corresponds the null function are those that are contained in all prime ideals. They have a very simple characterization:

Proposition. *An element of A belongs to all prime ideals if and only if it is nilpotent.*

Clearly the proof requires only that an element belonging to all prime ideals is nilpotent. Let f be a non-nilpotent element. Then we can construct the ring A_f defined in §1.1. The image of the points of the spectrum of this ring under the embedding

$$\operatorname{Spec} A_f \to \operatorname{Spec} A$$

consists of the prime ideals of A that do not contain any power of f, in particular, f itself. This proves the proposition.

Thus, the inapplicability in the general case of the "functional" point of view is connected with the presence of nilpotent elements in the ring. The set of all nilpotent elements of a ring forms an ideal, the so-called *nil-radical* of the ring.

With every point $x \in \operatorname{Spec} A$ there is connected the *local ring* $\mathcal{O}_x$: this is the local ring of the corresponding prime ideal. For example, if $A = \mathbb{Z}$, then for $x = (p)$ the ring $\mathcal{O}_x$ consists of the rational numbers whose denominator is prime to p (the p are prime numbers) and $\mathcal{O}_x = \mathbb{Q}$ for $x = (0)$.

This invariant of a point of a spectrum enables us to carry over to our general case a number of new geometrical concepts.

For example, the definition of simple points is connected with a purely algebraic property of their local rings. This suggests the following definition:

A point $x \in \operatorname{Spec} A$ is called *simple* or *regular* if the local ring $\mathcal{O}_x$ is Noetherian and regular.

We recall that, in general, $\operatorname{Spec} A \neq \operatorname{Specm} A$. Let $A = k[X]$ and suppose that a point of $\operatorname{Spec} A$ corresponds to a non-maximal prime ideal, that is, to an irreducible subvariety $Y \subset X$ of positive dimension. What is the geometrical meaning of regularity of this point of the spectrum? As the reader can easily verify, regularity in this case means that Y is not contained in the subvariety of singular points of X. Let $\mathfrak{m}_x$ be the maximal ideal of the local ring $\mathcal{O}_x$ of a point $x \in \operatorname{Spec} A$. Clearly

$$\mathcal{O}_x/\mathfrak{m}_x = k(x),$$

and the group $\mathfrak{m}_x/\mathfrak{m}_x^2$ is a vector space over $k(x)$. If $\mathcal{O}_x$ is a Noetherian ring (for example, if A is Noetherian), then this space is finite-dimensional. The vector space

$$\Theta_x = \operatorname{Hom}_{k(x)}(\mathfrak{m}_x/\mathfrak{m}_x^2, k(x))$$

is called the *tangent space* at the point x of the spectrum.

Example 1. If A is the ring of all integers of some algebraic number field K (for example, $A = \mathbb{Z}$, $K = \mathbb{Q}$), then $\operatorname{Spec} A$ consists of the maximal ideals and zero. For $x = 0$ we have $\mathcal{O}_x = K$, hence x is a regular point, and its tangent space is zero-dimensional. But if $x = \mathfrak{p} \neq (0)$, then $\mathcal{O}_x$ is known to be a principal ideal ring. Therefore these points are also regular, and their tangent spaces are one-dimensional.

Example 2. In order to meet with non-regular points we consider the ring $A = \mathbb{Z}[mi] = \mathbb{Z} + \mathbb{Z}mi$, where m is an integer greater than 1. The embedding $\varphi : A \to A'$, where $A' = \mathbb{Z}[i]$, determines the mapping

$$^a\varphi : \operatorname{Spec} A' \to \operatorname{Spec} A . \tag{1}$$

It is one-to-one if in both rings we restrict ourselves to prime ideals that are relatively prime to m. It is easy to check that the local rings of the corresponding ideals coincide. Therefore, a point $x \in \operatorname{Spec} A'$ can be non-regular only if the corresponding prime ideal divides m. Of such ideals there are just as many as they are integral prime divisors of m: if $p|m$, then $\mathfrak{p} = (p, mi)$ is the corresponding ideal. In this case $k(x) = \mathbb{F}_p$ is the field of p elements and $\mathfrak{m}_x/\mathfrak{m}_x^2 = \mathfrak{p}/\mathfrak{p}^2$ is two-dimensional over F_p. Therefore, the ideal $\mathfrak{m}_x$ is not principal. Since $\mathfrak{p}^2 \subset (p)$, the local ring $\mathcal{O}_x$ is non-regular. Hence all the prime ideals $\mathfrak{p} = (p, mi)$, $p|m$, determine the singular points of Spec A. The mapping (1) acounts for these singular points.

To determine the tangent spaces we could naturally go over to differential forms. The algebraic description of differential forms explained in Ch. III, § 4.2 enables us to carry them over to arbitrary rings. We do not need this construction later and do not study it in detail.

3. The Spectral Topology. The topological concepts we have used in connection with algebraic varieties suggest how to introduce a topology in the set Spec A. For this purpose we connect with every set $E \subset A$ the subset $V(E) \subset \operatorname{Spec} A$ consisting of the prime ideals $\mathfrak{p} \supset E$. The relations

$$V\left(\bigcup_\alpha E_\alpha\right) = \bigcap_\alpha V(E_\alpha),$$

$$V(I) = V(E') \cup V(E''),$$

where I is the intersection of the ideals generated by E' and E'', are obvious. They show that the sets $V(E)$ corresponding to arbitrary subsets $E \subset A$ satisfy the axioms for the system of all closed sets of a topological space.

The topology in which $V(E)$ are all the closed subsets of Spec A is called *spectral*.

Henceforth, if we talk of Spec A as a topological space, we always have in mind the spectral topology.

For any homomorphism $\varphi : A \to B$ and any set $E \subset A$ we have the relation

$$(^a\varphi)^{-1}(V(E)) = V(\varphi(E)),$$

from which it follows that the inverse image of a closed set under $^a\varphi$ is closed. This shows that $^a\varphi$ is a continuous mapping.

As an example we consider the natural homomorphism $\varphi : A \to A/\mathfrak{a}$, where $\mathfrak{a}$ is an ideal of A. Clearly, $^a\varphi$ is a homeomorphism of Spec $(A/\mathfrak{a})$ onto the closed set $V(\mathfrak{a})$. Every closed set in Spec A is of the form

$V(E)=V(\mathfrak{a})$, where $\mathfrak{a}$ is the ideal generated by E. Therefore all closed subsets of Spec A are homeomorphic to spectra of rings.

Let us consider another example. Let $S\subset A$ be a multiplicative system, $\varphi: A\to A_S$, $U_S={}^a\varphi(\operatorname{Spec} A_S)$, and $\psi: U_S\to \operatorname{Spec} A_S$ be the mappings and sets introduced in §1.1. We equip U_S with the topology of a subspace of Spec A, that is, we take the closed sets to be $V(E)\cap U_S$. A simple verification shows that not only ${}^a\varphi$, but also ψ is continuous. In other words, Spec A_S is homeomorphic to the space $U_S\subset \operatorname{Spec} A$.

Of particular importance is the special case when $S=\{f^n, n=1,2,\ldots\}$, where $f\in A$ is a non-nilpotent element. Here $U_S=\operatorname{Spec} A - V(f)$, where $V(f)$ is that $V(E)$ in which E consists of the single element f.

The open sets Spec $A-V(f)$ are called *principal* and are denoted by $D(f)$. It is easy to check that they form a basis of the spectral topology. As in the case of affine varieties, the significance of the principal open sets lies in the fact that they are homeomorphic to the spectra of rings A_f. Using these sets we can prove an important property of spectra:

Proposition. *The space* Spec A *is compact.*

What we have to show is that from every covering by open sets we can extract a finite subcovering. Since the principal open sets form a basis of Spec A, it is sufficient to show this for the covering Spec $A=\bigcup D(f_\alpha)$. The latter condition indicates that $\bigcap V(f_\alpha)=V(\mathfrak{a})=\emptyset$, where $\mathfrak{a}$ is the ideal generated by all the elements f_α. In other words, there are no prime ideals containing $\mathfrak{a}$, and this means that $\mathfrak{a}=A$. But then there exist elements $f_{\alpha_1},\ldots,f_{\alpha_r}$ and $g_1,\ldots,g_r\in A$ such that

$$f_{\alpha_1}g_1+\cdots+f_{\alpha_r}g_r=1\,.$$

From this, in turn, it follows that $(f_{\alpha_1},\ldots,f_{\alpha_r})=A$, that is,

$$\operatorname{Spec} A=D(f_{\alpha_1})\cup\ldots\cup D(f_{\alpha_r})\,.$$

The spectral topology is very "non-classical", to be precise, very non-Hausdorff. With such properties of affine varieties we have already become acquainted in Ch. I, for example: on an irreducible variety any two non-empty open sets intersect.

This property indicates that the Hausdorff separation axiom is not satisfied – there exist two distinct points for which any two neighbourhoods intersect. But owing to the fact that Spec A consists not only of the maximal, but of all prime ideals, this space is even "less Hausdorff" – non-closed points exist in it.

Let us find the closure of a point of Spec A. If our point is a prime ideal $\mathfrak{p}\subset A$, then its closure is $\bigcap\limits_{\mathfrak{p}\supset E} V(E)=V(\mathfrak{p})$, that is, it consists of the prime ideals $\mathfrak{p}'\supset\mathfrak{p}$ and is homeomorphic to Spec $A/\mathfrak{p}$. In particular, a prime ideal $\mathfrak{p}\subset A$ is a closed point of Spec A if and only if it is maximal.

If A has no divisors of zero, then the ideal (0) is prime and is contained in every prime ideal. Therefore its closure is the whole space – it is an everywhere dense point.

The existence of non-closed points in a topological space determines in it a certain hierarchy, which is laid down in the following definitions:

A point x is called a *specialization of a point* y if x is contained in the closure of y.

An everywhere dense point is called a *generic point of the space.*

When does the space Spec A have a generic point? As we have seen in § 1.2, the intersection of all prime ideals $\mathfrak{p} \subset A$ consists of all nilpotent elements of A, that is, it coincides with the nil-radical of the ring. If it is prime, then it determines a generic point of Spec A. But every prime ideal must contain all nilpotent elements, that is, the nil-radical. Therefore Spec A has a generic point if and only if its nil-radical is prime. This generic point is unique and is determined by the nil-radical.

4. Irreducibility, Dimension. The existence of a generic point is connected with an important geometrical property of X.

Namely, a topological space X necessarily fails to have a generic point if it can be represented in the form $X = X_1 \cup X_2$, where X_1 and X_2 are closed, $X_1 \neq X$, $X_2 \neq X$. Such a space is called reducible. For spectra of rings irreducibility is not only a necessary, but also a sufficient condition for the presence of a generic point. For it is enough to show that if Spec A is irreducible, then the nil-radical of A is prime; as we have shown above, it then follows that Spec A has a generic point. Suppose that the nil-radical N of A is not prime, and let $fg \in N, f \notin N, g \notin N$. Then

$$\operatorname{Spec} A = V(f) \cup V(g),$$
$$V(f) \neq \operatorname{Spec} A \neq V(g),$$

that is, Spec A is reducible.

Since all the closed subsets of Spec A are also homeomorphic to spectra of rings, this result carries over to arbitrary closed subsets. Thus, there exists a one-to-one correspondence between points and irreducible closed subsets of Spec A. It is determined by assigning to a point its closure.

The concept of a reducible space leads us at once to a decomposition into irreducible components. If A is a Noetherian ring, then there exists a representation

$$\operatorname{Spec} A = X_1 \cup \ldots \cup X_r,$$

where the X_i are irreducible closed subsets and $X_i \not\subset X_j$ for $i \neq j$, and this representation is unique. The proof of this fact repeats word for word the proof of the corresponding proposition for affine varieties, which was only based on the fact that the ring $k[X]$ is Noetherian.

Example 1. The simplest example of a decomposition of Spec A into irreducible components is the case of a ring A that is a product of finitely many rings without divisors of zero:

$$A = A_1 \times \cdots \times A_r .$$

In this case, as is easy to verify, Spec A is the union of its disjoint irreducible connected components Spec A_i.

Example 2. To analyse a somewhat less trivial example we take the group ring $\mathbb{Z}[\sigma]$ of a cyclic group of order 2:

$$A = \mathbb{Z}[\sigma] = \mathbb{Z} + \mathbb{Z}\sigma, \sigma^2 = 1 .$$

The nil-radical of this ring is equal to (0), but it is not prime, because A has divisors of zero: $(1+\sigma)(1-\sigma) = 0$. Therefore

$$\operatorname{Spec} A = X_1 \cup X_2 , \tag{1}$$

where $X_1 = V(1+\sigma)$, $X_2 = V(1-\sigma)$. The homomorphisms $\varphi_1, \varphi_2 : A \to \mathbb{Z}$ with the kernels $(1+\sigma)$ and $(1-\sigma)$ determine homeomorphisms

$$^a\varphi_1 : \operatorname{Spec} \mathbb{Z} \to V(1+\sigma) ,$$
$$^a\varphi_2 : \operatorname{Spec} \mathbb{Z} \to V(1-\sigma) ,$$

which show that X_1 and X_2 are irreducible and hence that (1) is the decomposition of Spec A into irreducible components.

Let us find the intersection $X_1 \cap X_2$. Clearly

$$X_1 \cap X_2 = V(1+\sigma, 1-\sigma) = V(\mathfrak{a})$$

where $\mathfrak{a}$ is the ideal $(1+\sigma, 1-\sigma) = (2, 1-\sigma)$. Since $A/\mathfrak{a} = \mathbb{Z}/(2)$, we see that $\mathfrak{a}$ is a maximal ideal, hence X_1 and X_2 intersect in a single point $x_0 = X_1 \cap X_2$. It is easy to check that if $x \neq x_0$, for example $x \in X_1$, $x \notin X_2$, then the homomorphism φ_1 establishes an isomorphism of the local rings of x and $\varphi_1(x)$. Therefore all the points $x \neq x_0$ are simple. But x_0 is singular, $\dim \Theta_{x_0} = 2$ and for $y_1 = (^a\varphi_1)^{-1}(x_0)$, $y_2 = (^a\varphi_2)^{-1}(x_0)$ we find that $d_{y_1}\Theta_{y_1}$ and $d_{y_2}\Theta_{y_2}$ are two distinct lines in $\Theta_{x_0} : x_0$ is a "double point with separated tangents".

It is convenient to represent Spec A by using the mapping

$$^a\varphi : \operatorname{Spec} A \to \operatorname{Spec} \mathbb{Z} ,$$

where $\varphi : \mathbb{Z} \to A$ is the natural embedding (similar to the way in which we have looked at Spec $\mathbb{Z}[i]$ in § 1.1). Then we get the picture of Fig. 14.

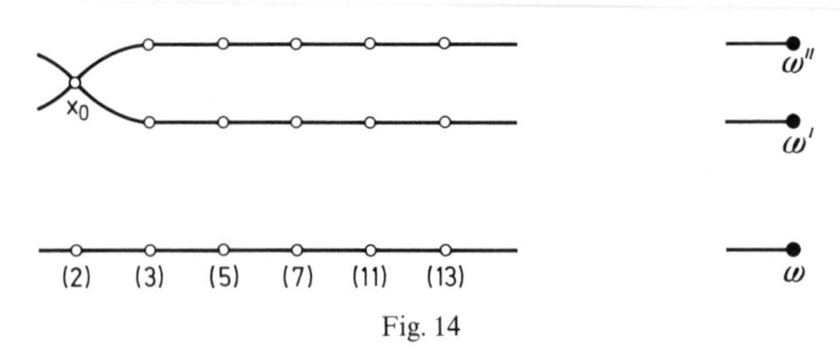

Fig. 14

Among the purely topological concepts, that is, those that can be defined by the spectral topology of the ring $k[X]$, there is also the dimension of an affine variety X. Of course, the definition we have given in Ch. 1 as the transcendence degree of the field $k(X)$ uses very specific properties of the ring $k[X]$: that it is an algebra over k, that it is embedded in a field, and that this field has finite transcendence degree over k. However, Corollary 3 in Ch. I, §6, puts it into a form that is applicable to any topological space.

The dimension of a topological space X is the integer n such that in X there exist chains of distinct irreducible closed sets

$$X_0 \subset X_1 \subset \ldots \subset X_n$$

and there do not exist chains with more terms.

Of course, not every topological space has finite dimension; this is not even true for spectral rings, even Noetherian ones. Nevertheless, for a number of important types of ring the dimension of Spec A is finite. In that case it is called the *dimension* of A. Without proof we quote the three main results.

A. If A is a Noetherian local ring, then the dimension of Spec A is finite and is the same as the dimension of A as it was defined in Ch. II, §2.1.

B. A finitely generated ring over a ring of finite dimension is itself of finite dimension.

C. If A is a Noetherian ring, then

$$\dim A[T_1, \ldots, T_n] = \dim A + n .$$

Proofs can be found in [29].

Example 3. The ring $\mathbb{Z}$ is of dimension 1. More generally, the ring of algebraic integers of some algebraic number field is of dimension 1, because in this ring every prime ideal other than (0) is maximal.

Example 4. To give an example of a ring of greater dimension we consider the case $A = \mathbb{Z}[T]$. In the hope that the reader has already analysed the structure of the spectrum of this ring in connection with § 1.1 we assume the structure of Spec $\mathbb{Z}[T]$ to be known. It is very simple: the maximal ideals are of the form $(p, f(T))$, where p is a prime number and $f \in \mathbb{Z}[T]$ is a polynomial which on reduction modulo p becomes irreducible; the non-maximal prime ideals other than (0) are principal and are of the form (p) or $(f(T))$, where f is a primitive polynomial. From this it follows that the chains of prime ideals of maximal length are:

$$(p, f(T)) \supset (g(T)) \supset (0)$$

or

$$(p, f(T)) \supset (p) \supset (0).$$

Thus, $\dim \mathbb{Z}[T] = 2$ in accordance with Proposition C.

Exercises

1. Let N be the nil-radical of a ring A. Show that the natural embedding ${}^a\varphi : \operatorname{Spec} A/N \to \operatorname{Spec} A$ is a homeomorphism.

2. Show that an element $f \in A$ is a divisor of zero if and only if there exists a decomposition $\operatorname{Spec} A = X \cup X'$ into closed subsets X, $X' \neq \operatorname{Spec} A$ such that $f(x) = 0$ for all points $x \in X$.

3. Let $\varphi : A \to B$ be an embedding of rings and let B be integral over A. Show that ${}^a\varphi$ is an epimorphism.

4. Let $\varphi : A \to B$ be a homomorphism of rings. Does the mapping ${}^a\varphi$ always carry closed points into closed ones? Is this true under the conditions of Exercise 3?

5. Show that $\overline{{}^a\varphi(V(E))} = V(\varphi^{-1}(E))$. The bar denotes closure.

6. Let X_1 and X_2 be closed subsets of $\operatorname{Spec} A$ and let $u_1, u_2 \in A$ be such that $u_1 + u_2 = 1$, $u_1 u_2 = 0$, $u_i(x) = 0$ for all points $x \in X_i$, $i = 1, 2$. Show that then $A = A_1 \times A_2$, with $X_i = {}^a\varphi_i(\operatorname{Spec} A_i)$, where $\varphi_i : A \to A_i$ is the natural homomorphism.

7. Let $\operatorname{Spec} A = X_1 \cup X_2$ be a decomposition into closed disjoint sets. Show that then $A = A_1 \times A_2$, $X_i = {}^a\varphi_i(\operatorname{Spec} A_i)$. *Hint*: Representing X_i in the form $V(E_i)$ find elements v_i, $i = 1, 2$, such that $v_1 + v_2 = 1$, $v_i(x) = 0$ for all $x \in X_i$. Using the proposition in § 1.2 construct functions u_1 and u_2 satisfying the conditions of Exercise 6.

8. Show that if A is a ring of finite type over an algebraically closed field, then the proposition of § 1.2 remains valid if in its statement prime ideals are replaced by maximal ideals. *Hint:* Use Hilbert's Nullstellensatz.

   Deduce that closed points are everywhere dense in every closed subset of $\operatorname{Spec} A$.

9. Let $A = \mathbb{Z}[T]/(F(T))$, where $F(T) \in \mathbb{Z}[T]$, p is a prime number, $F(0) \equiv 0(p)$ and $\mathfrak{p} \in A$ is the maximal ideal of A generated by p and the image of T. Show that the point $x \in \operatorname{Spec} A$ corresponding to $\mathfrak{p}$ is singular if and only if $F(0) \equiv 0(p^2)$, $F'(0) \equiv 0(p)$. *Hint:* Consider the homomorphism $M/M^2 \to \mathfrak{p}/\mathfrak{p}^2$, where $M = (p, F) \in \mathbb{Z}[T]$.

10. Show that in $\operatorname{Spec} \mathbb{Z}[T_1, \ldots, T_n]$ every closed subset each of whose components is of codimension 1 (that is, of dimension equal to $\dim \operatorname{Spec} \mathbb{Z}[T_1, \ldots, T_n] - 1$) is of the form $V(F)$, where $F \in \mathbb{Z}[T_1, \ldots, T_n]$.

11. Prove the following universal property of the ring of fractions A_S relative to a multiplicative system $S \subset A$ (Example 4 of § 1.1): if $f: A \to B$ is a homomorphism for which all the $f(s)$, $s \in S$, are invertible in B, then there exists a homomorphism $g : A_S \to B$ for which $f = gh$, where h is the natural homomorphism $A \to A_S$.

§2. Sheaves

1. Presheaves. The concept of the spectrum of a ring is only one of two elements from which the definition of a scheme is made up. The second element is the concept of a sheaf. In the preceding section we have used the fact that an affine variety is given by the ring of regular functions on it, and starting out from an arbitrary ring we have arrived at the corresponding geometrical concept – the spectrum. In the definition of the general concept of a scheme we also take as our basis the regular functions of a variety. But there may be too few of them if we consider functions that are regular on the whole variety. It is therefore natural to consider for any open set $U \subset X$ the ring of regular functions on it. In this way we obtain not one ring, but a system of rings, between which, as we shall see, there are various connections. We base our definition of a scheme on a similar object. However, to begin with we must analyse some definitions and the simplest facts relating to this kind of object.

Definition. Let X be a topological space and suppose that a certain set $\mathscr{F}(U)$ is associated with every open set U of it, and that for any two sets $U \subset V$ there is a mapping

$$\varrho_U^V : \mathscr{F}(V) \to \mathscr{F}(U).$$

This system of sets and mappings is called a *presheaf* if the following conditions are satisfied:

1) if U is empty, then the set $\mathscr{F}(U)$ consists of a single element;
2) ϱ_U^U is the identity mapping;
3) for any open sets $U \subset V \subset W$ we have

$$\varrho_U^W = \varrho_U^V \varrho_V^W . \tag{1}$$

Sometimes such a presheaf is simply denoted by the letter $\mathscr{F}$. If it is important to emphasize that the mapping ϱ_U^V refers precisely to a presheaf $\mathscr{F}$, then it is denoted by $\varrho_{U,\mathscr{F}}^V$.

If all the sets $\mathscr{F}(U)$ are groups, modules over a ring A, or rings, and if the mappings ϱ_U^V are homomorphisms of these structures, then we speak of a presheaf of groups, modules over A, or rings.

Evidently a presheaf $\mathscr{F}$ does not depend on a choice of the element $\mathscr{F}(\emptyset)$ (more accurately, for distinct choices we obtain isomorphic presheaves relative to a concept of isomorphism that the reader can easily supply). Therefore, to specify a presheaf it is sufficient to indicate the sets $\mathscr{F}(U)$ for non-empty sets U. If $\mathscr{F}$ is a presheaf of groups, then $\mathscr{F}(\emptyset)$ is the group consisting of one element only.

If $\mathscr{F}$ is a presheaf on X and $U \subset X$ is an open set, then the assignation $V \to \mathscr{F}(V)$ for all open sets $V \subset U$ obviously determines a presheaf on U. It is called the restriction of $\mathscr{F}$ and is denoted by $\mathscr{F}|_U$.

Examples. 1. Let M be a set, let $\mathscr{F}(U)$ consist of all functions on U with values in M, and for $U \subset V$ let ϱ_U^V be the restriction of a function given on V to U. The relation (1) is obvious. $\mathscr{F}$ is called the presheaf of all functions on X in M.

In order to carry over the intuitive picture of this example to the case of an arbitrary presheaf, the ϱ_U^V are called *restriction mappings*.

Example 1 can be varied.

2. Let M be a topological space, let $\mathscr{F}(U)$ consist of the continuous functions on U with values in M, and let the ϱ_U^V be as in Example 1. Then $\mathscr{F}$ is called a presheaf of continuous functions.

3. X is a differentiable manifold, $\mathscr{F}(U)$ the set of differentiable functions on U (with real values). The mappings ϱ_U^V are again as in Example 1.

4. X is an irreducible quasiprojective variety in which a topology is defined by the fact that closed subsets are algebraic subvarieties (so that the "topological" terminology of Ch. I turns into the usual topological concepts). For an open set $U \subset X$, let $\mathscr{F}(U)$ be the set of all rational functions on X that are regular at all points of U. The mappings ϱ_U^V are as in Example 1. Then $\mathscr{F}$ is a presheaf of rings, which is called the *presheaf of regular functions*.

2. The Structure Presheaf. Now we go over to the construction of the presheaf that will play the decisive role in what follows. It is defined on the topological space $X = \operatorname{Spec} A$. The presheaf we are about to define is called the *structure presheaf* on $\operatorname{Spec} A$ and is denoted by $\mathcal{O}$. To make the logic of our definition clearer we give it first in a more special form.

Let us assume, to begin with, that the ring A has no divisors of zero, and let K be its field of fractions. Now we can copy faithfully the Example 4 above. For an open set $U \subset \operatorname{Spec} A$ we denote by $\mathcal{O}(U)$ the set of those elements $u \in K$ which for every point $x \in U$ have a representation $u = a/b$, $a, b \in A$, $b(x) \neq 0$, (that is, b is not contained in the prime ideal x).

Clearly $\mathcal{O}(U)$ is a ring. Since all the rings $\mathcal{O}(U)$ are contained in K, we can compare them as subsets of a single set. Evidently, if $U \subset V$, then $\mathcal{O}(V) \subset \mathcal{O}(U)$. We denote by ϱ_U^V the embedding of $\mathcal{O}(V)$ in $\mathcal{O}(U)$. A trivial verification shows that we obtain a presheaf of rings.

Before completing the discussion of this case, let us compute $\mathcal{O}(\operatorname{Spec} A)$. Our arguments repeat the proof of Theorem 4 in Ch. I, § 3. The condition $u \in \mathcal{O}(\operatorname{Spec} A)$ means that for every point $x \in \operatorname{Spec} A$ there exist elements a_x and $b_x \in A$ such that

$$u = a_x/b_x, \qquad b_x(x) \neq 0. \tag{1}$$

Consider the ideal $\mathfrak{a}$ generated by all the elements b_x, $x \in \operatorname{Spec} A$. By (1) it is not contained in any prime ideal of A, so that $\mathfrak{a} = \mathrm{A}$. Thus, there exist points $x_1, \ldots, x_r$ and elements $c_1, \ldots, c_r \in A$ such that

$$c_1 b_{x_1} + \cdots + c_r b_{x_r} = 1 .$$

Multiplying (1) for $x = x_i$ by $c_i b_{x_i}$ and adding up we find that

$$u = \Sigma a_{x_i} c_i \in A .$$

Thus, $\mathcal{O}(\operatorname{Spec} A) = A$.

Now we go over to the case of an arbitrary ring A. The preceding arguments suggest that it would be natural to set $\mathcal{O}(\operatorname{Spec} A) = A$. But there also exist other open sets U that have natural candidates for the ring $\mathcal{O}(U)$, namely the principal open sets $D(f)$, $f \in A$. For we have seen in § 1.3 that $D(f)$ is homeomorphic to Spec A_f, and it is equally plausible to set

$$\mathcal{O}(D(f)) = A_f .$$

Thus, so far we have defined the presheaf $\mathcal{O}(U)$ on principal open sets $D(f)$. Before defining it on all open sets we introduce the homomorphisms ϱ_U^V, of course, only for principal open sets V and U.

To begin with, let us find out when $D(f) \subset D(g)$. This is equivalent to the fact that $V(f) \supset V(g)$, that is, every prime ideal containing g also contains f. In other words, the image $\bar{f}$ of f in the ring $A/(g)$ is contained in every prime ideal of this ring. In § 1.2 we have seen that this is equivalent to $\bar{f}$ being nilpotent, that is, to $f^n \in (g)$ for some $n > 0$. Thus, $D(f) \subset D(g)$ if and only if for some $n > 0$ and $u \in A$

$$f^n = gu . \tag{2}$$

In this case we can construct the homomorphism $\varrho_{D(f)}^{D(g)}$ of A_g into A_f by setting

$$\varrho_{D(f)}^{D(g)}(a/g^l) = au^l/f^{nl} .$$

An obvious verification shows that this mapping does not depend on the representation of an element $t \in A_g$ in the form $t = a/g^l$ and that it is a homomorphism. (This homomorphism can be written down in a more invariant form, by making use of the universal ring of fractions A_S, see Exercise 11 to § 1. In our case g and its powers are invertible in A_f owing to (2), from which the existence of the homomorphism $\varrho_{D(f)}^{D(g)}$ follows.)

Before we formulate the final definition we return for a moment to the case already considered when A has no divisors of zero. Then we can indicate a method of computing $\mathcal{O}(U)$ for every open set U in terms of $\mathcal{O}(V)$, where V are the various principal open sets. Namely, if $\{D(f)\}$ are all possible principal open sets contained in U, then, as is very easy to check,

$$\mathcal{O}(U) = \bigcap \mathcal{O}(D(f)) .$$

In the general case one might wish to take this equation as definition, but this is impossible, because the $\mathcal{O}(D(f))$ are not contained entirely in one common set. But they are connected among each other by the homomorphisms $\varrho_{D(g)}^{D(f)}$ if $D(g) \subset D(f)$. In this situation a natural generalization of the intersection is the projective limit of sets. Let us recall its definition. Let I be a partially ordered set, $\{E_\alpha, \alpha \in I\}$ a system of sets, and f_α^β for any α, $\beta \in I$, $\alpha \leqslant \beta$, a mapping of E_β into E_α satisfying the following conditions:

1) f_α^α is the identity map of E_α, and 2) for $\alpha \leqslant \beta \leqslant \gamma$ we have $f_\alpha^\gamma = f_\alpha^\beta f_\beta^\gamma$. Consider the subset of the product $\prod_{\alpha \in I} E_\alpha$ of the sets E_α consisting of those elements $x = \{x_\alpha; x_\alpha \in E_\alpha\}$ for which $x_\alpha = f_\alpha^\beta x_\beta$ for all $\alpha \leqslant \beta$. This subset is called the *projective limit of the system of sets* E_α *relative to the system of homomorphisms* f_α^β and is denoted by $\varprojlim E_\alpha$. The mappings $x \to x_\alpha$, $x \in \varprojlim E_\alpha$ of the projective limit are said to be *natural.*

If the E_α are rings, modules, or groups, and if the f_α^β are homomorphisms of these structures, then $\lim E_\alpha$ is a structure of the same type. The reader can find a more detailed description of this construction in [12], Ch. VIII, §3. One has to keep in mind here that the condition on the set I to be directed is not essential for the definition of the projective limit.

Now we are prepared for the final definition:

$$\mathcal{O}(U) = \varprojlim \mathcal{O}(D(f))$$

where the projective limit is taken over all $D(f) \subset U$ relative to the system of homomorphisms $\varrho_{D(g)}^{D(f)}$ for $D(g) \subset D(f)$ we have constructed above.

By definition, $\mathcal{O}(U)$ consists of families $\{u_\alpha\}, u_\alpha \in A_{f_\alpha}$, where f_α are all those elements for which $D(f_\alpha) \subset U$, and

$$u_\alpha = \varrho_{D(f_\alpha)}^{D(f_\beta)} u_\beta \quad \text{if} \quad D(f_\beta) \supset D(f_\alpha) . \tag{3}$$

For $U \subset V$ every family $\{v_\alpha\} \in \mathcal{O}(V)$ consisting of $v_\alpha \in A_{f_\alpha}$, $D(f_\alpha) \subset V$, determines a subfamily $\{v_\beta\}$ consisting of the v_β with index β such that $D(f_\beta) \subset U$. Evidently $\{v_\beta\} \in \mathcal{O}(U)$. *We set*

$$\varrho_U^V(\{v_\alpha\}) = \{v_\beta\} .$$

A trivial verification shows that $\mathcal{O}(U)$ and ϱ_U^V determine a presheaf $\mathcal{O}$, the so-called *structure presheaf*, on Spec A.

If $U = \operatorname{Spec} A$, then $D(1) = U$, so that 1 is one of the f_α, say f_0. The mapping

$$\{u_\alpha\} \to u_0$$

determines, as is easy to check, an isomorphism $\mathcal{O}(\operatorname{Spec} A) \simeq A$.

In particular, if $u=\{u_\alpha;\ D(f_\alpha)\subset U\}\in\mathcal{O}(U)$, then by definition $\varrho^U_{D(f)}u=\{u_\beta;D(f_\beta)\subset D(f)\}$. By what was said above, the assignation $\{u_\beta, D(f_\beta)\subset D(f)\}\to u_\alpha$ if $f=f_\alpha$ determines an isomorphism between $\mathcal{O}(D(f_\alpha))$ and A_{f_α} by which

$$u_\alpha=\varrho^U_{D(f_\alpha)}u\,. \tag{4}$$

This formula allows us to recover u_α from the element $u\in\mathcal{O}(U)$ determined by it.

3. Sheaves. We assume that the topological space X is the union of open sets U_α. Every function f on X is uniquely determined by its restrictions to the sets U_α, and if on each of the U_α a function f_α is given for which the restrictions of f_α and f_β agree on $U_\alpha\cap U_\beta$, then there exists a function f on X such that every f_α is its restriction to U_α. Continuous functions, differentiable functions on a differentiable manifold, and regular functions on a quasiprojective algebraic variety have the same property, which expresses the local character of the concept of being a continuous, differentiable, regular function. It can be rephrased for any presheaf and singles out an exceptionally important class of presheaves.

Definition. A presheaf $\mathcal{F}$ on a topological space is called a *sheaf* if for every open set $U\subset X$ and any open covering of it $U=\bigcup U_\alpha$ the following conditions are satisfied:

1) if $\varrho^U_{U_\alpha}s_1=\varrho^U_U s_2$ for $s_1, s_2\in\mathcal{F}(U)$ and all U_α, then $s_1=s_2$;

2) if $s_\alpha\in\mathcal{F}(U_a)$ are such that $\varrho^{U_\alpha}_{U_\alpha\cap U_\beta}s_\alpha=\varrho^{U_\beta}_{U_\alpha\cap U_\beta}s_\beta$, then there exists an $s\in\mathcal{F}(U)$ for which $s_\alpha=\varrho^U_{U_\alpha}s$ for all U_α.

We have already given a number of examples of sheaves before defining this concept. We now indicate the simplest example of a presheaf that is not a sheaf. Let X be a topological space, M a set, $\mathcal{F}(U)=M$ for all $U\subset X$, and ϱ^V_U the identity mapping. Evidently $\mathcal{F}$ is a presheaf. Suppose that in X there exist disconnected open sets and that $U=U_1\cup U_2$, $U_1\cap U_2=\emptyset$ is a representation of one of them as the union of disjoint open sets. Let m_1 and m_2 be distinct elements of M and $s_1=m_1\in\mathcal{F}(U_1)$, $s_2=m_2\in\mathcal{F}(U_2)$. Evidently the condition $\varrho^{U_1}_{U_1\cap U_2}s_1=\varrho^{U_2}_{U_1\cap U_2}s_2$ holds automatically, however, there does not exist an $s\in\mathcal{F}(U)$ such that $\varrho^U_{U_1}s=s_1$, $\varrho^U_{U_2}s=s_2$, because $m_1\neq m_2$.

Theorem 1. *The structure presheaf on* Spec A *is a sheaf.*

To begin with, we check the conditions 1) and 2) in the definition of a sheaf for the case when U and the U_α are principal open sets.

First of all, we note that both conditions need only be verified for the case $U=\operatorname{Spec}A$. For if $U=D(f)$, $U_\alpha=D(f_\alpha)$, then, as the reader can easily verify, the conditions 1) and 2) hold for U and U_α if they hold for Spec A_f and the sets $\bar U_\alpha=D(\bar f_\alpha)$, where $\bar f_\alpha$ is the image of f_α under the natural homomorphism $A \to A_f$. Now we proceed to the verification of 1) and 2) for $U_\alpha = D(f_\alpha)$, $\bigcup U_\alpha = \operatorname{Spec} A$.

1. Since $\mathcal{O}$ is a presheaf of groups, it is sufficient to show that if $u \in \mathcal{O}(\operatorname{Spec} A) = A$ and $\varrho_{U_\alpha}^{\operatorname{Spec} A} u = 0$ for all U_α, then $u = 0$.

The condition $\varrho_{U_\alpha}^{\operatorname{Spec} A} u = 0$ means that

$$f_\alpha^{n_\alpha} u = 0 \tag{1}$$

for all α and certain $n_\alpha \geqslant 0$. Since $D(f_\alpha) = D(f_\alpha^{n_\alpha})$, we have $\bigcup D(f_\alpha^{n_\alpha}) = \operatorname{Spec} A$. We have already seen that this leads to an identity

$$f_{\alpha_1}^{n_1} g_1 + \dots + f_{\alpha_r}^{n_r} g_r = 1$$

for suitable $g_1, \dots, g_r \in A$. Multiplying (1) for $\alpha = \alpha_1, \dots, \alpha_r$ by $g_1, \dots, g_r$ and adding up we find that $u = 0$.

2. Since the space $\operatorname{Spec} A$ is compact, we may restrict ourselves to the case of a finite covering. For the reader can easily verify that if the assertion is true for subcoverings, then it is also true for the whole covering.

Let $\operatorname{Spec} A = D(f_1) \cup \dots \cup D(f_r)$ and $u_i \in A_{f_i}$, $u_i = v_i / f_i^n$ (a single n may be chosen because the covering is finite). First of all, we observe that $D(f) \cap D(g) = D(fg)$ (the verification is trivial). By definition

$$\varrho_{D(f_i f_j)}^{D(f_i)} u_i = \frac{v_i f_j^n}{(f_i f_j)^n},$$

and by hypothesis,

$$(f_i f_j)^m (v_i f_j^n - v_j f_i^n) = 0 .$$

Setting $v_j f_j^m = w_j$, $m + n = k$, we find that

$$u_i = w_i / f_i^k, \quad w_i f_j^k = w_j f_i^k. \tag{2}$$

As in the verification of 1), we see that

$$\Sigma f_i^k g_i = 1 .$$

We set $u = \Sigma w_j g_j$. By hypothesis,

$$f_i^k u = \sum_j w_j g_j f_i^k = \sum_j w_i g_j f_j^k = w_i . \tag{2}$$

Therefore $\varrho_{D(f_i)}^{\operatorname{Spec} A} u = w_i / f_i^k = u_i$.

The verification of 1) and 2) for arbitrary open sets is a formal consequence of what we have already proved. Our situation can be described in general terms as follows. On the topological space X a basis $\mathscr{V} = \{V_\tau\}$ of open sets is given that is closed under intersections. We assume that the presheaf $\mathscr{F}$ of groups on X satisfies the conditions: a) $\mathscr{F}(U) = \varprojlim \mathscr{F}(V_\alpha)$, where the limit is taken over all $V_\alpha \in \mathscr{V}$, $V_\alpha \subset U$, relative to the homomorphisms $\varrho^{V_\alpha}_{V_\beta}$, and b) the $\varrho^U_{V_\alpha}$ coincide with the natural homomorphisms of the projective limit. Both these properties are satisfied for the presheaf $\mathcal{O}$ – the first by definition, and the second by (4) in § 2.2. We show that under these conditions $\mathscr{F}$ is a sheaf if 1) and 2) hold for the sets $V_\tau \in \mathscr{V}$.

1. Let $U = \bigcup_\xi U_\xi = \bigcup_{\xi,\lambda} V_{\xi,\lambda}$, $V_{\xi,\lambda} \in \mathscr{V}$. If $\varrho^U_{U_\xi} u = 0$ for all U_ξ, then $\varrho^U_{V_{\xi,\lambda}} u = 0$. Introducing new indices $(\xi,\lambda) = \gamma$, we obtain $U = \bigcup V_\gamma$, $\varrho^U_{V_\gamma} u = 0$ for all V_γ. To show that $u = 0$ we need only verify by condition b) that $\varrho^U_{V_\alpha} u = 0$ for all $V_\alpha \subset U$. This follows at once from an analysis of the homomorphisms corresponding to the sets

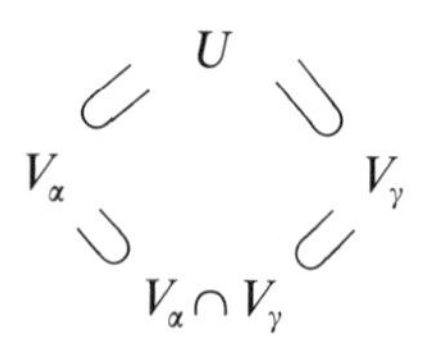

Indeed,

$$\varrho^{V_\alpha}_{V_\alpha \cap V_\gamma} \varrho^U_{V_\alpha} u = \varrho^U_{V_\alpha \cap V_\gamma} u = \varrho^{V_\gamma}_{V_\alpha \cap V_\gamma} \varrho^U_{V_\gamma} u = 0$$

for all V_γ, hence $\varrho^U_{V_\alpha} u = 0$ because $V_\alpha = \bigcup (V_\alpha \cap V_\gamma)$, and 1) holds by hypothesis for the sets V_α.

2. Let $u_\xi \in \mathscr{F}(U_\xi)$, $\varrho^{U_{\xi_1}}_{U_{\xi_1} \cap U_{\xi_2}} u_{\xi_1} = \varrho^{U_{\xi_2}}_{U_{\xi_2} \cap U_{\xi_1}} u_{\xi_2}$, $U_\xi = \bigcup_\lambda V_{\xi,\lambda}$. Setting $v_{\xi,\lambda} = \varrho^{U_\xi}_{V_{\xi,\lambda}} u_\xi$ and $\gamma = (\xi,\lambda)$ we verify that

$$\varrho^{V_{\gamma_1}}_{V_{\gamma_1} \cap V_{\gamma_2}} v_{\gamma_1} = \varrho^{V_{\gamma_2}}_{V_{\gamma_1} \cap V_{\gamma_2}} v_{\gamma_2}. \tag{3}$$

This follows from the analysis of the homomorphisms ϱ corresponding to the sets

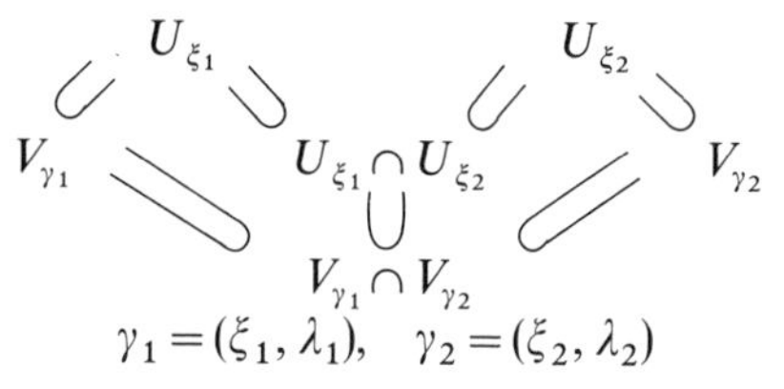

$\gamma_1 = (\xi_1, \lambda_1)$, $\gamma_2 = (\xi_2, \lambda_2)$

The left-hand side of (3) is equal to

$$\varrho^{U_{\xi_1}}_{V_{\gamma_1} \cap V_{\gamma_2}} u_{\xi_1} = \varrho^{U_{\xi_1} \cap U_{\xi_2}}_{V_{\gamma_1} \cap V_{\gamma_2}} \varrho^{U_{\xi_1}}_{U_{\xi_1} \cap U_{\xi_2}} u_{\xi_1}.$$

Evidently it is equal to the right-hand side. By (3), for every $V_\alpha \in \mathscr{V}$, $V_\alpha \subset U$ the elements $\varrho^{V_\gamma}_{V_\alpha \cap V_\gamma} v_\gamma$ satisfy a similar relation, hence by hypothesis there exist elements $v_\alpha \in \mathscr{F}(V_\alpha)$ for which $\varrho^{V_\alpha}_{V_\alpha \cap V_\gamma} v_\alpha = \varrho^{V_\gamma}_{V_\alpha \cap V_\gamma} v_\gamma$. An obvious verification shows that these elements determine an element u of the projective limit $\varprojlim \mathscr{F}(V_\alpha)$. For it $\varrho^U_{V_\alpha} u = v_\alpha$. Hence for $u'_\xi = \varrho^U_{U_\xi} u$

we have $\varrho_{V_\tau}^{U_\xi} u'_\xi = \varrho_{V}^{U_\xi} u_\xi$, for all $V_\tau \subset U_\xi$, $V_\tau \in \mathscr{V}$, and so $u'_\xi = u_\xi$. This completes the proof of the theorem.

4. The Stalks of a Sheaf. We return to the analysis of our concepts of a sheaf and a presheaf. We consider, to begin with, a presheaf for which all the sets $\mathscr{F}(U)$ are subsets of a single set and the mappings ϱ_U^V are embeddings of $\mathscr{F}(V)$ in $\mathscr{F}(U)$; this is so, for example, for the sheaf $\mathscr{O}$ on the irreducible space Spec A. Then we may consider the union $\mathscr{F}_x$ of sets $\mathscr{F}(U)$ for all open sets U containing the given point x. For the sheaf of continuous functions $\mathscr{F}_x$ these are the germs of the functions continuous on some neighbourhood of x, that is, the result of identifying functions that coincide in some such neighbourhood. For the sheaf of regular functions on an irreducible quasiprojective variety $\mathscr{F}_x$ this is the local ring of x.

In the general case not all the sets $\mathscr{F}(U)$ are contained in a single one, but they are connected by the homomorphisms ϱ_U^V, and this enables us to replace the union by the inductive limit. The definition, similar to that of the projective limit, can be read in the book [12], Ch. VIII, §4.

Definition. The *stalk* $\mathscr{F}_x$ of a presheaf $\mathscr{F}$ at a point $x \in X$ is the inductive limit of the sets $\mathscr{F}(U)$ for all U containing x relative to the system of mappings ϱ_U^V for $U \subset V$.

By definition, an element of $\mathscr{F}_x$ is given by an element of any of the $\mathscr{F}(U)$, $x \in U$, but here two elements $u \in \mathscr{F}(U)$ and $v \in \mathscr{F}(V)$, $x \in U$, $x \in V$, are identified if there exists a $W \subset U \cap V$, $x \in W$, such that $\varrho_W^U u = \varrho_W^V v$.

Applying this definition to the case of the structure sheaf $\mathscr{O}$ on Spec A we see that the stalk $\mathscr{O}_x$ is the same as the local ring of the prime ideal $x \in$ Spec A.

In the general case, for every open set U containing x there is defined the natural homomorphism

$$\varrho_x^U : \mathscr{F}(U) \to \mathscr{F}_x .$$

If $\mathscr{F}$ is a sheaf and if $\varrho_x^U u_1 = \varrho_x^U u_2$ for two elements $u_1, u_2 \in \mathscr{F}(U)$ and all points $x \in U$, then $u_1 = u_2$. For by definition this means that every point $x \in U$ has a neighbourhood W, $W \subset U$, such that $\varrho_W^U u_1 = \varrho_W^U u_2$. By definition of a sheaf it then follows that $u_1 = u_2$.

Thus, for a sheaf $\mathscr{F}$ the elements of $\mathscr{F}(U)$ can be given by families $\{u_x; u_x \in \mathscr{F}_x, x \in U\}$. Of course, as a result we do not obtain all families of this form. The following condition is clearly necessary:

For every point $x \in U$ there exists a neighbourhood $W \subset U$ and an element $w \in \mathscr{F}(W)$ such that $u_y = \varrho_y^W w$ for all points $y \in W$.

The reader can easily verify that, conversely, any family satisfying this condition corresponds to some element $u \in \mathscr{F}(U)$.

This is true, of course, only when $\mathscr{F}$ is a sheaf. But if $\mathscr{F}$ is an arbitrary presheaf, we may nevertheless consider the set $\mathscr{F}'(U)$ of all families $\{u_x\}$ satisfying the condition above. For $U \subset V$ the mapping

$$\varrho_U^V : \{v_x; v_x \in \mathscr{F}_x, x \in V\} \to \{v_y; v_y \in \mathscr{F}_y, y \in U\}$$

turns $\mathscr{F}'(U)$ into a presheaf. It is easy to see that, in fact, we obtain in this way a sheaf. It is called the *sheaf associated with the presheaf* $\mathscr{F}$. It is the sheaf "closest" to $\mathscr{F}$. For example, if $\mathscr{F}$ is a presheaf for which $\mathscr{F}(U) = M$ for all U (so that it consists of the constant functions on U with values in M), then $\mathscr{F}'(U)$ consists of all functions on U that are constant on each connected component of U.

Exercises

1. Let X be a discrete topological space, $F(U)$ the collection of those mappings $f: U \to M$ for which $f(U)$ is finite, and for $U \subset V$ let ϱ_U^V be the restriction. Is $\mathscr{F}$ a presheaf? Is it a sheaf?

2. Let X be a smooth quasiprojective variety in which a topology is introduced as in Example 4 of §2.1. For an open set $U \subset X$ we set $\mathscr{F}(U) = \Omega^p[U]$, and for $U \subset V$ we define ϱ_U^V as the restriction of differential forms. Is $\mathscr{F}$ a presheaf? Is it a sheaf?

3. Let A be a ring $\mathfrak{a} \subset A$ an ideal. For every non-nilpotent $f \in A$ we denote by $\mathfrak{a}_f$ the ideal generated in A_f by the images of the elements of $\mathfrak{a}$ under the homomorphism $A \to A_f$. By analogy with what we have done in §2.2, construct the presheaf $\mathscr{F}$ for which $\mathscr{F}(U) = \mathfrak{a}_f$ if $U = D(f)$, and show that $\mathscr{F}$ is a sheaf. A simpler variant: examine the case when A has no divisors of zero.

4. Let X be a topological space, M an Abelian group, $\mathscr{F}(U)$ the factor group of all locally constant functions on U with values in M by the constant functions, and let ϱ_U^V be defined by means of restrictions. Show that $\mathscr{F}$ is a presheaf and determine the sheaf $\mathscr{F}'$ associated with it.

5. Show that the structure sheaf $\mathcal{O}$ on Spec A can be defined as follows. For the elements $u \in \mathcal{O}(U)$ we take families of elements $\{u_x; u_x \in \mathcal{O}_x, x \in U\}$ ($\mathcal{O}_x$ is the local ring of the prime ideal x) that satisfy the condition: for any point $y \in U$ there exists a principal neighbourhood $y \in D(f) \subset U$ and an element $u \in A_f$ such that all the u_x for $x \in D(f)$ are images of u under the natural homomorphisms $A_f \to \mathcal{O}_x$. If $U \subset V$, then by taking in the family $v = \{v_x; x \in V\}$ those v_x for which $x \in U$ we obtain $\varrho_U^V v$.

6. Let A be a one-dimensional local ring and $\xi \in \operatorname{Spec} A$ a generic point. Show that ξ is an open set and find $\mathcal{O}(\xi)$.

7. Let A be the local ring of the origin of coordinates in $\mathbb{A}^2$. Find $\mathcal{O}(U)$, where $U = \operatorname{Spec} A - x$ and x is a closed point.

§3. Schemes

1. Definition of a Scheme

Definition. A *ringed space* is a pair $(X, \mathcal{O})$ consisting of a topological space X and a sheaf of rings $\mathcal{O}$.

A *morphism of ringed spaces* $\varphi: (X, \mathcal{O}_X) \to (Y, \mathcal{O}_Y)$ is a collection of a continuous mapping $\varphi: X \to Y$ and homomorphisms

$$\psi_U: \mathcal{O}_Y(U) \to \mathcal{O}_X(\varphi^{-1}(U))$$

for every open set $U \subset Y$ for which $\varphi^{-1}(U)$ is non-empty. It is required that the diagram

$$\begin{array}{ccc} \mathcal{O}_X(\varphi^{-1}(V)) & \xrightarrow{\varrho^{\varphi^{-1}(V)}_{\varphi^{-1}(U)}} & \mathcal{O}_X(\varphi^{-1}(U)) \\ \uparrow{\scriptstyle \psi_V} & & \uparrow{\scriptstyle \psi_U} \\ \mathcal{O}_Y(V) & \xrightarrow{\varrho^V_U} & \mathcal{O}_Y(U) \end{array}$$

is commutative for any U and V, $U \subset V$.

Sometimes the sheaf $\mathcal{O}$ is denoted by $\mathcal{O}_X$. It is called the *structure sheaf* of the ringed space.

Example 1. Any topological space X is a ringed space if we take for $\mathcal{O}_X$ the sheaf of continuous functions. Any continuous mapping $\varphi: X \to Y$ defines a morphism if we set $\psi_U(f) = \varphi^*(f)$ for $f \in \mathcal{O}_Y(U)$.

Example 2. Any differentiable manifold is a ringed space if we take for $\mathcal{O}_X$ the sheaf of differentiable functions. Just as above, any differentiable mapping determines a morphism.

Example 3. Any ring A determines the ringed space $(\operatorname{Spec} A, \mathcal{O})$ where $\mathcal{O}$ is the structure sheaf. Henceforth we denote this ringed space by $\operatorname{Spec} A$. Let us show that a homomorphism $\lambda: A \to B$ determines a morphism $\varphi: \operatorname{Spec} B \to \operatorname{Spec} A$. We set $\varphi = {}^a\lambda$. For $U = D(f) \subset \operatorname{Spec} A$ we have $\varphi^{-1}(U) = D(\lambda(f))$. The mapping $a/f^n \to \lambda(a)/\lambda(f)^n$ determines a homomorphism ψ_U of the ring $A_f = \mathcal{O}_{\operatorname{Spec} A}(U)$ into the ring $B_{\lambda(f)} = \mathcal{O}_{\operatorname{Spec} B}(\varphi^{-1}(U))$. The reader can easily verify that these homomorphisms extend to homomorphisms $\psi: \mathcal{O}_{\operatorname{Spec} A}(U) \to \mathcal{O}_{\operatorname{Spec} B}(\varphi^{-1}(U))$ for every open set $U \subset \operatorname{Spec} A$ and define a morphism φ of ringed spaces.

In what follows we frequently denote a ringed space $(X, \mathcal{O}_X)$ by a single letter X and a morphism $X \to Y$ given by mappings φ and ψ_U by a single letter φ.

A simple check shows that by combining two morphisms $\varphi: X \to Y$ and $\varphi': Y \to Z$ (φ as well as ψ_U) we obtain a morphism $\varphi'\varphi: X \to Z$. A morphism having an inverse is called an isomorphism.

If $(X, \mathcal{O}_X)$ is a ringed space and $U \subset X$ an open set, then by restricting the sheaf $\mathcal{O}_X$ to U we obtain the ringed space $(U, \mathcal{O}_X|_U)$. In this sense we shall henceforth often regard an open set $U \subset X$ as a ringed space.

As regards the examples we have given we make two remarks.

1. Whereas in Examples 1 and 2 the morphism is uniquely determined by the mapping $\varphi: X \to Y$ because the corresponding homomorphisms ψ_U are given by inverse images of functions, in Example 3 this is not so. For instance, if the ring A has a non-zero nil-radical N, $B = A/N$, and $\lambda: A \to B$ is the natural projection, then $\operatorname{Spec} A = \operatorname{Spec} B$ and $\varphi = {}^a\lambda$ is the identity mapping; however, even for $U = \operatorname{Spec} A$ we see

that $\psi_U = \lambda$ is not an isomorphism. Thus, a morphism of ringed spaces does not reduce to a mapping of the corresponding topological spaces. To emphasize this fact the word "mapping" is here replaced by the term "morphism".

2. The concept of a ringed space gives a convenient principle of classifying geometric objects. For example, let us take differentiable manifolds. They can be defined as ringed vector spaces, namely those in which every point has a neighbourhood U such that the ringed space $(U, \mathcal{O}|_U)$ is isomorphic to $(\bar{U}, \bar{\mathcal{O}})$, where $\bar{U}$ is a domain in n-dimensional Euclidean space and $\bar{\mathcal{O}}$ is the sheaf of differentiable functions on it. Precisely this definition is contained, for example, in the book [26], the only difference being that the terminology of sheaves is not used.

The general idea of this method of determining geometric objects is the following: we impose restrictions on the local structure of the ringed space by requiring each point to have a neighbourhood isomorphic, as a ringed space, to one of the ringed spaces of a previously fixed class.

The last remark leads us to the basic definition:

A *scheme* is a ringed space $(X, \mathcal{O})$ in which every point has a neighbourhood U such that the ringed space $(U, \mathcal{O}|_U)$ is isomorphic to Spec A, where A is a ring*.

A *morphism of schemes* $\varphi: X \to Y$ is defined as a morphism of corresponding ringed spaces, satisfying the following condition: if $U \subset Y$ is open, $x \in f^{-1}(U)$, $u \in \mathcal{O}_Y(U)$ and $u(\varphi(x)) = 0$, then $\varphi_u(u)(x) = 0$.

If X is a scheme and A a ring, then the morphism $X \to \operatorname{Spec} A$ determines a homomorphism $A \to \mathcal{O}_X(U)$ for every open set $U \subset X$, that is, it turns $\mathcal{O}_X$ into a sheaf of algebras over A. It is not hard to show that, conversely, if $\mathcal{O}_X$ is a sheaf of algebras over A, then it specifies the canonical morphism $X \to \operatorname{Spec} A$. A scheme X for which a morphism $X \to \operatorname{Spec} A$ is given is called a *scheme over* A. A morphism of schemes over A is determined by the condition that the diagram should be commutative,

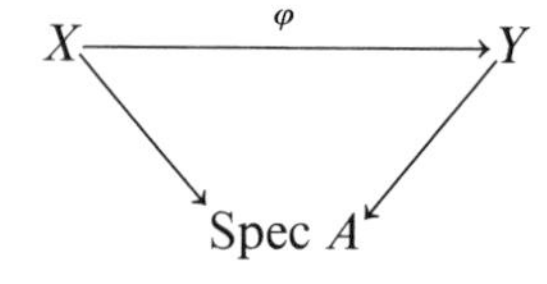

and this is equivalent to the fact that all the ψ_U are algebra homomorphisms over A.

Since every ring is an algebra over the ring of integers $\mathbb{Z}$, every scheme is a scheme over $\mathbb{Z}$. In this sense the concept of a scheme over A generalizes that of a scheme.

---

* If $(X, \mathcal{O})$ is scheme, $U \subset X$ open, $x \in U$ and $u \in \mathcal{O}(U)$, we say that $v(x) = 0$ if, for a neighbourhood $U' \subset U$ of x, for which there is an isomorphism $\varphi: (U', \mathcal{O}|_{U'}) \to \operatorname{Spec} A$, $p^U_{U'}(v)$ and x are mapped by φ to such clements $v' \in A'$ and $x' \in \operatorname{Spec} A$ that $v'(x') = 0$. Obviously this does not depend on the choice of U'.

Here are two very simple examples of schemes.

Example 4. Example 3 of a ringed space shows that Spec A is a scheme for every ring A. Such schemes are called *affine*. There is a one-to-one correspondence between ring homomorphisms $\lambda: A \to B$ and morphisms $\lambda: \operatorname{Spec} B \to \operatorname{Spec} A$.

Example 5. Let us find out how the notion of a quasiprojective variety fits into the language of schemes. We begin with the case of an affine variety X over an algebraically closed field k. The scheme $\operatorname{Spec} k[X]$ defined in Example 4 does not coincide with X even as a set: $\operatorname{Spec} k[X]$ consists of all prime ideals of the ring $k[X]$, which in turn corresponds to all irreducible subvarieties of X, and not only to its points. Nevertheless, the variety X and the scheme $\operatorname{Spec} k[X]$ are very naturally connected with each other, if only because regular mappings of affine varieties $X \to Y$ and morphisms of schemes $\operatorname{Spec} k[X] \to \operatorname{Spec} k[Y]$ are one and the same thing: both correspond to algebra homomorphisms $k[Y] \to k[X]$. Thus, we have here an isomorphism of categories.

Now let us consider an arbitrary quasiprojective variety X over a field k and associate with it in a similar fashion a scheme $\tilde{X}$ over k. For the set $\tilde{X}$ we take the collection of all irreducible subvarieties of X (including X). Let $U \subset X$ be an open subset and $\tilde{U}$ the set of its irreducible subvarieties. By associating with a subvariety $Z \subset U$ its closure $\bar{Z} \subset X$ we define an embedding of $\tilde{U}$ into $\tilde{X}$. The subsets $\tilde{U} \subset \tilde{X}$ define a topology in $\tilde{X}$. Finally, we define the sheaf $\mathcal{O}_{\tilde{X}}$ by the condition $\mathcal{O}_{\tilde{X}}(\tilde{U}) = k[U]$ with the natural restriction mapping. The reader is invited to verify that in this way we turn $\tilde{X}$ into a scheme over k.

A regular mapping $f: X \to Y$ determines a mapping of sets $\tilde{f}: \tilde{X} \to \tilde{Y}$ in which an irreducible variety $Z \subset X$ corresponds to $\overline{f(Z)}$, the closure of $f(Z)$ in Y. Finally, for $U \subset Y$ we define the homomorphism $\tilde{f}_U: \mathcal{O}_{\tilde{Y}}(\tilde{U}) \to \mathcal{O}_{\tilde{X}}(\tilde{f}^{-1}(U))$ as the homomorphism $f^*: k[U] \to k[f^{-1}(U)]$. The reader can easily verify that $\tilde{f}$ is a morphism of $\tilde{X}$ in $\tilde{Y}$, and that the assignation $f \to \tilde{f}$ defines a one-to-one correspondence between regular mappings $X \to Y$ and morphisms $\tilde{X} \to \tilde{Y}$ as schemes over k. Again we have an isomorphism of categories.

In what follows we frequently do not distinguish between a quasiprojective variety and the scheme corresponding to it.

In conclusion we make a few obvious remarks concerning the definition of a scheme.

The structure sheaf $\mathcal{O}$ of a scheme X has one important property: its stalk $\mathcal{O}_x$ over any point $x \in X$ is a local ring. For the stalk $\mathcal{F}_x$ of an arbitrary sheaf $\mathcal{F}$ on a space X does not change if we replace X by an open subset U containing x. For the structure sheaf on an affine scheme Spec A we have already seen that $\mathcal{O}_x$ is the local ring of a prime ideal

$x \in \operatorname{Spec} A$. Owing to this the local properties considered in §1 for affine schemes, such as regularity of a point, tangent space, etc. automatically carry over to arbitrary schemes.

Properties stated in §1 in terms of a topological space such as irreducibility, dimension, etc. are also applicable to an arbitrary scheme. Finally, certain concepts introduced earlier for quasiprojective varieties at once carry over to schemes. A *rational morphism* of a scheme X into Y is a class of equivalent morphisms $\varphi: U \to Y$, where U is an open dense set in X, and where morphisms $\varphi: U \to Y$ and $\psi: V \to Y$ are called equivalent if they agree on $U \cap V$. Two schemes X and Y are called *birationally isomorphic* if they have isomorphic open dense subsets (see the proposition in Ch. I, §4.3).

2. Pasting of Schemes. By definition, every scheme is covered by open sets isomorphic to affine schemes (or, as we say briefly, by affine open sets). Can we perhaps recover the scheme X by knowing such a covering $X = \bigcup U_\alpha$? We consider this problem in a somewhat greater generality, without assuming that the open sets U_α are necessarily affine.

First of all, we may remark that every open subset U of a scheme X is a scheme; this follows from the fact that every point has an affine neighbourhood V and that the $D(f) \subset V$ from a basis of open sets.

If $X = \bigcup U_\alpha$ is an open covering, then the schemes U_α are not independent: U_α and U_β have the isomorphic open subset $U_\alpha \cap U_\beta$. Therefore we start out from the following data; a system of schemes $U_\alpha, \alpha \in I$, in each of them a system of open subsets $U_{\alpha,\beta} \subset U_\alpha$, $\alpha, \beta \in I$, $U_{\alpha,\alpha} = U_\alpha$, and a system of isomorphisms of schemes $\varphi_{\alpha,\beta}: U_{\alpha,\beta} \to U_{\beta,\alpha}$. Let us find out how we can construct a scheme X, an open covering $X = \bigcup V_\alpha$, and a system of isomorphisms $\psi_\alpha: U_\alpha \to V_\alpha$, such that ψ_α restricted to $U_{\alpha,\beta}$ determines an isomorphism of the schemes $U_{\alpha,\beta}$ and $V_\alpha \cap V_\beta$, and $\psi_\beta \varphi_{\alpha,\beta} \psi_\alpha^{-1}$ is the identity mapping of $V_\alpha \cap V_\beta$. If such a scheme X exists, then we say that it is obtained by a pasting of the schemes U_α.

For a pasting to be possible it is necessary, as is easy to verify, that the following conditions hold:

$$\varphi_{\alpha,\alpha} = 1, \quad \alpha \in I, \quad \varphi_{\alpha,\beta}\varphi_{\beta,\alpha} = 1, \quad \alpha, \beta \in I. \tag{1}$$

The restriction $\varphi'_{\alpha,\beta}$ of the morphism $\varphi_{\alpha,\beta}$ to $U_{\alpha,\beta} \cap U_{\alpha,\gamma}$ is an isomorphism of $U_{\alpha,\beta} \cap U_{\alpha,\gamma}$ and $U_{\beta,\alpha} \cap U_{\beta,\gamma}$, and these isomorphisms are connected by the relations

$$\varphi'_{\alpha,\gamma} = \varphi'_{\beta,\gamma}\varphi'_{\alpha,\beta}, \quad \alpha, \beta, \gamma \in I. \tag{2}$$

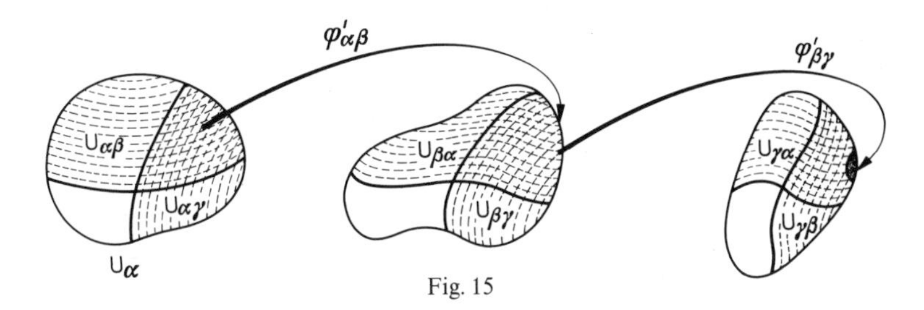

Fig. 15

The morphisms and schemes occurring in the conditions (1) and (2) are illustrated in Fig. 15.

Next we show that if the conditions (1) and (2) are satisfied, then a pasting is possible. First of all, we define X as a set. For this purpose we introduce in T, the disjoint union of all the U_α, a relation by setting $x \sim y$ if $x \in U_{\alpha,\beta}$, $y \in U_{\beta,\alpha}$, $y = \varphi_{\alpha,\beta}(x)$. The conditions (1) and (2) guarantee this relation is an equivalence relation. We denote by X the factor set of this equivalence relation, and let $p: T \to X$ be the canonical projection.

We introduce a topology in X by taking as open those sets $U \subset X$ for which $p^{-1}(U)$ is open. (The topology in T is determined by the open sets $\bigcup W_\alpha$, where W_α is open in U_α.) It is easy to see that p establishes a homeomorphism ψ_α of the sets U_α with open subsets $V_\alpha \subset X$ and that $X = \bigcup V_\alpha$.

Finally, we define the sheaf $\mathcal{O}_X$ on X as follows. For a W contained in some V_α we set $\mathcal{O}_X(W) = \mathcal{O}_{U_\alpha}(\psi_\alpha^{-1}(W))$, choosing an arbitrary $V_\alpha \supset W$. The choice of another $V_\beta \supset W$ replaces $\mathcal{O}_X(W)$ by an isomorphic ring. The homomorphisms $\varrho^W_{W'}$, where $W' \subset W \subset V_\alpha$, are defined in the obvious manner. Therefore the presheaf $\mathcal{O}_X$ is defined not on all the sets, but those W on which it is defined form a basis of open sets. The situation is the same as in the definition of the structure sheaf on Spec A. In this way we can extend the definition of $\mathcal{O}_X(U)$ to all open sets $U \subset X$ as the projective limit $\mathcal{O}_X(W)$, where $W \subset U$ are those open sets on which it was previously defined. There remains a standard verification of a large number of properties (that $\mathcal{O}_X$ is a sheaf, X a scheme, etc.), which we omit.

As a first application of this construction we define a scheme $\mathbb{P}^n(A)$, a so-called projective space over the ring A. For this purpose we consider $n+1$ independent variables $T_0, \ldots, T_n$, and in the ring of fractions $A[T_0, \ldots, T_n]_{(T_0 \ldots T_n)}$ the subrings $A_i = A[T_0/T_i, \ldots, T_n/T_i]$. We set $U_i = \operatorname{Spec} A_i$, $U_{ij} = D(T_j/T_i) \subset U_i$. By definition, $U_{ij} = \operatorname{Spec} A_{ij}$, where $A_{ij} = (A_i)_{(T_j/T_i)}$ consists of the elements

$$F(T_0, \ldots, T_n)/T_i^p T_j^q \in A[T_0, \ldots, T_n]_{(T_0 \ldots T_n)}$$

for which F is a form of degree $p+q$. Hence it follows that A_{ij} and A_{ji} agree in $A[T_0, \ldots, T_n]_{(T_0 \ldots T_n)}$, and so we have the natural isomorphism $\varphi_{ij}: U_{ij} \to U_{ji}$. It is easy to check that the conditions (1) and (2) both hold. As a result of pasting we obtain the scheme $\mathbb{P}^n(A)$.

3. Closed Subschemes. If the ring homomorphism $\lambda: A \to B$ is an epimorphism, then the mapping ${}^a\lambda: \operatorname{Spec} B \to \operatorname{Spec} A$ determines a homeomorphism of Spec B onto the closed subset $V(\mathfrak{a}) \subset \operatorname{Spec} A$, where $\mathfrak{a}$ is the kernel of λ. In that case Spec B is called a closed subscheme of Spec A, and the morphism ${}^a\lambda$ a closed embedding. We generalize these concepts at once to arbitrary schemes.

A morphisms of schemes $\varphi: Y \to X$ is called a *closed embedding* (or *closed immersion*) if every point $x \in X$ has an affine neighbourhood U such that the scheme $\varphi^{-1}(U)$ is affine and the homomorphism $\psi_U: \mathcal{O}_X(U) \to \mathcal{O}_Y(\varphi^{-1}(U))$ is an epimorphism.

In that case Y is called a *closed subscheme* of X.

Since closedness is a local property, $\varphi(Y)$ then is a closed subset of the topological space X.

To make this definition consistent with the example from which we have started out we prove the following proposition.

Proposition. *If X is an affine scheme, $X = \operatorname{Spec} A$, and $\varphi: Y \to X$ a closed embedding, then Y is also affine: $Y = \operatorname{Spec} B$, and $\varphi = {}^a\lambda$, where $\lambda: A \to B$ is an epimorphism of rings.*

We can find a covering $X = \bigcup U_i$, $U_i = D(f_i)$, $f_i \in A$, such that $\varphi^{-1}(U_i) = \operatorname{Spec} A_i$ and $\psi_i: A_{f_i} \to A_i$ is an epimorphism. We set $\ker \psi_i = \mathfrak{a}_i \subset A_{f_i}$, $\varrho^X_{U_i} = {}^a\lambda_i$, $\bigcap \lambda_i^{-1} \mathfrak{a}_i = \mathfrak{a}$. The morphism φ makes Y into a scheme over A. But since $\mathfrak{a} \subset \lambda_i^{-1} \mathfrak{a}_i$, under the action of A on $\mathcal{O}_Y(\varphi^{-1}(U_i))$ the ideal $\mathfrak{a}$ acts trivially. In other words, Y is a scheme over $A/\mathfrak{a}$. This shows that the diagram

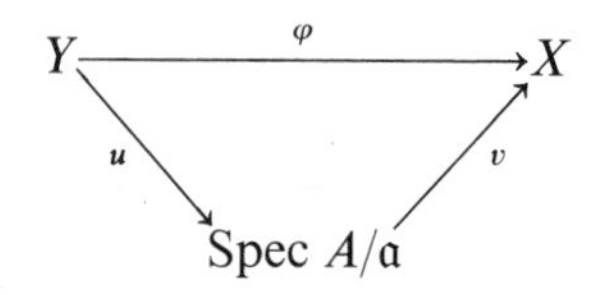

is commutative, where v is a closed embedding.

The proposition will be proved if we can verify that u is an isomorphism. Locally u is given by homomorphisms [in the sets $\varphi^{-1}(U_i)$ and $v^{-1}(U_i)$]

$$u_i: (A/\mathfrak{a})_{\bar{f}_i} \to A_{f_i}/\mathfrak{a}_i,$$

where $\bar{f}_i$ is the image of f_i in $A/\mathfrak{a}$. It is sufficient to show that all the u_i are isomorphisms.

That u_i is an epimorphism follows at once from the fact that $\mathfrak{a} \subset \lambda_i^{-1}\mathfrak{a}_i$. To prove that it is a monomorphism we use the following remarks. The ring $\mathcal{O}_Y(\varphi^{-1}(U_i \cap U_j))$ can be described in two ways:

$$\mathcal{O}_Y(\varphi^{-1}(U_i \cap U_j)) = (A_i)_{\psi_i(\lambda_i f_j)} = (A_j)_{\psi(\lambda_j f_i)}. \tag{1}$$

We consider the localization homomorphisms

$$\lambda_j^i : A_{f_i} \to (A_{f_i})_{\lambda_i(f_j)} = A_{(f_i f_j)}.$$

From (1) it follows at once that

$$\lambda_j^i \mathfrak{a}_i = \lambda_i^j \mathfrak{a}_j, \tag{2}$$

where, for example, $\lambda_j^i \mathfrak{a}_i$ is the ideal generated by the elements

$$\lambda_j^i \alpha, \quad \alpha \in \mathfrak{a}_i \quad \text{in} \quad A_{(f_i f_j)}.$$

Suppose that $a \in A$ determines an element of the kernel of the homomorphism u_i. Then $\lambda_i(a) \in \mathfrak{a}_i$. By (2) it then follows that

$$\lambda_j^i \lambda_i(a) \in \lambda_i^j \mathfrak{a}_j.$$

The left-hand side is the image of a under the localization $A \to A_{(f_i f_j)}$ and is therefore equal to $\lambda_i^j \lambda_j(a)$, and the elements on the right-hand side are of the form $\lambda_i^j(a_j)/\lambda_j(f_i)^l$. Thus,

$$\lambda_i^j(\lambda_j(f_i)^l \lambda_j(a) - a_j) = 0.$$

Therefore

$$\lambda_j(f_i)^{l+m} \lambda_j(a) = \lambda_j(f_i)^m a_j \in \mathfrak{a}_j$$

for some m. So we see that

$$\lambda_j(f_i^{l+m} a) \in \mathfrak{a}_j, \tag{3}$$

where l and m can be chosen to be the same for all j. The relation (3), which has been proved for all j, shows that $f_i^{l+m} a \in \mathfrak{a}$, that is $(\bar{f}_i)^{l+m}\bar{a} = 0$, where $\bar{a}$ is the image of $A/\mathfrak{a}$. This means that $\bar{a}$ determines the zero element in $(A/\mathfrak{a})_{\bar{f}_i}$. The proposition is now proved.

Closed subschemes provide us with new examples of schemes. Thus, the closed subschemes of the scheme $\mathbb{P}^n(A)$ give us a new extensive type of schemes over the ring A, generalizing projective varieties. Even our customary quasiprojective varieties contain vastly more closed subschemes than closed subvarieties.

For example, on the affine line $X = \operatorname{Spec} k[T]$ a closed subscheme other than X is of the form $\operatorname{Spec} k[T]/(F)$, where $F(T)$ is an arbitrary polynomial, whereas closed subvarieties correspond only to the collection of roots of these polynomials without reflecting the multiplicities of the roots.

If $\varphi: X \to Y$ is a morphism of schemes and Y' a closed subscheme of Y, then we can determine its inverse image $\varphi^{-1}(Y')$, which is a closed subscheme of X. We only treat the case when X and Y are affine schemes, $X = \operatorname{Spec} A$, $Y = \operatorname{Spec} B$, and $\varphi = {}^a\lambda$, $\lambda: B \to A$. Then a closed embedding of Y' in Y is determined by the natural homomorphism $B \to B/\mathfrak{b}$. If $\lambda(\mathfrak{b})A = A$, then the set $\varphi^{-1}(Y')$ is empty. If this is not so, then clearly $X' = \operatorname{Spec} A/\lambda(\mathfrak{b})A$ is a closed subscheme of X, which is called the inverse image of Y'. As a topological space it is actually the inverse image of the subspace $Y' \subset Y$.

For example, if X and Y are isomorphic to the affine line $\mathbb{A}^1$ over a field k, if φ is given by the mapping $\varphi(x) = x^2$, and if the characteristic of k is not 2, then $\varphi^{-1}(y)$ for $y \neq 0$ consists of two connected components isomorphic to $\operatorname{Spec} k$ (that is, two "ordinary" points), and $\varphi^{-1}(y) = \operatorname{Spec} k[T]/(T^2)$ for $y = 0$. This example shows that schemes with nilpotent elements in rings $\mathcal{O}(U)$ can arise in the most classical situations. We have already had occasion to comment that it is natural to define the inverse images of subvarieties of codimension 1 as divisors, that is, as subvarieties with multiplicities. In the simplest cases these multiplicities prove sufficient to specify a sheaf on these subvarieties. In the general case this is a palliative: it is clear that the inverse image under a morphism of two objects must be an object of the same kind, in our case a scheme, but here we very often arrive at schemes with nilpotent elements.

Even more extreme is the situation in the example when $X = Y = \mathbb{A}^1$ is an affine line over a field k of characteristic p and $\varphi(x) = x^p$. This mapping is one-to-one, but not an isomorphism. Applying our concept of inverse image we see that $\varphi^{-1}(x) = \operatorname{Spec} k[T]/(T^p)$, that is, the inverse image of every point contains nilpotent elements in its sheaf. It is interesting that in this case X and Y are algebraic groups under addition, and φ is a homomorphism. Therefore it is natural to expect that $\varphi^{-1}(0)$ is also a "group" of some new type. In the next section we shall see that this is so.

4. Reducibility and Nilpotents. If the rings $\mathcal{O}_X(U)$ have no nilpotent elements, then the scheme X is called *reduced*. With every scheme X there is associated a reduced closed subscheme X' whose topological space coincides with X. For an open set $U \subset X$ the ring $\mathcal{O}_{X'}(U)$ is defined as the quotient ring of $\mathcal{O}_X(U)$ by its nil-radical (the ideal generated by all nilpotent elements). This scheme is denoted by X_{red}.

Let us consider some examples, with the object of illustrating the role of non-reduced schemes in the study of "classical" quasiprojective varieties. Let X be a quasiprojective variety defined over a field k, and $I_0 = \operatorname{Spec} k$. What are the morphisms $\varphi: I_0 \to X$ (as a scheme over k)? Since I_0 is a quasiprojective variety consisting of a single point y_0 and since morphisms of quasiprojective varieties are the same thing as regular mappings, φ is completely specified by the image $x=\varphi(y_0)$ of y_0. Of these morphisms there are just as many as there are closed points $x \in X$.

Now we complicate this example somewhat by introducing nilpotent elements.

We set $I_1 = \operatorname{Spec} T_1$, where $T_1 = k[T]/(T^2)$, and we investigate the morphisms $\varphi: I_1 \to X$. Since I_0 is a closed subscheme in I_1 $(I_0 = (I_1)_{\mathrm{red}})$, we see that φ determines a morphism $\varphi': I_0 \to X$ specified by the point $x \in X$. It is sufficient for us to give the homomorphism $\psi_U: \mathcal{O}_X(U) \to T_1$ for affine open sets $U \subset X$. If $x \notin U$, then $\psi_U = 0$. Let $x \in U$. Since $\mathcal{O}_X(U) = k + \mathfrak{m}_x$, where $\mathfrak{m}_x$ is the maximal ideal of the point x, we see that ψ_U is uniquely determined by its action on $\mathfrak{m}_x$ and by $\psi_U(\mathfrak{m}_x) \subset k\varepsilon$, where ε is the image of T in T_1. As $\varepsilon^2 = 0$, we have $\psi_U(\mathfrak{m}_x^2) = 0$, hence ψ_U determines a linear mapping of $\mathfrak{m}_x/\mathfrak{m}_x^2$ into $k\varepsilon$, that is, into k.

The space of linear functions on $\mathfrak{m}_x/\mathfrak{m}_x^2$ is the tangent space Θ_x to X at x. Therefore, every morphism $\varphi: I_1 \to X$ for which $\varphi(I_1) = x$ determines a tangent vector at x. It is easy to verify that this establishes a one-to-one correspondence between these morphisms and vectors in Θ_x.

This result has the following geometrical interpretation. We consider an affine neighbourhood U of x and in it a closed subscheme $T_x = \operatorname{Spec} k[U]/\mathfrak{m}_x^2$, where $\mathfrak{m}_x$ is the maximal ideal of x in $k[U]$. The homomorphism $k[U] \to k[U]/\mathfrak{m}_x^2$ determines a closed embedding $T_x \to U$. It is easy to see that T_x is also a closed subscheme in X and does not depend on the choice of the neighbourhood U. The arguments above show that every morphism $\varphi: I_1 \to X$ is of the form $\varphi: j\psi$, where ψ is a morphism $I_1 \to T_x$, and j a closed embedding of T_x in X. Thus, the morphisms $I_1 \to X$ carrying I_0 into x, are in one-to-one correspondence with the morphisms $I_1 \to T_x$. The scheme T_x is fairly large: the tangent space to it at x is the same as that to X. But it is also small enough for the morphism $\varphi: I_1 \to T_x$ to be uniquely determined by its differential $d_{x_0}\varphi$, $x_0 = I_0$. This is the geometrical interpretation of our computations. It justifies the name "*infinitely small neighbourhood of x of the first order*" for the subscheme T_x. Similarly we can define an "*infinitely small neighbourhood of x of order n*". These definitions lead to a "*theory of jets*" on algebraic varieties.

As a second example we consider the concept of a family of subvarieties of a variety X, which was defined in Ch. I, §6.5, as a subvariety Γ in the product $X \times S$, where S is an algebraic variety. For a point $s \in S$ the subvariety Y of X determined by $Y \times s = (X \times s) \cap \Gamma$ is determined the variety of this family corresponding to s.

Assuming now that S is an arbitrary scheme, the same definition can be preserved verbatim. We only have to make use of the concept of products of schemes, which will be defined in the next section. For example, if $S=I_1$, then a family with base S is a subscheme Γ of $X\times I_1$. It is easy to verify that if s_0 is the closed point of I_1, then the embedding $s_0\to I_1$ determines the closed embedding $X=X\times s_0\to X\times I_1$. The inverse image of Γ determines a subscheme Γ_{s_0} in X—the unit subscheme of the family. The concept of a family of divisors as defined in Ch. III, §3.5, can also be generalized. Although the family at which we arrive is purely "infinitesimal", its presence can be very important. The subscheme (or divisor) Γ_{s_0} can turn out to be reduced, that is, a quasi-projective subvariety in the variety X, and the possibility of including it in this family points to the existence of a subvariety of infinitely small deformations. Every family in the previous sense containing the subvariety Y, of course, also determines an infinitely small deformation of this subvariety. However, there exist, for example, a surface X and on it a curve Y for which the set of infinitely small deformations is isomorphic to k and which cannot be included in any family of curves other than Y (that is, it has no finite deformations). An example is contained in the book [24], Lecture 22. This example makes it particularly clear how schemes with nilpotent elements naturally arise in connection even with "classical" problems of the geometry of algebraic varieties.

5. Finiteness Conditions. Two properties of schemes, which we shall analyse presently, have the character of "finite-dimensionality".

A scheme X is called *Noetherian* if it has a finite covering of affine open sets

$$X=\bigcup U_i,\qquad U_i=\operatorname{Spec} A_i, \tag{1}$$

in which the rings A_i are Noetherian.

A scheme X over a ring B is called a *scheme of finite type over* B if it has a finite covering (1) in which the A_i are algebras of finite type over B.

Obviously, a scheme of finite type over a Noetherian ring is Noetherian.

We shall prove propositions of one and the same kind referring to both these concepts.

Proposition 1. *If an affine scheme* Spec A *is Noetherian, then the ring* A *is Noetherian.*

By hypothesis, there exists a finite covering (1) such that the rings A_i are Noetherian. Let $\mathfrak{a}_1\subset\mathfrak{a}_2\subset\dots$ be a chain of ideals in A. As was shown in §2.2, $A=\mathcal{O}(X)$, where $\mathcal{O}$ is the structure sheaf on X. We consider the ideals $\mathfrak{a}_n^{(i)}=\varrho_{U_i}^X\mathfrak{a}_n A_i\subset A_i$. Since the rings A_i are Noetherian and there are finitely many of them, there exists an N such that

$$\mathfrak{a}_{n+1}^{(i)}=\mathfrak{a}_n^{(i)} \tag{2}$$

for all i and all $n \geqslant N$. Let us show that then $\mathfrak{a}_{n+1} = \mathfrak{a}_n$ for $n \geqslant N$. Indeed, since the U_i form a covering of X, it follows from (2) that

$$(\varrho_x^X \mathfrak{a}_{n+1})\mathcal{O}_x = (\varrho_x^X \mathfrak{a}_n)\mathcal{O}_x$$

for all points $x \in X$ and all $n \geqslant N$. It now remains to repeat the arguments in §2.2. If $u \in \mathfrak{a}_{n+1}$, then

$$u = a_x/b_x,\ a_x \in \mathfrak{a}_n,\ b_x \in A,\ b_x(x) \neq 0\,.$$

There exist points $x_1, \ldots, x_r$ and elements $c_1, \ldots, c_r \in A$ such that $c_1 b_{x_1} + \cdots + c_r b_{x_r} = 1$. Then

$$u = \Sigma a_{x_i} c_i \in \mathfrak{a}_n\,,$$

that is, $\mathfrak{a}_n = \mathfrak{a}_{n+1}$.

Proposition 2. *If an affine scheme* Spec A *has finite type over a ring* B, *then* A *is an algebra of finite type over* B.

By hypothesis, there exists a covering (1) such that the algebras A_i have finite type over B. Since the space Spec A_i is compact, it has a finite covering by principal open sets $D(f)$, $f \in A$. The corresponding algebras $(A_i)_f = A_f$ have finite type over B. Therefore, we may assume at once that in (1) $U_i = D(f_i)$. Suppose that the generators of the algebra A_i over B are of the form $x_{ij}/f_i^{n_{ij}}$. On the other hand, as $\bigcup D(f_i) = \operatorname{Spec} A$, there exist elements $g_i \in A$ such that

$$\Sigma f_i g_i = 1\,. \tag{3}$$

We denote by $A' \subset A$ the subalgebra generated over B by the elements x_{ij}, f_i, and g_i, and we show that $A' = A$. Let $x \in A$. By hypothesis, $x \in A_{f_i}$ for all f_i. This means that there exists an n (by taking it sufficiently large we may assume it to be independent of i) such that $f_i^n x$ belongs to the subalgebra generated over B by the elements x_{ij} and f_i. In particular,

$$f_i^n x \in A' \tag{4}$$

for all f_i. By raising (3) to a sufficiently high power we obtain a relation $\Sigma f_i^n g_i^{(n)} = 1$, where the $g_i^{(n)}$ belong to the subalgebra generated over B by the elements f_j and g_j. In particular, $g_i^{(n)} \in A'$. Multiplying the relations (4) by $g_i^{(n)}$ and adding up we see that $x \in A'$.

Exercises

1. Let X be a ringed space and G a group consisting of automorphisms of X. Define a set Y as the factor set of the points of X with respect to G, and let $p: X \to Y$ be the natural projection. Introduce in Y the topology in which a set $U \subset Y$ is open if and only if $p^{-1}(U) \subset X$ is open. Finally, define a presheaf $\mathcal{O}_Y$ by the condition: $\mathcal{O}_Y(U) = \mathcal{O}_X(p^{-1}(U))^G$. Here A^G denotes the set of G-invariant elements of A [it must be checked that G is in a natural sense a group of automorphisms of the ring $\mathcal{O}_X(p^{-1}(U))$].

Show that $(Y, \mathcal{O}_Y)$ is a ringed space. It is called the factor space of X with respect to G and is denoted by X/G.

2. Let k be an infinite field, $\mathbb{A}^2$ an affine plane over k, $X = \mathbb{A}^2 - (0,0)$, and let G consist of the automorphisms $(x, y) \to (\alpha x, \alpha y)$, $\alpha \in k, \alpha \neq 0$. Show that in the notation of Exercise 1 the ringed space Y coincides with the projective line $\mathbb{P}^1$ over k.

3. Let X be the same as in Exercise 2, but let G consist of the automorphisms $(x, y) \to (\alpha x, \alpha^{-1} y)$, $\alpha \in k, \alpha \neq 0$. Show that Y is a scheme. Show also that if $X = \mathbb{A}^2$, and G the same as above, then Y is not a scheme.

4. Investigate the inverse images of the points $x \in \operatorname{Spec} \mathbb{Z}$ in the morphism ${}^a\varphi$ in Example 3 of § 1.1.

5. Investigate the inverse images of points under the morphism $X \to Y$ projecting the circle $x^2 + y^2 = 1$ onto the x-axis: $f(x, y) = x$, where all the varieties are defined over the field $\mathbb{R}$ of real numbers. In other words,

$$X = \operatorname{Spec} \mathbb{R}[T_1, T_2]/(T_1^2 + T_2^2 - 1), \; Y = \operatorname{Spec} \mathbb{R}[T_1].$$

6. Show that in Example 5 of § 3.1 the points of the varieties coincide with the closed points of the scheme $\tilde{X}$.

7. Let Γ be a homogeneous ring, $\Gamma = \bigoplus_{n \geq 0} \Gamma_n$, $\Gamma_n \cdot \Gamma_m \subset \Gamma_{n+m}$. An ideal $\mathfrak{a} \subset \Gamma$ is called homogeneous if $\mathfrak{a} = \bigoplus_{n \geq 0} (\mathfrak{a} \cap \Gamma_n)$. Denote by $\operatorname{Proj} \Gamma$ the collection of homogeneous prime ideals $\mathfrak{p} \subset \Gamma$ that do not contain the ideal $\bigoplus_{n>0} \Gamma_n$, and introduce in this set the topology induced by the embedding $\operatorname{Proj} \Gamma \subset \operatorname{Spec} \Gamma$. For a homogeneous element $f \in \Gamma_m$, $m > 0$, let $\Gamma_{(f)}$ denote the subring of Γ_f consisting of the quotients $g/f^k, g \in \Gamma_{mk}$, $k \geqq 0$. Set

$$G_+(f) = D(f) \cap \operatorname{Proj} \Gamma.$$

Let ψ_f be the composition of mappings $G_+(f) \to D(f) \to \operatorname{Spec} \Gamma_f \to \operatorname{Spec} \Gamma_{(f)}$. Show that ψ_f is a homeomorphism between $G_+(f)$ and $\operatorname{Spec} \Gamma_{(f)}$. Show that the structure sheaves over $\operatorname{Spec} \Gamma_{(f)}$ (for all homogeneous f), carried over by means of ψ_f to $\operatorname{Proj} \Gamma$, determine a single sheaf $\mathcal{O}$ and that $(\operatorname{Proj} \Gamma, \mathcal{O})$ is a scheme. This scheme is also denoted by $\operatorname{Proj} \Gamma$.

8. Show that if in the notation of Exercise 7, Γ is a graded algebra over a ring A, that is, $\Gamma_n \cdot A \subset \Gamma_n$, then in this way a natural structure of a scheme over A is defined in the scheme $\operatorname{Proj} \Gamma$.

9. In the notation of Exercise 7, let $\Gamma = A[T_0, \ldots, T_n]$ with the usual grading by degrees. Show that the scheme $\operatorname{Proj} \Gamma$ is isomorphic to $\mathbb{P}^n(A)$.

10. Let Y be an affine n-dimensional variety over a field k, let y be a simple point of it and $\mathfrak{m}_y \subset k[Y]$ the corresponding maximal ideal. In the notation of Exercise 7, set $\Gamma = \bigoplus_{n \geq 0} \mathfrak{m}_y^n$, $\mathfrak{m}_y^0 = k[Y]$. Show that $\operatorname{Proj} \Gamma = \tilde{X}$, where X is the variety obtained from Y by the σ-process centred at y, and that the morphism $\tilde{\sigma} : \operatorname{Proj} \Gamma \to \operatorname{Spec} k[Y]$ corresponding to the σ-process is determined by the natural algebra structure over $k[Y]$ which exists in Γ (see Exercise 8).

§ 4. Products of Schemes

1. Definition of a Product. It would hopeless to define the product of two schemes X and Y in terms of the set of pairs (x, y), $x \in X$, $y \in Y$. For when $X = Y = \mathbb{A}^1$, we have $X \times Y = \mathbb{A}^2$, and to the points $X \times Y$ there correspond irreducible subvarieties of the plane $\mathbb{A}^2$. Consequently, among them there are all irreducible curves which, of course, cannot be represented in the form of pairs (x, y). Therefore, first of all we must try to clarify what properties we expect from a product of schemes, and then we must tackle the problem whether a scheme with these properties exists. This was the way in which we arrived in Ch. I at the definition of the product of quasiprojective varieties.

We consider schemes over an arbitrary ring A. By definition, this means a scheme X and a morphism $X \to \operatorname{Spec} A$. We even consider a more general situation: a morphism of two arbitrary schemes $X \to S$. Such an object is called a scheme over S. It is clear how to define a morphism of schemes $\varphi: X \to S$ and $\psi: Y \to S$ over S; this is a morphism $f: X \to Y$ for which $\varphi = \psi \cdot f$.

If $\varphi: X \to S$ and $\psi: Y \to S$ are two schemes over S, then evidently their product over S (which we denote by $X \times_S Y$) must have projections onto the factors, that is, two morphisms of schemes over S, namely $p_X: X \times_S Y \to X$ and $p_Y: X \times_S Y \to Y$ in a commutative diagram.

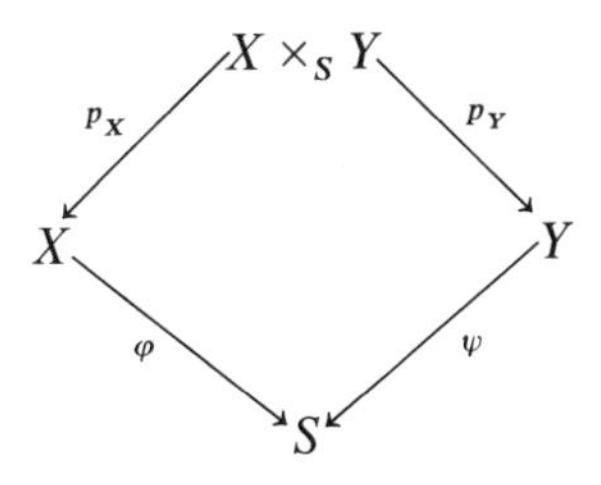

Furthermore, it is natural to require universality of the product. This means that for any scheme Z and morphisms $u: Z \to X$ and $v: Z \to Y$ for which the diagram

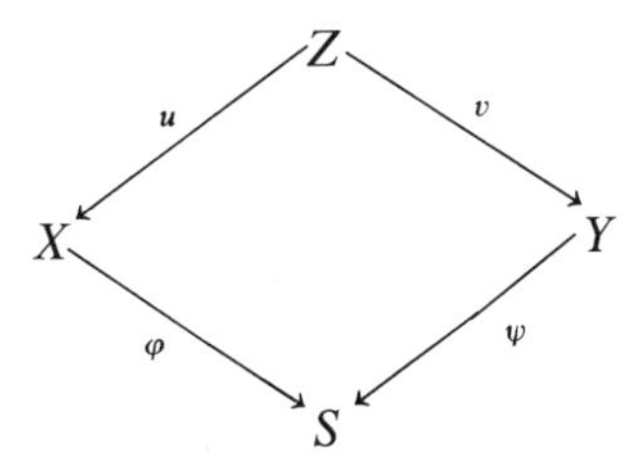

commutes there exists a morphism $h: Z \to X \times_S Y$ such that $p_X h = u$, $p_Y h = v$, and the morphism h with these properties must be unique. The morphism h is denoted by (u, v).

If a scheme $X \times_S Y$ satisfying these conditions exists, then evidently it is unique to within an isomorphism. It is called the *product of the schemes X and Y over S*. Occasionally, instead of using the term "scheme X over S" one simply talks of the morphism $\varphi: X \to S$, and then $X \times_S Y$ is called the fibred product of the morphisms φ and ψ.

The definition is compatible with the definition of the product of two objects of a category. In our case we consider the category of schemes over S.

In the category of sets the fibred product of two mappings $\varphi: X \to S$ and $\psi: Y \to S$ exists and coincides with the subset $Z \subset X \times Y$ consisting of those pairs (x, y), $x \in X$, $y \in Y$ for which $\varphi(x) = \psi(y)$. The situation is similar in the category of quasiprojective varieties over an algebraically closed field k.

The product of two schemes over a scheme S exists. The proof of this assertion is essentially elementary, but somewhat lengthy. It can be found in the book [15], Ch. I. We confine ourselves to some remarks, which may help the reader to reconstruct a proof.

If X, Y, and S are affine schemes, $X = \operatorname{Spec} A$, $Y = \operatorname{Spec} B$, $S = \operatorname{Spec} C$, then the specification of X and Y as schemes over S determines in A and B algebra structures over C. In this case the scheme $Z = \operatorname{Spec}(A \otimes_C B)$ is the product of X and Y over S if we endow it with the projections $p_X = {}^a f: Z \to X$ and $p_Y = {}^a g: Z \to Y$, corresponding to the homomorphisms $f: A \to A \otimes_C B$, $f(a) = a \otimes 1$, and $g: B \to A \otimes_C B$, $g(b) = 1 \otimes b$. This proposition is a simple consequence of the definition of the tensor product.

In the general case we have to consider a covering $S = \bigcup W_\alpha$, $X = \bigcup U_{\alpha\beta}$, $Y = \bigcup V_{\alpha\gamma}$ by affine sets such that $\varphi(U_{\alpha\beta}) \subset W_\alpha$, $\psi(V_{\alpha\gamma}) \subset W_\alpha$. Then $\varphi: U_{\alpha\beta} \to W_\alpha$ and $\psi: V_{\alpha\gamma} \to W_\alpha$ are affine schemes over W_α, and by the preceding the products $U_{\alpha\beta} \times_{W_\alpha} V_{\alpha\gamma}$ exist. It is not hard to verify that these schemes satisfy the conditions (1) and (2) in § 3.2 (for a suitable choice of open subsets and isomorphisms, which are easy to indicate), so that they can be pasted together into a single scheme. After this we have to define the projections of this scheme onto X and Y and verify that the condition of universality holds.

From the definition of the product it easily follows that it is associative: $(X \times_S Y) \times_S Z = X \times_S (Y \times_S Z)$.

If S is an affine scheme: $S = \operatorname{Spec} A$, then $X \times_{\operatorname{Spec} A} Y$ is also denoted by $X \times_A Y$.

An arbitrary scheme can be regarded as a scheme over $\mathbb{Z}$. Therefore the product of any two schemes is defined over $\mathbb{Z}$: $X \times_{\mathbb{Z}} Y$. It is simply called the *product of the schemes* and is denoted by $X \times Y$.

As a first application of the concept of a product we define the inverse image of a closed subscheme (in § 3.3 this definition had only been given for affine schemes). If Y is a closed subscheme of a scheme X, if $j: Y \to X$ is a closed embedding, and if $\varphi: X' \to X$ is any morphism, then for the scheme $Y' = Y \times_X X'$ by definition the morphism $j': Y \times_X X' \to X'$ exists. It is not hard to verify that j' is a closed embedding, so that Y' is a closed subscheme of X'. It is called the inverse image of Y under the morphism φ. It is easy to check that for the case of affine schemes this definition is the same as that given earlier.

The advantage of the new definition consists in that it may be applied to certain other situations. For example, let x be a point of a scheme X, not necessarily closed. We set $T = \operatorname{Spec} k(x)$ and define a morphism $T \to X$ by the fact that $\varphi(T) = x$ and $\psi_U(\mathcal{O}(U)) = 0$ if the open affine set U does not contain x. But if $x \in U$, $U = \operatorname{Spec} A$, then x is a prime ideal of A, and we define ψ_U as the natural homomorphism $A \to k(x)$ into the field of fractions of A/x. The homomorphisms ψ_U automatically extend to all open sets $U \subset X$ and define a morphism $\varphi: T \to X$.

If $\varphi: X' \to X$ is another morphism, then the scheme $X' \times_X T$ is called the inverse image of x or the *fibre* of the morphism φ over x. It has a morphism $X' \times_X T \to T$, that is, it is a scheme over $k(x)$. If the point x is not closed, then its inverse image is not closed, in general.

2. Group Schemes. The concept of products makes it possible to carry over to schemes the definition of an algebraic group. For this purpose we have to reformulate the definition of an algebraic group given in Ch. III, § 3.2, so that we do not talk of points but only of morphisms.

Let $\varphi: X \to S$ be a scheme over S. A group law is defined by a morphism

$$\mu: X \times_S X \to X\,.$$

The role of the unit element is played by a morphism

$$\varepsilon: S \to X$$

such that $\varphi \cdot \varepsilon = 1$ (we have already seen several times that, for example, for schemes over a field k the morphism $\operatorname{Spec} k \to X$ determines a point in X).

The association of every element with its inverse is replaced by the specification of a morphism

$$i: X \to X\,.$$

The property of the unit element is expressed by the fact that

$$\mu \cdot (\varepsilon\varphi, 1) = \mu \cdot (1, \varepsilon\varphi) = 1 \tag{1}$$

(1 is the identity morphism).

The property of the inverse element is expressed by the condition

$$\mu \cdot (i, 1) = \mu \cdot (1, i) = \varepsilon\varphi\,. \tag{2}$$

It remains to write down the conditions of associativity. For this purpose we observe that by the associativity of the product of schemes we have two morphisms $X \times_S X \times_S X \to X \times_S X : (\mu, 1)$ and $(1, \mu)$. Our condition has the form

$$\mu \cdot (\mu, 1) = \mu \cdot (1, \mu)\,. \tag{3}$$

If the conditions (1), (2), and (3) are satisfied, then a scheme X over S with the morphisms μ, ε and i is called a *group scheme* over S.

We leave it to the reader to make the natural definition of homomorphism and isomorphism of group schemes.

Here is a typical example, which shows the usefulness of extending the concept of an algebraic group to that of a group scheme. Let $X=Y=G_a$ be the scheme of $\mathbb{A}^1$ over an algebraically closed field k of characteristic p in which the group law is defined as $\mu(x,y)=x+y$. These are even algebraic groups with which we are already acquainted. We consider the homomorphism of these groups: $f(x)=x^p$. As a point mapping it is a monomorphism and as a mapping of abstract groups it is an isomorphism, but as a regular mapping of varieties it is not an isomorphism. This is a serious difference from the usual situation in the theory of groups.

In the preceding subsection we have seen that if we regard f as a morphism of schemes, then the inverse image of every point is a nontrivial scheme (that is, not Spec k). It is natural to try and turn $f^{-1}(0)$ into a group scheme. For this purpose we denote the scheme by Z and its closed embedding in X by j. We consider the morphism

$$\mu\cdot(j,j): Z\times_k Z\to X.$$

The reader is advised to show as an exercise that there is a morphism

$$\mu': Z\times_k Z\to Z$$

such that $\mu\cdot(j,j)=j\cdot\mu'$ and that μ' turns Z into a group scheme.

It can be shown that Z is the kernel of the homomorphism f in the sense as this is understood in the theory of categories. Generally speaking, the category of commutative algebraic groups over a field k of finite characteristic is not an Abelian category, however, by extending it to the category of commutative group schemes over k we arrive at an Abelian category.

3. Separation. Finally we explain what is perhaps the most important application of the concept of a product, namely to the question of separation of schemes.

The image of the morphism $\Delta=(1,1)$; $X\to X\times_S X$ is called the *diagonal.*

A scheme X over S is called *separated* if its diagonal is closed. A scheme X is called separated if it is separated as a scheme over $\mathbb{Z}$.

The analogous condition for topological spaces defines Hausdorff spaces ([8], §8.1). In the case of schemes the meaning of this condition is slightly different: the topological space connected with the scheme is in any case almost never Hausdorff. To get a feeling for the meaning of the condition of separation we give an example of a non-separated scheme.

Let U_1 and U_2 be two copies of an affine line over a field k and let $U_{12} \subset U_1$, $U_{21} \subset U_2$ be open sets obtaining by deleting a point o (for some fixed choice of coordinates T_1 on U_1 and T_2 on U_2). The mapping φ that associates with every point $x \in U_{12}$ the point $u' \in U_{21}$ with the same coordinate is, of course, an isomorphism. Evidently the conditions necessary for pasting together U_1 and U_2 with respect to U_{12} and U_{21} are satisfied. As a result we obtain a scheme X over k, which is called an affine line with o as a branch point. In fact, it has two points o_1 and o_2 that are obtained from o in U_1 and U_2, respectively. Let us show that this scheme is not separated over k.

The closed points of the scheme $X \times_k X$ are of the form (x_1, x_2), where x_1 and x_2 are closed points in X, and the mapping Δ is given by $\Delta(x) = (x, x)$. Since the scheme X, by construction, is covered by two affine sets V_1 and V_2 isomorphic to U_1 and U_2, we see that $X \times X$ is covered by the four sets $V_1 \times V_1$, $V_1 \times V_2$, $V_2 \times V_1$, and $V_2 \times V_2$.

Consider, for example, the set $V_1 \times V_2$. It is isomorphic to $\mathbb{A}^1 \times \mathbb{A}^1$, and its intersection with $\Delta(X)$ consists of the points (x, x), $x \in V_1 \cap V_2 = U_{12}$. From this it is already clear that $\Delta(X)$ is not closed, in fact, not even its intersection with $V_1 \times V_2$ is closed. To complete the picture we can compute the closure of $\Delta(X)$. The closure of $\Delta(X) \cap (V_1 \times V_2)$ in $V_1 \times V_2$ is evidently obtained by adding the point (o_1, o_2). Considering in the same way all the four open sets $V_i \times V_j$, $i, j = 1, 2$, we find that the closure of $\Delta(X)$ is obtained by adding two points: (o_1, o_2) and (o_2, o_1). Hence it follows that the closure of $\Delta(X)$ is isomorphic to the line $\mathbb{A}^1$ in which the point o is "split" into four points: (o_1, o_1), (o_2, o_2), (o_1, o_2), (o_2, o_1), of which the first two belong to $\Delta(X)$, but the second two do not.

To give a clearer idea of the influence of non-separation on properties of a scheme we analyse this example of a scheme X in somewhat more detail. The fields $k(V_1)$ and $k(V_2)$ are isomorphic and determine a field, which we naturally call a field of rational functions on X. The local rings $\mathcal{O}_x$ of points $x \in X$ are subrings of this field. Let us find out what the rings $\mathcal{O}_{o_1}$ and $\mathcal{O}_{o_2}$ are.

Clearly $\mathcal{O}_{o_1}$ is the same as the local ring of o_1 on V_1. Since the isomorphism between U_{12} and U_{21} extends to the identity isomorphism between U_1 and U_2, here the functions in $\mathcal{O}_{o_1}$ correspond to functions in $\mathcal{O}_{o_2}$, and this means that $\mathcal{O}_{o_1} = \mathcal{O}_{o_2}$. Thus, two distinct points have one and the same local ring. Furthermore, any function in this ring takes at o_1 and o_2 the same value: these two points are not separable by means of rational functions. It can be shown that in the general case non-separation is connected with a similar phenomenon.

Let us now proceed to the general analysis of the concept of separation.

Proposition 1. *An affine scheme X over a ring B is separated and $\Delta: X \to X \times_B X$ is a closed embedding.*

Let $X = \operatorname{Spec} A$ and let A be a B-algebra. Since $X \times_B X = \operatorname{Spec}(A \otimes_B A)$, the morphism $\Delta: X \to X \times_B X$ is associated with a homomorphism $\lambda: A \otimes_B A \to A$. According to the definition λ is specified by the fact that

$$\lambda u = 1, \lambda v = 1 \tag{1}$$

for homomorphisms $u, v: A \to A \otimes_B A$

$$u(a) = a \otimes 1, v(a) = 1 \otimes a .$$

Hence it is easy to see that $\lambda(a \otimes b) = ab$. From this, as in fact already from (1), it follows that λ is an epimorphism, and this means that Δ is a closed embedding.

Since every scheme is covered by affine sets, which are separated, non-separation must be connected with certain properties of the pasting of affine schemes. This is corroborated by the following result in which we consider only the case when X is a scheme over an affine scheme $S = \operatorname{Spec} B$.

Proposition 2. *Let $X = \bigcup U_\alpha$ be an affine covering for which the following conditions hold:*

1) *all the sets $U_\alpha \cap U_\beta$ are affine and*

2) *the ring $\mathcal{O}_X(U_\alpha \cap U_\beta)$ is generated by the rings $\varrho^{U_\alpha}_{U_\alpha \cap U_\beta} \mathcal{O}_X(U_\alpha)$ and $\varrho^{U_\beta}_{U_\alpha \cap U_\beta} \mathcal{O}_X(U_\beta)$. Then the scheme X is separated over B.*

Let $u, v: X \times_B X \to X$ be the standard morphisms of a product. Then

$$\Delta^{-1}(u^{-1}(U_\alpha) \cap v^{-1}(U_\beta)) = \Delta^{-1}(u^{-1}(U_\alpha)) \cap \Delta^{-1}(v^{-1}(U_\beta)) = U_\alpha \cap U_\beta . \tag{2}$$

On the other hand, from the definition of a product it follows easily that for any open sets $U, V \subset X$ the open set $u^{-1}(U) \cap v^{-1}(V) \subset X \times X$ is isomorphic to $U \times V$. Together with (2) this shows that for separation of the scheme X it is sufficient that the restrictions $\Delta_{\alpha,\beta}$ of the morphism Δ to $U_\alpha \cap U_\beta$

$$\Delta_{\alpha,\beta}: U_\alpha \cap U_\beta \to U_\alpha \times_B U_\beta$$

should have a closed image. But by 1) $U_\alpha \cap U_\beta$ is affine: $U_\alpha \cap U_\beta = \operatorname{Spec} C_{\alpha,\beta}$, and by 2) the corresponding ring homomorphism $A_\alpha \otimes_B A_\beta \to C_{\alpha,\beta}$, $U_\alpha = \operatorname{Spec} A_\alpha$ is an epimorphism. But this means that $\Delta_{\alpha,\beta}$ is a closed embedding.

It is not hard to show that the converse is also true. We verify only one perfectly obvious, but useful, part of it: *in a separated scheme the intersection of two open affine sets is affine.* In fact,

$$U \cap V = \Delta^{-1}(U \times V) .$$

If U and V are affine, then so is $U \times V$, and if X is separated, then Δ is a closed embedding, therefore $U \cap V$ is a closed subscheme of an affine scheme. According to the proposition in §3.3, it is itself affine.

Now we turn our attention to an interesting feature of the criterion stated in Proposition 2: it does not depend on the morphism $X \to S$. Thus, the property of separation of a scheme X over an affine scheme S does not depend on the choice of S and the morphism $X \to S$. It could have been formulated by regarding X, for example, as a scheme over $\mathbb{Z}$.

An important application of Proposition 2 is the verification that the projective space $\mathbb{P}^n(A)$ over an arbitrary ring A is separated. In this case $\mathbb{P}^n(A) = \bigcup_{i=0,\ldots,n} U_i$, $U_i = \operatorname{Spec} A[T_0/T_i, \ldots, T_n/T_i]$. Since

$$U_i \cap U_j = \operatorname{Spec} A[T_0/T_i, \ldots, T_n/T_i]_{(T_j/T_i)},$$

this set is obviously affine. Now $\mathcal{O}_x(U_i \cap U_j)$ consists of elements $F(T_0, \ldots, T_n)/T_i^p T_j^q$, where $F \in A[T_0, \ldots, T_n]$ is a form of degree $p+q$. The rings $\varrho^{U_i}_{U_i \cap U_j} \mathcal{O}_X(U_i)$ and $\varrho^{U_j}_{U_i \cap U_j} \mathcal{O}_X(U_j)$ consist of elements F/T_i^p and G/T_j^q, where F and G are forms of degree p and q, respectively. Evidently they generate $\mathcal{O}_X(U_i \cap U_j)$.

It is easy to verify that in a separated scheme a closed subscheme and an open subset are separated. Hence it follows that quasiprojective varieties are separated.

We turn our attention to those properties of quasiprojective varieties that are connected with their separation. Particularly often we have used the fact that a regular mapping is uniquely determined by its restriction to any open dense subset. The corresponding property of schemes is closely connected with separation. Namely, if a scheme X is separated, then for any scheme Y and morphisms $f: Y \to X$, $g: Y \to X$ the set $Z \subset Y$ consisting of the points for which $f(y) = g(y)$ is closed. For we have the morphisms $(f, g): Y \to X \times X$, and Z is the inverse image of the diagonal under this morphism.

This shows that only for separated schemes are rational morphisms a natural generalization of morphisms. If a scheme X is not separated, then two distinct morphisms $Y \to X$ may define one and the same rational morphism.

Another frequently occurring property is that a regular mapping has a closed graph. If $f: Y \to X$ is a morphism of schemes, its graph is defined as the image of the morphism $(1, f): Y \to Y \times X$. It is the inverse image of the diagonal in $X \times X$ relative to the morphism

$$f \times 1 : Y \times X \to X \times X \, (f \times 1) = (f \circ p_Y, p_X),$$

where $p_Y : Y \times X \to Y$ and $p_X : Y \times X \to X$ are the natural projections. Thus, the graph of a morphism is closed if X is a separated scheme.

Exercises

1. Let X and Y be schemes over an algebraically closed field k. Show that the correspondence $u \to (p_X(u), p_Y(u))$ establishes a one-to-one correspondence between the closed points of the scheme $X \times_k Y$ and the pairs (x, y) where x and y are closed points in X and Y, respectively.

2. Find all the points of the scheme $\operatorname{Spec} \mathbb{C} \times_{\mathbb{R}} \operatorname{Spec} \mathbb{C}$, where $\mathbb{C}$ and $\mathbb{R}$ are the fields of real and complex numbers.

3. Let X be an affine group scheme over an affine scheme $S = \operatorname{Spec} B$, $X = \operatorname{Spec} A$, A an algebra over B. Show that the group law determines a homomorphism $\mu: A \to A \otimes_B A$, the unit morphism determines a homomorphism $\varepsilon: A \to B$, and the inverse element an automorphism $i: A \to A$. Formulate condition (1), (2), and (3) of §4.2 in terms of these homomorphisms.

4. Show that the kernel Y of the homomorphism $G_a \to G_a: x \to x^p$, as constructed in §4.2, is an affine group scheme, $Y = \operatorname{Spec} A$, $A = k[T]/(T^p)$. Compute in this case all the homomorphisms introduced in Exercise 3.

5. Consider the analogous homomorphism of the multiplicative groups $G_m \to G_m$, $x \to x^p$, and compute its kernel Y'. Show that the group schemes Y (see Exercise 4) and Y' are not isomorphic.

6. Let k be a field of characteristic 2. Show that to within an isomorphism there exist only two group schemes $X = \operatorname{Spec} A$, where $A = k[T]/T^2$, namely the schemes Y and Y' (Exercises 4 and 5).

7. Show that the non-separated scheme considered in §4.3 coincides with the scheme of Exercise 3 in §3.

8. Show that the scheme $\operatorname{Proj} \Gamma$ is always separated (Exercise 8 to §3).

Chapter VI. Varieties

§ 1. Definition and Examples

1. Definitions. In this chapter we consider schemes that are more closely connected with quasiprojective varieties. These schemes are called algebraic varieties. It is precisely this concept that we arrive at in trying to give an invariant definition of an algebraic variety.

Definition. *A variety* over an algebraically closed field is a reduced separated scheme of finite type over k.

A *morphism* of varieties is a morphism as schemes over k.

A variety X which is an affine scheme is called an *affine variety*.

As we have seen in Ch. V, § 3, every quasiprojective variety determines a scheme. This scheme is a variety, which we also call quasiprojective.

By definition, every variety X has a finite covering $X=\bigcup U_i$, where the U_i are affine varieties. From this it follows that X is of finite dimension. If X is irreducible, then all the U_i are dense in X and $\dim X=\dim U_i$. Furthermore, they all are birationally isomorphic, because $U_i \cap U_j$ is open and dense both in U_i and U_j. Therefore the fields of rational functions $k(U_i)$ are isomorphic to each other and can be identified. The field so obtained is called the *field of rational functions* on X and is denoted by $k(X)$. The dimension of X is equal to the transcendence degree of $k(X)$.

A closed point of a variety X that belongs to an affine open subset U is also closed in U and is a point of the corresponding affine variety with coordinates in k. There are plenty of such points on X.

Proposition. *The closed points are dense in any closed subset.*

Observe, first of all, that in an affine variety (and even in an affine scheme) every non-empty closed subset has a closed point. For a non-empty closed subset Z of the space $\operatorname{Spec} A$ is of the form $\operatorname{Spec} B$, where B is a quotient ring of A. Since every ring has a maximal ideal, Z has a closed point.

If X is an arbitrary variety, $Z \subset X$ a closed subset, and $z \in Z$, then it is sufficient to show that $Z \cap U$ contains a closed point for every neighbourhood U of z. We may restrict ourselves to affine U because they form a basis of all open sets. For affine U, by the preceding, $Z \cap U$ has a closed point. But here is a hidden danger: the point may be closed in U, but not in X. This actually occurs, for example, in the case of the set $U = \operatorname{Spec} \mathcal{O} - \{x\}$, where $\mathcal{O}$ is the local ring of a closed point x of a curve. Fortunately, in the case of varieties all is well: if a point $z \in X$ is closed in some neighbourhood U of it, then it is also closed in X. This follows from the fact that closed points x of varieties are characterized by the property $k(x) = k$. For a point x is closed in X if and only if it is closed in all affine open sets containing it, and for an affine variety the condition $k(x) = k$ evidently characterizes the closed points. The field $k(x)$ depends only on the local ring of x, hence does not change when X is replaced by an open set U containing x. This proves the proposition.

Since a variety is a reduced scheme, the elements $f \in \mathcal{O}_X(U)$ are uniquely characterized by their values $f(x) \in k(x)$ for all $x \in U$. According to the proposition they are characterized by their values at closed points. But then $k(x) = k$, so that the elements $f \in \mathcal{O}_X(U)$ can be interpreted as functions on the set of closed points with values in k.

If $\varphi: X \to Y$ is a morphism of varieties, $x \in X$ and $y = \varphi(x)$, then the homomorphism of local rings $\varphi^*: \mathcal{O}_y \to \mathcal{O}_x$ determines a field embedding $k(y) \to k(x)$. If x is a closed point, then $k(x) = k$, hence $k(y) = k$, that is, y is also closed. Therefore the image of a closed point is closed. Thus, by interpreting the elements $f \in \mathcal{O}_Y(U)$ as functions on closed points we can define a homomorphism $\psi_U: \mathcal{O}_Y(U) \to \mathcal{O}_X(\varphi^{-1}(U))$ by the condition $\psi_U(f)(x) = f(\varphi(x))$. In other words, the specification of the mapping $\varphi: X \to Y$ and even of its restriction to the set of closed points determines a morphism.

Of course, a variety X has masses of non-reduced closed subschemes. But every closed subset $Z \subset X$ can be made into a reduced scheme, or as we shall say henceforth, into a *closed subvariety*. If X is an affine variety, $X = \operatorname{Spec} A$, $Z = V(\mathfrak{a})$, then we set $Z = \operatorname{Spec} A/N'$, where N' consists of all the elements $a \in A$, a power of which is contained in $\mathfrak{a}$ (the inverse image of the nil-radical of the ring $A/\mathfrak{a}$). The general case is obtained by pasting together.

All this shows how close varieties are to quasiprojective varieties. Indeed, all the local concepts and properties analysed in Ch. II, the concept of a simple point, the theorem that the set of singular points is closed, the properties of normal varieties, are preserved verbatim for algebraic varieties. The same applies to properties of divisors and differential forms.

The only properties whose transfer to varieties is not obvious are those that are connected with projectiveness. Let us clarify at once what conditions replace projectiveness in the case of arbitrary varieties.

The property of being projective is, of course, far from "abstract". But there is one proposition at our disposal, namely Theorem 3 in Ch. I, § 5, that gives an "intrinsic" characterization of projective varieties. We recall the definition:

A variety X is called *complete* if for every variety Y the projection morphism $p: X \times Y \to Y$ carries closed sets into closed sets.

The basic properties of projective varieties: closure of the image, absence of non-constant everywhere regular functions ($\mathcal{O}_X(X)=k$), were derived from Theorem 3 in Ch. I, § 5.2, and are therefore true for complete varieties. Note that in proving closure of the image we made use of the fact that the graph is closed. As we have seen in Ch. V, § 3, this follows from the fact that varieties are separated.

Among all properties of projective varieties proved in Ch. I–IV there is only one in which projectiveness is used directly and not via an application of Theorem 3 in Ch. I, § 5: this is the very important Theorem 3 of Ch. II, § 3. Next we show how it generalizes to arbitrary complete varieties.

Theorem 1. *If X is a smooth irreducible variety and $\varphi: X \to Y$ a rational morphism of it into a complete variety, then the set of points at which φ is not defined is of codimension not less than* 2.

Let $V \subset X$ be the set of points at which φ is defined, Γ_φ the graph of the morphism $\varphi: V \to Y$ in $V \times Y$, and Z its closure in $X \times Y$, which we regard as a closed subset of $X \times Y$. The image of Z under the projection $p: X \times Y \to X$ is closed because Y is complete. Since $p(Z) \supset V$, we see that $p(Z)=X$. The restriction $p: Z \to p(Z)$ is a birational isomorphism: it is an isomorphism between Γ_φ and V. The theorem is a consequence of the following result:

Lemma. *If $p: Z \to X$ is an epimorphism and a birational isomorphism, and if X is a smooth variety, then the set of points at which the rational morphism p^{-1} is not defined is of codimension not less than two.*

For $\varphi = qp^{-1}$, where q is the restriction to Z of the projection $X \times Y \to Y$. Therefore φ is defined at those points at which p^{-1} is defined.

Proof of the Lemma. We assume that there exists a subvariety $T \subset X$ of codimension 1 such that p^{-1} is not defined at any of its points. By replacing Z, X, and T by affine open subsets we may assume that they are affine and that $T \subset p(Z) \subset X$. Let $Z \subset \mathbb{A}^m$ and let $u_1, \dots, u_m$ be coordinates in $\mathbb{A}^m$ as elements of $\mathcal{O}_Z(Z)$. We consider a point $t \in T$ and represent the rational functions $(p^{-1})^*(u_i)$ in the form

$$(p^{-1})^*(u_i) = g_i/h,$$

where $g_1, \ldots, g_m, h \in \mathcal{O}_t$ are coprime. Then

$$h \cdot (p^{-1})^* (u_i) = g_i , \qquad p^*(h) \cdot u_i = p^*(g_i) .$$

Therefore $g_i(\tau) = 0$ for all points $\tau \in T$ at which $h(\tau) = 0$, and this contradicts the fact that the elements $g_1, \ldots, g_m, h \in \mathcal{O}_t$ are coprime.

The varieties we have introduced turn out to be so close in their properties to quasiprojective varieties that the question arises: are perhaps the two concepts identical? A little later we shall show in § 2.3 that this is not so: there exist varieties that cannot be embedded in any projective space. However, what is much more important, owing to its invariant, intrinsic character, the concept of a variety turns out to be a vastly more flexible instrument. Many constructions can be carried out very simply and naturally within the framework of this concept. Post factum it can be shown sometimes that we need not go beyond the framework of quasiprojective or projective varieties; however, often this is only of secondary interest. In the following three subsections we give some important examples of such constructions.

For the simplest example we can point to the definition of the product of varieties. Within the context of varieties the definition is very simple: the arguments of Ch. V, § 4.1, simplify considerably if we make use of the fact that the set of closed points of the variety $X \times Y$ is of the form (x, y), where x and y are closed points in X and Y, respectively (see Exercises 1 and 2). In Ch. I, § 5, we had to spend a considerable effort on this definition, because we had to convince ourselves that the product of quasiprojective varieties is again quasiprojective.

Another example we can consider here is the concept of *normalization* of a variety. Let X be an irreducible variety, and K a finite extension of the field $k(X)$. We show that there exists a normal irreducible variety and a morphism ν_K of it, $\nu_K : X^\nu_K \to X$, such that $k(X^\nu_K) = K$ and that the embedding $\nu^*_K : k(X) \to k(X^\nu_K)$ coincides with the given embedding $k(X) \to K$. This variety is unique: for two normalizations X^ν_K and $\tilde{X}^\nu_K$ there exists an isomorphism $f : X^\nu_K \to \tilde{X}^\nu_K$ such that the diagram

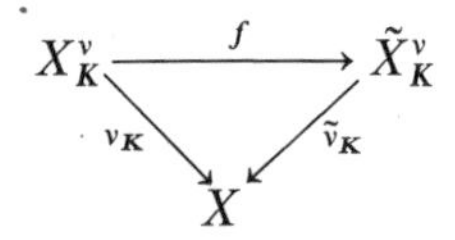

commutes. The variety X^ν_K is called the normalization of X in K.

The uniqueness is proved word-for-word as in Ch. II, § 5.2, where we considered the case $K = k(X)$. To prove the existence we consider an affine covering $X = \bigcup U_i$. The integral closure A^ν_i of the ring $k[U_i]$ in K is a finitely generated algebra, as we have seen in Ch. II, § 5.2.

Therefore, in the field K the normalization $\nu_{K,i}: U^\nu_{K,i} \to U_i$ of the affine variety U_i exists and is affine. From the uniqueness of the normalization it follows that $\nu^{-1}_{K,i}(U_i \cap U_j)$ and $\nu^{-1}_{K,j}(U_i \cap U_j)$ are isomorphic. This makes it possible to paste the varieties $U^\nu_{K,i}$ into a single scheme X^ν_K, which obviously is a reduced irreducible scheme of finite type over k. Let us show that the scheme X^ν_K is separated.

We have to show that in $X^\nu_K \times X^\nu_K$ the diagonal is closed, and for this purpose it is sufficient to verify that the diagonal is closed in a neighbourhood of every point $\xi \in X^\nu_K \times X^\nu_K$. Suppose that the morphism $\nu \times \nu: X^\nu_K \times X^\nu_K \to X \times X$ carries ξ into $\eta \in X \times X$ and that U' is an affine neighbourhood of η such that $(\nu \times \nu)^{-1}(U') = V'$ is affine. Its existence follows from that of the normalization in the affine case. Since X is separated, the scheme $U = \Delta \cap U'$ is closed in U', hence affine. From this it follows that the scheme $(\nu \times \nu)^{-1}(U)$ is also affine and a fortiori so is its irreducible component V containing ξ. Let $\delta^\nu: X^\nu_K \to X^\nu_K \times X^\nu_K$ and $\delta: X \to X \times X$ be the diagonal morphisms. We set $W = (\delta^\nu)^{-1}(V) = \nu^{-1}\delta^{-1}(U)$. So we obtain the commutative diagram

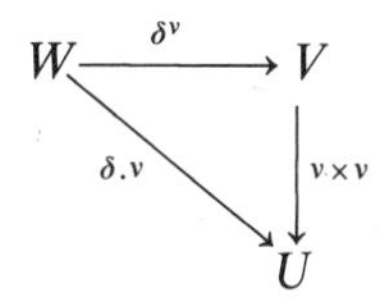

in which the morphism δ^ν corresponds to a finite regular mapping of affine varieties. All the more this is true for the morphism $\delta^\nu: W \to V$ (a module that is finite over a ring is a fortiori finite over a larger ring). Applying Theorem 4 of Ch. I, § 5 we see that $\delta^\nu(W) = V$, and this shows that the diagonal is closed in the neighbourhood V' of ξ.

Thus, the scheme X^ν_K is an irreducible variety, and as a trivial verification shows, it is the required normalization.

So we see that in the context of arbitrary varieties the construction of the normalization is trivial. The question remains whether the normalization of a quasiprojective variety is quasiprojective. This is true, but we shall not give the proof here, which naturally is based on purely projective arguments. It can be found, for example, in [20] Ch. V, § 4.

In the case of curves we can repeat the proofs of Theorem 6 and 7 in Ch. II, § 5.3. They show that the normalization (in its function field) of any irreducible curve is quasiprojective, and for a complete curve projective. In particular, if the curve is smooth, then it is quasiprojective. In fact, this is true for arbitrary curves, but the proof is more complicated, and we omit it here.

2. Vector Bundles. One of the most important constructions of algebraic varieties having a typical non-projective character is a vector bundle. We recall that the general concept of a fibration does not differ at all from a morphism of varieties $p: X \to S$ or from that of a variety over S. We are interested in fibrations whose fibres are vector spaces. In formulating this concept we have to bear in mind that an n-dimensional vector space over a field k has the natural structure of an algebraic variety isomorphic to $\mathbb{A}^n$.

Definition. A *family of vector spaces* is a fibration $p: E \to X$ in which every fibre $p^{-1}(x)$, $x \in X$, has the structure of a vector space over $k(x)$, and the corresponding structure of an algebraic variety is the same as the structure of $p^{-1}(x)$ as inverse image of the point x under the morphism p.

The fibre $p^{-1}(x)$ is denoted by E_x.

A *morphism* f of a family $p: E \to X$ into a family $q: F \to X$ is a morphism $f: E \to F$ for which the diagram

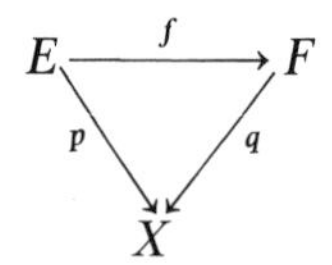

is commutative (in particular, f maps E_x into F_x) and the mapping $f_x: E_x \to F_x$ is linear over $k(x)$.

It is obvious how to define an isomorphism of families.

The simplest example of a family is the direct product $E = X \times V$, where V is a vector space over k, and p is the projection of $X \times V$ onto X. This family and any family isomorphic to it are called *trivial*.

Example 1. Let V and W be two linear spaces of dimension m and n, respectively. Let us find the common form of a morphism of the trivial families $f: X \times V \to X \times W$. We choose bases $v_1, \ldots, v_m$ in V and $w_1, \ldots, w_n$ in W and denote the corresponding coordinates by $\xi_1, \ldots, \xi_m$ and $\eta_1, \ldots, \eta_n$. The projections $p: X \times V \to V$ and $q: X \times W \to W$ determine elements $x_i = p^* \xi_i \in \mathcal{O}_{X \times V}(X \times V)$ and $y_j = q^* \eta_j \in \mathcal{O}_{X \times W}(X \times W)$. It is obvious that closed points $\alpha \in X \times V$ and $\beta \in X \times W$ are uniquely determined by the values $x_i(\alpha) \in k$ and $y_j(\beta) \in k$. Therefore the morphism f is uniquely determined by the elements $f^* y_j \in \mathcal{O}_{X \times V}(X \times V)$.

The composition of the isomorphism $X \to X \times v_i$ and the embedding $X \times v_i \to X \times V$ determines a morphism $\varphi_i: X \to X \times V$. We set $a_{ij} = \varphi_i^* f^* y_j \in \mathcal{O}_X(X)$. Then

$$f^* y_j = \Sigma a_{ij} x_i \,. \tag{1}$$

For it is sufficient to verify this equality at all closed points $\alpha \in X \times V$, and there it follows at once from the definition of a morphism of families (linearity of the mapping f_x).

Conversely, every matrix (a_{ij}), $a_{ij} \in \mathcal{O}_X(X)$, leads by means of (1) to a morphism $f: X \times V \to X \times W$. Clearly we obtain an isomorphism if and only if $m=n$ and the determinant $\det(a_{ij})$ is an invertible element of the ring $\mathcal{O}_X(X)$.

If $p: E \to X$ is a family of vector spaces, then for every open set $U \subset X$ the fibration $p: p^{-1}(U) \to U$ is a family of vector spaces. It is called the *restriction* of the family E to U and is denoted by $E|_U$.

Definition. A family of vector spaces $p: E \to X$ is called a *vector bundle* if every point $x \in X$ has a neighbourhood U such that the restriction of the family E to U is trivial.

Obviously the dimension of the fibre E_x of a vector bundle is a locally constant function on X, in particular, is constant when X is connected. In that case the number $\dim E_x$ is called the rank of the bundle and is denoted by $\operatorname{rk} E$.

Example 2. Let V be an $(n+1)$-dimensional vector space, and $\mathbb{P}^n$ the projective space consisting of lines $l \subset V$. The line corresponding to a point $x \in \mathbb{P}^n$ is denoted by l_x. In $\mathbb{P}^n \times V$ we consider the set E of pairs (x, v) for which $x \in \mathbb{P}^n$ and $v \in V$, v being a closed point of l_x. Obviously, this is the set of closed points of some quasiprojective subvariety of $\mathbb{P}^n \times V$, which we also denote by E. The projection $\mathbb{P}^n \times V \to \mathbb{P}^n$ determines a morphism $p: E \to \mathbb{P}^n$. We show that $p: E \to \mathbb{P}^n$ is a vector bundle. We introduce in V a coordinate system $(x_0, \ldots, x_n)$, which we also take to be a system of homogeneous coordinates for $\mathbb{P}^n$. The restriction of E to the open set $\mathbb{A}^n_j \times V \neq 0$ consists of the points

$$\xi = ((x_0 : x_1 : \ldots : x_n);\ (y_0, \ldots, y_n)),\ t_i = x_i/x_j,\ y_i = t_i y_j\,,$$

and the mapping $\xi \to ((t_1, \ldots, t_n), y_0)$ determines an isomorphism of this family with $U_0 \times k$.

The rank of the bundle we have constructed is 1. The projection $\mathbb{P}^n \times V \to V$ determines a morphism $q: E \to V$. The reader can easily verify that this morphism is the same as the σ-process at the point

$$O = (0, \ldots, 0) \in V \quad \text{and} \quad q^{-1}(O) = \mathbb{P}^n \times 0\,.$$

We consider the vector bundle $p: E \to X$ and a morphism $f: X' \to X$. The product $E' = E \times_X X'$ has a morphism $p': E' \to X'$. This morphism determines a vector bundle. For $E|_U \simeq U \times V$, $U \subset X$, and if $U' = f^{-1}(U)$, then $E'|_{U'} = E \times_U U' \simeq U'X$. This bundle is also denoted by $f^*(E)$. Evidently, $\operatorname{rk} f^*(E) = \operatorname{rk} E$.

Example 3. Let X be a projective variety, $f: X \to \mathbb{P}^n$ a closed embedding of it in a projective space, and $p: E \to \mathbb{P}^n$ the bundle of Example 2. Then $f^*(E)$ is a bundle over X of rank 1. Generally speaking, it depends on the embedding f and is its most important invariant.

Since a vector bundle is locally trivial, it is pasted together from several trivial bundles. This leads to an effective method of constructing bundles.

Let $X = \bigcup U_\alpha$ be a covering for which the bundle $p: E \to X$ is trivial on each of the U_α. We fix an isomorphism

$$\varphi_\alpha : p^{-1}(U_\alpha) \simeq U_\alpha \times V.$$

For the intersection $U_\alpha \cap U_\beta$ we have two isomorphisms:

$$\varphi_{\alpha\beta} = \varphi_\alpha|_{p^{-1}(U_\alpha \cap U_\beta)}, \varphi_{\beta\alpha} = \varphi_\beta|_{p^{-1}(U_\alpha \cap U_\beta)} : p^{-1}(U_\alpha \cap U_\beta) \to (U_\alpha \cap U_\beta) \times V.$$

Therefore $\varphi_{\beta\alpha}\varphi_{\alpha\beta}^{-1}$ is an automorphism of the bundle $(U_\alpha \cap U_\beta) \times V$.

We now use the result of Example 1. Choosing a basis in the vector space V we write the automorphism $\varphi_{\beta\alpha}\varphi_{\alpha\beta}^{-1}$ in the form of a matrix $C_{\alpha,\beta}$ with coefficients in $\mathcal{O}_X(U_\alpha \cap U_\beta)$. Obviously these matrices satisfy the pasting conditions:

$$C_{\alpha,\alpha} = 1, \quad C_{\alpha,\gamma} = C_{\alpha,\beta} C_{\beta,\gamma} \quad \text{on} \quad U_\alpha \cap U_\beta \cap U_\gamma. \tag{2}$$

Conversely, the specification for arbitrary α and β of a matrix $C_{\alpha,\beta}$ with elements in $\mathcal{O}_X(U_\alpha \cap U_\beta)$ determines a vector bundle provided that the matrices satisfy the conditions (2).

The matrices $C_{\alpha,\beta}$ are called the *transition matrices* of the bundle.

It is easy to clarify the dependence of the matrices $C_{\alpha,\beta}$ on the choice of the isomorphisms φ_α. Other isomorphisms φ'_α have the form $\varphi'_\alpha = f_\alpha \varphi_\alpha$, where f_α is an automorphism of the trivial bundle $U_\alpha \times V$.

The automorphism f_α can be described by a matrix B_α with coefficients in $\mathcal{O}_X(U_\alpha)$ having an inverse of the same form. So we arrive at the new matrices

$$C'_{\alpha,\beta} = B_\alpha C_{\alpha,\beta} B_\beta^{-1}.$$

Conversely, every such change of the matrices $C_{\alpha,\beta}$ leads to an isomorphic bundle.

3. Bundles and Sheaves. Vector bundles are generalizations of vector spaces. Now we introduce an analogue of the points of a vector space.

Definition. A *section* of a vector bundle $p: E \to X$ is a morphism $s: X \to E$ for which $p \cdot s = 1$ on X.

In particular, $s(x) = o_x$ (the null vector of E_x) is a section, the so-called null section.

The set of sections of a bundle E is denoted by $\mathscr{L}(E)$.

Example 1. A section f of a trivial bundle $X \times k \to X$ of rank is simply a morphism of X into $\mathbb{A}^1$, that is, $f \in \mathcal{O}_X(X)$. Thus, $\mathscr{L}(X \times k) = \mathcal{O}_X(X)$. In particular, $\mathscr{L}(\mathbb{P}^n \times k) = k$, and similarly $\mathscr{L}(\mathbb{P}^n \times V) = V$.

Consider the bundle E of Example 2 above. Each of its sections $s: \mathbb{P}^n \to E$ determines, in particular, a section $s: \mathbb{P}^n \to \mathbb{P}^n \times V$, hence by Cor. 2, Ch. I, § 5.2, has the form $s(x) = (x, v)$ for some fixed $v \in V$. But since $s(x) \in E$, we see that $v \in l_x$ for all $x \in \mathbb{P}^n$, consequently $v = 0$. Thus, $\mathscr{L}(E) = 0$. This shows, in particular, that E is not isomorphic to the trivial bundle.

In terms of transition matrices a section s is given by associating with every set U_α a vector $s_\alpha = (f_{\alpha,1}, \ldots, f_{\alpha,n})$, $f_{\alpha,i} \in \mathcal{O}_X(U_\alpha)$, with $s_\alpha = C_{\alpha,\beta} s_\beta$ in $U_\alpha \cap U_\beta$.

Starting from the definition of a vector bundle it is easy to verify that for two sections s_1 and s_2 there exists a section $s_1 + s_2$ such that

$$(s_1 + s_2)(x) = s_1(x) + s_2(x)$$

for every point $x \in X$. The sum on the right-hand side has a meaning because $s_1(x), s_2(x) \in E_x$, and E_x is a vector space.

Similarly the equation

$$(f \cdot s)(x) = f(x)\, s(x)$$

determines a product of a section with an element $f \in \mathcal{O}_X(X)$.

Thus, the set $\mathscr{L}(E)$ is a module over $\mathcal{O}_X(X)$. Let us associate with every open set $U \subset X$ the collection $\mathscr{L}(E, U)$ of sections of the bundle E restricted to U. An obvious verification shows that we obtain a sheaf, which is denoted by $\mathscr{L}_E$. This is a sheaf of Abelian groups, but it also has a more delicate structure, which we shall now determine in the general case.

Definition. Let $\mathscr{F}$ be a sheaf of Abelian groups and $\mathscr{G}$ a sheaf of rings on a topological space X; suppose also that every open set $U \subset X$ determines a module structure over $\mathscr{G}(U)$ in $\mathscr{F}(U)$. Under these conditions $\mathscr{F}$ is called a *sheaf of modules* over $\mathscr{G}$ if the multiplication $\mathscr{F}(U) \otimes \mathscr{G}(U) \to \mathscr{F}(U)$ commutes with the restriction homomorphisms ϱ_U^V, that is, if the diagram

$$\begin{array}{ccc} \mathscr{F}(V) \otimes \mathscr{G}(V) & \longrightarrow & \mathscr{F}(V) \\ {\scriptstyle \varrho^V_{U,\mathscr{F}} \otimes \varrho^V_{U,\mathscr{G}}} \downarrow & & \downarrow {\scriptstyle \varrho^V_{U,\mathscr{F}}} \\ \mathscr{F}(U) \otimes \mathscr{G}(U) & \longrightarrow & \mathscr{F}(U) \end{array}$$

commutes for $U \subset V$. In that case every fibre $\mathscr{F}_x$ of the sheaf $\mathscr{F}$ is a module over the fibre $\mathscr{G}_x$ of $\mathscr{G}$.

A *homomorphism of two sheaves of modules* $\mathscr{F}'$ and $\mathscr{F}''$ over one and the same sheaf of rings $\mathscr{G}$ is a system of homomorphisms $\varphi_U : \mathscr{F}'(U) \to \mathscr{F}''(U)$ of modules over $\mathscr{G}(U)$ for which the diagram

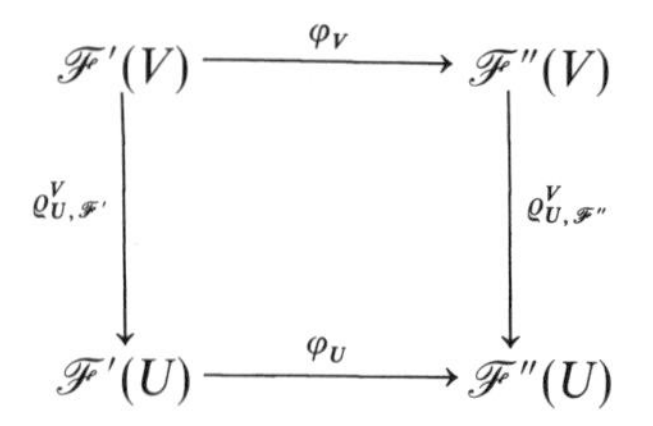

commutes for all $U \subset V$.

Clearly, the sheaf $\mathscr{L}_E$ corresponding to the bundle is a sheaf of modules over the structure sheaf $\mathcal{O}_X$.

Every operation that is invariantly defined over modules can be extended to sheaves of modules. In particular, for arbitrary modules over a ring A the following operations are defined:

$$M \oplus M_1, \; M \otimes_A M_1, \; M^* = \mathrm{Hom}(M, A), \; \Lambda_A^p M .$$

Applying them to modules $\mathscr{F}(U)$ and $\mathscr{F}_1(U)$ over rings, and taking the associated sheaves, we arrive at the sheaves $\mathscr{F} \oplus \mathscr{F}_1$, $\mathscr{F} \otimes_{\mathscr{G}} \mathscr{F}_1$, $\mathscr{F}^*$, $\Lambda_{\mathscr{G}}^p \mathscr{F}$, which are called, respectively, *the direct sum*, *the tensor product*, *the dual sheaf*, and *the exterior power*.

The sheaf of a trivial bundle of rank n is given by the fact that $\mathscr{L}_E(U) = \mathcal{O}(U)^n$, that is, $\mathscr{L}_E$ is the direct sum of n copies of $\mathcal{O}_X$. Such a sheaf is called *free* of rank n. Let $\mathscr{F}$ be a sheaf of modules over the structure sheaf $\mathcal{O}$. If each point has a neighbourhood U such that the sheaf $\mathscr{F}_{|U}$ is free and of finite rank, then $\mathscr{F}$ is called a *locally free* sheaf of finite rank. Evidently, if a sheaf $\mathscr{F}$ is locally free, then each of its fibres $\mathscr{F}_x$ is a free $\mathcal{O}_x$-module. The sheaf $\mathscr{L}_E$ corresponding to any vector bundle E is locally free of finite rank, because E is locally isomorphic to a trivial bundle.

Theorem 2. *The association $E \to \mathscr{L}_E$ establishes a one-to-one correspondence between vector bundles and locally free sheaves of finite rank (both considered to within an isomorphism).*

We now show how to associate a vector bundle to a locally free sheaf $\mathscr{F}$ of finite rank. Clearly X can be assumed to be connected. Suppose that $X = \bigcup U_\alpha$ is a covering such that $\mathscr{F}|_{U_\alpha}$ is a free sheaf and $\varphi_\alpha : \mathscr{F}|_{U_\alpha} \xrightarrow{\sim} \mathcal{O}_{U_\alpha}^{n_\alpha}$ the corresponding isomorphism. Then

$$\varphi_\beta \varphi_\alpha^{-1} : \mathcal{O}_{U_\alpha \cap U_\beta}^{n_\alpha} \to \mathcal{O}_{U_\alpha \cap U_\beta}^{n_\beta} \tag{1}$$

is an isomorphism of sheaves of modules. Since X is connected, it follows that all the numbers n_α are equal. We set $n_\alpha = n$. Every endomorphism of the sheaf of modules $\mathcal{O}_U^n$ is given by a matrix $C=(c_{ij})$, $c_{ij}\in\mathcal{O}_U(U)$. Thus, the isomorphism (1) determines a matrix $C_{\alpha,\beta}$, and it is evident that these matrices satisfy the relations (2) in § 1.2. Therefore they determine a vector bundle E. A trivial verification, which we omit, shows that $\mathcal{L}_E=\mathcal{F}$. This proves the theorem.

It is easy to check that the correspondence $E\to\mathcal{L}_E$ between vector bundles and locally free sheaves enables us to associate with every homomorphism of bundles a homomorphism of sheaves of modules over $\mathcal{O}_X$. In other words, this is an equivalence of the two catagories.

We observe that the fibre of the bundle and the stalk of the corresponding sheaf are totally different objects. For example, if $E=X\times k$, then $\mathcal{L}_E=\mathcal{O}_X$, $E_x=k$, and $(\mathcal{L}_E)_x=\mathcal{O}_x$. In the general case the fibre E_x can be recovered from the stalk $(\mathcal{L}_E)_x$ by means of the relations

$$E_x\simeq(\mathcal{L}_E)_x/\mathfrak{m}_x(\mathcal{L}_E)_x, \tag{2}$$

where $\mathfrak{m}_x$ is the maximal ideal of $\mathcal{O}_x$. It is sufficient to verify this locally, assuming that $E=U\times k^n$, $\mathcal{L}_E=\mathcal{O}_U^n$, and then it is obvious.

Theorem 2 yields a convenient method of constructing bundles.

Example 2. Let E and F be vector bundles, $\mathcal{L}_E$ and $\mathcal{L}_F$ the corresponding locally free sheaves. It is clear that the sheaves $\mathcal{L}_E\oplus\mathcal{L}_F$, $\mathcal{L}_E\otimes\mathcal{L}_F$, $\mathcal{L}_E^*$, $\Lambda_{\mathcal{O}}^p\mathcal{L}_E$ are locally free. Their corresponding bundles are denoted by $E\oplus F$, $E\otimes F$, E^*, $\Lambda^p E$. In the case $p=\operatorname{rk}E$, $\Lambda^p E$ is denoted by $\det E$.

If the bundles E and F in the covering $X=\bigcup U_\alpha$ are defined by the matrices $C_{\alpha,\beta}$ and $D_{\alpha,\beta}$, then the bundles $E\oplus F$, $E\otimes F$, E^* and $\Lambda^p E$ are given in the same covering by the matrices

$$\begin{pmatrix} C_{\alpha,\beta} & 0 \\ 0 & D_{\alpha,\beta}\end{pmatrix},\ C_{\alpha,\beta}\otimes D_{\alpha,\beta},\ (C_{\alpha,\beta})^{*-1},\ \Lambda^p C_{\alpha,\beta}. \tag{3}$$

For $p=\operatorname{rk}E$ the bundle $\Lambda^p E$ is given by the one-dimensional matrices $\det C_{\alpha,\beta}$.

From the relations (2) it follows that under these operations on the bundles the corresponding operation on vector spaces is performed in every fibre.

Example 3. Let X be a smooth variety. By assigning to an open set U the group $\Omega^p[U]$ of differentiable forms regular on U we evidently determine a sheaf of modules over $\mathcal{O}_X$. It is called the *sheaf of p-dimensional differential forms.*

Theorem 2 of Ch. III, § 4.3, asserts that this sheaf is locally free. Consequently, by Theorem 2 it determines a bundle, which is denoted by Ω^p. In particular, Ω^1 is called the *cotangent bundle*.

The stalk of the sheaf $\mathscr{F}$ of one-dimensional differential forms at a point $x \in X$ is of the form $\mathscr{F}_x = \mathcal{O}_x\, dt_1 + \cdots + \mathcal{O}_x\, dt_n$, where $t_1, \ldots, t_n$ are local parameters at x, and the sum is direct. The homomorphism $\mathscr{F}_x \to \mathscr{F}_x / \mathfrak{m}_x \mathscr{F}_x$ can be written in the form

$$u_1\, dt_1 + \cdots + u_n\, dt_n \to u_1(x)\, dt_1 + \cdots u_n(x)\, dt_n\,,$$

from which it follows by (2) that

$$\Omega_x^1 \simeq \mathscr{F}_x / \mathfrak{m}_x \mathscr{F}_x \simeq \mathfrak{m}_x / \mathfrak{m}_x^2\,. \tag{4}$$

Obviously $\Lambda^p \Omega^1 = \Omega^p$, $\det \Omega^1 = \Omega^n$, $n = \dim X$.

Example 4. The bundle dual to the cotangent bundle is called the *tangent bundle* and is denoted by Θ. By virtue of (4), for every point $x \in X$

$$\Theta_x = (\mathfrak{m}_x / \mathfrak{m}_x^2)^*\,,$$

that is, it is the tangent space at x.

The last general question we wish to discuss in connection with vector bundles are the concepts of subbundle and factor bundle.

Definition. If a morphism of bundles $\varphi: F \to E$ is a closed embedding of varieties, then it is called an *embedding of the bundles*. In this case $\varphi(F)$ is called a *subbundle* of E.

Proposition. *A subbundle $F \subset E$ of a vector bundle is locally a direct summand.*

The assertion is that every point $x \in X$ has a neighbourhood U and a fibration G over U such that

$$E|_U \simeq F|_U \oplus G\,. \tag{5}$$

By assumption we can suppose that both F and G are trivial on U, so $E \cong U \times W$, $F \cong U \times V$ and $f: U + V \to U \times W$ is a subbundle. Let $x_1, \ldots, x_m$ be coordinates in V and $y_1, \ldots, y_n$ in W. Since f is a closed embedding the associated $f^*: \mathcal{O}_{U \times W} \to \mathcal{O}_{U \times V}$ is an epimorphism. Let (a'_{ij}), $a'_{ij} \in \mathcal{O}(U)$, be the associated matrix. There is a matrix (b_{jk}), $b_{jk} \in \mathcal{O}(U)$, such that

$$X_k = \sum_{j=1}^{n} b_{jk} f^* y_j\,, \qquad k = 1, \ldots, n\,.$$

The matrix b_{jk} defines a map $g: U \times W \to U \times V$. Since $(a_{ij})(b_{jk}) = 1$, the composition $gf: U \times V \to U \times V$ is the identity. This shows that the associated sheaves of modules satisfy the relation

$$\mathscr{L}_E|_U \cong \mathscr{L}_F|_U \oplus \mathscr{F},$$

where $\mathscr{F}$ is a free sheaf of $\mathcal{O}_U$ modules.

Now we can define the *factor bundle* E/F with respect to a subbundle $F \subset E$.

As a set, of course,

$$E/F = \bigcup_{x \in X} E_x/F_x .$$

To introduce in it the structure of a variety we consider an open set U on which (5) holds, and we identify $\bigcup_{x \in U} E_x/F_x$ with an algebraic variety G. It is easy to verify that these structures are compatible on distinct open sets U and define E/F as a vector bundle.

The translation into the language of transition matrices is obvious. If we take a covering $X = \bigcup U_\alpha$ so that (5) is true for all U_α, then the matrices $C_{\alpha,\beta}$ defining E can be written in the form

$$C_{\alpha,\beta} = \begin{pmatrix} D_{\alpha,\beta} & 0 \\ * & D'_{\alpha,\beta} \end{pmatrix},$$

where $D_{\alpha,\beta}$ determines the bundle F, and $D'_{\alpha,\beta}$ the bundle E/F. Hence it follows at once that

$$\det E = \det F \det E/F . \tag{6}$$

Example 5. Let X be a smooth variety, and $Y \subset X$ a smooth closed subvariety. We define the *normal bundle* $N_{X/Y}$ to Y in X. The definition given in differential geometry is not applicable in the algebraic situation, because it is connected with the notion of the orthogonal complement $W^\perp$ of a linear subspace $W \subset V$. However, the space $W^\perp$ is defined so that it is isomorphic to V/W, and this we can utilize.

We denote by Θ'_X the restriction of the bundle Θ_X to the subvariety Y. It is defined as $j^*\Theta_X$, where $j: Y \to X$ is a closed embedding. The bundle Θ_Y is a subbundle of Θ'_X. For by definition $\Theta'_X = j^*\Theta_X = j^*((\Omega^1_X)^*) = (j^*\Omega^1_X)^*$. The restrictions of differential forms from X to Y determines a homomorphism $\varphi: j^*\Omega^1_X \to \Omega^1_Y$, and

$$\varphi^*: \Theta_Y = (\Omega^1_Y)^* \to (j^*\Omega^1_X)^* = \Theta'_X .$$

By definition,

$$N_{X/Y} = \Theta'_X/\Theta_Y .$$

Let us compute the transition matrices of the normal bundle. The homomorphism $\Theta'_X \to N_{X/Y}$ determines a homomorphism

$$\psi: N^*_{X/Y} \to j^*\Omega^1_X$$

of the dual bundles. It is easy to see that ψ determines a closed embedding, so that $N^*_{X/Y}$ can be regarded as a subbundle in $j^*\Omega^1_X$, and that Ω^1_Y is a factor with respect to this bundle. It is enough to verify these assertions on open sets, on which our bundles are trivial, and then they are obvious.

As we have seen, the forms $du_1, \dots, du_n$ are a basis of the $\mathcal{O}_X(U)$-module $\Omega^1_X[U]$, when the functions $u_1, \dots, u_n$ determine local parameters at an arbitrary point $x \in U$. This basis defines a basis $\eta_1, \dots, \eta_n$ of the $\mathcal{O}_Y(U \cap Y)$-module given by the sheaf corresponding to the bundle $j^*\Omega^1_X$. Here $\varphi(\eta_i)$ is the restriction of the form du_i to Y.

According to Theorem 5 of Ch. II, § 3, we can choose functions $u_1, \dots, u_n$ such that $u_1, \dots, u_m$ are local equations of Y in U. According to the same theorem the restrictions of the forms $du_{m+1}, \dots, du_n$ determine a basis in $\Omega^1_Y[U \cap Y]$, hence $\eta_1, \dots, \eta_m$ is a basis of the $\mathcal{O}_Y(U \cap Y)$-module $N^*_{X/Y}(U \cap Y)$.

Suppose that systems $u_{\alpha,1}, \dots, u_{\alpha,n}$ and $u_{\beta,1}, \dots, u_{\beta,n}$ are chosen in the sets U_α and U_β in the manner indicated. The transition matrix for the bundle Ω^1_X is determined by the expansion

$$du_{\alpha,i} = \sum_{j=i}^{n} c_{i,j}\, du_{\beta,j}, \quad i = 1, \dots, n, \quad c_{i,j} \in \mathcal{O}_X(U), \tag{7}$$

and the transition matrix for $j^*\Omega^1_X$ in the basis $\eta_1, \dots, \eta_n$ is obtained by restricting the elements of this matrix to $U \cap Y$.

Since $u_{\alpha,i} \in (u_{\beta,1}, \dots, u_{\beta,m})$, $i = 1, \dots, m$, on $U_\alpha \cap U_\beta$, we have

$$u_{\alpha,i} = \sum_{j=1}^{m} f_{i,j} u_{\beta,j}, \quad i = 1, \dots, m, \quad f_{i,j} \in \mathcal{O}_X(U_\alpha \cap U_\beta).$$

Hence

$$du_{\alpha,i} = \sum_{j=1}^{m} f_{i,j}\, du_{\beta,j} + \sum_{j=1}^{m} u_{\beta,j}\, df_{i,j}. \tag{8}$$

To make these formulae compatible with (7) we have to express $df_{i,j}$ in terms of $du_1, \dots, du_n$. But we are interested in the formulae for η_i that are obtained by restricting all the functions occuring in it to Y. Since $u_{\beta,j} = 0$ on Y, $j = 1, \dots, m$, the second group of terms in (8) disappears. Thus,

$$\eta_{\alpha,i} = \sum_{j=1}^{m} \bar{f}_{i,j} \eta_{\beta,j}, \quad i = 1, \dots, m,$$

where $\bar{f}_{i,j}$ is the restriction of $f_{i,j}$ to $U_\alpha \cap U_\beta \cap Y$. As we have seen, these are the transition matrices of the bundle $N^*_{X/Y}$. The matrices for $N_{X/Y}$ are obtained by transposition and inversion. The transition to the inverse matrix is equivalent to an interchange of the order of α and β.

Finally, we obtain the simple formula

$$C_{\alpha,\beta} = (h_{i,j}|_Y) \tag{9}$$

if

$$u_{\beta,j} = \Sigma\, h_{i,j} u_{\alpha,i} \quad \text{in} \quad U_\alpha \cap U_\beta\,.$$

Almost all constructions of this subsection are based fundamentally on the possibility of specifying the bundle abstractly, without embedding in a projective space. It can be shown, however, that a vector bundle over a quasiprojective variety is itself quasiprojective. We do not prove this here.

4. Divisors and Line Bundles. To every divisor D on an irreducible variety X there corresponds a linear space $\mathscr{L}(D)$ (we do not assume X to be smooth and we consider locally principal divisors). This association can be turned into a sheaf on the variety X. For this purpose we observe that a divisor D on X determines a divisor on any open subset $U \subset X$: we have to restrict the local equations of the divisor D to U. We denote the divisors so obtained by D_U and we set

$$\mathscr{L}_D(U) = \mathscr{L}(U, D_U)\,,$$

where $\mathscr{L}(U, D_U)$ is the space associated with the divisor D_U on the variety U. Obviously $\mathscr{L}_D(U) \subset k(X)$, and $\mathscr{L}_D(V) \subset \mathscr{L}_D(U)$ if $U \subset V$. We denote the inclusion of $\mathscr{L}_D(V)$ in $\mathscr{L}_D(U)$ by ϱ_U^V. The system $\{\mathscr{L}_D(U), \varrho_U^V\}$ determines a presheaf, in fact, a sheaf, as is easy to verify, which we denote by $\mathscr{L}_D$.

Multiplication of elements $f \in \mathscr{L}_D(U)$ by $h \in \mathscr{O}_X(U)$ turns $\mathscr{L}_D$ into a sheaf of modules over $\mathscr{O}_X$. This sheaf is locally free. For if D is defined on an open set U_α by the local equation f_α, then the elements $g \in \mathscr{L}_D(U_\alpha)$ are characterized by the condition $g f_\alpha \in \mathscr{O}_X(U_\alpha)$. This shows that the mapping $g \to g f_\alpha$ determines an isomorphism

$$\varphi_\alpha : \mathscr{L}_{D|U_\alpha} \to \mathscr{O}_{X|U_\alpha}\,. \tag{1}$$

In § 1.3 we have seen that this sheaf determines a vector bundle E_D, and from (1) it follows that $\operatorname{rk} E_D = 1$. Bundles of rank 1 are called *line bundles* (their fibres are straight lines). Let us write down the transition functions of the bundle E_D. Since the isomorphism (1) is given on U_α by multiplication by f_α, the automorphism $\varphi_\beta \varphi_\alpha^{-1}$ is given on $U_\alpha \cap U_\beta$ by multiplication by $f_\alpha^{-1} f_\beta$. Note that $f_\alpha^{-1} f_\beta \in \mathscr{O}_X(U_\alpha \cap U_\beta)$ in view of the consistency of the system f_α. Similarly $(f_\alpha^{-1} f_\beta)^{-1} = f_\beta^{-1} f_\alpha \in \mathscr{O}_X(U_\alpha \cap U_\beta)$. Thus, in this case the transition matrix $\varphi_{\alpha,\beta}$ of order one can be written in the form

$$\varphi_{\alpha,\beta} = f_\alpha^{-1} f_\beta\,. \tag{2}$$

If the divisor D is replaced by an equivalent divisor $D' = D + (f)$, $f \in k(X)$, then multiplication by f determines an isomorphism of the modules $\mathscr{L}(U, D_U)$ and $\mathscr{L}(U, D'_U)$. We have verified this in Ch. III, § 1.5. Evidently we obtain in this way an isomorphism of the sheaves $\mathscr{L}_D$ and $\mathscr{L}_{D'}$. The bundles E_D and $E_{D'}$ even have identical transition matrices. Thus, both the sheaf $\mathscr{L}_D$ and the bundle E_D correspond to an integral class of divisors.

Theorem 3. *The association $D \to \mathscr{L}_D \to E_D$ determines a one-to-one correspondence between* 1) *divisor classes,* 2) *classes (to within isomorphism) of sheaves of $\mathcal{O}_X$-modules, locally isomorphic to $\mathcal{O}_X$, and* 3) *classes of vector bundles of rank* 1.

The correspondence between the sets 2) and 3) was established in Theorem 2. Therefore it is enough for us to show that $D \to E_D$ determines a one-to-one correspondence between the sets 1) and 3). To show this we construct the inverse mapping.

Suppose that in the covering $X = \bigcup U_\alpha$ the line bundle E is given by transition matrices $\varphi_{\alpha,\beta}$ of order one, where $\varphi_{\alpha,\beta} \in \mathcal{O}_X(U_\alpha \cap U_\beta)$, $\varphi_{\alpha,\beta}^{-1} \in \mathcal{O}_X(U_\alpha \cap U_\beta)$. From the relations (2) in § 1.2 it follows that $\varphi_{\beta,\alpha} = \varphi_{\alpha,\beta}^{-1}$ and

$$\varphi_{\alpha,\beta} = \varphi_{\gamma,\alpha}^{-1} \varphi_{\gamma,\beta} \quad \text{on} \quad U_\alpha \cap U_\beta \cap U_\gamma . \tag{3}$$

We fix an index γ, which we denote by 0, and we set $\gamma = 0$ in (3). The embedding $\mathcal{O}_X(U_\alpha \cap U_\beta) \to k(X)$ enables us to regard the $\varphi_{\alpha,\beta}$ as elements of $k(X)$, and (3) holds for them, as before. We set $f_\alpha = \varphi_{0,\alpha}$. The system of elements f_α on the sets U_α is consistent, since

$$f_\alpha^{-1} f_\beta = \varphi_{\alpha,\beta} , \tag{4}$$

therefore determines a divisor D. Comparison of (2) and (4) shows that $E = E_D$.

We now prove that the divisor class D depends only on the bundle E and not on the choice of the covering and the transition matrices $\varphi_{\alpha,\beta}$. Two systems $\{\varphi_{\alpha,\beta}, U_\alpha\}$ and $\{\varphi'_{\lambda,\mu}, U'_\lambda\}$ can be compared on the covering $\{U_\alpha \cap U'_\lambda\}$ by setting $\tilde{\varphi}_{\alpha,\beta,\lambda,\mu} = \varphi_{\alpha,\beta}$, $\tilde{\varphi}'_{\alpha,\beta,\lambda,\mu} = \varphi'_{\lambda,\mu}$ on $U_\alpha \cap U_\beta \cap U'_\lambda \cap U'_\mu$. Therefore we may assume from the very beginning that the covering is common to the two cases: $X = \bigcup U_\alpha$. As was shown in § 1.2, we then have

$$\varphi'_{\alpha,\beta} = \psi_\alpha^{-1} \varphi_{\alpha,\beta} \psi_\beta, \; \psi_\alpha, \; \psi_\alpha^{-1} \in \mathcal{O}_X(U_\alpha) . \tag{5}$$

By definition of the functions f_α and f'_α,

$$f'_\alpha = \psi_0^{-1} \varphi_{0\alpha} \psi_\alpha = \psi_0^{-1} f_\alpha \psi_\alpha ,$$

and by (5), $D' = D - (\psi_0)$.

So we have actually constructed a mapping of the set 3) into 1). An obvious substitution shows that it is inverse to the mapping $D \to E_D$. This proves the theorem.

For any morphism $f: X \to Y$ the following relation holds:

$$f^* E_D = E_{f^* D}, \tag{6}$$

whose simple verification is left to the reader.

The class of divisors corresponding by Theorem 3 to the line bundle E is called its *characteristic class* and is denoted by $c(E)$.

Example 1. If $\dim X = n$, and Ω^n is the bundle introduced in § 1.2, then $c(\Omega^n) = K$ is the canonical class.

Example 2. Let X be a smooth variety and $Y \subset X$ a smooth hypersurface. In this case the normal bundle $N_{X/Y}$ is linear. Let us compute its characteristic class.

Suppose that Y is given in an affine covering $X = \bigcup U_\alpha$ by local equations f_α. Then $f_\alpha^{-1} f_\beta = f_{\alpha,\beta}$, where $f_{\alpha,\beta}, f_{\alpha,\beta}^{-1} \in \mathcal{O}(U_\alpha \cap U_\beta)$. According to the formulae (9) of § 1.3, the transition matrices of the bundle $N_{X/Y}$ have the form $f_{\alpha,\beta}|_Y = (f_\alpha^{-1} f_\beta)|_Y$. But we have just seen that $f_\alpha^{-1} f_\beta$ are the transition matrices for the bundle E_Y. So we have proved the formula

$$N_{X/Y} = E_Y|_Y .$$

By (6) it then follows that

$$c(N_{X/Y}) = \varrho_Y(C_Y),$$

where C_Y is the divisor class on X containing Y, and $\varrho_Y: \mathrm{Cl}(X) \to \mathrm{Cl}(Y)$ is the homomorphism of restriction to Y. We recall the explicit way of obtaining ϱ_Y: we have to replace Y by an equivalent divisor Y' not containing Y as a component, and then restrict Y' to Y.

Since the divisor classes form a group, the correspondence established in Theorem 3 determines a group operation also on the set of line bundles or sheaves locally isomorphic to $\mathcal{O}$. From (2) it is clear that addition of divisors corresponds to multiplication of one-dimensional transition matrices. In a more invariant form this operation is given as the tensor product of bundles or sheaves (see Theorem 2). Here multiplication of sheaves by $\mathcal{O}$ plays the role of the unit element, and the inverse to the sheaf $\mathscr{L}_D$ is $\mathscr{L}_{-D}$. Therefore locally free sheaves of $\mathcal{O}$-modules of rank 1 are also called *invertible sheaves.*

Although invertible sheaves and divisor classes correspond to each other in a one-to-one manner, the former are technically more convenient to use. For example, the inverse image $f^*\mathscr{F}$ is defined for every morphism f and every sheaf $\mathscr{F}$. It is easy to verify that if a sheaf $\mathscr{F}$ is invertible, then so is $f^*\mathscr{F}$. The corresponding operation on divisor classes requires for its definition arguments connected with a shift of the support of a divisor.

The technical advantages of invertible sheaves are connected with a fundamental phenomenon: in a closely related situation in the theory of complex analytic manifolds the concepts of an invertible sheaf and a divisor class are already inequivalent, and invertible sheaves give more information and lead to more natural problems. On this point see Exercises 6, 7, and 8 to Ch. VIII, § 2.

As an application of the preceding concepts we derive a relation which we have stated and used in Ch. IV, § 2.3.

Theorem 4. *The genus g_Y of a smooth curve Y on a smooth complete surface X can be expressed by the formula*

$$g_Y = \tfrac{1}{2}(Y+K, Y) + 1\,, \tag{7}$$

where K is the canonical class of X.

For a smooth subvariety $Y \subset X$ in a smooth variety X formula (6) of § 1.3 gives

$$\varrho_Y(\det \Theta_X) = \det \Theta'_X = \det \Theta_Y \det N_{X/Y}\,.$$

From formula (3) of § 1.3 it follows that $\det(E^*) = (\det E)^{-1}$ for every bundle. Since

$$\det \Omega_X^1 = \Lambda^n \Omega_X^1 = \Omega_X^n\,,$$

we obtain

$$\varrho_Y(c(\Omega_X^n)) = c(\Omega_Y^m) - c(\det N_{X/Y})\,,$$

where $\dim X = n$, $\dim Y = m$. Now let $m = n-1$. We make use of the results obtained in the discussion of Examples 1 and 2 and arrive at the relation

$$\varrho_Y(K_X) = K_Y - \varrho_Y(C_Y)\,. \tag{8}$$

Finally, if $n=2$, $m=1$, then the equality of the degrees of the divisors on the two sides of the equation follows from (8).

Recalling that $\deg \varrho_Y(D) = (Y, D)$ and that by the Riemann-Roch theorem $\deg K_Y = 2g_Y - 2$, we obtain

$$(Y, K) = 2g_Y - 2 - (Y^2)\,,$$

and the theorem follows.

Exercises

1. Let k be an algebraically closed field. We define a pseudovariety over k as a ringed space in which every point has a neighbourhood isomorphic to Specm A, where A is an algebra over k of finite type and without nilpotent elements, and where the topology and the sheaf on Specm A are defined word for word as in Ch. V. Show that by associating with every variety the set of its closed points we determine an isomorphism of the categories of varieties and pseudovarieties.

2. Define the product of two pseudovarieties X and Y starting out from the fact that $X \times Y$ consists of pairs (x, y), $x \in X$, $y \in Y$, constructing an affine covering of this set from affine coverings of X and Y, and using the definition of the product of affine varieties in Ch. I.

3. Prove that a variety is complete if and only if its irreducible components are complete.

4. A fibering $X \to S$ (not necessarily a vector bundle!) is called locally trivial if every point $s \in S$ has a neighbourhood U such that the restriction of X to U is isomorphic to $F \times U$ (as a scheme over U). Show that if the basis of S and the fibre of a locally trivial fibering X are complete, then so is X.

5. Determine the transition matrices of the bundle in Example 2 of § 1.2 corresponding to a covering of $\mathbb{P}^n$ by the sets $\mathbb{A}^n_i$. Find the characteristic class of this bundle.

6. Let D be a divisor on a variety X for which the space $\mathscr{L}(D)$ is finite dimensional, $\mathscr{F} = \mathscr{F}_D$ its corresponding invertible sheaf, and f a rational mapping in $\mathbb{P}^n$, $n = l(D) - 1$, that is associated with $\mathscr{L}(D)$ in accordance with Ch. III, § 1.5. Show that f is regular at those and only those points $x \in X$ for which the stalk $\mathscr{F}_x$ is generated over $\mathcal{O}_X$ by the space $\varrho_x \mathscr{L}(D)$.

7. Let X be a smooth affine variety: $X = \operatorname{Spec} A$. Show that the module $\Theta_x(X)$ over A is isomorphic to the module of derivations of A, that is, the k-linear mappings $d: A \to A$ for which $d(xy) = d(x) \cdot y + x \cdot d(y)$, $x, y \in A$.

8. Show that the normal bundle to a line C in $\mathbb{P}^n$ is a sum of $n-1$ isomorphic 1-dimensional bundles E. Find $c(E)$.

9. Suppose that $n-1$ hypersurfaces $C_1, \ldots, C_{n-1}$ of degree $m_1, \ldots, m_{n-1}$ in $\mathbb{P}^n$ intersect transversally with respect to a curve X. Find its genus.

10. Let $f: E \to X$ be a vector bundle and $X = \bigcup U_\alpha$ a covering over the elements of which E is trivial: $E|_{U_\alpha} \simeq U_\alpha \times k^n$. We embed k^n in $\mathbb{P}^n$ as points with $x_0 \neq 0$ and paste together the varieties $U_\alpha \times \mathbb{P}^n$ by means of the transition matrices $C_{\alpha,\beta}$ of E, which are now regarded as matrices of projective transformations in $\mathbb{P}^n$. Show that we can obtain in a suitable manner a variety $\tilde{E}$ in which E is an open subset such that $\tilde{E}$ is smooth, $f: \tilde{E} \to X$ is regular, and its fibre is isomorphic to $\mathbb{P}^n$.

11. In the notation of Exercise 10, let $X = \mathbb{P}^1$, E_n the bundle of rank 1 corresponding to the divisor nx_∞ on $\mathbb{P}^1$, $n > 0$. Show that $\tilde{E}_n - E_n$ is a curve which f maps isomorphically onto $\mathbb{P}^1$. Let C_0 be the null section of E_n, which evidentally is contained in $\tilde{E}_n$, and F the fibre of E_n. Show that on the surface $\tilde{E}_n: C_0 - C_\infty \sim nF$. Find (C_0^2) and (C_∞^2).

12. Show that in the notation of Exercise 11 the restriction of divisors $D \in \operatorname{Div} \tilde{E}_n$ to the common fibre determines a homomorphism $\operatorname{Cl} \tilde{E} \to \mathbb{Z}$ whose kernel is $\mathbb{Z} \cdot F$. Show that $\operatorname{Cl} \tilde{E}_n$ is a free group with the two generators C_0 and F.

13. In the notation of Exercises 10–12 find the canonical class of the surface $\tilde{E}_n$.

14. Show that the surfaces $\tilde{E}_n$ corresponding to distinct $n \geqslant 0$ are non-isomorphic. *Hint:* Show that on $\tilde{E}_n$ there is a unique irreducible curve with a negative square and that this square is $-n$.

§ 2. Abstract and Quasiprojective Varieties

1. Chow's Lemma. We prove a result which throws some light on the connections between complete and projective varieties. Of course, every irreducible variety is birationally isomorphic to a projective variety, for example, the projective closure of any of its open affine subsets. However, in this direction one can prove significantly more:

Chow's Lemma. *For every complete irreducible variety X there exists a projective variety $\bar{X}$ and an epimorphism $f: \bar{X} \to X$ that is a birational isomorphism.*

The idea of the proof is the same as that which we used to construct a projective embedding of a normalization of a curve.

Let $X = \bigcup U_i$ be a finite affine covering. For every affine variety $U_i \subset \mathbb{A}^{n_i}$ we denote by Y_i its closure in the projective space $\mathbb{P}^{n_i} \supset \mathbb{A}^{n_i}$. Obviously, the variety $Y = \Pi Y_i$ is projective.

We set $U = \bigcap U_i$. The embeddings $\psi: U \to X$ and $\psi_i: U \to U_i \subset Y_i$ determine a morphism

$$\varphi: U \to X \times Y, \qquad \varphi = \psi \times \Pi \psi_i .$$

We denote by $\bar{X}$ the closure of the set $\varphi(U)$ in $X \times Y$.

The projection $p_X: X \times Y \to X$ determines a morphism $f: \bar{X} \to X$. We show that it is a birational isomorphism. For this purpose it is sufficient to verify that

$$f^{-1}(U) = \varphi(U) . \tag{1}$$

For $p_X \varphi = 1$ on U, and by (1) f coincides on $f^{-1}(U)$ with the isomorphism φ^{-1}. The equation (1) is equivalent to the relation

$$(U \times Y) \cap \bar{X} = \varphi(U) , \tag{2}$$

that is, to the fact that $\varphi(U)$ is closed in $U \times Y$. But this is obvious, because $\varphi(U)$ coincides in $U \times Y$ with the graph of the morphism $\Pi \psi_k$. Here f is an epimorphism because $f(\bar{X}) \supset U$, and U is dense in X.

It remains to show that $\bar{X}$ is projective. To do this we use the projection $g: X \times Y \to Y$ and show that its restriction $\bar{g}: \bar{X} \to Y$ is a closed embedding. Since the notion of a closed embedding is local, it is sufficient to find open sets $V_i \subset Y$ such that $\bar{X} \subset \bigcup g^{-1}(V_i)$ and g determines a closed embedding of $\bar{X} \cap g^{-1}(V_i)$ in V_i. We set

$$V_i = p_i^{-1}(U_i) ,$$

where $p_i: Y \to Y_i$ are the projections. First of all, the $g^{-1}(V_i)$ cover $\bar{X}$. To see this it is enough to show that

$$g^{-1}(V_i) = f^{-1}(U_i) , \tag{3}$$

because $\bigcup U_i = X$ and $\bigcup f^{-1}(U_i) = \bar{X}$. In its turn, (3) follows from the fact that

$$f = p_i g \quad \text{on} \quad f^{-1}(U_i). \tag{4}$$

It is sufficient to verify (4) on some open subset $W \subset f^{-1}(U_i)$. In particular, we may take $W = f^{-1}(U) = \varphi(U)$ [in accordance with (1)], and then (4) is obvious.

Thus, it remains to verify that

$$g: \bar{X} \cap g^{-1}(V_i) \to V_i$$

determines a closed embedding. We recall that

$$V_i = p_i^{-1}(U_i) = U_i \times \hat{Y}_i, \quad \hat{Y}_i = \prod_{j \neq i} Y_j,$$
$$g^{-1}(V_i) = X \times U_i \times \hat{Y}_i.$$

We denote by Z_i the graph of the morphism $U_i \times \hat{Y}_i \to X$, which is the composition of the projection onto U_i and the embedding in X. The set Z_i is closed in $X \times U_i \times \hat{Y}_i = g^{-1}(V_i)$, and its projection onto $U_i \times \hat{Y}_i = V_i$ is an isomorphism. On the other hand, $\varphi(U) \subset Z_i$, and since Z_i is closed, we see that $\bar{X} \cap g^{-1}(V_i)$ is closed in Z_i. Therefore, the restriction of the projection to this set is a closed embedding, and Chow's lemma is proved.

Similar arguments show the analogous fact for arbitrary varieties, where $\bar{X}$ in this case is quasiprojective (see Exercise 7).

2. The σ-Process Along a Subvariety. Chow's lemma shows that an arbitrary variety is fairly close to a quasiprojective one. Nevertheless these are distinct concepts. Simple examples of non-quasiprojective varieties will be constructed in § 2.3. This construction makes use of a generalization of the σ-process defined in Ch. II, § 4. The difference consists in the fact that now we construct a morphism $\sigma: X' \to X$ for which the rational morphism σ^{-1} blows up not a point $x_0 \in X$ but a whole smooth subvariety. The construction follows very closely the case we have already discussed.

a) *The Local Construction.* According to Theorem 5 of Ch. II, § 3, for every closed point of a smooth subvariety Y of a smooth variety X there exists a neighbourhood U and functions $u_1, \ldots, u_m \in \mathcal{O}_X(U)$, $m = \operatorname{codim}_X Y$, such that $\mathfrak{a}_Y = (u_1, \ldots, u_m)$ in $\mathcal{O}_X(U)$ and that $d_x u_1, \ldots, d_x u_m$ are linearly independent at any closed point $x \in U$ (the latter condition means $u_1, \ldots, u_m$ can be included in a system of local parameters). If these conditions are satisfied, then we say that $u_1, \ldots, u_m$ are local parameters of the subvariety Y in U.

Suppose that X is affine and that Y has local parameters $u_1, \dots, u_m$ in the whole of X. We consider the product $X \times \mathbb{P}^{m-1}$, and in it the closed subvariety X' defined by the equations $t_i u_j(x) = t_j u_i(x)$, $i, j = 1, \dots, m$, where $(t_1, \dots, t_m)$ are homogeneous coordinates in $\mathbb{P}^{m-1}$. The projection $X \times \mathbb{P}^{m-1} \to X$ determines a morphism $\sigma: X' \to X$. Here $\sigma^{-1}(Y) = Y \times \mathbb{P}^{m-1}$, and σ determines an isomorphism

$$X' - (Y \times \mathbb{P}^{m-1}) \to X - Y.$$

If $x' = (y, z)$ is a closed point on X', $y \in X$, $z \in \mathbb{P}^{m-1}$, $z = (z_1 : \dots : z_m)$ and $z_i \neq 0$, then in a neighbourhood of x' we have $u_j = u_i s_j$, $s_j = t_j/t_i$. Let $v_1, \dots, v_{n-m}, u_1, \dots, u_m$ be a system of local parameters at a point y on X. Then the maximal ideal of the point x' on X' has the form

$$\begin{aligned} \mathfrak{m}_{x'} &= (v_1, \dots, v_{n-m}, u_1, \dots, u_m, s_1 - s_1(x'), \dots, s_m - s_m(x')) \\ &= (v_1, \dots, v_{n-m}, s_1 - s_1(x'), \dots, \widehat{s_i - s_i(x')}, u_i, \dots, s_m - s_m(x')). \end{aligned}$$

Hence, as in Ch. II, § 4.2, it follows that X' is smooth, n-dimensional, and irreducible. Just as there, so we have here:

Lemma. *If the σ-process $\tau: \bar{X} \to X$ is determined by another system of parameters $v_1, \dots, v_m$ of the same subvariety $Y \subset X$, then there exists an isomorphism $\varphi: X' \to \bar{X}$ for which*

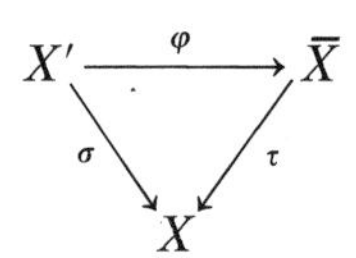

is commutative. This isomorphism is unique.

On the open sets $X' - \sigma^{-1}(Y)$ and $\bar{X} - \tau^{-1}(Y)$ we have $\varphi = \tau^{-1}\sigma$, and the uniqueness follows from this. By definition, in these sets

$$\varphi(x; t_1 : \dots : t_m) = (x; v_1(x) : \dots : v_m(x)),$$
$$\psi(x; t'_1 : \dots : t'_m) = (x; u_1(x) : \dots : u_m(x)),$$

where $\psi = \varphi^{-1}$.

By hypothesis,

$$v_l = \sum_j h_{l,j} u_j, \qquad h_{l,j} \in k[X]. \tag{1}$$

In the open set $t_i \neq 0$ we write $s_j = t_j/t_i$, we express (1) in the form

$$v_l = u_i g_l, \qquad g_l = \sum_j (\sigma^* h_{l,j}) s_j, \tag{2}$$

and we set

$$\varphi(x; t_1 : \dots : t_m) = (x; g_1 : \dots : g_m). \tag{3}$$

The same simple verification as in the proof of the analogous lemma in Ch. II, § 4.2, shows that φ is a morphism and coincides with the one already constructed on $X' - \sigma^{-1}(Y)$. The construction of ψ is similar.

b) *The Global Construction.* Let $X = \bigcup U_\alpha$ be an affine covering such that Y is defined in U_α by the equations $u_{\alpha,1}, \ldots, u_{\alpha,m}$. By applying to U_α and $Y \cap U_\alpha$ the construction under a) we obtain a system of varieties X'_α and morphisms $\sigma_\alpha: X'_\alpha \to U_\alpha$. The relation $X'_\alpha \supset \sigma_\alpha^{-1}(U_\alpha \cap U_\beta)$ holds for any α and β, and by the lemma there exist uniquely determined isomorphisms

$$\varphi_{\alpha,\beta}: \sigma_\alpha^{-1}(U_\alpha \cap U_\beta) \to \sigma_\beta^{-1}(U_\alpha \cap U_\beta).$$

It is easy to check that they satisfy the conditions for pasting together and determine a variety X' and a morphism $\sigma': X' \to X$. The so constructed morphism is called the σ-process with centre in Y. From the lemma it follows in an obvious way that neither X' nor σ depend on the covering $\{U_\alpha\}$ or the system of parameters $u_{\alpha,i}$.

c) *The Exceptional Subvariety.* The subvariety $\sigma^{-1}(Y)$ is known to us locally:

$$\sigma^{-1}(Y \cap U_\alpha) = (Y \cap U_\alpha) \times \mathbb{P}^{m-1}. \tag{4}$$

Globally we are concerned here with a fibering of a new type: $\sigma^{-1}(y)$, $y \in Y$, is a projective space. The relation (4) shows in what sense our fibering is locally trivial.

With every vector bundle $p: E \to X$ we can connect a fibering $\varphi: \mathbb{P}(E) \to X$ of this type. For this purpose we define $\mathbb{P}(E)$ as the set

$$\mathbb{P}(E) = \bigcup_{x \in X} \mathbb{P}(E_x),$$

where $\mathbb{P}(E_x)$ is the projective space of lines of the vector space E_x. To equip $\mathbb{P}(E)$ with the structure of an algebraic variety we consider a covering $X = \bigcup U_\alpha$ in which E is given by transition matrices $C_{\alpha,\beta}$. Having fixed an isomorphism $p^{-1}(U_\alpha) \simeq U_\alpha \times V$, where V is a vector space, we obtain a mapping

$$\bigcup_{x \in U_\alpha} \mathbb{P}(E_x) \xrightarrow{\sim} U_\alpha \times \mathbb{P}(V),$$

by means of which we can introduce in this space the structure of an algebraic variety. Obviously all these structures are compatible with each other and determine on $\mathbb{P}(E)$ a unique structure of an algebraic variety. This is called the *projectivization* of the vector bundle E.

Specifically, $\mathbb{P}(E)$ is pasted together from open sets

$$\varphi^{-1}(U_\alpha) \simeq U_\alpha \times \mathbb{P}(V)$$

by means of a rule for pasting that is determined by the automorphisms of the variety $(U_\alpha \cap U_\beta) \times \mathbb{P}(V)$:

$$\varphi_{\alpha,\beta}(u, \xi) = (u, \mathbb{P}(C_{\alpha,\beta})\, \xi), \tag{5}$$

where $u \in U_\alpha \cap U_\beta$, $\xi \in \mathbb{P}(V)$, and $\mathbb{P}(C_{\alpha,\beta})$ is the projective transformation with the matrix $C_{\alpha,\beta}$.

We return to the variety $\sigma^{-1}(Y)$ that arises in the σ-process $\sigma: X' \to X$. It is pasted together from the open sets $(Y \cap U_\alpha) \times \mathbb{P}^{m-1}$, and the rule of pasting is given by the formulae (1). This rule falls exactly under the type (5) if for $C_{\alpha,\beta}$ we take the matrix

$$C_{\alpha,\beta} = (h_{i,j}|_Y).$$

The functions $h_{i,j}$ are defined by (1), and a single glance at the transition matrices of the normal bundle—formulae (9) in § 1.3—is sufficient to convince us that the $C_{\alpha,\beta}$ correspond to the bundle $N_{X/Y}$. Thus, we may express the result of our discussion by the simple formula

$$\sigma^{-1}(Y) \simeq \mathbb{P}(N_{X/Y}).$$

d) *The Behaviour of Subvarieties.*

Proposition. *Let Z be a closed irreducible smooth subvariety of X, transversal to Y at each of their points of intersection, $\sigma: X' \to X$ the σ-process with centre in Y. Then the subvariety $\sigma^{-1}(Z)$ consists of two irreducible components:*

$$\sigma^{-1}(Z) = \sigma^{-1}(Y \cap Z) \cup Z',$$

and $\sigma: Z' \to Z$ determines the σ-process of the variety Z with centre in $Y \cap Z$.

The proof follows very closely the arguments in Ch. II, § 4.3.

Our problem is local, therefore we may take it that $Y \cap Z$ has in X the local parameters $u_1, \ldots, u_l$, and that among them $u_1, \ldots, u_r, \ldots, u_m$ are parameters for Y, and $u_{r+1}, \ldots, u_m, \ldots, u_l$ for Z. Then X' is determined in $X \times \mathbb{P}^{m-1}$ by the equations

$$t_i u_j = t_j u_i, \quad i, j = 1, \ldots, m. \tag{6}$$

We denote by $\bar{Z}$ the closure of the set $\sigma^{-1}(Z - (Y \cap Z))$. Evidently $\sigma^{-1}(Z) = \sigma^{-1}(Y \cap Z) \cup \bar{Z}$. Since $u_{r+1} = \cdots = u_l = 0$ at every point of $\sigma^{-1}(Z - (Y \cap Z))$ and at least one of $u_1, \ldots, u_r \neq 0$, we have on $\bar{Z}$

$$t_{r+1} = \cdots = t_m = 0.$$

Therefore

$$\bar{Z} \subset Z \times \mathbb{P}^{r-1},$$

where $t_1, \ldots, t_r$ are homogeneous coordinates in $\mathbb{P}^{r-1}$ and on $\bar{Z}$ the following relations hold:

$$t_i u_j = t_j u_i\,, \qquad i, j = 1, \ldots, r\,.$$

These relations determine the σ-process $\sigma: Z' \to Z$ with centre in $Y \cap Z$. So we see that $\bar{Z} \subset Z'$, and since both varieties have one and the same dimension and Z' is irreducible, we have $\bar{Z} = Z'$.

The subvariety $Z' \subset X'$ is called the *proper inverse image* of the subvariety $Z \subset X$ under the σ-process.

In conclusion we make a few remarks in connection with the notion of a σ-process.

1. It can be shown that a σ-process does not lead us out of the class of quasiprojective varieties. We do not give a proof here.

2. The existence of σ-processes whose centres are not points creates a whole range of new difficulties in the theory of birational isomorphisms of varieties of dimension greater than 2. In particular it is not known to what degree the results we have obtained in Ch. IV, § 3.4 for surfaces can be carried over to them. It is only known that not every morphism $X \to Y$ that is a birational isomorphism splits into a product of σ-processes. A relevant example was constructed by Hironaka. Whether it is true that every birational isomorphism is a product of σ-processes and their inverse morphisms is unknown at present. On the other hand, the theorem on the elimination of points of inderminacy by means of σ-processes is true in any dimension if k is a field of characteristic 0; this was also proved by Hironaka.

3. Example of a Non-Quasiprojective Variety. The variety we are going to construct by way of example is complete. If a complete variety were isomorphic to a quasiprojective variety, then by the theorem on the closure of the image it would be projective. Consequently, it is sufficient to construct an example of a complete but non-projective variety.

The proof that it is non-projective is based on the fact that the intersection indices on projective varieties have certain specific properties. Therefore we begin with some general remarks on intersection indices.

We shall make use of concepts that are a very special case of the ring of classes of cycles of which we have talked in Ch. IV, § 2.3. In our special case the required definitions are easy to give independently. Let X by a three-dimensional smooth complete variety, C an irreducible curve, and D a divisor on X. We assume that $C \not\subset \operatorname{Supp} D$. Then the restriction $\varrho_C(D)$ defines a locally principal divisor on C (we do not assume C to be smooth) for which the intersection index is defined (see the remark in connection with the definition of intersection index, Ch. IV, § 1.1). In this case the intersection index is denoted by $\deg \varrho_C(D)$ and is also called the intersection index of the curve C and the divisor D:

$$(C, D) = \deg \varrho_C(D)\,.$$

The arguments of Ch. IV, § 1 show that this index as a function of D is additive and invariant under equivalence. In particular, the index (C, Δ) is defined, where Δ is a divisor class containing D. Besides, in the application we need only the case when C is a smooth curve, and then both these properties are obvious.

We consider the free Abelian group A^I generated by all the curves $C \subset X$. For an element $a \in A^I$ the index (a, Δ), $\Delta \subset Cl\,X$, is defined by additivity. On A^I we introduce an equivalence relation: $a \approx b$ if $(a, \Delta) = (b, \Delta)$ for any divisor class Δ. In that case a and b are called *numerically equivalent*.

We consider an example, which is basic for what follows. If $a = \Sigma n_i C_i$, $a' = \Sigma n'_j C'_j$, and if all the curves C_i, C'_j lie on a smooth surface $Y \subset X$ and $a \sim a'$ as divisors on Y, then $a \approx a'$. Indeed, for every divisor D on X the restriction operation $\varrho^X_{C_i}(D)$ can be carried out in two stages:

$$\varrho^X_{C_i} = \varrho^Y_{C_i} \cdot \varrho^X_Y\,,$$

hence for $a \in \operatorname{Div} Y$

$$(a, D)_X = (a, \varrho^X_Y D)_Y.$$

From this our assertion follows by virtue of the invariance of the intersection index on Y under equivalence of divisors.

The preceding arguments referred to any complete variety X. The projectiveness of a variety X implies an important property: if $a = \Sigma n_i C_i$, $n_i > 0$, then $a \not\approx 0$. In fact, for the intersection of an irreducible curve C with a hyperplane section H of X the formula

$$(C, H) = \deg C$$

is obvious, in particular, $(C, H) > 0$. Therefore also $(a, H) = \Sigma n_i (C_i, H) > 0$.

Before proceeding to the construction of the example we consider an auxiliary construction. Let C_1 and C_2 by two smooth curves in a smooth three-dimensional variety V, where C_1 and C_2 intersect transversally at x_0. Suppose that the curves C_1 and C_2 are rational. Although our results are true independently of this fact, this assumption simplifies the deductions somewhat. We denote by $\sigma: V' \to V$ the σ-process with centre in C_1. According to the proposition in § 2.2 $\sigma^{-1}(C_2)$ consists of two components:

$$\sigma^{-1}(C_2) = \sigma^{-1}(x_0) \cup C'_2\,,$$

where $\sigma: C_2' \to C_2$ is the σ-process at $x_0 \in C_2$. Hence in our case it is an isomorphism. We denote the surface $\sigma^{-1}(C_1)$ by S_1. As a very simple exercise on the formula defining the σ-process we recommend the reader to verify that S_1 and C_2' intersect in the single point $\bar{x}_0$, $\sigma(\bar{x}_0) = x_0$, and transversally. We denote the fibre of the morphism

$$\sigma: S_1 \to C_1$$

at each point $x \in C_1$ by k_x. Since we have assumed that C_1 is a rational curve, any two points on it are equivalent: $x' \sim x''$, hence on S_1

$$k_{x'} \sim k_{x''} .$$

The situation in which we find ourselves is illustrated in Fig. 16.

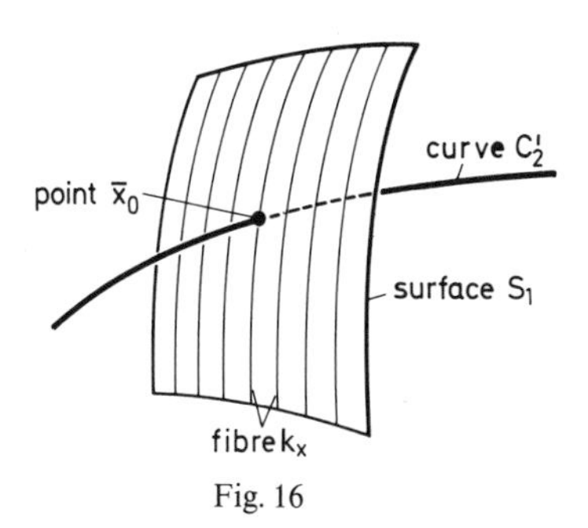

Fig. 16

We now consider the σ-process of the variety V' with centre in C_2':

$$\bar{\sigma}: \bar{V} \to V' .$$

The inverse image $(\bar{\sigma})^{-1}(S_1)$ of the surface S_1 is irreducible: according to the proposition in §2.2, $(\bar{\sigma})^{-1}(S_1) = (\bar{\sigma})^{-1}(\bar{x}_0) \cup S_1'$, and $\bar{\sigma}: S_1' \to S_1$ is the σ-process of the surface S_1 with centre at $\bar{x}_0$. Hence it follows that $(\bar{\sigma})^{-1}(\bar{x}_0) \subset S_1'$. On the surface S_1' we have $(\bar{\sigma})^{-1}(k_{x_0}) = \bar{l} \cup \bar{l}'$, where $\bar{l} = (\bar{\sigma})^{-1}(\bar{x}_0)$, and $\bar{\sigma}: \bar{l}' \to k_{x_0}$ is an isomorphism. For $x \neq x_0$ the fibre $\bar{\sigma}^{-1}(k_x)$ is irreducible. We denote it by $\bar{l}_x'$. By the preceding we have on S_1'

$$\bar{k}_x \sim \bar{l}' + \bar{l} . \tag{1}$$

We denote by S_2 the surface $(\bar{\sigma})^{-1}(C_2')$. Like S_1, it is stratified over C_2' into fibres $\bar{l}_y'$, $y \in C_2'$, and on S_2

$$\bar{k}_{y_1} \sim \bar{k}_{y_2} \quad \text{and} \quad \bar{k}_{\bar{x}_0} = \bar{l} . \tag{2}$$

The surfaces S_1' and S_2 intersect on the line $\bar{l}$. Their mutual disposition is illustrated in Fig. 17.

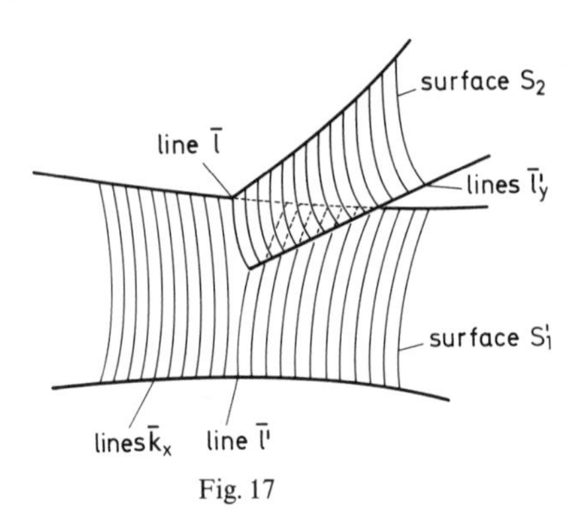

Fig. 17

Now let us go over to numerical equivalence. Substituting (1) in (2) we find that

$$\bar{k}_x \approx \bar{l}' + \bar{k}_y . \tag{3}$$

The basic feature of this relation is its asymmetry relative to $\bar{k}_x$ and $\bar{l}'$, which is connected with the order in which we have carried out the σ-processes. We utilise this in the example, which we are now about to construct.

We consider a smooth three-dimensional variety V and in it two smooth rational curves C_1 and C_2, which intersect transversally in two points x_0 and x_1 (for example, $V \supset \mathbb{P}^2$, C_1 and C_2 are a line and a conic in $\mathbb{P}^2$). In the variety $V_0 = V - x_1$ we carry out, as before, σ-processes first in $C_1 - x_1$, and then in the proper inverse image of the curve $C_2 - x_1$. So we obtain a morphism

$$\sigma_0 : \bar{V}_0 \to V - x_1 .$$

In $V_1 = V - x_0$ we carry out the σ-processes in the opposite order: first with centre in $C_2 - x_0$, and then in the proper inverse image of the curve $C_1 - x_0$. So we obtain a morphism

$$\sigma_1 : \bar{V}_1 \to V - x_0 .$$

Evidently, the varieties $\sigma_0^{-1}(V - x_0 - x_1)$ and $\sigma_1^{-1}(V - x_0 - x_1)$ are isomorphic, and the morphisms σ_0 and σ_1 agree on them. For the curve $C_1 \cup C_2 - \{x_0, x_1\}$ is disconnected, therefore, both $\sigma_0^{-1}(V - x_0 - x_1)$ and $\sigma_1^{-1}(V - x_0 - x_1)$ can be obtained by carrying out in $V - x_0 - x_1$ the σ-process with centre in $C_1 - x_0 - x_1$ in the open set $V - C_2$, and the σ-process with centre in $C_2 - x_0 - x_1$ in the open set $V - C_1$, and then pasting together the resulting varieties with respect to the set $V - (C_1 \cup C_2)$, on which the two σ-processes agree.

Thus, we may paste the varieties $\bar{V}_0$ and $\bar{V}_1$ with respect to the open subsets $\sigma_0^{-1}(V-x_0-x_1)$ and $\sigma_1^{-1}(V-x_0-x_1)$ and obtain a variety $\bar{V}$ and a morphism

$$\sigma: \bar{V} \to V.$$

In $\bar{V}$ the relation (3) holds, which we have derived using the existence of a common point x_0 on the curves C_1 and C_2. Similarly, the existence of the point x_1 leads to the relation

$$\bar{k}_y \approx \bar{\bar{l}}' + \bar{k}_x, \tag{4}$$

where $\bar{\bar{l}}'$ is an irreducible curve. Substituting one relation in the other we find that

$$\bar{k}_x \approx \bar{l}' + \bar{\bar{l}}' + \bar{k}_x,$$

hence

$$\bar{l}' + \bar{\bar{l}}' \approx 0. \tag{5}$$

To arrive at a contradiction to the assumption that $\bar{V}$ is projective it remains to show that it is complete. For any variety Z the projection $\bar{V} \times Z \to Z$ can be split up into the composition of the mappings $(\sigma, 1): \bar{V} \times Z \to V \times Z$ and the projection $V \times Z \to Z$. Since V is projective, the image of a closed set under the second projection is closed, and we need only show the analogous property for $(\sigma, 1)$. We know that V is the union of the two open sets $V-x_0$ and $V-x_1$, and since the notion of closure is of local character, it is sufficient to verify that

$$(\sigma, l): (\sigma, l)^{-1}((V-x_i) \times Z) \to (V-x_i) \times Z, \quad i=0,1,$$

carries closed sets into closed sets. On the sets $V-x_i$ the morphism σ coincides with the composition of σ-processes, and it remains to show that for any σ-process $\sigma: U' \to U$ and any Z the morphism

$$(\sigma, 1): U' \times Z \to U \times Z$$

carries closed sets into closed sets. Again the local character of the problem allows us to assume that σ is given by the construction under a) in § 2.2, that is, $U' \subset U \times \mathbb{P}^{m-1}$ and σ is induced by the projection $U \times \mathbb{P}^{m-1} \to U$. But then our assertion follows from the fact that a projective space is complete: Theorem 3 of Ch. I, § 5.

Thus, if X were quasiprojective, then it would be projective, but this is impossible, because the relation (5) cannot hold in a projective variety.

The foundation of the argument on which the example is constructed are, of course, the relations (3) and (4). They lead to (5), which cannot hold on projective varieties. Perhaps these relations

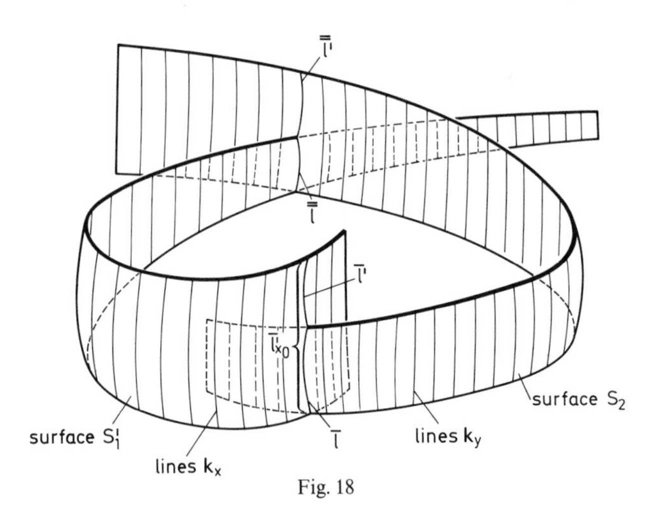

Fig. 18

become clearer if they are illustrated in a very primitive way (Fig. 18). Here the splitting of the fibre k_{x_0} into two components is shown as the composition of the segment k_{x_0} from two segments: $\bar{l}$ and $\bar{l}'$.

Remarks. 1. The dimension 3 in this example is not accidental. It can be shown that a 2-dimensional smooth complete variety is projective. On the other hand, there exist examples of complete, but not projective, two-dimensional varieties with singular points.

2. In our example we consider an affine open subset $U \subset \bar{V}$. If both the curves $\bar{l}'$ and $\bar{\bar{l}}'$ in (5) were to intersect U, then we could find a divisor D for which $(\bar{l}' \cdot D) > 0$, $(\bar{\bar{l}}' \cdot D) > 0$, which contradicts (5). For D we could take the closure of a hyperplane section of that affine space in which U lies. Thus, $\bar{l}'$ and $\bar{\bar{l}}'$ lie "very far apart" in $\bar{V}$: if an open affine subset contains at least one point of $\bar{l}'$, then it does not intersect $\bar{\bar{l}}'$.

4. Criteria for Projectiveness. In conclusion we give some criteria, which characterize projective varieties among arbitrary complete varieties. We do not state them in the greatest possible generality. In particular, in the first two we assume the variety to be smooth. This could be omitted, but it would require some additional explanations.

1. *Criterion of Chevalley-Kleiman.* A smooth complete variety is projective if and only if any finite set of its points is contained in an affine open subset.

Evidently, on a projective variety X there always exists a hyperplane section H that does not contain a given finite set S, so that $S \subset X - H$, and $X - H$ is affine. Therefore one half of the criterion is obvious. In the example of the non-projective variety we have constructed above this criterion obviously does not hold (see Remark 2 after the example).

2. *Criterion of Nakai-Moishezon.* A smooth complete variety X is projective if and only if on it there exists a divisor H such that for every closed subvariety Y

$$((\varrho_Y(H))^m)_Y > 0\,, \quad m = \dim Y.$$

Here $\varrho_Y(H)$ denotes the restriction of H to Y.

For projective varieties H can be taken to be a hyperplane section. In that case

$$(\varrho_Y(H)^m)_Y = \deg Y.$$

Therefore the criterion evidently holds for projective varieties.

In formulating the last criterion we recall that on the projective space $\mathbb{P}^n$ there is defined a line bundle $E \subset \mathbb{P}^n \times V$, where V is the vector space whose lines are represented by points of $\mathbb{P}^n$ (Examples 1 and 2 of § 1.4). Here the projection $\mathbb{P}^n \times V \to V$ determines a morphism $E \to V$, which coincides with the σ-process in V with centre at the origin of coordinates. In this representation the only exceptional subvariety is the null section of the bundle E. Let $X \subset \mathbb{P}^n$ be a closed subvariety. The bundle $E' = \varrho_X(E)$, the restriction of E to X, is a closed subset of E, and the σ-process $\sigma: E \to V$ determines a morphism $\sigma': E' \to V$. From the completeness of a projective space it follows that σ carries closed set into closed sets. Therefore, $V' = \sigma'(E')$ is an affine variety. Obviously, the only exceptional subvariety for σ' is the null section.

These arguments establish the "only if" part of the following criterion.

3. *Criterion of Grauert.* A complete variety is projective if and only if on it there exists a line bundle E and a morphism $f: E \to V$ onto an affine variety V such that f is a birational isomorphism and its only exceptional subvariety is the null section of the bundle E.

More briefly, the condition of Grauert's criterion can be stated as contractibility of the null section of E to a point.

Exercises

1. Give a new proof of Theorem 1 in § 1, using Chow's lemma and a reduction to Theorem 3 in Ch. II, § 3.

2. Show that if X is a complete variety and $\sigma: X' \to X$ a σ-process, then X' is also a complete variety.

3. Show that $\mathbb{P}(E) \simeq \mathbb{P}(E')$ if E is a vector bundle, and $E' = E \otimes L$, where L is a vector bundle of rank 1.

4. Let X be a smooth complete variety, $\dim X = 3$, $Y \subset X$ a smooth curve, $\sigma: X' \to X$ a σ-process with centre in Y, $y_0 \in Y$, $l = \sigma^{-1}(y_0)$. Show that $(l, \sigma^* D) = 0$, where D is any divisor on X and σ^* its inverse image on X'.

5. Under the conditions of Exercise 4, let $S = \sigma^{-1}(Y)$. Show that $(l, S) = -1$. *Hint:* Consider a surface D passing through Y that is smooth at y_0, and apply to it the result of Exercise 4.

6. Show that for every smooth projective 3-dimensional variety there exists a non-projective variety birationally isomorphic to it.

7. Show that for any irreducible variety X there exists a quasiprojective variety $\bar{X}$ and an epimorphism $f: \bar{X} \to X$ that is a birational isomorphism. There exists an embedding $\bar{X} \subset \mathbb{P}^n \times X$ such that f is the restriction to $\bar{X}$ of the projection $\mathbb{P}^n \times X \to X$.

§ 3. Coherent Sheaves

1. Sheaves of Modules. In connection with vector bundles we have come across sheaves of bundles over a sheaf of rings $\mathcal{O}_X$. Such sheaves are a particularly convenient tool in the investigation of algebraic varieties. One example will be in this section. We now begin with some general properties of these sheaves.

Let us consider a very general situation: a ringed space, that is, a topological space X on which a sheaf $\mathcal{O}$ of rings is given. Later we shall analyse sheaves on X that are sheaves of modules over $\mathcal{O}$. We do not say so explicitly and simply speak of sheaves of modules. It is clear that every sheaf of Abelian groups over a topological space X can be regarded as a sheaf of modules over a sheaf of rings $\mathcal{O}$, if we take for $\mathcal{O}$ the sheaf of locally constant functions with values in $\mathbb{Z}$.

The definition of a homomorphism $f: \mathcal{F} \to \mathcal{G}$ of sheaves of modules was given in § 1.3. We recall that this is a system of homomorphisms $f_U: \mathcal{F}(U) \to \mathcal{G}(U)$ of modules over $\mathcal{O}(U)$ satisfying certain conditions of compatibility.

Example 1. Let X be a smooth algebraic variety over a field k, $\mathcal{O}_X$ the sheaf of regular functions, Ω^1 the sheaf of one-dimensional regular differential forms. By assigning to $f \in \mathcal{O}_X(U)$ the differential $df \in \Omega^1(U)$ we define a homomorphism of sheaves

$$d: \mathcal{O}_X \to \Omega^1 .$$

It is a homomorphism of sheaves of modules over the sheaf of locally constant functions with values in k, but not over the sheaf $\mathcal{O}_X$.

Our next aim is to define the kernel and image of a homomorphism of sheaves of modules. The first definition is perfectly obvious. Let $f: \mathcal{F} \to \mathcal{G}$ be a homomorphism of sheaves of modules. We set $\mathcal{K}(U) = \operatorname{Ker} f_U$. From the definition of a homomorphism it follows that for $U \subset V$ we have $\varrho_U^V \mathcal{K}(V) \subset \mathcal{K}(U)$. Therefore the system $\{\mathcal{K}(U), \varrho_U^V\}$ determines a presheaf. A simple verification shows that it is a sheaf of modules. By definition this is the *kernel of the homomorphism* f.

The kernel of a homomorphism is an example of a *subsheaf* of $\mathscr{F}$. This is the name for a sheaf of modules $\mathscr{F}'$ for which $\mathscr{F}'(U) \subset \mathscr{F}(U)$ for all open sets $U \subset X$, where the homomorphisms $\varrho^V_{U,\mathscr{F}'}$ are the restrictions of $\varrho^V_{U,\mathscr{F}}$ to the modules of $\mathscr{F}'(V)$.

The matter is somewhat more complicated with the concept of the image of a homomorphism $f: \mathscr{F} \to \mathscr{G}$. The fact is that the $\mathscr{O}(U)$-modules $\mathscr{I}(U) = \operatorname{Im} f_U$, together with the homomorphisms $\varrho^V_{U,\mathscr{G}}$, determine a presheaf, which, speaking generally, is not a sheaf.

Example 2. Let X be a topological space, for the time being quite arbitrary, $\mathscr{O}$ the sheaf of locally constant functions with values in the field of real numbers $\mathbb{R}$, $\mathscr{F}$ and $\mathscr{G}$ sheaves of continuous functions with values in $\mathbb{R}$ and in the circle $\mathrm{G} = \mathbb{R}/\mathbb{Z}$ respectively. A homomorphism $f: \mathscr{F} \to \mathscr{G}$ is given by the fact that for $U \subset X$ and $\varphi \in \mathscr{F}(U)$

$$f_U(\varphi)(x) \equiv \varphi(x) \pmod{\mathbb{Z}}, \qquad x \in U.$$

For every element $\psi \in \mathscr{G}(U)$ and every point $x \in U$ there exists a neighbourhood V_x of x such that the set $\psi(V_x) \subset \mathrm{G}$ is not the whole of G. Then there exists a set $T \subset \mathbb{R}$ such that the projection $p: \mathbb{R} \to \mathbb{R}/\mathbb{Z} = \mathrm{G}$ determines a homeomorphism $p: T \to \psi(V_x)$. The function $\varphi = p^{-1}\psi$ then belongs to $\mathscr{F}(V_x)$, and $f_{V_x}(\varphi) = \psi$. In other words, $\operatorname{Im} f_{V_x} = \mathscr{G}(V_x)$.

On the other hand, in general, $\operatorname{Im} f_U \neq \mathscr{G}(U)$. For example, let $X = \mathrm{G}$ and let $\xi \in \mathscr{G}(X)$ be the identity mapping. It is easy to see that this cannot be "lifted" to $\mathbb{R}$, that is, there is no continuous function $\varphi: \mathrm{G} \to \mathbb{R}$ such that $\varphi(x) \equiv x \pmod{\mathbb{Z}}$. As we have seen, there exists a covering $\mathrm{G} = \bigcup V_\alpha$ such that $\varphi_\alpha = \varrho^G_{V_\alpha}\varepsilon \in \operatorname{Im} f_{V_\alpha}$. Evidently these functions are compatible on the intersections $V_\alpha \cap V_\beta$. However, there is no function $\varphi \in \operatorname{Im} f_G$ for which $\varrho^G_{V_\alpha}\varphi = \varphi_\alpha$. Condition 2) in the definition of a sheaf (Ch. V, § 2.3) is not satisfied.

There is a natural way of defining the image of a homomorphism $f: \mathscr{F} \to \mathscr{G}$ of sheaves of modules. We defined a presheaf $\mathscr{I}'$ by the condition

$$\mathscr{I}'(U) = f_U(\mathscr{F}(U)), \qquad U \subset X.$$

The sheaf $\mathscr{I}$ associated with the presheaf $\mathscr{I}'$ (Ch. V, § 2.4) is called the *image of the homomorphism* f and is denoted by $\operatorname{Im} f$.

Recalling the definition of the sheaf associated with a presheaf we see that $\operatorname{Im} f$ is a subsheaf of $\mathscr{G}$ and that $(\operatorname{Im} f)(U)$ consists of the elements $a \in \mathscr{G}(U)$ such that every point $x \in U$ has a neighbourhood U_x for which

$$\varrho^U_{U_x}(a) \in f_{U_x}(\mathscr{F}(U)_x)).$$

It is obvious that f determines a homomorphism

$$\mathscr{F} \to \operatorname{Im} f.$$

From the definition it follows at once that a homomorphism $f: \mathscr{F} \to \mathscr{G}$ for which $\operatorname{Ker} f = 0$ and $\operatorname{Im} f = \mathscr{G}$ is an isomorphism.

A sequence of homomorphisms $\mathscr{F}_1 \xrightarrow{f_1} \mathscr{F}_2 \cdots \xrightarrow{f_n} \mathscr{F}_{n+1}$ is said to be *exact* If $\operatorname{Im} f_i = \operatorname{Ker} f_{i+1}$, $i = 1, \ldots, n$. If the sequence $0 \to \mathscr{F} \xrightarrow{f} \mathscr{G} \xrightarrow{g} \mathscr{H} \to 0$ is exact, then $\mathscr{F}$ can be regarded as a subsheaf of $\mathscr{G}$. Consequently

$$(\operatorname{Im} f)(U) = f(\mathscr{F}(U)),$$

that is, in the construction of the sheaf $\operatorname{Im} f$ the transition from the presheaf to its associated sheaf is superfluous. Therefore the sequence

$$0 \to \mathscr{F}(U) \xrightarrow{f_U} \mathscr{G}(U) \xrightarrow{g_U} \mathscr{H}(U) \tag{1}$$

is exact for every open set U.

Example 2 shows that the sequence

$$0 \to \mathscr{F}(U) \xrightarrow{f_U} \mathscr{G}(U) \xrightarrow{g_U} \mathscr{H}(U) \to 0,$$

is not exact, in general (for example, when $U = X$). This fact is the reason why there is a non-trivial cohomology theory of sheaves.

For every subsheaf $\mathscr{F}$ of a sheaf $\mathscr{G}$ we can construct a homomorphism $f: \mathscr{G} \to \mathscr{H}$ such that $\operatorname{Ker} f = \mathscr{F}$, $\operatorname{Im} f = \mathscr{H}$. To construct it we set

$$\mathscr{H}'(U) = \mathscr{G}(U)/\mathscr{F}(U)$$

and define the homomorphisms $\varrho^V_{U,\mathscr{H}'}$ as the result of the action of the homomorphisms $\varrho^V_{U,\mathscr{G}}$ on these factor groups. This defines the presheaf $\mathscr{H}'$. For $\mathscr{H}$ we choose its associated sheaf.

It is easy to verify that for the stalks of these sheaves we have the relation

$$\mathscr{H}_x = \mathscr{G}_x/\mathscr{F}_x.$$

Therefore, an element $a \in \mathscr{G}(U)$ determines elements $a_x \in \mathscr{H}_x$ for all points $x \in U$. An obvious verification shows that all the elements $\{a_x\}$ give an element $a' \in \mathscr{H}(U)$, and $f: a \to a'$ is a homomorphism with the required properties. The sheaf $\mathscr{H}$ is called the factor sheaf of $\mathscr{G}$ by $\mathscr{F}$. Evidently the sequence $0 \to \mathscr{F} \to \mathscr{G} \to \mathscr{H} \to 0$ is exact.

Example 3. Let X be an irreducible algebraic variety over a field k, and $\mathscr{K}^*$ the sheaf of locally constant functions with values in the multiplicative group of $k(X)$. The sheaf $\mathcal{O}^*$ is defined by the fact that $\mathcal{O}^*(U)$ is the group of invertible elements of the ring $\mathcal{O}(U)$. It easy to verify that for the factor sheaf $\mathscr{D} = \mathscr{K}^*/\mathcal{O}^*$ the group $\mathscr{D}(U)$ is isomorphic to the group of locally principal divisors of the variety U. In this example $\mathscr{K}^*$ and $\mathcal{O}^*$ are regarded as sheaves of Abelian groups.

Definition. The *support* of a sheaf $\mathscr{F}$ is defined as the set $X-W$, where W is the union of all open sets $U\subset X$ for which $\mathscr{F}(U)=0$.

This set is closed and is denoted by $\operatorname{Supp}\mathscr{F}$.

Proposition. *If S is the support of a sheaf $\mathscr{F}$ and if U, V, $U\subset V$, are two open sets for which $U\cap S=V\cap S$, then the homomorphism $\varrho_U^V:\mathscr{F}(V)\to\mathscr{F}(U)$ is an isomorphism.*

Let $a\in\mathscr{F}(V)$, $\varrho_U^V a=0$. By definition of S, every point $x\in V$, $x\notin S$, has a neighbourhood V_x, which we may regard as contained in V, such that

$$\varrho_{V_x}^V(a)=0\,.$$

By hypothesis, for points $x\in S$ such a neighbourhood is U. From the definition of a sheaf it follows that $a=0$.

Let $a\in\mathscr{F}(U)$. We consider a covering $V=\bigcup U_\alpha$ in which $U_0=U$ and $U_\alpha\cap S=\emptyset$ for $\alpha\neq 0$ (for example, U_α for $\alpha\neq 0$ are sufficiently small neighbourhoods of points $x\in V$, $x\notin S$).

We set $a_0=a$, $a_\alpha=0$ for $\alpha\neq 0$. From the conditions of the proposition it follows that

$$\varrho_{U_\alpha\cap U_\beta}^{U_\alpha}(a_\alpha)=\varrho_{U_\alpha\cap U_\beta}^{U_\beta}(a_\beta)\,.$$

Hence, according to the definition of a sheaf, there exists an element $a'\in\mathscr{F}(V)$ for which

$$\varrho_{U_\alpha}^V a'=a_\alpha\,,$$

in particular, $\varrho_U^V(a')=a$ for $\alpha=0$. This proves the proposition.

From the proposition it follows that if S is the support of the sheaf, then the modules $\mathscr{F}(U)$ for all sets U having a given intersection with S are canonically isomorphic. Therefore we can define a sheaf $\bar{\mathscr{F}}$ on S, by setting

$$\bar{\mathscr{F}}(\bar{U})=\mathscr{F}(U)\quad\text{if}\quad U\cap S=\bar{U}$$

for open sets $\bar{U}\subset S$.

Example 4. Let X be a scheme and $Y\subset X$ a closed subscheme. We define a subsheaf $\mathscr{J}_Y$ of the structure sheaf $\mathcal{O}_X$ by the condition $\mathscr{J}_Y(U)=\mathfrak{a}_Y$ if U is an affine open set, $U=\operatorname{Spec}A$, and $\mathfrak{a}_Y\subset A$ is the ideal of the subscheme $Y\cap U$. Obviously, if U does not intersect Y, then $\mathscr{J}_Y|_U=\mathcal{O}_X|_U$. Therefore the sheaf $\mathscr{F}=\mathcal{O}_X/\mathscr{J}_Y$ is 0 on these open sets, that is, its support is contained in Y. The corresponding sheaf $\bar{\mathscr{F}}$ is the same as the structure sheaf $\mathcal{O}_Y$ of Y.

Note. Our definition of the support of a sheaf is not the one generally accepted, however, it is somewhat more convenient for our purposes. Besides, the two definitions coincide in those situations in which they are going to be applied later.

2. Coherent Sheaves. In discussing vector bundles we have already come across locally free sheaves. Now we consider a class of sheaves that is in the same relation to an arbitrary module of finite type as locally free sheaves are to free modules of finite rank.

The concepts introduced in § 3.1 are now going to be applied to the case when $(X, \mathcal{O}_X)$ is an arbitrary scheme. We begin with the local analysis and assume that $X = \operatorname{Spec} A$, where A is an arbitrary ring.

For every module M over A and every multiplicative system S of elements of A we define the localization of M relative to S by setting

$$M_S = M \otimes_A A_S .$$

The module M_S can be described in the same way as the localization A_S of a ring in Ch. V, § 1.1: it consists of pairs (m, s), $m \in M$, $s \in S$, with the same rules of identification, addition and multiplication by elements of A_S as in the case of rings. We write the pair (m, s) in the form m/s. In particular, by taking for S the system of powers of a non-nilpotent element $f \in A$ we obtain a module M_f over the ring A_f.

The homomorphisms $A_S \to A_{S'}$, which are defined for $S \subset S'$, give rise to homomorphisms $M_S \to M_{S'}$. This enables us to associate with a module M over a ring A a sheaf $\tilde{M}$ on $\operatorname{Spec} A$. Its definition simply copies that of the sheaf $\mathcal{O}$, into which it turns in case $M = A$. By virtue of this we omit some verifications, which in the general case do not differ at all from those made in Ch. V, § 2.2.

For an open set $U = D(f), f \in A$, we set

$$\tilde{M}(U) = M_f .$$

For any open set U we consider all the $f \in A$ for which $D(f) \subset U$. For them there are defined the homomorphisms

$$M_g \to M_f$$

if $D(g) \supset D(f)$. Using these homomorphisms we can define the projective limit of the groups M_f. We set

$$\tilde{M}(U) = \varprojlim_{D(f) \subset U} M_f .$$

The group $\tilde{M}(U)$ is a module over the ring $\mathcal{O}(U) = \varprojlim A_f$; this is a general property of a projective limit. The inclusion $U \subset V$ defines a homomorphism $\varrho_U^V : \tilde{M}(V) \to \tilde{M}(U)$, just as in the case $M = A$. The system $(\tilde{M}(U), \varrho_U^V)$ defines a sheaf of modules $\tilde{M}$ over the sheaf of rings $\mathcal{O}_X$.

Every homomorphism of A-modules $\varphi : M \to N$ determines homomorphisms $\varphi_f : M_f \to N_f$ for all $f \in A$, and after passage to the limit a homomorphism of sheaves $\tilde{\varphi} : \tilde{M} \to \tilde{N}$. If $\varphi : M \to N$ and $\psi : N \to L$ are two such homomorphisms, then

$$\widetilde{\varphi\psi}=\tilde{\varphi}\cdot\tilde{\psi}\,.$$

The module M can be recovered from the sheaf $\tilde{M}$. In fact, there is a generalization of the relation proved in Ch. V, § 2.2,

$$\tilde{M}(\operatorname{Spec} A)=M\,,$$

which is proved word-for-word in the same way. Hence it follows that the correspondence $M\to\tilde{M}$ between modules M and sheaves obtained from them is one-to-one. Furthermore, a simple verification enables us to deduce that the correspondence $\varphi\to\tilde{\varphi}$ is a homomorphism of the groups

$$\operatorname{Hom}(M,N)\sim\operatorname{Hom}(\tilde{M},\tilde{N})\,,$$

where on the left-hand side we have the group of homomorphisms of modules over A, and on the right-hand side that of sheaves of modules over $\mathcal{O}_{\operatorname{Spec} A}$.

Now we can go over to a globalization of these concepts.

Let X be a Noetherian scheme.

Definition. A sheaf $\mathcal{F}$ on X is called *coherent* if every point $x\in X$ has an affine neighbourhood U such that $U=\operatorname{Spec} A$, A is a Noetherian ring, and the sheaf $\mathcal{F}|_U$ is isomorphic to a sheaf of the form $\tilde{M}$, where M is a module of finite type over A.

Proposition. *If X is an affine Noetherian scheme and $X=\operatorname{Spec} A$, then every coherent sheaf $\mathcal{F}$ on X is of the form $\tilde{M}$, where M is a module of finite type over A.*

Proof. Since the open sets of the form $D(f)$ determine a basis of the spectral topology, there exist elements $f_i\in A$ such that $\bigcup D(f_i)=X$ and on $D(f_i)$ the sheaf $\mathcal{F}$ is isomorphic to one of the form $\tilde{M}_i$, where M_i is a module of finite type over the ring A_{f_i}. The space Spec A being compact, we may assume that the elements f_i are finite in number. We set $\mathcal{F}(X)=M$ and show that $\mathcal{F}=\tilde{M}$.

For every non-nilpotent element $g\in A$ there is defined a homomorphism $\varrho^X_{D(g)}: M\to\mathcal{F}(D(g))$, which owing to the fact that $\mathcal{F}(D(g))$ is a module over A_g can be extended uniquely to a homomorphism of A_g-modules

$$\varphi_g:\tilde{M}(D(g))\to\mathcal{F}(D(g))\,.$$

An obvious verification shows that this system of homomorphisms determines a unique homomorphism of sheaves of modules $\tilde{M}\to\mathcal{F}$. We show that φ is an isomorphism.

Everything reduces to proving that the homomorphism φ_g is an isomorphism. To do this we consider the sequence of homomorphisms

$$0\to M\xrightarrow{\lambda}\oplus M_i\xrightarrow{\mu}\oplus M_{i,j}\,, \tag{1}$$

where

$$M_{i,j}=(M_i)_{f_j}=(M_j)_{f_i}=\mathscr{F}(D(f_i f_j)), \qquad \lambda(m)=(\ldots\varrho^X_{D(f_i)}(m)\ldots),$$
$$\mu(\ldots m_i\ldots m_j\ldots)=(\ldots(\varrho^{D(f_i)}_{D(f_i f_j)}(m_i)-\varrho^{D(f_j)}_{D(f_i f_j)}(m_j)\ldots).$$

In the sequence (1) M_i and $M_{i,j}$ are regarded as A-modules. From the definition of a sheaf it follows that this sequence is exact. Now we use the important but trivially verifiable property of the functor $M\to M_g$: it carries exact sequences into exact sequences. In particular, the sequence

$$0\to M_g\xrightarrow{\lambda_g}\oplus(M_i)_g\xrightarrow{\mu_g}\oplus(M_{i,j})_g$$

is exact. On the other hand, consider the sheaf $\mathscr{F}_{|D(g)}$. It has a similar exact sequence

$$0\to\mathscr{F}(D(g))\xrightarrow{\lambda'_g}\bigoplus_i\mathscr{F}(D(gf_i))\xrightarrow{\mu'_g}\bigoplus_{i,j}\mathscr{F}(D(gf_i f_j)).$$

But $\mathscr{F}(D(gf_i))\simeq(M_i)_g$, $\mathscr{F}(D(gf_i f_j))\simeq(M_{i,j})_g$. These isomorphisms induce an isomorphism $\varphi'_g: M_g\simeq\mathscr{F}(D(g))$. It is easy to check that φ'_g coincides with φ_g on the images of elements of M, and therefore on the whole of M_g. So we have shown that φ is an isomorphism and hence that $\mathscr{F}=\tilde{M}$.

It remains to show that the module M is Noetherian when it is known to us that the modules $M_i=M_{f_i}$ are Noetherian. Let $\{M_n\}$ be an increasing sequence of submodules of M. Then $(M_n)_{f_i}=(M_{n+1})_{f_i}$ for all f_i and sufficiently large n. Hence it follows that $M_n=M_{n+1}$.

3. Dévissage of Coherent Sheaves. Here we give an account of a method that enables us to reduce, although only in very coarse problems, arbitrary coherent sheaves to free sheaves.

Proposition 1. *For every coherent sheaf $\mathscr{F}$ on a Noetherian irreducible and reduced scheme X there exists an open dense set W such that the sheaf $\mathscr{F}|_W$ is free.*

The assertion has local character, therefore we may restrict ourselves to the case when $X=\operatorname{Spec} A$, where A is a Noetherian ring without nilpotent elements, and $\mathscr{F}=\tilde{M}$, where M is a module of finite type over A. Furthermore, it is obvious that we may take X to be irreducible. Under this assumption it follows that A has no divisors of zero.

We recall that the rank of an A-module is the maximal number of linearly independent elements over A. By hypothesis, the rank of M is finite. We denote it by r and let $x_1,\ldots,x_r$ be linearly independent elements over A. By definition, the submodule M' generated by them is free. Let $y_1,\ldots,y_m$ be a system of generators of M. Then there exist elements $d_i \neq 0$, $d_i \in A$, such that

$$d_i y_i \in M' . \tag{1}$$

We consider the open set $W = D(d)$, $d = d_1 \dots d_m$. The sheaf $\mathscr{F}|_W$ is isomorphic to $\tilde{M}_d$. But by (1) $M_d = M'_d$, hence

$$\mathscr{F}|_W = \tilde{M}'_d .$$

The module M'_d over A_d is free, because M' is free, and the proposition is proved.

Proposition 2. *For every coherent sheaf $\mathscr{F}$ over an irreducible Noetherian reduced scheme X there exists a coherent sheaf $\mathscr{G}$ containing a free subsheaf $\mathcal{O}^r$ and a homomorphism $\varphi : \mathscr{F} \to \mathscr{G}$ such that the supports of the sheaves* Ker φ *and $\mathscr{G}/\mathcal{O}^r$ are different from X.*

As will be clear from the proof, we can construct a homomorphism $\varphi : \mathscr{F} \to \mathscr{G}$ for which the support not only of the sheaf $\operatorname{Ker} \varphi$ but also of $\mathscr{G}/\operatorname{Im} \varphi$ is different from the whole scheme X. Since the support of $\mathscr{G}/\mathcal{O}^r$ is different from X, Proposition 2 shows that every coherent sheaf is free "to within sheaves of support other than X".

Proof. Let W be the open set whose existence is established in Proposition 1, and let $f : \mathscr{F}|_W \to \mathcal{O}^r|_W$ be the isomorphism that exists according to the same proposition. We may assume that W is a principal open set and will do this in what follows. We define the sheaf $\mathscr{G}$ by the condition:

$$\mathscr{G}(U) = f_{U \cap W}\, \varrho^U_{U \cap W} \mathscr{F}(U) + \varrho^U_{U \cap W} \mathcal{O}^r(U) . \tag{2}$$

Since $\varrho^U_{U \cap W} \mathcal{O}^r(U) \subset \mathcal{O}^r(U \cap W)$, and $f_{U \cap W} \varrho^U_{U \cap W} \mathscr{F}(U) \subset \mathcal{O}^r(U \cap W)$, both terms on the right-hand side of (2) are contained in one and the same group. We consider the sum of these subgroups, which obviously is a submodule of $\mathcal{O}^r(U \cap W)$ over the ring $\mathcal{O}(U)$, provided that for $x \in \mathcal{O}^r(U \cap W)$, $a \in \mathcal{O}(U)$, we set $a \cdot x = \varrho^U_{U \cap W}(a) \cdot x$. Since the modules $\mathscr{F}(U)$ and $\mathcal{O}^r(U)$ are of finite type over $\mathcal{O}(U)$, so is the module $\mathscr{G}(U)$ over $\mathcal{O}(U)$.

The definition of the homomorphisms $\varrho^U_{V, \mathscr{G}}$ is self-evident. From what we have said above it follows at once that the sheaf $\mathscr{G}$ we have constructed is coherent.

The homomorphism $\varrho^U_{U \cap W}$ for $\mathcal{O}^r$ is an embedding. It is sufficient to verify this for an affine open set $U = \operatorname{Spec} A$. We consider a principal open set $D(f) \subset U \cap W$. The kernel of the homomorphism $\varrho^U_{D(f)}$ consists of those elements $x \in A$ such that $f^n x = 0$ for some $n \geqslant 0$. Since the scheme X is irreducible, the ring A has no divisors of 0, hence $x = 0$. A fortiori $\operatorname{Ker} \varrho^U_{U \cap W} = 0$. Thus, we may identify $\mathcal{O}^r$ with a subsheaf of $\mathscr{G}$ by means of the homomorphism $\varrho^U_{U \cap W}$.

We define the homomorphism $\varphi:\mathscr{F}\to\mathscr{G}$ by the conditions

$$\varphi_U=f_{U\cap W}\varrho^U_{U\cap W}.$$

If $U\subset W$, then

$$\mathscr{G}(U)=f_U\varrho^U_{U\cap W}\mathscr{F}(U)=f_U\mathscr{F}(U)=\mathcal{O}^r(U)=\varrho^U_{U\cap W}\mathcal{O}^r(U)$$

and f_U is an isomorphism. Therefore φ_U is an isomorphism, and $\mathscr{G}(U)=\mathcal{O}^r(U)$. This shows that the sheaves $\operatorname{Ker}\varphi$ and $\mathscr{G}/\mathcal{O}^r$ vanish on W, hence the supports of these sheaves are contained in $X-W$.

Proposition 2 leads to the question of the structure of coherent sheaves whose supports differ from the whole sheaf. If the support of a sheaf $\mathscr{F}$ is a closed set $Y\subset X$, then according to § 3.1 on Y a sheaf $\bar{\mathscr{F}}$ is defined by the condition

$$\bar{\mathscr{F}}(\bar{U})=\mathscr{F}(U)\quad\text{if}\quad U\cap Y=\bar{U}.$$

We regard Y as a reduced closed subscheme of X. Is perhaps $\bar{\mathscr{F}}$ a coherent sheaf over this scheme or at least over the sheaf of $\mathcal{O}$-modules on it? Generally speaking, this is not true, as the following example shows. Let $X=\operatorname{Spec}\mathbb{Z}$, and suppose that $\mathscr{F}$ corresponds to the module $\mathbb{Z}/p^2\mathbb{Z}$, where p is a prime number. The support of $\mathscr{F}$ is the ideal (p), and the reduced scheme corresponding to it is $\operatorname{Spec}(\mathbb{Z}/p\mathbb{Z})$. Obviously it is impossible to define on $\mathbb{Z}/p^2\mathbb{Z}$ the structure of a module over the ring $\mathbb{Z}/p\mathbb{Z}$.

However, we can show that in a weaker form the sheaf $\bar{\mathscr{F}}$ can be made into a coherent sheaf over Y.

Proposition 3. *If $\mathscr{F}$ is a coherent sheaf on a Noetherian scheme X, with support $Y\neq X$, then in the sheaf $\bar{\mathscr{F}}$ on the reduced subscheme Y there exists a sequence of subsheaves $\bar{\mathscr{F}}=\bar{\mathscr{F}}_0\supset\bar{\mathscr{F}}_1\supset\cdots\supset\bar{\mathscr{F}}_m=0$ whose factors $\bar{\mathscr{F}}_i/\bar{\mathscr{F}}_{i+1}$ are all coherent $\mathcal{O}_y$-modules.*

Proof. In § 3.2 we have given an example of a sheaf $\mathscr{J}_Y$ connected with the subscheme Y. Clearly, the sheaf $\bar{\mathscr{F}}$ is coherent if the following relation holds:

$$\bar{\mathscr{F}}\cdot\mathscr{J}_Y=0. \tag{3}$$

For in this case all the $\mathcal{O}_X(U)$-modules $\mathscr{F}(U)$ are modules over $\mathcal{O}_X(U)/\mathscr{J}_Y(U)=\mathcal{O}_Y(U)$. Therefore, if in the affine open set $U=\operatorname{Spec}A$ the sheaf $\mathscr{F}$ is of the form $\tilde{M}$, where M an A-module, then $M\cdot\mathfrak{a}_Y=0$, hence M is a $A/\mathfrak{a}_Y$-module. Here $\bar{\mathscr{F}}=\tilde{M}$ if we now regard M as a module over $A/\mathfrak{a}_Y$.

Let us show that a somewhat weaker assertion is true: there exists a number $m>0$ for which

$$\mathscr{F}\cdot\mathscr{J}_Y^m=0. \tag{4}$$

We consider the affine open set $U = \operatorname{Spec} A$, for which the restriction of $\mathscr{F}$ to it is of the form $\tilde{M}$, where M is a module of finite type over A. Let $\mathfrak{a}_Y \subset A$ be the ideal of the subset $Y \cap U$. If $f \in \mathfrak{a}_Y$, then $D(f) \subset U - (U \cap Y)$, and by hypothesis the restriction of $\mathscr{F}$ to $D(f)$ is zero. This means that $M_f = 0$, hence for every $m \in M$ there exists a $j(m) > 0$ such that $f^{j(m)} m = 0$. Since M is a module of finite type, it follows from this that $f^j M = 0$ for some $j > 0$. From the fact that this relation holds for every $f \in \mathfrak{a}_Y$ and that the ideal $\mathfrak{a}_Y$ has a finite basis we deduce that

$$\mathfrak{a}_Y^l M = 0 \tag{5}$$

for some $l > 0$. In other words, (4) holds for the open set U. Choosing a finite covering of X by such open sets and taking for m the maximum of those numbers l for which (5) holds on each of these factors, we obtain (4) on the whole of X.

We set $\mathscr{F}_i = \mathscr{F} \mathscr{J}_Y^i$, $i = 0, \ldots, m$, $\mathscr{F}_0 = \mathscr{F}$. Obviously the supports of all the sheaves $\mathscr{F}_i$ are contained in Y. We denote by $\bar{\mathscr{F}}_i$ the sheaves determined by the $\mathscr{F}_i$ on Y. By (4) $\bar{\mathscr{F}}_m = \mathscr{F}_m = 0$. Since

$$(\mathscr{F}_i / \mathscr{F}_{i+1}) \cdot \mathscr{J}_Y = 0 ,$$

the relation (3) holds for the sheaves $\bar{\mathscr{F}}_i / \bar{\mathscr{F}}_{i+1}$, hence they are coherent. This proves Proposition 3.

In conclusion we show that by the method we have used all the time we can reduce the study of sheaves to the case of irreducible schemes.

Proposition 4. *Let X be a Noetherian reduced scheme, $X = \bigcup X_i$ a representation in the form of a union of irreducible components, and $\mathscr{F}$ a coherent sheaf on X. There exist coherent sheaves $\mathscr{F}_i$ on X and a homomorphism $\varphi : \mathscr{F} \to \oplus \mathscr{F}_i$ such that the support of $\mathscr{F}_i$ is contained in X_i, the sheaf $\bar{\mathscr{F}}_i$ determined on X_i by $\mathscr{F}_i$ is coherent, and the support of the kernel of φ is contained in $\bigcup_{i \neq j} (X_i \cap X_j)$.*

Proof. We set $\mathscr{F}_i = \mathscr{F} / \mathscr{F} \mathscr{J}_{X_i}$, let $\varphi_i : \mathscr{F} \to \mathscr{F}_i$ be the natural projection, and $\varphi = \oplus \varphi_i$. We have seen in § 3.2 that the support of $\mathscr{F}_i$ is contained in X_i, and since $\mathscr{F}_i \mathscr{J}_{X_i} = 0$, the sheaf $\bar{\mathscr{F}}_i$ is coherent.

We consider the open set

$$U_i = X_i - \bigcup_{i \neq j} (X_i \cap X_j) .$$

On it $\mathscr{J}_{X_j} = \mathcal{O}_X$ for $j \neq i$, and $\mathscr{J}_{X_i} = 0$, therefore $\mathscr{F}_j|_{U_i} = 0$ for $j \neq i$, and $\mathscr{F}_i|_{U_i} = \mathscr{F}|_{U_i}$. Thus, $\varphi_j = 0$ for $j \neq i$, and $\varphi = \varphi_i$ is an isomorphism. Therefore the kernel of φ is 0 on $\bigcup U_i$, and this is what we have claimed.

4. The Finiteness Theorem

Theorem. *If X is a complete variety over a field k and $\mathscr{F}$ a coherent sheaf on X, then the vector space $\mathscr{F}(X)$ is finite-dimensional over k.*

The basis of the proof is the following remark. If we are given a homomorphism of sheaves over k

$$\varphi:\mathscr{F}\to\mathscr{G}, \quad \mathscr{H}=\operatorname{Ker}\varphi, \quad \dim\mathscr{H}(X)<\infty, \quad \dim\mathscr{G}(X)<\infty, \tag{1}$$

then the space $\mathscr{F}(X)$ is finite-dimensional. This follows from the definition of the kernel, according to which $\mathscr{H}(X)$ is the kernel of the homomorphism $\varphi_X:\mathscr{F}(X)\to\mathscr{G}(X)$. Hence we obtain by induction that the space $\mathscr{F}(X)$ is finite-dimensional if there exist subsheaves

$$\mathscr{F}=\mathscr{F}_0\supset\mathscr{F}_1\supset\cdots\supset\mathscr{F}_m=0, \tag{2}$$

for which the spaces $\mathscr{F}_i/\mathscr{F}_{i+1}(X)$ are finite-dimensional.

We prove the theorem by induction on the dimension of the variety X. If $\dim X=0$, then X consists of finitely many points, the coherent sheaf $\mathscr{F}$ on X is, by definition, a finite-dimensional vector space over k, and the theorem is proved.

We now assume that the theorem is true for complete varieties whose dimension is less than that of X. We show that we can then derive the theorem for all sheaves $\mathscr{F}$ on X whose support is contained in a closed subvariety Y, $\dim Y<\dim X$.

Indeed, by the definition of $\bar{\bar{\mathscr{F}}}$ we have $\mathscr{F}(X)=\bar{\bar{\mathscr{F}}}(Y)$, and we may apply the theorem to the coherent sheaves on Y. Here we are faced with the difficulty that, in general, the sheaf $\bar{\bar{\mathscr{F}}}$ is not coherent on Y; but the position is saved by Proposition 3 of § 3.3. It gives us a sequence $\bar{\bar{\mathscr{F}}}=\bar{\bar{\mathscr{F}}}_0\supset\bar{\bar{\mathscr{F}}}_1\supset\cdots\supset\bar{\bar{\mathscr{F}}}_m=0$, in which the sheaves $\bar{\bar{\mathscr{F}}}_i/\bar{\bar{\mathscr{F}}}_{i+1}$ are coherent on Y, so that the inductive hypothesis is applicable to them. We obtain a sequence of sheaves (2), and from its existence it follows that the space $\bar{\bar{\mathscr{F}}}(Y)$, hence also $\mathscr{F}(X)$, is finite-dimensional.

The next step of the proof consists in the reduction of the theorem to the case of irreducible varieties. Let $X=\bigcup X_i$ be the decomposition into irreducible components. Here we can apply Proposition 4 of § 3.3. Our homomorphism φ has a kernel whose support is contained in the subvariety $\bigcup\limits_{i\neq j}(X_i\cap X_j)$, which is of smaller dimension than X. Therefore it is enough for us to show that the space $(\oplus\mathscr{F}_i)(X)$ is finite-dimensional. But

$$(\oplus\mathscr{F}_i)(X)=\oplus(\bar{\bar{\mathscr{F}}}_i(X_i)),$$

and since $\bar{\bar{\mathscr{F}}}_i$ is a coherent sheaf on X_i, this reduces the assertion to the case of irreducible varieties X_i.

Finally, we come to the central part of the proof, assuming X to be irreducible. Here we rely on Proposition 2 of § 3.3. Since X is complete, we have $\mathcal{O}(X)=k$, hence $\dim \mathcal{O}^r(X)=r$. Since the support of $\mathcal{G}/\mathcal{O}^r$ is different from X, the theorem is true for this sheaf, hence we have for $\mathcal{G}$ the homomorphism $\psi: \mathcal{G} \to \mathcal{G}/\mathcal{O}^r$ which satisfies (1). Therefore the space $\mathcal{G}(X)$ is finite-dimensional. On the other hand, the homomorphism $\varphi: \mathcal{F} \to \mathcal{G}$ constructed in Proposition 2 of § 3.3 again satisfies (1), hence the space $\mathcal{F}(X)$ is also finite-dimensional, as required.

The theorem we have proved has many important applications. Some of these were mentioned earlier. First of all, in § 1.4 we have associated with every divisor D on a variety X a sheaf $\mathcal{L}_D$ such that $\mathcal{L}_D(X)$ is isomorphic to the space $\mathcal{L}(D)$ introduced in Ch. III, § 1.5. We have seen in § 3.4 that the sheaf $\mathcal{L}_D$ is locally free of rank 1, hence coherent. Thus, our theorem is applicable to it, and we obtain the result, which has already been used several times:

Corollary 1. *The dimension $l(D)$ of a locally principal divisor D over a complete variety is finite.*

Applying the theorem to the sheaf corresponding to the cotangent bundle Ω^1 and its exterior powers Ω^p we obtain:

Corollary 2. *On a complete smooth variety the dimensions h^p of the spaces $\Omega^p[X]$ of regular differential forms are finite.*

This result was also stated in Ch. III, where we have seen that it gives a number of birational invariants of a variety.

As a further example we consider the sheaf $\mathcal{T}$ corresponding to the tangent bundle. The elements of the group $\mathcal{T}(X)$ are called *regular vector fields* on X. Such an element can be regarded as a function associating with every point $x \in X$ the tangent vector $t_x \in \Theta_x$ at that point. In this case our theorem yields:

Corollary 3. *The dimension of the space of regular vector fields on a smooth complete variety is finite.*

Exercises

1. A coherent sheaf $\mathcal{F}$ is called a torsion sheaf if $\mathcal{F}(U)$ is a torsion module over $\mathcal{O}_X(U)$ for every open set U. Show that $\mathcal{F}$ is a torsion sheaf if and only if its support is different from X (the scheme X is assumed to be irreducible).

2. Find the general form of torsion sheaves on a smooth curve.

3. Let $E \to X$ be a vector bundle over an affine variety $X=\operatorname{Spec} A$. Show that the set M_E of sections of E is a module of finite type over A.

4. Show that the module M_E introduced in Exercise 3 is projective over the ring A (for the definition of a projective module see [10], Ch. I, § 2).

5. Show that the modules M_E and $M_{E'}$ are isomorphic if and only if E and E' are.

6. Show that every vector bundle over an affine line $\mathbb{A}^1$ is trivial.

7. Let $E \to X$ be a vector bundle over a complete variety X. Show that the set of sections M_E is a finite-dimensional vector space.

8. Show that the set of morphisms $f: E_1 \to E_2$ of vector bundles $E_i \to X$, $i = 1, 2$, over a complete variety X forms a finite-dimensional space.

9. Let A be a one-dimensional regular local ring, $\mathscr{K}$ its field of fractions, $X = \operatorname{Spec} A$, $x \in X$ a generic point, $U = \{x\}$.

A sheaf $\mathscr{F}$ of $\mathcal{O}$-modules over X is given by an A-module M, a linear space L over $\mathscr{K}$, and an A-homomorphism $\varphi: M \to L$. Express in these terms the fact that $\mathscr{F}$ is a coherent sheaf. Construct an example of a subsheaf of a coherent sheaf that is not coherent.

10. Let X be an irreducible variety, $x_0 \in X$ a closed point. We define a presheaf $\mathscr{F}$ on X by setting $\mathscr{F}(U) = \mathcal{O}(U)$ if U does not contain x_0, $\mathscr{F}(U) = 0$ if U contains x_0. Show that $\mathscr{F}$ is a sheaf, that it is not coherent, and that it is a subsheaf of $\mathcal{O}$.

Part III. Algebraic Varieties over the Field of Complex Numbers and Complex Analytic Manifolds

Chapter VII. Topology of Algebraic Varieties

§ 1. The Complex Topology

1. Definitions. In Ch. II, § 2.3, we have seen that the set of complex points of an algebraic variety defined over the field of complex numbers is a topological space. In Ch. II this was shown for quasiprojective varieties, the only ones at our disposal at that moment. But these arguments remain valid for arbitrary varieties. Presently we shall give a general definition. The topology that determines on X a given structure of a scheme is called the *spectral* topology.

First we introduce some notation. For a variety X defined over the field $\mathbb{C}$ of complex numbers we denote by $X(\mathbb{C})$ the set of its closed points. We consider a set $U \subset X$ that is open in the spectral topology, finitely many functions $f_1, \dots, f_m$ that are regular on U, and a number $\varepsilon > 0$. We denote by $V(U; f_1, \dots, f_m; \varepsilon)$ the set of those points $x \in U(\mathbb{C})$ for which

$$|f_i(x)| < \varepsilon, \quad i = 1, \dots, m.$$

We turn the set $X(\mathbb{C})$ into a topological space, by taking the sets $V(U; f_1, \dots, f_n; \varepsilon)$ as a basis of open sets.

The topology so defined is called the *complex* topology. Let us compare it with the spectral topology, which we have considered earlier. If $Y \subset X$ is a closed subset in the spectral topology, then $Y(\mathbb{C}) \subset X(\mathbb{C})$. From the definition it follows that $Y(\mathbb{C})$ is closed in $X(\mathbb{C})$ in the complex topology and that the complex topology of the set $Y(\mathbb{C})$ is the same as its topology as a subset of $X(\mathbb{C})$. However, not every set that is closed in the complex topology is of the form $Y(\mathbb{C})$, where Y is closed in X in the spectral topology. An example is the set of points $x \in \mathbb{A}^1(\mathbb{C})$ for which $|t(x)| \leqslant 1$, where t is a coordinate on $\mathbb{A}^1$. The morphism $f: X \to Y$ of algebraic varieties evidently determines a continuous mapping $f: X(\mathbb{C}) \to Y(\mathbb{C})$.

In some respects the complex topology is simpler than the spectral topology. As a very simple example we show that $(X_1 \times X_2)(\mathbb{C})$ in the complex topology is the product of $X_1(\mathbb{C})$ and $X_2(\mathbb{C})$. It is clear that

$$V(U_1; f_1, \ldots, f_m; \varepsilon) \times V(U_2; g_1, \ldots, g_n; \varepsilon)$$
$$= V(U_1 \times U_2; p_1^* f_1, \ldots, p_1^* f_m, p_2^* g_1, \ldots, p_2^* g_m; \varepsilon),$$

where p_1 and p_2 are the projections of $X_1 \times X_2$ onto X_1 and X_2. Therefore products of open sets in $X_1(\mathbb{C})$ and $X_2(\mathbb{C})$ are open in $(X_1 \times X_2)(\mathbb{C})$. To verify that they form a basis of open sets it is sufficient to do this for affine X_1 and X_2. Embedding them in affine spaces we reduce the verification to the case $X_1 = \mathbb{A}^{n_1}$, $X_2 = \mathbb{A}^{n_2}$, where it is obvious.

In the complex topology $X(\mathbb{C})$ is a Hausdorff space. For by the definition of a variety the diagonal Δ is closed in $X \times X$ in the spectral topology. Therefore $\Delta(\mathbb{C})$ is closed in $(X \times X)(\mathbb{C})$ in the complex topology. As we have just seen, $(X \times X)(\mathbb{C}) = X(\mathbb{C}) \times X(\mathbb{C})$, and $\Delta(\mathbb{C})$ coincides with the diagonal of this space: the set of points of the form (x, x), $x \in X(\mathbb{C})$. The fact that the diagonal is closed is equivalent to the space being Hausdorff ([8], § 8.1).

The topological space $\mathbb{P}^n(\mathbb{C})$ is compact, hence so are all its closed subsets. In particular, this refers to the spaces $X(\mathbb{C})$, where X is a projective variety. If X is a complete variety, then by using Chow's lemma in Ch. VI, § 2.1, we construct a morphism $f: X' \to X$, where X' is a projective variety. This morphism is birational, hence $f(X')$ is dense in X, and since X' is projective, $f(X') = X$. In particular, $f(X'(\mathbb{C})) = X(\mathbb{C})$. Since f is a continuous mapping and $X'(\mathbb{C})$ is compact, it follows that $X(\mathbb{C})$ is compact. It can be shown that this property characterizes complete varieties over the field of complex numbers: if the space $X(\mathbb{C})$ is compact, then the variety X is complete (see Exercises 1 and 2 to Ch. VII, § 2). It is evident that for an arbitrary variety X the space $X(\mathbb{C})$ is locally compact.

The arguments in Ch. II, § 2.3, can now be applied to the investigation of the complex topology of any (not only quasiprojective) smooth variety defined over the field of complex numbers. They show that in this case $X(\mathbb{C})$ in the complex topology is a topological manifold of dimension $2 \dim X$.

The preceding definition admits the following generalization. We consider an arbitrary field k and denote by k' its algebraic closure. Let X be a scheme over k such that the scheme $X \times_k \operatorname{Spec} k'$ is an algebraic variety over k'. This scheme is called an algebraic variety *defined over* k. An example is an affine or a projective variety over k' in which an ideal has a basis consisting of polynomials with coefficients in k.

If X is an algebraic variety over k, then $X(k)$ denotes the set of those closed points $x \in X$ for which $k(x) = k$.

If k is the field $\mathbb{R}$ of real numbers or the field $\mathbb{Q}_p$ of p-adic numbers, then we can define in the set of points $X(k)$ a topology in exactly the same way as we have done this in the case $k=\mathbb{C}$.

If $k=\mathbb{R}$ and X is a smooth variety, then $X(\mathbb{R})$ is a topological manifold of dimension $\dim X$. Henceforth we only consider the topological space $X(\mathbb{C})$, except in § 4, where we study the space $X(\mathbb{R})$ in the case when X is a curve.

We always consider the space $X(\mathbb{C})$ with the complex topology. For the remainder of this section we consider a space $X(\mathbb{C})$, where X is smooth. We use here a somewhat larger topological apparatus than in the remaining parts of the book. But on the other hand, the results we obtain are nowhere used later.

2. Algebraic Varieties as Differentiable Manifolds. Orientation. Let X be an n-dimensional smooth variety over the field $\mathbb{C}$ of complex numbers, $x \in X$ a point (from now on we only consider closed points) and $t_1, \ldots, t_n$ a system of local parameters in it. As was shown in Ch. II, § 2.3, there exists a neighbourhood U of x at $X(\mathbb{C})$ which by means of the functions $t_1, \ldots, t_n$ is mapped homeomorphically onto a domain of the space $\mathbb{C}^n$. In view of this, every function on U can be regarded as a function of the variables $t_1(y), \ldots, t_n(y)$, or of the real variables $u_1, \ldots, u_n$, $v_1, \ldots, v_n$, if $t_j = u_j + i v_j$.

Definition. A real-valued function on U *belongs to the class* C^∞ if it is infinitely differentiable as a function of $u_1, \ldots, u_n$; $v_1, \ldots, v_n$.

As was shown in Ch. II, § 2.3, other local parameters $t'_1, \ldots, t'_n$ are analytic functions of $t_1, \ldots, t_n$. Therefore the definition we have given does not depend on the choice of local parameters, and the concept is well-defined.

It is easy to verify that our definition introduces in the topological space $X(\mathbb{C})$ the structure of a differentiable manifold ([26], § 1).

There is a natural connection between the properties of the algebraic variety X and the differentiable manifold $X(\mathbb{C})$. Differential forms $\omega \in \Omega^p[X]$ are (complex-valued) differential forms on $X(\mathbb{C})$. If $E \to X$ is a vector bundle, then $E(\mathbb{C}) \to X(\mathbb{C})$ is a topological vector bundle. Here we only have to ignore that E_x is a vector space over $\mathbb{C}$ and have to regard it as a space (of double the dimension) over $\mathbb{R}$. In this correspondence a tangent bundle $\Theta \to X$ corresponds to a tangent bundle of the differentiable manifold $X(\mathbb{C})$.

Presently we consider the question of orientation of the differentiable manifold $X(\mathbb{C})$. First we recall the relevant definitions.

An *orientation of the one-dimensional vector space* $\mathbb{R}$ is one of the two connected components of the set $\mathbb{R} - 0$; an *orientation of an n-dimensional vector space* F is an orientation of the one-dimensional space $\Lambda^n F$.

An *orientation of a* (locally trivial) *vector bundle* $f: E \to X$ is a collection of orientations ω_x of the fibres E_x such that every point has a neighbourhood U and an isomorphism $f^{-1}(U) \to U \times F$ carrying all the orientations ω_x, $x \in U$, into one and the same orientation of F. An *orientation of an differentiable manifold* is an orientation of its tangent bundle.

Proposition. *If X is a smooth variety over $\mathbb{C}$, then the differentiable manifold $X(\mathbb{C})$ is orientable.*

The reason for this is very simple: if an n-dimensional vector space F over $\mathbb{C}$ is regarded as a $2n$-dimensional space over $\mathbb{R}$, then it has some canonical orientation. To define it we choose a basis $e_1, \dots, e_n$ in F over $\mathbb{C}$. Then the vectors $u_1, \dots, u_{2n} = e_1, ie_1, \dots, e_n, ie_n$ form a basis of F over $\mathbb{R}$ and have the orientation $u_1 \wedge \dots \wedge u_{2n}$ of this space. Let us verify that this orientation does not depend on the choice of basis $e_1, \dots, e_n$. Let $f_1, \dots, f_n$ be another basis of F over $\mathbb{C}$. We denote by φ the linear transformation over $\mathbb{C}$ that carries $e_1, \dots, e_n$ into $f_1, \dots, f_n$, and by $\tilde{\varphi}$ the same transformation φ in F, regarded as a linear space over $\mathbb{R}$. We have to show that $\det \tilde{\varphi} > 0$, and this follows from the identity

$$\det \tilde{\varphi} = |\det \varphi|^2 . \tag{1}$$

To prove it we consider the space $\tilde{F} = F \otimes_{\mathbb{R}} \mathbb{C}$, and in it the transformation I defined by

$$I(f \otimes \alpha) = if \otimes \alpha, f \in F, \alpha \in \mathbb{C} . \tag{2}$$

Then $\tilde{F} = F_1 \oplus F_2$, where F_1 and F_2 are the eigenspaces for I corresponding to the eigenvalues i and $-i$. By analogy with (2) we extend $\tilde{\varphi}$ to $\tilde{F}$. Of course, its determinant remains unchanged. It is easy to see that F_1 and F_2 are invariant under $\tilde{\varphi}$ and that the matrix of $\tilde{\varphi}$ in F_1 is the same as the matrix of φ in F, and in F_2 is its complex conjugate. Equation (1) follows from this.

Now let $f: \Theta \to X(\mathbb{C})$ be the tangent bundle and ω_x the canonical orientation in $\Theta_x, x \in X$. We show that by this the orientation in X is determined. If $U \subset X$ is such that

$$\psi: f^{-1}(U) \simeq U \times F \tag{3}$$

is an isomorphism of algebraic vector bundles, then *a fortiori* this is true for the corresponding differentiable bundles. But in (3)

$$\psi_x : \Theta_x \to F$$

is an isomorphism of complex vector spaces. Therefore it carries the canonical orientation ω_x in Θ into the canonical orientation ω in F. This proves the proposition.

From now on the orientation we have just constructed is called the *canonical orientation of the manifold* $X(\mathbb{C})$.

We have obtained a first restriction, which shows that not every even-dimensional manifold can be represented in the form $X(\mathbb{C})$, where X is some smooth algebraic variety. For example, such a representation does not exist for the real projective plane.

3. The Homology of Smooth Projective Varieties. The orientability of a differentiable manifold can be expressed in terms of its homology. We recall this connection (see, for example, [19], 251). The orientation ω of an n-dimensional vector space E over $\mathbb{R}$ determines an element of the relative homology group $\omega \in H_n(E, E-0, \mathbb{Z})$. If U is a chart containing the point x of the manifold M, and $\varphi : U \to E$ a diffeomorphism carrying U into a neighbourhood of 0 in E, then we have the excision isomorphisms

$$H_n(U, U-x, \mathbb{Z}) \to H_n(M, M-x, \mathbb{Z}),$$
$$H_n(\varphi(U), \varphi(U)-0, \mathbb{Z}) \to H_n(E, E-0, \mathbb{Z})$$

and the isomorphism

$$\varphi_{n*} : H_n(U, U-x, \mathbb{Z}) \to H_n(\varphi(U), \varphi(U)-0, \mathbb{Z}).$$

Finally, $d_x\varphi$ is an isomorphism of the tangent spaces

$$d_x\varphi : \Theta_x \to E.$$

Using this system of isomorphisms we can associate with the orientation ω_x of the tangent space Θ_x a homology class for which we preserve the same notation:

$$\omega_x \in H_n(M, M-x, \mathbb{Z}).$$

The orientation of the compact manifold M then determines a class $\omega_M \in H_n(M, \mathbb{Z})$, which is uniquely characterized by the fact that under the homomorphism

$$H_n(M, \mathbb{Z}) \to H_n(M, M-x, \mathbb{Z}),$$

corresponding to any point $x \in M$ it is mapped into ω_x. The class ω_M is called the *orienting class of the manifold M*.

The proposition in § 1.2 shows that if X is a smooth complete algebraic variety, then $X(\mathbb{C})$ has a canonically defined orienting class $\omega_{X(\mathbb{C})} \in H_{2n}(X(\mathbb{C}), \mathbb{Z})$, $n = \dim X$. In what follows we talk occasionally of X itself as of a $2n$-dimensional homology class, meaning by this the class $\omega_{X(\mathbb{C})}$.

The preceding arguments construct the class $\omega_{X(\mathbb{C})} \in H_{2n}(X(\mathbb{C}), \mathbb{Z})$, which is necessarily non-zero, because it determines a non-zero class in the group $H_{2n}(X(\mathbb{C}), X(\mathbb{C})-x, \mathbb{Z})$. For the same reason this class is of infinite order. Hence for a smooth complete variety X

$$H_{2n}(X(\mathbb{C}),\mathbb{C}) \neq 0 .$$

This is a special case of the following more general result.

Proposition. *For a smooth projective variety X of dimension n*

$$H_{2l}(X(\mathbb{C}),\mathbb{C}) \neq 0, \quad l \leqslant n .$$

We can indicate $2l$-dimensional cycles on $X(\mathbb{C})$ that are not homologous to 0. For this purpose we consider a smooth subvariety $Y \subset X$ of dimension l, for example, the corresponding section by a linear subspace of a projective space $\mathbb{P}^N$ containing X. Let j denote the morphism of embedding Y in X and also the embedding $Y(\mathbb{C}) \to X(\mathbb{C})$. The homology class we consider is $j_* \omega_Y$. We show that it is not homologous to 0. Somewhat inaccurately, but more intuitively, we can express this by saying that smooth subvarieties are not homologous to 0 in an enveloping variety.

First of all, we can make a simple reduction. By combining the embedding $Y \to X$ with the embedding $X \to \mathbb{P}^N$ we see that it is sufficient to prove our proposition for the composite embedding $Y \to \mathbb{P}^N$. In other words, we may assume that $X = \mathbb{P}^N$, $N = n$.

The only result we use is Stokes' formula ([26], 34), from which it follows that if φ is a closed m-dimensional differential form on some manifold, ω an n-dimensional cycle, and

$$\int_\omega \varphi \neq 0 ,$$

then ω is not homologous to 0.

The form we construct is defined over the manifold $\mathbb{P}^N(\mathbb{C})$. To write it down we make use of the fact that a differential form over $\mathbb{C}^n$ can be expressed in terms of the differentials $dz_1, \ldots, dz_n, d\bar{z}_1, \ldots, d\bar{z}_n$ ([9], 171). Consequently such an expression is also possible in local coordinates on an algebraic variety. If φ is the form, we denote by $d'\varphi$ its partial differential with respect to the variables $z_1, \ldots, z_n$ only, and by $d''\varphi$ with respect to the variables $\bar{z}_1, \ldots, \bar{z}_n$. Then $d\varphi = d'\varphi + d''\varphi$.

Let $\zeta_0, \ldots, \zeta_n$ be homogeneous coordinates in $\mathbb{P}^n$ and ξ a linear form. In the domain $\xi \neq 0$ we set

$$\varphi = i d' d'' H = i \sum_{k,j} \frac{\partial^2 H}{\partial z_k \partial \bar{z}_j} dz_k \wedge d\bar{z}_j ,$$

$$H = \log \sum_{l=0}^{n} \left| \frac{\zeta_l}{\xi} \right|^2 .$$

Replacing ξ by another form η does not change φ in the domain where $\xi \neq \eta \neq 0$, because

$$d' d'' \log \left| \frac{\xi}{\eta} \right|^2 = 0 .$$

Thus, the form φ does not depend on the choice of ξ, and by considering various ξ we can define φ as a form on the whole of $\mathbb{P}^n(\mathbb{C})$. This form is closed owing to the fact that $d d' d'' = (d')^2 d'' + d'(d'')^2 = 0$.

Let us show that if Y is a smooth l-dimensional subvariety of $\mathbb{P}^n$, then

$$\int_{\omega_Y} \varphi^l > 0 , \tag{1}$$

where φ^l is the l-th exterior power of φ. This then gives us the required result. The inequality (1) is a consequence of the following sharper assertion.

Let $t_1, \ldots, t_l$ be local parameters in a neighbourhood of some point $y \in Y$, $t_j = u_j + i v_j$, $j = 1, \ldots, l$. We treat $u_1, v_1, \ldots, u_l, v_l$ as local parameters over the differentiable manifold $Y(\mathbb{C})$, and we let $j: Y(\mathbb{C}) \to \mathbb{P}^n(\mathbb{C})$ be the embedding. Then the form $j^* \varphi^l$ on Y has in these local parameters the form

$$j^* \varphi^l = F\, du_1 \wedge dv_1 \wedge \cdots \wedge du_l \wedge dv_l ,$$

where F is a real function and

$$F > 0 . \tag{2}$$

Compared with (1) the inequality (2) has the advantage that it bears purely local character. We now proceed to verify it.

The property of φ on which the proof of the inequality (2) is based is the following. We write φ in local coordinates, and assuming, for example, that $\zeta_0 \neq 0$, we set $z_i = \zeta_i / \zeta_0$. Then

$$\varphi = i d' d'' \log \left(1 + \sum_{i=1}^{n} |z_i|^2 \right) = i \sum c_{l,j}\, dz_l \wedge d\bar{z}_j ,$$

and here $(c_{l,j})$ is the matrix of a Hermitian positive-definite form.

The property of being Hermitian is evident from the fact that

$$c_{l,j} = \frac{\partial^2 H}{\partial z_l \partial \bar{z}_j} , \quad H = \log \left(1 + \sum_{i=1}^{n} |z_i|^2 \right) .$$

Positive-definiteness means that

$$\sum_{l,j} c_{l,j} \xi_l \bar{\xi}_j > 0 \tag{3}$$

for any $\xi_1, \ldots, \xi_n$ not vanishing simultaneously. The left-hand side of (3) can be rewritten in the form

$$d'_\xi d''_\xi \log\left(1 + \sum_{i=1}^{n} |z_i|^2\right),$$

where d'_ξ is the differential with respect to the variables $z_1, \ldots, z_n$ in the direction $\xi = (\xi_1, \ldots, \xi_n)$, that is,

$$d'_\xi f = \sum \frac{\partial f}{\partial z_i} \xi_i ,$$

and similarly

$$d''_\xi f = \sum \frac{\partial f}{\partial \bar{z}_i} \bar{\xi}_i .$$

Using this, the left-hand side of (3) can be rewritten in the form

$$\frac{1}{(1+\sum |z_i|^2)^2} \left(\left(\sum |\xi_i|^2\right)\left(1 + \sum |z_i|^2\right) - \left|\sum z_i \bar{\xi}_i\right|^2\right),$$

and by Cauchy's inequality this is not less than

$$\frac{\sum |\xi_i|^2}{(1+\sum |z_i|^2)^2} .$$

The remaining computation is simple linear algebra. If $t_1, \ldots, t_l$ are local parameters on Y, then by substituting the expression

$$dz_j = \sum \frac{dz_j}{dt_l} dt_l$$

in the expression for φ we see that

$$j^* \varphi = i \sum_{\alpha, \beta = 1}^{l} g_{\alpha,\beta} dt_\alpha \wedge d\bar{t}_\beta .$$

Here $g_{\alpha,\beta}$ is obtained from $c_{l,j}$ by means of the formula for the restriction of a bilinear form from the complex linear space $\Theta_{y,\mathbb{P}^n}$ to the complex linear subspace $\Theta_{y,Y}$. Therefore $(g_{\alpha,\beta})$ is also the matrix of a positive definite Hermitian form. Finally, a simple calculation shows that

$$j^* \varphi^l = i^l \det(g_{\alpha\beta})\, dt_1 \wedge d\bar{t}_1 \wedge \ldots \wedge dt_l \wedge d\bar{t}_l$$

and if $t_j = u_j + i v_j$, then $dt_j \wedge d\bar{t}_j = -2i\, du_j \wedge dv_j$. Therefore in (2)

$$F = 2^l \det(g_{\alpha\beta}) ,$$

and the inequality (2) follows from the fact that the determinant of a positive definite Hermitian form is a positive real number. This completes the proof of the proposition.

We give a sketch of another proof of this proposition. Although it uses more topological tools, its idea is very lucid. We recall that if M is a compact n-dimensional oriented manifold, then $H^p(M,\mathbb{C})\otimes H^{n-p}(M,\mathbb{C}) \to H^n(M,\mathbb{C})$ defines a duality between the spaces $H^p(M,\mathbb{C})$ and $H^{n-p}(M,\mathbb{C})$. Since $H^p(M,\mathbb{C})$ is dual to $H_p(M,\mathbb{C})$, the spaces $H_p(M,\mathbb{C})$ and $H_{n-p}(M,\mathbb{C})$ are also dual to each other. The corresponding scalar product is called the *intersection index* or the *Kronecker index* of the two cycles. Let V and W be two smooth oriented subvarieties of dimension p and $n-p$ in M, intersecting transversally. This means that the intersection consists of finitely many points and in each such point $x\in V\cap W$ we have $\Theta_{x,M} = \Theta_{x,V}\oplus\Theta_{x,W}$. In this case we can look at the embeddings $j_V: V\to M$, $j_W: W\to M$ and exhibit for the intersection index $(j_{V*}\omega_V\cdot j_{W*}\omega_W)$ the simple formula

$$(j_{V*}\omega_V\cdot j_{W*}\omega_W)=\sum_{x\in V\cap W} c(V,W,x)\,, \tag{4}$$

where $c(V,W,x)$ is equal to $+1$ or -1 according as the natural orientation of $\Theta_{x,V}\oplus\Theta_{x,W}$ is the same as the orientation of $\Theta_{x,M}$ or opposite. (For these results see [13], Ch. II.7.)

Now let $M=X(\mathbb{C})$, $V=Y(\mathbb{C})$, $W=Z(\mathbb{C})$, where X, Y and Z are smooth complete algebraic varieties, $Y\subset X$, $Z\subset X$, and Y and Z intersect in X transversally. Then all the terms c in (4) are $+1$. For in this case $\Theta_{x,X}$, $\Theta_{x,Y}$, and $\Theta_{x,Z}$ are complex linear spaces. The transition from a complex basis of $\Theta_{x,Y}\oplus\Theta_{x,Z}$ to a complex basis $\Theta_{x,X}$ is effected by a complex linear transformation. The corresponding real transformation carrying a real basis of $\Theta_{x,Y}\oplus\Theta_{x,Z}$ into a real basis of $\Theta_{x,X}$ has positive determinant, as we have seen in § 1.2. Therefore

$$c(Y(\mathbb{C}),Z(\mathbb{C}),x)=+1\,.$$

If X is a smooth projective variety, $Y\subset X$, and Y is smooth, then there exists a smooth subvariety Z intersecting Y transversally with respect to a non-empty subset. For Z we can take the section of X by the corresponding linear subspace. In that case (4) shows that

$$(j_{Y*}\omega_Y\cdot j_{Z*}\omega_Z)=\deg Y$$

is equal to the degree of Y in the enveloping projective space. From this it follows, of course, that $j_{Y*}\omega_Y\neq 0$.

We can establish a connection between these arguments and the earlier proof of the proposition. Namely, if $X=\mathbb{P}^n$, and if Z is a linear subspace of dimension $n-l$, then the homology class $j_{Z*}\omega_Z$ determines by means of the intersection index a linear form on $H_{2l}(X(\mathbb{C}),\mathbb{C})$. It can be shown that the form φ we have constructed corresponds, to within a factor, to the cohomology class determined by a hyperplane section, and φ^l to the class determined by Z. More accurately, the class of hyperplane sections determines the form $\frac{1}{2\pi}\varphi$. This means that the degree of a smooth l-dimensional subvariety $Y \subset \mathbb{P}^n$ can be expressed as an integral over the oriented cycle ω_Y:

$$\deg Y = \frac{1}{(2\pi)^l} \int_{\omega_Y} \varphi^l . \tag{5}$$

In conclusion we remark that the proposition is probably true for arbitrary complete varieties. However, the author does not know how to prove it.

Exercises

1. Show that the formula (5) above remains valid when Y is a curve, possibly non-smooth. *Hint:* Consider the normalization morphism.

2. Show that if X is a smooth projective curve, $\omega \in \Omega^1[X]$, and $\int_\sigma \omega = 0$ for all $\sigma \in H_1(X, \mathbb{Z})$, then $\omega = 0$. *Hint:* Consider the function $\varphi(x) = \int_{x_0}^{x} \omega$, where x_0 is a fixed and x an arbitrary point on X; show that the integral does not depend on the path of integration, that the function φ is continuous and as a function of the local parameter at x holomorphic. Show that the existence of a maximum of $|\varphi(x)|$ leads to a contradiction to the maximum modulus principle for analytic functions.

3. Show that the assertion of Exercise 2 is true for a smooth projective variety of arbitrary dimension.

4. Show that if G is an algebraic group acting on a projective smooth variety X and if the space $G(\mathbb{C})$ is connected, then $g^*\omega = \omega$ for $g \in G$, $\omega \in \Omega^1[X]$. *Hint:* Show that the cycles σ and $g_*\sigma$ are homologous.

§ 2. Connectedness

The object of this section is to show that if X is an irreducible algebraic variety over the field of complex numbers, then the space $X(\mathbb{C})$ is connected. By covering X with affine open sets (in the spectral topology) it is easy to reduce the theorem to the case when X is affine. In that case the proof is based on the fact that according to Theorem 10 in Ch. I, § 5, there exists a finite mapping $f: X \to \mathbb{A}^n$ onto an affine space. Using the terminology introduced in Ch. V we shall speak in what follows of finite morphisms. In § 2.1 we deduce some simple topological properties of algebraic varieties, and § 2.2 is devoted to a proof of the main result: the fact that $X(\mathbb{C})$ is connected. Here we make use of some simple properties of analytic functions of several complex variables, which are proved in § 2.3.

1. Auxiliary Lemmas

Lemma 1. *If X is an irreducible algebraic variety, and Y a proper subvariety, then the set $X(\mathbb{C}) - Y(\mathbb{C})$ is everywhere dense in $X(\mathbb{C})$.*

We consider first the case when X is an algebraic curve. Then Y consists of finitely many closed points. Let $v: X^v \to X$ be the normalization morphism, $Y' = v^{-1}(Y)$. Since X^v is a smooth curve, every point $y' \in Y'$ has a neighbourhood U homeomorphic to the disc $|z| < 1$ in the plane of the complex variable z. Obviously $U - y'$ is everywhere dense in U, and therefore also $X^v(\mathbb{C}) - Y'$ is everywhere dense in $X^v(\mathbb{C})$. Since v is an epimorphism, it follows that $X(\mathbb{C}) - Y$ is everywhere dense in $X(\mathbb{C})$.

The general case can be reduced to the one just considered by a simple induction on the dimension n of the variety X. Suppose that $n > 1$. For every point $y \in Y(\mathbb{C})$ there exists an irreducible subvariety X' of codimension 1 in X that contains y and does not contain any irreducible component of Y passing through y. For we take in an affine neighbourhood U of y a point $y_i \neq y$ on every irreducible component Y_i of the variety Y passing through y, and we consider the section of U by a hyperplane L in the enveloping affine space such that $y \in L$, $y_i \notin L$, for all the chosen points y_i. The closure in X of any irreducible component of this section passing through y can be taken for the variety X'. We set $Y' = X' \cap Y$. By the inductive hypothesis $X'(\mathbb{C}) - Y'(\mathbb{C})$ is everywhere dense in $X'(\mathbb{C})$. In particular, the point y lies in the closure of $X'(\mathbb{C}) - Y'(\mathbb{C})$. Therefore, a fortiori, it lies in the closure of $X(\mathbb{C}) - Y(\mathbb{C})$. Since y can be taken to be any point of $Y(\mathbb{C})$, this proves the lemma.

Corollary. *If X is an irreducible algebraic variety, $Y \neq X$ an algebraic subvariety, and $X(\mathbb{C}) - Y(\mathbb{C})$ is open in $X(\mathbb{C})$ and connected, then $X(\mathbb{C})$ is also connected.*

For if $X(\mathbb{C}) = M_1 \cup M_2$ is a decomposition into two closed disjoint sets, then $X(\mathbb{C}) - Y(\mathbb{C})$ splits into its intersections with M_1 and M_2. Since $X(\mathbb{C}) - Y(\mathbb{C})$ is connected, it must coincide with one of these intersections, hence be contained in M_1 or in M_2. But then also its closure is contained in the corresponding set. According to Lemma 1 this closure coincides with $X(\mathbb{C})$, and this means that one of the sets M_1 or M_2 is empty.

Lemma 2. *If $V \subset \mathbb{A}^n$ is closed in the spectral topology, then $V(\mathbb{C})$ is connected.*

Proof. Let $\mathbb{A}^n - V = Y$, $x_1, x_2 \in V(\mathbb{C})$. Through x_1 and x_2 we draw a line L that does not contain any irreducible component of Y. The space $L(\mathbb{C})$ is homeomorphic to $\mathbb{C}$, and $L(\mathbb{C}) \cap V(\mathbb{C})$ to the space $\mathbb{C} - \{y_1, \ldots, y_m\}$, where $\{y_1, \ldots, y_m\} = L \cap Y$. It follows that $L(\mathbb{C}) \cap V(\mathbb{C})$ is connected, hence that x_1 and x_2 belong to one and the same connected component of $V(\mathbb{C})$. Since x_1 and x_2 are arbitrary, $V(\mathbb{C})$ is connected.

2. The Main Theorem. The proof that the space $X(\mathbb{C})$ is connected is based on the following result, which reduces the problem to a simpler situation.

Lemma. *For every irreducible variety X there exists a set $U\subset X$, open in the spectral topology, and a finite morphism $f:U\to V$ onto an open subset V of the affine space $\mathbb{A}^n$ (in the spectral topology) such that the following conditions hold:*

1) *U is isomorphic to a subvariety $V(F)$ in $V\times\mathbb{A}^1$, and*

$$F(T)\in\mathbb{C}[\mathbb{A}^n][T]\subset\mathbb{C}[V\times\mathbb{A}^1]$$

is an irreducible polynomial over $\mathbb{C}[\mathbb{A}^n]$ with highest coefficient 1, *and $f:U\to V$ is induced by the projection $V\times\mathbb{A}^1\to V$;*

2) *the continuous map $f:U(\mathbb{C})\to V(\mathbb{C})$ is an unramified covering.*

Proof. Let $X'\subset X$ be an affine set that is open in the spectral topology, and let $f:X'\to\mathbb{A}^n$ be the finite morphism whose existence is guaranteed by Theorem 9 in Ch. I, § 5. In the ring $\mathbb{C}[X']\supset\mathbb{C}[\mathbb{A}^n]$ we consider a primitive element θ of the extension $\mathbb{C}(X')/\mathbb{C}(\mathbb{A}^n)$, and we denote by $F(T)$, $F\in\mathbb{C}[\mathbb{A}^n][T]$, its minimal polynomial over $\mathbb{C}(\mathbb{A}^n)$. Evidently, $\mathbb{C}[\mathbb{A}^n][\theta]\subset\mathbb{C}[X']$, and from the fact that θ is a primitive element it follows that for every element $\xi\in\mathbb{C}[X']$ there exists an element $a_\xi\in\mathbb{C}[\mathbb{A}^n]$ such that $\alpha_\xi\cdot\xi\in\mathbb{C}[\mathbb{A}^n][\theta]$. If $\xi_1,\dots,\xi_s$ is a basis of the module $\mathbb{C}[X']$ over $\mathbb{C}[\mathbb{A}^n]$ and $a=a_{\xi_1}\dots a_{\xi_s}$, then

$$a\cdot\mathbb{C}[X']\subset\mathbb{C}[\mathbb{A}^n][\theta]\,. \tag{1}$$

We denote by $d\in\mathbb{C}[\mathbb{A}^n]$ the discriminant of the polynomial $F(T)$, we set $c=a\cdot d$, $V=\mathbb{A}^n-V(c)$, $U=X'-V(c)$, and as before, we denote by $f:U\to V$ the restriction of the morphism f. We show that for U, V, and f the conditions 1) and 2) of the lemma are satisfied.

Obviously, by virtue of (1) $\mathbb{C}[U]=\mathbb{C}[X']_c=\mathbb{C}[\mathbb{A}^n][\theta]_c=\mathbb{C}[V][\theta]$. Therefore U is isomorphic to the subset $V(F)$ in $V\times\mathbb{A}^1$, and 1) holds. Let $m=\deg F$, $v_0\in V$. Since $d(v_0)\neq 0$ (d is the discriminant of F), the equation $F(v_0)(T)=0$ has m distinct roots $\mu_1,\dots,\mu_m$, and

$$F'_T(v_0,\mu_i)\neq 0\,,\quad i=1,\dots,m\,. \tag{2}$$

We denote by $z_1,\dots,z_n$ coordinates in $\mathbb{A}^n$. They determine local parameters at $v_0\in\mathbb{A}^n$. By (2) they [more accurately, $f^*(z_1),\dots,f^*(z_n)$] determine local parameters also at $u_i=(v_0,\mu_i)$. Therefore, there exist neighbourhoods $\tilde U_i$ of u_i and $\tilde V$ of v_0 in the complex topology such that the mappings $u\to(f^*(z_1),\dots,f^*(z_n))$ and $v\to(z_1,\dots,z_n)$ are homeomorphisms of $\tilde U$ and $\tilde V$ onto one and the same domain of $\mathbb{C}^n$. This shows that $f:\tilde U_i\to\tilde V$ is a homeomorphism. We can take $\tilde V$ so small that the $\tilde U_i$ do not intersect for $i\neq j$. Then $f^{-1}(\tilde V)=\bigcup\tilde U_i$. For the number of inverse images of a point $v \in \tilde{V}$ cannot exceed m, but it has already m inverse images in $\tilde{U}_1, \ldots, \tilde{U}_m$. This completes the verification of condition 2).*

Theorem. *If X is an irreducible algebraic variety over the field of complex numbers, then the space $X(\mathbb{C})$ is connected.*

Proof. Let U be the set whose existence is guaranteed by the lemma. According to the Corollary to Lemma 1 we need only prove that the set $U(\mathbb{C})$ is connected. Let $U(\mathbb{C}) = M_1 \cup M_2$ be a decomposition into two closed disjoint subsets. The mapping $f: U(\mathbb{C}) \to V(\mathbb{C})$ constructed in the lemma carries open sets into open sets and closed sets into closed sets. Since M_1 and M_2 are open and closed in $U(\mathbb{C})$, we see that $f(M_1)$ and $f(M_2)$ are open and closed in $V(\mathbb{C})$. By Lemma 2, $V(\mathbb{C})$ is connected, hence $f(M_1) = f(M_2) = V(\mathbb{C})$.

Clearly the restriction of f to M_1 determines an unramified covering $f_1: M_1 \to V(\mathbb{C})$. From the fact that $V(\mathbb{C})$ is connected it follows easily that the number of inverse images in M_1 of a point $v \in V(\mathbb{C})$ is one and the same for all v. We denote this number, the so-called degree of the covering $f: M_1 \to V(\mathbb{C})$, by r. Since also $f(M_2) = V(\mathbb{C})$, we have $r < m$, where m is the degree of the covering f.

For a point $v \in V(\mathbb{C})$ we take a neighbourhood V_v such that $f_1^{-1}(V_v) = U_1 \cup \cdots \cup U_r$, $U_i \cap U_j = \emptyset$ for $i \neq j$ and that the restriction of f_1 to U_i is a homeomorphism $f_i: U_i \to V_v$, $i = 1, \ldots, r$.

For every function $\theta \in \mathbb{C}[U]$ that is integral over $\mathbb{C}[\mathbb{A}^n]$ we consider its restrictions $\theta_1, \ldots, \theta_r$ to the sets $U_1, \ldots, U_r$, and we denote by $g_1, \ldots, g_r$ their elementary symmetric functions. The idea of the next argument is as follows: we show that there exist polynomials $p_1, \ldots, p_r \in \mathbb{C}[\mathbb{A}^n]$ such that at all points $v \in f(M_1)$ their restrictions to V_v coincide with $g_1, \ldots, g_r$. Hence it follows easily that θ satisfies a relation

$$\theta^r - f^*(p_1)\theta^{r-1} + \cdots + (-1)^r f^*(p_r) = 0 \tag{3}$$

at all points $x \in M_1$.

Since a similar relation (with another $r' < m$) holds at the points $x \in M_2$ there exist polynomials $P_1, P_2 \in \mathbb{C}[\mathbb{A}^n][T]$ of degree less than m such that $P_i(\theta) = 0$ in M_i, $i = 1, 2$. Therefore $P_1(\theta) P_2(\theta) = 0$ in $\mathbb{C}[U]$, and since $\mathbb{C}[U]$ has no divisors of zero, we see that the function $\theta \in \mathbb{C}[U]$ satisfies an equation over $\mathbb{C}[\mathbb{A}^n]$ of degree less than m. This contradicts the fact that by definition $m = [\mathbb{C}(U) : \mathbb{C}(\mathbb{A}^n)]$.

Proceeding to putting this plan into effect we observe, first of all, that the functions $(f_i^{-1})^*(\theta)$ are analytic on the set V_v in the coordinates $z_1, \ldots, z_n$ in $\mathbb{A}^n(\mathbb{C})$. For according to the lemma, the local parameters at $u_i = f_i^{-1}(v)$ can be expressed as analytic functions of $f^*(z_1), \ldots, f^*(z_n)$,

* The verification of condition 2) could, of course, be replaced by a reference to Ch. II, § 5.3. We have preferred to give an independent simple proof, to avoid using the more difficult Theorem 8 in Ch. II, § 5.

and θ is in a sufficiently small neighbourhood an analytic function of the local parameters at v. Thus, $g_1, \ldots, g_r$ are also analytic functions of $z_1, \ldots, z_n$ in V_v. Consequently, each of the functions g_i is analytic in $z_1, \ldots, z_n$ on the whole set $V(\mathbb{C})$. We recall that $V(\mathbb{C})$ is obtained from the whole space $\mathbb{A}^n(\mathbb{C})$ by excluding points of an algebraic subvariety $S \subset \mathbb{A}^n$. We consider the behaviour of the functions g_i in a neighbourhood of a point $s \in S(\mathbb{C})$. By the choice of θ it satisfies an equation

$$\theta^m + f^*(a_1)\,\theta^{m-1} + \cdots + f^*(a_m) = 0\,, \quad a_i \in \mathbb{C}[\mathbb{A}^n]\,. \tag{4}$$

The values of the functions $(f_i^{-1})^*(\theta)$ are roots of this equation. Therefore, in every compact neighbourhood of s the functions g_i are bounded. From this it follows that they may be extended to functions analytic in the whole of $\mathbb{A}^n(\mathbb{C})$ (Lemma 1 in § 2.3).

We show that the so obtained analytic functions g_i on $\mathbb{A}^n(\mathbb{C})$ are polynomials in the coordinates $z_1, \ldots, z_n$.

For this purpose we estimate the order of their growth in dependence of the growth of $\max|z_i|$. For a point $z = (z_1, \ldots, z_n) \in \mathbb{A}^n(\mathbb{C})$ we set $|z| = \max|z_i|$. Applying to the equation (2) the standard estimate for the modulus of a root of an algebraic equation we find that

$$|\theta(x)| < 1 + \max_i |a_i(f(x))|\,.$$

By hypothesis, the functions a_i are polynomials in $z_1, \ldots, z_n$. If the maximum of the degrees of these polynomials is equal to l, then there exists a constant C such that

$$|\theta(x)| < C|z|^l\,.$$

Hence it follows that $(f_i^{-1})^*(\theta)$ for every $i = 1, \ldots, r$ satisfies the same inequality and therefore that

$$|g_i(z)| < C|z|^{il}\,, \quad i = 1, \ldots, r\,.$$

So we see that the $g_i(z)$ are analytic functions in the whole of $\mathbb{A}^n(\mathbb{C}) = \mathbb{C}^n$ having polynomial growth. Consequently they are polynomials in $z_1, \ldots, z_n$ (Lemma 2 in § 2.3).

So we have established the relation (3), and hence have concluded the proof of the theorem.

3. Analytic Lemmas. We prove here the two lemmas on analytic functions of complex variables that were used in the preceding subsection.

Lemma 1. *Let $S \subset \mathbb{A}^n$, $S \neq \mathbb{A}^n$ be an algebraic subvariety and g an analytic function on the set $\mathbb{A}^n(\mathbb{C}) - S(\mathbb{C})$ that is bounded in the neighbourhood of every point $s \in S(\mathbb{C})$. Under these conditions g can be extended to a* *holomorphic function on the whole of* $\mathbb{A}^n(\mathbb{C}) = \mathbb{C}^n$, *and there is only one such extension.*

The uniqueness of the extension follows at once from the uniqueness theorem for analytic functions.

Clearly it is sufficient for us to find for every point $s \in S(\mathbb{C})$ a neighbourhood U such that g can be extended from $U - (U \cap S(\mathbb{C}))$ to U as an analytic function. Then we obtain a single extension by virtue of its uniqueness.

To prove the existence of the extension we observe that we may replace S by a larger algebraic subvariety and may therefore assume that S is given by a single equation $f(z_1, \ldots, z_n) = 0$. Making the transformation $z_i' = z_i + c_i z_n$, $i = 1, \ldots, n-1$, $z_n' = z_n$, for suitably chosen numbers $c_1, \ldots, c_{n-1}$ we can achieve that f as a polynomial in z_n has highest coefficient 1:

$$f(z_1, \ldots, z_n) = z_n^l + h_1(z')\, z_n^{l-1} + \cdots + h_l(z'),$$

where $z' = (z_1, \ldots, z_{n-1})$.

Suppose that s is the origin of coordinates. Then

$$f(0, \ldots, 0, z_n) = z_n^m (z_n - \lambda_1), \ldots, (z_n - \lambda_{n-m}).$$

By the theorem on the continuity of the roots of an algebraic equation the roots of $f(z'; z_n) = 0$ tend, as $z' \to 0$, either to 0 or to $\lambda_1, \ldots, \lambda_{n-m}$. Therefore, we can find a real number $r > 0$ and an $\varepsilon > 0$ such that for $|z'| < \varepsilon$ the equation $f(z'; z_n) = 0$ does not have roots with $|z_n| = r$. We define $|z'| = \max\limits_{1, \ldots, n-1} |z_i|$.

We set

$$G(z_1, \ldots, z_n) = \frac{1}{2\pi i} \int\limits_{|w| = r} \frac{g(z'; w)}{w - z_n}\, dw$$

and show that G is an analytic function for $|z'| < \varepsilon$, $|z_n| < r$, and is an extension of g to this domain.

The fact that G is analytic can be verified directly by integration. By assumption, g is analytic at every point $(\alpha_1, \ldots, \alpha_{n-1}, w)$, $|\alpha_i| < \varepsilon$, $|w| = r$. Therefore the function $\dfrac{g(z_1, \ldots, z_{n-1}; w)}{w - z_n}$ for every w with $|w| = r$ is analytic at the point $z_1 = \alpha_1, \ldots, z_{n-1} = \alpha_{n-1}$, $z_n = \beta$, $|\beta| < r$. Integrating its expansion in a Taylor series

$$\frac{g(z_1, \ldots, z_{n-1}; w)}{w - z_n} = \sum c_{l_1, \ldots, l_n}(w)\, (z_1 - \alpha_1)^{l_1}, \ldots, (z_{n-1} - \alpha_{n-1})^{l_{n-1}} (z_n - \beta)^{l_n} \qquad (1)$$

with respect to the circle $|w| = r$, we obtain the Taylor expansion for G.

We now show that G agrees with g where g is defined. For this purpose we set $z_i = \alpha_i$, $i = 1, \ldots, n-1$, $|\alpha_i| < \varepsilon$, and we consider the functions $g(\alpha; z_n)$ and $G(\alpha; z_n)$. The preceding arguments show that $G(\alpha; z_n)$ is analytic for $|z_n| < r' < r$, and $g(\alpha; z_n)$ is by hypothesis analytic at all points z_n with $|z_n| < r'$, except possibly the finitely many roots of the equation $f(\alpha; z_n) = 0$, but is bounded in their neighbourhood. Therefore, it has no poles for $|z_n| < r$, and formula (1) shows that $G(\alpha; z_n) = g(\alpha; z_n)$ by Cauchy's integral formula.

Lemma 2. *Let $f(z_1, \ldots, z_n)$ be an analytic function in the whole of $\mathbb{C}^n$ and suppose that there exists a constant C such that*

$$|f(z)| < C|z|^l, \ z = (z_1, \ldots, z_n), \ |z| = \max |z_i| . \tag{2}$$

Then f is a polynomial of degree $\leqslant l$.

Suppose that in the expansion of f in a Taylor series

$$f = F_0 + F_1 + \cdots$$

a homogeneous constituent F_j for some $j > l$ is not identically 0. We can find numbers $\alpha_1, \ldots, \alpha_n$ such that $F_j(\alpha_1, \ldots, \alpha_n) \neq 0$. Then for $g(w) = f(\alpha_1 w, \ldots, \alpha_n w)$ the coefficient of w^j in the Taylor series is non-zero, and, as before, the estimate (2) holds. Subtracting from g the sum of the first l terms of the Taylor series we obtain the function

$$g_1(w) = a_l w^l + \cdots ,$$

for which $a_l \neq 0$ and (2) is also true.

By hypothesis, the function g_1/w^l is bounded in the whole plane, hence constant. This contradicts the fact that $a_l \neq 0$.

Exercises

1. Show that if X is a quasiprojective variety and the space $X(\mathbb{C})$ is compact, then X is projective.
2. Show that if the space $X(\mathbb{C})$ is compact, then the variety X is complete. *Hint:* Use Exercise 7 to Ch. VI, § 2.
3. Let X be a reduced and irreducible scheme of finite type over $\mathbb{C}$, and $X(\mathbb{C})$ the set of its closed points equipped with the same topology as in § 1.1. Show that if $X(\mathbb{C})$ is a Hausdorff space, then X is separated.
4. Show that the automorphism group of a non-hyperelliptic smooth projective curve of genus $\neq 0$ is finite. *Hint:* Show that if $\varphi: X \to \mathbb{P}^{g-1}$ is the embedding corresponding to the canonical class, then the automorphisms of the curve $\varphi(X)$ are induced by projective transformations in P^{g-1} and therefore form an algebraic group G. Apply Exercise 4 to § 1.
5. Extend the result of Exercise 4 to hyperelliptic curves.

§ 3. The Topology of Algebraic Curves

The association of the topological space $X(\mathbb{C})$ with an algebraic variety X leads to problems of two types. Firstly, it is interesting to clarify what topological spaces we obtain in this manner, and if possible, to go as far as a topological classification of them. Secondly, to investigate which of the invariants of the topological space $X(\mathbb{C})$ have an algebraic meaning. In other words, it is a question of constructing for a variety X defined over an arbitrary field k invariants which for $k=\mathbb{C}$ would turn into prescribed invariants of $X(\mathbb{C})$.

In this section we indicate how to solve problems of the two types in the simplest case, when X is a smooth projective curve. However, this is almost the only case in which a topological classification of the spaces $X(\mathbb{C})$ and the algebraic meaning of the resulting topological invariants are completely known.

1. The Local Structure of Morphisms. Let X be a smooth algebraic curve. Every point $x\in X(\mathbb{C})$ has a neighbourhood U (in the complex topology) that is homeomorphic to a neighbourhood of the origin of coordinates in the complex plane $\mathbb{C}$. This homeomorphism can be given by any local parameter t at x:

$$t: U \to \mathbb{C}. \tag{1}$$

From now on we assume U to be chosen so that $t(U)$ is the domain $|z|<\varepsilon$ in $\mathbb{C}$. The mapping (1) allows us to regard t as a coordinate in U so that the number $t(x)$, $x\in U$, determines the point x uniquely.

Let $f: X\to Y$ be a morphism of smooth curves, $f(X)$ dense in Y, $x\in X$, $f(x)=y$. We show that the points x and y have neighbourhoods U and V and in them coordinates u and v in which the mapping f has a very simple expression.

We choose local parameters t and v at the points x and y and neighbourhoods U of x and V of $f(x)$ such that $f(U)\subset V$ and the open embeddings (1) hold:

$$t: U\to\mathbb{C},$$
$$v: V\to\mathbb{C}.$$

Regarding t as coordinate on U and v on V, we can say that f is defined in U by prescribing $v(f(x'))$ as a function of $t(x')$ for points $x'\in U$. In other words, to specify f we have to represent $f^*(v)$ as a function of t.

Since $f(X)$ is dense in Y, the function $f^*(v)$ does not vanish identically on X, is regular, and vanishes at x. We set

$$f^*(v)=t^l\varphi, \quad \varphi\in\mathcal{O}_x, \quad \varphi(x)\neq 0. \tag{2}$$

To φ there corresponds a formal power series $\Phi(T)$, $\Phi(0)\neq 0$, with a positive radius of convergence. Therefore, by taking, if necessary, for U a smaller neighbourhood of x we may assume that

$$\varphi(x')=\Phi(t(x')), \quad x'\in U .$$

Since $\Phi(0)\neq 0$, there exists a power series $\psi(T)=\Phi(T)^{1/l}$, also having a positive radius of convergence. Therefore the function $u(x')=\psi(t(x'))\cdot t(x')$ is defined for points x' in a sufficiently small neighbourhood of x, which we denote again by U. So we have constructed a mapping

$$u: U\to\mathbb{C},$$

with

$$t^l\varphi=u^l \text{ in } U .$$

The function $u=t\cdot\Psi(t)$ is no longer rational on X and is defined only in a sufficiently small complex neighbourhood U of x. However, in this neighbourhood it is evidently continuous. Like t, it determines a homeomorphism of some neighbourhood of x onto an open set in $\mathbb{C}$. For by the implicit function theorem the analytic function $z\Psi(z)$ has in some neighbourhood of 0 an inverse function and therefore determines a homeomorphism of some neighbourhood of 0 in $\mathbb{C}$.

By construction, $f^*(v)=u^l$ in the open set U. Thus, we have reached the end of our analysis by obtaining a simple local representation for the mapping f.

Theorem 1. *For every morphism of smooth curves $f: X\to Y$ for which $f(X)$ is dense in Y, and for every point $x\in X$ there exist neighbourhoods U of x and V of $f(x)$ and homeomorphisms $u: U\to\mathbb{C}$ and $v: V\to\mathbb{C}$ onto neighbourhoods of 0 in $\mathbb{C}$ such that*

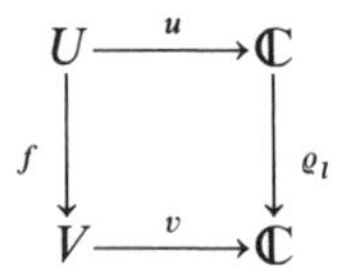

is commutative. Here $\varrho_l(z)=z^l$, and l is defined as the order of the zero of $f^*(t)$ at x if t is a local parameter at y.

If u and v are interpreted as coordinates in U and V, then Theorem 1 asserts that under restriction to these sets and in these coordinates the mapping f can be expressed in the very simple form

$$v=u^l . \tag{3}$$

Obviously, the open sets U and V can be chosen so that $u(U)$ and $v(V)$ coincide with the open disc $|z|<1$.

Such neighbourhoods are called *normal.*

The number l in (2) and (3) is called the *order of ramification* of f at the point $x \in X$. If the order of ramification of f in at least one point $x \in f^{-1}(y)$ is greater than 1, then y is called a ramification point of f.

Clearly, the order of ramification of f at x is the same as the multiplicity with which x occurs in the divisor $f^*(y)$. If the curves X and Y are not only smooth, but also projective, then Theorem 1 of Ch. III, § 2, shows that y is not a ramification point if and only if the number of inverse images of this point is equal to the degree $\deg f$ of the morphism f. In other words, our definition is compatible with the definition of a ramification point given in Ch. II, § 5.3. From Theorem 7 in Ch. II, § 5, it follows that a morphism of smooth projective curves has finitely many ramification points.

2. Triangulation of Curves. Here we show that the space $X(\mathbb{C})$ is triangularizable, where X is a smooth projective algebraic curve. For the convenience of the reader the definition of triangulation and the basic facts on the classification of triangularizable two-dimensional manifolds are compiled in § 3.4.

The triangularizability of the space $X(\mathbb{C})$ is obtained as a consequence of a more general fact. To state it we introduce the following definition.

A triangulation Φ of a topological space X is said to be *compatible* with a triangulation Ψ of a space Y relative to a continuous mapping $f: X \to Y$ if $f^{-1}(E) = \bigcup E_i$, $E_i \in \Phi$, for every simplex $E \in \Psi$, and if the mapping $f: E_i \to E$ is a homeomorphism.

Theorem 2. *If $f: X \to Y$ is a morphism of smooth projective curves and if the space $Y(\mathbb{C})$ is triangularizable, then $X(\mathbb{C})$ and $Y(\mathbb{C})$ have triangulations that are compatible relative to f. If $Y(\mathbb{C})$ is a combinatorial surface, then so is $X(\mathbb{C})$.*

Proof. We consider an arbitrary point $y \in Y(\mathbb{C})$, and we set $f^{-1}(y) = \{x_1, \ldots, x_l\}$. According to Theorem 1 we can take a neighbourhood V_y of y and neighbourhoods U_i of the points x_i that are normal and pairwise disjoint.

From the covering V_y of $Y(\mathbb{C})$ we select a finite subcovering. We obtain a finite set of points y_α in $Y(\mathbb{C})$, for each point y_α a neighbourhood V_α, and for $x_{\alpha,i} \in f^{-1}(y_\alpha)$ normal disjoint neighbourhoods $U_{\alpha,i}$. Obviously, if $y \in V_\alpha$, $y \neq y_\alpha$, then y is not a ramification point.

By assumption $Y(\mathbb{C})$ has a triangulation Ψ_0. By a proposition to be proved in § 3.4 there exists a finer triangulation Ψ such that all ramification points of the morphism f are vertices of it and that every simplex of it is contained in some neighbourhood V_α.

Let E be an arbitrary simplex of the triangulation Ψ. By hypothesis, E is contained in some open set V_α. If y_α is not a ramification point, then $f_i: U_{i,\alpha} \to V_\alpha$ for every $x_i \in f^{-1}(y_\alpha)$ is a homeomorphism. We denote by E_i the inverse image $f^{-1}(E)$ in $U_{i,\alpha}$. If $t: E \to \sigma$ is the mapping that occurs in the definition of the triangulation Ψ, then we set $t_i = t f_i: E_i \to \sigma$. We include the sets E_i and the homeomorphisms t_i in the triangulation Φ.

Suppose now that $E \subset V_\alpha$ and that y_α is a ramification point. We consider two cases.

a) $y_\alpha \notin E$. In suitable coordinates the mapping $f_i: U_{i,\alpha} \to V_\alpha$ has the form $v = f^*(u) = u^l$. Since $y_\alpha \notin E$ and the set E is simply-connected, we see that $v \neq 0$ on E and that every branch of the function $\sqrt[l]{v}$ determines there a single-valued function. From this it follows that $f_i^{-1}(E)$ splits into l connected components $E_1, \ldots, E_l$ and that the mapping $f_i: E_i \to E$ is a homeomorphism. We include the sets E_i and the mappings $t_j = t f_j: E_j \to \sigma$ in the triangulation Φ.

b) $y_\alpha \in E$. Now we can apply the same arguments to the set $E - y_\alpha$. We find that $f^{-1}(E - y_\alpha)$ splits into l connected components $\tilde{E}_1, \ldots, \tilde{E}_l$.

Setting

$$E_j = \tilde{E}_j \cup x_i \qquad (x_i \in f^{-1}(y_\alpha) \cap U_{i,\alpha}),$$

we see that $f_j: E_j \to E$ is a homeomorphism. Again we include all the E_j and the mappings $t_j = t f_j: E_j \to \sigma$ in the triangulation Φ.

A simple verification shows that the sets and mappings so constructed determine a triangulation Φ of the space $X(\mathbb{C})$, and also that this triangulation is compatible with Ψ relative to f and satisfies the definition of a combinatorial surface. This verification is left to the reader.

Theorem 3. *If X is a smooth projective curve, then the space $X(\mathbb{C})$ is triangularizable and is a combinatorial surface.*

We show first the triangularizability of the space $\mathbb{P}^1(\mathbb{C})$. We indicate a special triangulation, which is not very economical, but is useful for subsequent applications.

The decomposition of the surface of an octahedron into faces, edges, and vertices gives a triangulation of it. We imagine the octahedron to be inscribed in a two-dimensional sphere. By projecting this triangulation from an interior point of the octahedron we obtain a triangulation of the two-dimensional sphere. Since the projective line over the field of complex numbers is homeomorphic to a two-dimensional sphere, we obtain a triangulation of the projective line.

We identify $\mathbb{P}^1(\mathbb{C})$ with the plane of a complex variable, supplemented by the point at infinity. The triangulation we have constructed is given by the decomposition corresponding to the real axis, the imaginary axis, and the circle $|z|=1$.

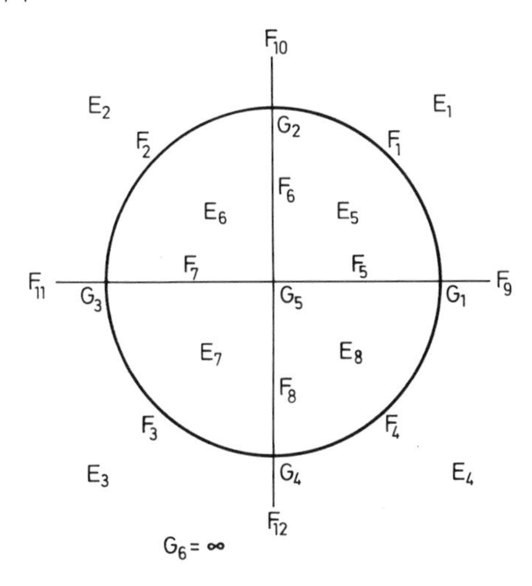

$$d(E_i)=2, i=1, \ldots, 8; d(F_i)=1, \\ i=1, \ldots, 12; d(G_i)=0, \quad i=1, \ldots, 6.$$

Now we consider a non-constant rational function f on X. It determines a morphism $f: X \to \mathbb{P}^1$. It remains to apply Theorem 2.

3. Topological Classification of Curves. We apply the topological classification of surfaces that will be explained in § 3.4 to the surfaces $X(\mathbb{C})$, where X are smooth projective curves. According to Theorem A of § 3.4, for this purpose we have to clarify whether the spaces $X(\mathbb{C})$ are orientable, and we have to compute their Euler characteristics.

Theorem 4. *If X is a smooth projective curve, then the space $X(\mathbb{C})$ is orientable.*

This is, of course, a special case of Proposition 1 in § 1.2; however, we give another much more elementary proof.

We use an arbitrary morphism $f: X \to \mathbb{P}^1$ and consider triangulations Φ and Ψ of the spaces $X(\mathbb{C})$ and $\mathbb{P}^1(\mathbb{C})$ that are compatible relative to f. Their existence follows from Theorem 2. Since the two-dimensional sphere is orientable, so is the triangulation Ψ.

From the fact that Φ and Ψ are compatible it follows that the simplexes of Φ are precisely those into which the inverse images of simplexes in Ψ decompose. Therefore every simplex $E \in \Phi$ is mapped by f homeomorphically onto a simplex $F \in \Psi$.

We fix an orientation of Ψ, that is, of all its triangles. To a triangle $E \in \Phi$ we give the orientation that is obtained from the orientation of $f(E) \in \Psi$ by means of the homeomorphism $f: E \to f(E)$. It remains to verify that in this way we obtain an orientation of the whole triangulation Φ. This is simple. Suppose that two triangles E' and E'' have a common edge E with the vertices b and c. We denote the vertices of E' and E'' by a, b, c and b, c, d. We set $f(a)=a'$, $f(b)=b'$, $f(c)=c'$, $f(d)=d'$. Then a', b', c' and b', c', d' are the vertices of the triangles $f(E')$ and $f(E'')$ of Ψ. Suppose that the chosen orientation gives for $f(E')$ the order (a', b', c') of the vertices. Then according to the definition of the orientation of a triangulation, the order in $f(E'')$ must be (c', b', d'). By hypothesis, the order of the vertices in E' and E'' is (a, b, c) and (c, b, d), hence it is clear that on E they determine opposite orientations. This proves the theorem.

Now we proceed with the second problem: the definition of the Euler characteristic of $X(\mathbb{C})$ for an irreducible smooth projective curve X.

Theorem 5. *The Euler characteristic of the space $X(\mathbb{C})$ is $2-2g$, where g is the genus of X.*

Again we use a regular mapping $f: X \to \mathbb{P}^1$ and compatible triangulations Φ and Ψ of the spaces $X(\mathbb{C})$ and $\mathbb{P}^1(\mathbb{C})$. We denote the numbers in the Definition (2) of § 3.4 of the Euler characteristic by c_0, c_1, c_2 for Φ and c'_0, c'_1, c'_2 for Ψ. Then

$$\chi(\mathbb{P}^1(\mathbb{C})) = c'_0 - c'_1 + c'_2\,, \quad \chi(X(\mathbb{C})) = c_0 - c_1 + c_2\,.$$

Let us see how these numbers are connected. By the definition of compatibility of triangulations, for every simplex $E \in \Psi$

$$f^{-1}(E) = \bigcup E_i\,, \tag{1}$$

where $f: E_i \to E$ is a homeomorphism. In the proof of Theorem 4 we have seen that by sorting out all the simplexes $E \in \Psi$ we obtain among the E_i all the simplexes of Φ.

How many simplexes E_i are there in (1)? If $\deg f = n$ and $d(E) > 0$, then there are n of them. For their number cannot be larger than n, because every point $y \in \mathbb{P}^1$ has at most n inverse images. But it cannot be smaller than n, because there are only finitely many points (the ramification points) having fewer than n inverse images. Let $d(E)=0$, so that E is a point $y \in \mathbb{P}^1(\mathbb{C})$. Denoting by $\bar{y}$ the divisor consisting of the single point y with multiplicity 1 we set

$$f^*(\bar{y}) = \sum_{i=1}^{r} l_i \bar{x}_i\,. \tag{2}$$

The number of inverse images of y is r, but since $\sum_{i=1}^{r} l_i = n$, we have $r = n - \Sigma(l_i - 1)$. Therefore we obtain

$$c_2 = nc_2',$$
$$c_1 = nc_1',$$
$$c_0 = nc_0' - \sum(l_i - 1),$$

where the last sum contains the ramification orders of f at all points. As a result we see that

$$\chi(X(\mathbb{C})) = \chi(\mathbb{P}^1(\mathbb{C})) \cdot n - \sum(l_i - 1).$$

On the other hand, from the triangulation (1) of § 3.2, for example, it is clear that $\chi(\mathbb{P}^1(\mathbb{C})) = 2$. Therefore finally

$$\chi(X(\mathbb{C})) = 2n - \sum(l_i - 1). \tag{3}$$

We now consider an arbitrary rational differential form $\omega \neq 0$ on $\mathbb{P}^1$ and compute the divisor $(f^*(\omega))$ of the inverse image of ω on X. Suppose that at $y \in \mathbb{P}^1$ we have $v_y(\omega) = m$. Then

$$\omega = t^m \cdot g \cdot dt,$$

where t is a local parameter at y, $g \in \mathcal{O}_y$, $g(y) \neq 0$. If the numbers l_i denote the same thing as in (3), then $v_{x_i}(f^*(t)) = l_i$, that is, $f^*(t) = \tau^{l_i} h_i$, where τ is a local parameter at x_i, $h_i \in \mathcal{O}_{x_i}$, $h_i(x_i) \neq 0$. From this it follows that

$$v_{x_i}(f^*(\omega)) = ml_i + l_i - 1.$$

In other words,

$$(f^*(\omega)) = f^*((\omega)) + \sum(l_i - 1)\,\bar{x}_i, \tag{4}$$

where, as in (3), the last sum is extended over all points $x_i \in X$ at which the ramification exponent of f is greater than 1.

Since $f^*(\omega)$ is a differential form on X, we have $\deg((f^*(\omega))) = 2g - 2$. In exactly the same way $\deg((\omega)) = -2$. Finally, for every divisor D on $\mathbb{P}^1$ the equation $\deg(f^*(D)) = n \deg D$ holds, because this is true for a divisor consisting of a single point, by Theorem 1 of Ch. III, § 2. Considering the degrees of the divisors on the two sides of (4) we therefore find

$$2g - 2 = -2n + \sum(l_i - 1).$$

Comparing this formula with (3) we obtain Theorem 5.

Theorems 4 and 5 in conjunction with the topological theorem A of § 3.4 give us a complete topological classification of smooth projective curves. They show that for two such curves the spaces $X(\mathbb{C})$ are homeomorphic if and only if the curves have the same genus.

No similar result is known for varieties X of dimension >1. We quote one of the simplest results on the connection between topological and algebraic properties of smooth complete varieties, which generalizes Theorem 5. Since $\chi(X(\mathbb{C}))=b_0-b_1+b_2$, where b_1 is the one-dimensional Betti number of $X(\mathbb{C})$ and $b_0=b_2=1$, Theorem 5 can be expressed by the equation

$$b_1=2g\,.$$

If X is an arbitrary projective smooth variety, then a similar result holds:

$$b_1=2h^1\,,$$

where b_1 is the one-dimensional Betti number of the space $X(\mathbb{C})$ and $h^1=\dim_{\mathbb{C}}\Omega^1[X]$. Using more delicate constructions we can express all the remaining Betti numbers of $X(\mathbb{C})$ in terms of algebraic invariants of the variety X.

In conclusion we mention that the topological classification of smooth projective curves can be obtained by other means, within the framework of the theory of differentiable manifolds. Then we have to use a less elementary topological apparatus, but on the other hand, the exposition is more invariant. We sketch only the general course of this exposition, omitting all details.

In § 1.2 we have seen how to prove orientability on the space $X(\mathbb{C})$ for every smooth variety X, by using concepts of the theory of differentiable manifolds. The combinatorial classification of surfaces (Theorem A of § 3.4) must be replaced by its "smooth" analogue, the theorem that every connected compact orientable surface can be obtained by attaching finitely many handles to the sphere. The proof of this theorem follows easily from Morse theory (see, for example [35]). It remains to prove Theorem 5. For this purpose we have to consider the tangent bundle θ of the two-dimensional surface $X(\mathbb{C})$. Its first Chern class $c_1(\theta)$ is an element of the group $H^2(X(\mathbb{C}),\mathbb{Z})$. This group has a canonical generator: the homology class φ for which $\varphi(w_X)=1$, where w_X is the orienting class of X. Therefore $c_1(\theta)=v\cdot\varphi$, $v\in\mathbb{Z}$, hence the integer v is determined. In our class c_1 is the Euler class, and therefore $v=\chi(X(\mathbb{C}))$, that is, $c_1(\theta)=\chi(X(\mathbb{C}))\cdot\varphi$, where $\chi(X(\mathbb{C}))$ is the Euler characteristic of $X(\mathbb{C})$ (see [19], 388).

On the other hand, as an algebraic vector bundle over X, θ is one-dimensional and corresponds to the divisor class $-K$, where K is the canonical class of X. Since $\deg(-K)=2-2g$, the relation

$$\chi(X(\mathbb{C}))=2-2g$$

follows from the general result.

If E is a one-dimensional vector bundle on a smooth projective curve X, D its characteristic class, and $c_1(E)$ the Chern class of the corresponding bundle on $X(\mathbb{C})$, then

$$c_1(E)=(\deg D)\varphi .$$

A proof of this for the case of a divisor D on a variety X of arbitrary dimension is given in [11].

4. Combinatorial Classification of Surfaces. For the convenience of the reader we recall here some elementary topological concepts and results.

Let V be an n-dimensional affine space over the field of real numbers. Then any two points $P, Q \in V$ determine the vector $\overrightarrow{PQ}$ belonging to the n-dimensional vector space $\mathbb{R}^n$, and every vector $x \in \mathbb{R}^n$ and point $P \in V$ determine a point $Q \in V$ such that $\overrightarrow{PQ}=x$; this is written as $P+x=Q$. For any points $P_1, \ldots, P_m \in V$ and numbers $\lambda_1, \ldots, \lambda_m \in \mathbb{R}$ such that $\Sigma\lambda_i=1$, the point $Q+\Sigma\lambda_i\overrightarrow{QP_i}$ does not depend on the choice of the auxiliary point Q and is written in the form $\Sigma\lambda_i P_i$. If the points $P_0, \ldots, P_r \in V$ do not lie in any affine subspace of dimension less than r, then the representation of R in the form $R=\sum_{i=0}^{r}\lambda_i P_i$, $\Sigma\lambda_i=1$, is unique.

A set of points $R \in V$ that are representable in the form

$$R=\sum_{i=0}^{r}\lambda_i P_i, \quad \sum\lambda_i=1, \quad \lambda_i \geqslant 0, \tag{1}$$

in terms of independent points $P_0, \ldots, P_r$ is called an r-dimensional *simplex*. The points P_i are called its *vertices*.

If $P_{i_1}, \ldots, P_{i_{r-s}}$ are any $r-s$ vertices of a simplex σ, then the points $R \in \sigma$ for which $\lambda_{i_1}=\cdots=\lambda_{i_{r-s}}=0$ in (1) themselves form an s-dimensional simplex with the vertices P_j, $j \neq i_1, \ldots, i_{r-s}$. This simplex is called a *face* of σ.

Let X be a Hausdorff space. We define a *triangulation* (more accurately, a finite triangulation) of X. This is the name for: a) a finite family Φ of closed subsets E_i of X, b) the assignment of a non-negative integer $d(E_i)$ to every subset $E_i \in \Phi$, and c) a homeomorphism $t_i: E_i \to \sigma_i$, where σ_i is a simplex of dimension $d(E_i)$. Here the following conditions must be satisfied:

1) $X = \bigcup E_i$;

2) if $E_i \in \Phi$ and $E_j \in \Phi$, then $E_i \cap E_j \in \Phi$ or is empty;

3) if $E_j \subset E_i$, then $t_i(E_j)$ is a face of σ_i; all the faces of this simplex are obtained in this way.

From the definition it follows that if $d(E_i)=0$, then E_i is a point $x \in X$. All these points are called the vertices of the triangulation. The subsets E_i are called a simplicial triangulation, and the vertices of the triangulation contained in a given simplex F_i are called its vertices. It is easy to show that if the set $K=\{x_1, \ldots, x_N\}$ of the vertices of a triangulation is known to us and also what subsets $S \subset K$ are the sets of vertices of a single simplex, then we can reconstruct X from this information. Thus, a triangulation of a space enables us to give a purely combinatorial scheme for it. Topological spaces admitting at least one triangulation are called *triangularizable*.

In connection with triangulations of the spaces $X(\mathbb{C})$, where X is a smooth projective curve, we need triangulations Φ having the following properties:

a) all the simplexes of the triangulation are of dimension $\leqslant 2$,

b) every simplex of dimension <2 is a face of some simplex of dimension 2,

c) every simplex of dimension 1 is a face of precisely two simplexes of dimension 2.

A topological space having a triangulation with these properties is called a *combinatorial surface*.

In what follows we make use of the operation of refinement or subdivision of a given triangulation. We give a simplified description of the operation of subdivision that fits triangulations for which $d(E_i) \leqslant 2$ for all $E_i \in \Phi$.

Let X be a topological space, Φ a triangulation of it with $d(E_i) \leqslant 2$ for all $E_i \in \Phi$, and E_r one of the simplexes of the triangulation for which $d(E_r)=1$. We choose any interior point ξ on the segment $t_r(E_r)$ and denote by Γ' and Γ'' the parts into which ξ divides this segments. We set $x=t_r^{-1}(\xi)$. Let E_i, $i \in I$, be those simplexes of Φ for which $d(E_i)=2$, $E_i \supset E_r$. We divide the triangle $t_i(E_i)$ into two triangles T_i' and T_i'' by joining the point $t_i(x)$ to the vertex opposite the side $t_i(E_r)$.

We consider the family Φ' consisting of the following closed subsets of X:

$d=0$: those $E_j \in \Phi$ for which $d(E_j)=0$, and the point x:

$d=1$: those $E_j \in \Phi$ for which $d(E_j)=1$, $j \neq r$, and $t_r^{-1}(\Gamma')$, $t_r^{-1}(\Gamma'')$;

$d=2$: $E_j \in \Phi$, $j \notin I$, and $t_i^{-1}(T')$, $t_i^{-1}(T'')$ for $i \in I$.

For example:

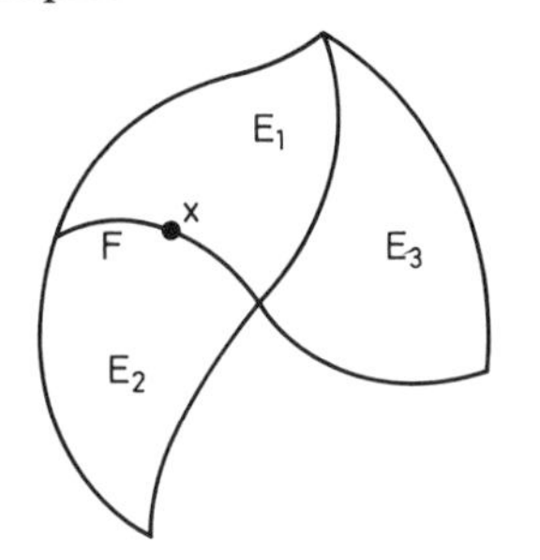

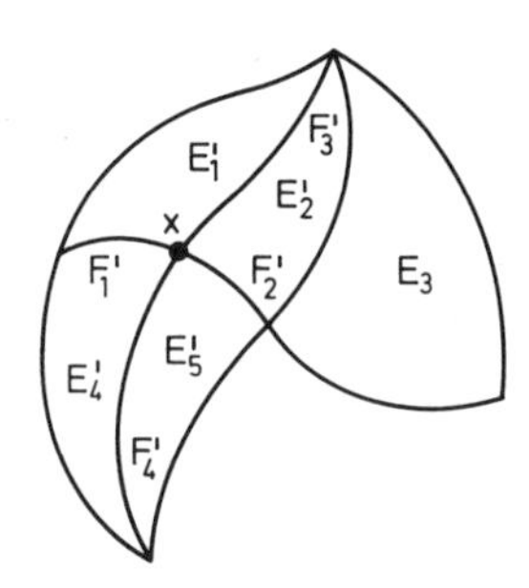

Thus, every set of the family Φ' either coincides with a simplex of the triangulation Φ or is part of such a simplex. We define a mapping t' as the corresponding mapping in Φ or its restriction.

It is easy to verify that the conditions 1), 2), and 3) hold, so that Φ' is a triangulation. It is called a *subdivision* of Φ.

The following property is an immediate consequence of the definition of subdivision.

Proposition. *Let $S \subset X$ be a finite set, $X = \bigcup U_i$ a finite covering, and Φ a triangulation of X. There exists another triangulation obtained from Φ by a sequence of subdivisions for which all the points of S are vertices and each of its simplexes is contained in one of the sets U_i.*

Such a triangulation is said to be *inscribed* in the covering $\{U_i\}$.

If a point $s \in S$ belongs to a simplex E_i with $d(E_1) = 1$, then by a single process of subdivision we make it into a vertex. But if $s \in E_i$, $d(E_i) = 2$, then we choose a simplex $E_j \subset E_i$, $d(E_j) = 1$, and the vertex P of the triangle $t_i(E_i)$ opposite the segment $t_i(E_j)$. For ξ we choose the point of intersection of the side $t_i(E_j)$ and the line joining $t_i(s)$ and P. By performing the corresponding subdivision we arrive at the previous case.

In order to inscribe a triangulation in a covering it is sufficient to do this for every triangle. We leave the details of this simple argument to the reader.

Now we recall the concept of orientation of some triangulation of a surface. An orientation of a triangle or a segment is a choice of one of the two possible directions of going around its vertices. Thus, every triangle or segment has two distinct orientations, which are called opposite. Every orientation of a triangle determines an orientation of is edges.

Let X be a combinatorial surface and Φ a corresponding triangulation. An *orientation* of Φ is a choice of orientations of all its triangles such that on every segment $E \in \Phi$, $d(E) = 1$, the triangles whose face E is determine opposite orientations. A triangulation having an orientation is said to be *orientable*. If X is connected, then a triangulation Φ has either precisely two orientations or none. It is easy to verify that the triangulation of the two-dimensional sphere [or of $\mathbb{P}^1(\mathbb{C})$] constructed in § 3.3 is orientable.

The property of the surface of being orientable does not depend on its triangulation. In other words, if Φ and Ψ are two distinct triangulations of one and the same surface, then they are either both orientable or both non-orientable. In invariant terms the condition of orientability of a surface can be written as $H_2(X, \mathbb{Z}) \neq 0$.

The last topological concept we need is that of the *Euler characteristic* of a surface. If a triangulation Φ has c_0 vertices, c_1 edges, and c_2 triangles, then its Euler characteristic is defined as:

$$\chi_\Phi(X) = c_0 - c_1 + c_2 . \tag{2}$$

Like the property of orientability, the Euler characteristic does not depend on the triangulation of the surface and is therefore denoted by $\chi(X)$. An invariant definition of it is:

$$\chi(X) = \dim_K H_0(X, K) - \dim_K H_1(X, K) + \dim_K H_2(X, K),$$

where K is an arbitrary field of characteristic 0. It is easy to check that the Euler characteristic of the two-dimensional sphere [that is, of $\mathbb{P}^1(\mathbb{C})$] is equal to 2.

The main result of the topology of surfaces is that the topological invariants we have introduced: the property of orientability and the Euler characteristic, form a complete system of topological invariants on connected triangularizable surfaces.

Theorem A. *Two connected triangularizable surfaces are homeomorphic if and only if they are simultaneously orientable or non-orientable and their Euler characteristics are the same.*

For a proof see [2] or [28].

The condition of triangularizibility here is superfluous: it can be shown that every surface is triangularizible, but we do not need this.

§ 4. Real Algebraic Curves

A *real algebraic curve* is a scheme X over the field of real numbers $\mathbb{R}$ such that $X \otimes_{\mathbb{R}} \mathbb{C}$ is an algebraic curve. Henceforth we assume that $X \otimes_{\mathbb{R}} \mathbb{C}$ is a smooth irreducible projective curve. As before, we denote by $X(\mathbb{R})$ the set of closed points $x \in X$ for which $k(x) = \mathbb{R}$. In simple terms, X is a projective smooth irreducible curve defined by an equation with real coefficients, and $X(\mathbb{R})$ is the set of points on it with real coordinates.

As we have seen, $X(\mathbb{R})$ is a compact one-dimensional manifold. However, it is not necessarily connected, so that the analogue of

Theorem 3 in § 2 is not true here. An example of a disconnected manifold $X(\mathbb{R})$ has occurred in Ch. II, § 3.1.

A connected one-dimensional compact manifold is homeomorphic to a circle. This is not hard to show directly, and for the connected components of the manifold $X(\mathbb{R})$ it follows at once from their triangularizibility, which we shall prove soon. Thus, $X(\mathbb{R})$ is homeomorphic to a certain number of disjoint circles, so that the only topological invariant of this space is the number of its connected components.

In this section we prove the main result, which links this topological invariant of $X(\mathbb{R})$ with algebraic properties of the curve X:

Harnack's Theorem. *If X is a smooth projective curve of genus g defined over the field of real numbers, then the number of connected components of the space $X(\mathbb{R})$ does not exceed $g+1$.*

There are several proofs of this theorem. One of them runs entirely in the real domain. It can be found in the book [20]. We give an outline of another proof, which is interesting in that a certain property of the space $X(\mathbb{R})$ is deduced from its embedding in the space $X(\mathbb{C})$.

1. Involutions. In the proof of Harnack's theorem which we shall set forth the main role is played by the mapping τ that associates with every point $x \in \mathbb{P}^n(\mathbb{C})$ the point $\tau(x)$ with the complex conjugate coordinates. Evidently τ determines a homeomorphism of the topological space $\mathbb{P}^n(\mathbb{C})$ (however, it is not an automorphism of the algebraic variety $\mathbb{P}^n$!). Since the curve X is given by equations with real coefficients, $\tau(X(\mathbb{C})) = X(\mathbb{C})$, and τ determines a homeomorphism of the space $X(\mathbb{C})$. In other words, τ is an automorphism of the scheme $X \otimes_{\mathbb{R}} \mathbb{C}$ induced by the automorphism of complex conjugation on $\mathbb{C}$.

As before, we shall use a triangulation of $X(\mathbb{C})$, however, now it is convenient to choose it so that it is invariant under τ, that is, together with a simplex E it also contains the simplex $\tau(E)$. Let us show that such a triangulation exists.

Here we have to repeat the whole process of constructing a triangulation of $X(\mathbb{C})$. We begin with a triangulation of $\mathbb{P}^1(\mathbb{C})$ that is invariant under τ. Such a triangulation was indicated in § 3.2. It is now quite easy to verify that the proposition on refinements of a triangulation proved in § 3.4 can be sharpened in the sense that if Φ is a triangulation that is invariant under τ, then the refinement Ψ we have constructed is also invariant under τ. For this purpose we need only arrange that by breaking up a simplex E into two subsets E' and E'' we break at the same time $\tau(E)$ into $\tau(E')$ and $\tau(E'')$.

Finally, we choose a non-constant function $f \in \mathbb{R}(X)$, that is, a rational function of the coordinates with real coefficients. The corresponding mapping $f: X(\mathbb{C}) \to \mathbb{P}^1(\mathbb{C})$ obviously has the property

$$f(\tau(x)) = \tau(f(x)).$$

It is easy to verify that the process of constructing a triangulation Ψ of $X(\mathbb{C})$ that is compatible with the triangulation Φ of $\mathbb{P}^1(\mathbb{C})$, which was described in the proof of Theorem 4 of § 3, leads to a τ-invariant triangulation Φ if Ψ was τ-invariant. Thus, we have the following result.

Proposition 1. *The space $X(\mathbb{C})$ has a triangulation Φ that is invariant under the homeomorphism τ.*

Clearly the set $X(\mathbb{R})$ then consists of simplexes of this triangulation.

Proposition 2. *Let E be a one-dimensional simplex of a triangulation Φ of the surface $X(\mathbb{C})$ contained in the set $X(\mathbb{R})$, and let E' and E'' be the two two-dimensional simplexes whose boundary it is. Then $\tau(E') = E''$.*

Since E' and E'' are the only simplexes of Φ having the boundary E, since Φ is invariant under τ and $\tau(E) = E$, we have either $\tau(E') = E'$, $\tau(E'') = E''$, or $\tau(E') = E''$, $\tau(E'') = E'$.

Let x be an interior point of E. At x we take a local parameter $t \in \mathbb{R}(X)$, for example, the equation of the hyperplane with real coefficients passing through x and transversal to X. Suppose that U contains x and that

$$t: U \to \mathbb{C}$$

is a homeomorphism of U onto the disc $|z| < 1$ in C. We take U sufficiently small so that of all the simplexes of Φ it intersects only E, E', and E''.

Since $t \in \mathbb{R}(X)$, we have

$$t(\tau(x)) = \tau(t(x)), \tag{1}$$

and therefore $t(E \cap U)$ coincides with the real diameter of the disc $|z| < 1$. So we see that $U - (U \cap E)$ splits into two connected components: $U \cap (E' - E)$ and $U \cap (E'' - E)$. Similarly $t(U) - t(E)$ splits into two components: the upper and the lower semicircle. Clearly distinct components of $U - (U \cap E)$ are mapped onto distinct components of the image. But the two semicircles are complex conjugates, hence it follows by (1) that

$$\tau(U \cap (E' - E)) = U \cap (E'' - E).$$

Therefore $\tau(E') \cap E''$ is not empty, hence $\tau(E') = E''$.

2. Proof of Harnack's Theorem. We use the homology groups with coefficients in $\mathbb{Z}/2\mathbb{Z}$. For an arbitrary surface F the group $H_1(F, \mathbb{Z}/2\mathbb{Z})$ is denoted by $H_1(F)$.

Let $T_1, \ldots, T_l$ be the connected components of the space $X(\mathbb{R})$. In the triangulation Φ they all consist of one-dimensional and zero-dimensional simplexes. Obviously, in the group $H_1(X(\mathbb{C}))$ they are cycles, which we denote by the same letters.

Proposition. *In the group $H_1(X(\mathbb{C}))$ the cycles $T_1, \ldots, T_l$ are either independent or are connected by the single relation*

$$T_1 + \cdots + T_l = 0 .$$

If the proposition were not true, then in $H_1(X(\mathbb{C}))$ there would exist a relation

$$T_1 + T_2 + \cdots + T_r = 0 , \quad r < l .$$

In other words,

$$T_1 + \cdots + T_r = \partial S ,$$

where S is a two-dimensional chain of the triangulation Φ (if necessary, we change the numbering of the cycles T_i).

The two-dimensional simplexes that do not occur in S form a chain $\bar{S}$, and since $\partial(S + \bar{S}) = 0$, we have

$$\partial \bar{S} = \partial S = T_1 + \cdots + T_r . \tag{1}$$

Thus, every one-dimensional simplex occurring in a cycle T_i $(i \leqslant r)$ is a face of one two-dimensional simplex occurring with the coefficient 1 in S and one occurring in $\bar{S}$.

Observe that τS is a chain of Φ and that

$$\partial \tau S = \tau \partial S = \partial S \tag{2}$$

by virtue of (1) and the fact that $\tau T_i = T_i$. Since $H_1(X(\mathbb{C}))$ is a module over $\mathbb{Z}/2\mathbb{Z}$, (2) shows that

$$\partial(S + \tau S) = 0 \tag{3}$$

Since $H_2(X(\mathbb{C})) = \mathbb{Z}/2\mathbb{Z}$, it follows from (3) that either

$$S + \tau S = S + \bar{S} ,$$

and hence

$$\bar{S} = \tau S , \tag{4}$$

or else

$$\begin{aligned} S + \tau S &= 0 , \\ S &= \tau S . \end{aligned} \tag{5}$$

We consider an arbitrary one-dimensional simplex E_1 of Φ contained in one of the T_i, $i = 1, \ldots, r$. Let E' and E'' be the two-dimensional simplexes whose boundary it is. We may then suppose that E' occurs in S, and E'' in $\bar{S}$, with the coefficient 1. By applying Proposition 2 of § 4.1 we see that (5) is not true, so that (4) is true.

We now consider the set T_{r+1} (by assumption, $r < l$) and take in it an arbitrary point t. Clearly $t \in S$ or $t \in \bar{S}$, but $t \notin S \cap \bar{S}$, because this intersection is $T_1 + \cdots + T_r$ and does not intersect T_{r+1}. If, say, $t \in S$, then $\tau(t) \in \tau(S)$. But $\tau(t) = t$, because $t \in T_{r+1} \subset X(\mathbb{R})$, and $\tau(S) = \bar{S}$, therefore $t \in S \cap \bar{S}$, which is not true, as we have seen. This proves the proposition.

To complete the proof of Harnack's theorem we need one further topological argument. If F is an arbitrary orientable surface, then the group $H_1(F)$ is a module of finite rank m over $\mathbb{Z}/2\mathbb{Z}$. The intersection index associates with two elements α, $\beta \in H_1(F)$ an element of $\mathbb{Z}/2\mathbb{Z}$, denoted by (α, β). The function (α, β) is linear in each of its arguments and skew-symmetric, that is, $(\alpha, \alpha) = 0$ for every $\alpha \in H_1(F)$. From Poincaré's duality law it follows that it is non-singular, that is, if $\xi_1, \ldots, \xi_m$ is a basis of $H_1(F)$, then

$$\det((\xi_i, \xi_j)) \neq 0 .$$

Hence it follows that arbitrary elements $\alpha_1, \ldots, \alpha_n \in H_1(F)$ for which $n > m/2$ and $(\alpha_i, \alpha_j) = 0$, $i, j = 1, \ldots, n$, are linearly dependent.

We apply this remark to the group $H_1(X(\mathbb{C}))$. As was proved in § 3, its rank is $2g$. The cycles $T_1, \ldots, T_l$, the connected components of the set $X(\mathbb{R})$, are by definition disjoint. Therefore $(T_i, T_j) = 0$, $i, j = 1, \ldots, l$, hence any $g + 1$ of them are linearly dependent in $H_1(X(\mathbb{C}))$. If the number l of components were greater than $g + 1$, we would arrive at a contradiction to the proposition. This completes the proof of Harnack's theorem.

3. Ovals of Real Curves. In his lecture on "Problems of Mathematics", Hilbert, in the context of Harnack's theorem, raised the question of the mutual disposition of the connected components of a plane real curve $X \subset \mathbb{P}^2$ (these components are called *ovals* of X). We indicate here a precise formulation of this problem, but only for the case of curves of even degree.

In that case it can be shown that any oval of X is homologous to 0 in the topology of the space $\mathbb{P}^2(\mathbb{R})$ and splits it into two components, of which one is homeomorphic to a disc and the other to a Moebius strip. The first component is called the interior of the oval. Therefore it makes sense to talk of one oval lying inside another or including it. The problem is to clarify the possible dispositions of ovals (in the sense of one being included in another) for all real plane smooth curves of a given degree.

At present the answer is known for curves of degree 2, 4, or 6; we shall describe the result in the case of maximal number of ovals. Using the formula for the genus of a smooth plane curve of degree $2n$ we see that the maximum number of ovals admissible by Harnack's theorem is $2n^2 - 3n + 2$.

A curve of degree 2 can consist of a single oval. A curve of degree 4 has not more than 4 ovals. Here only one disposition is possible: all 4 ovals lie outside each other. A curve of degree 6 has not more than eleven ovals.

In this case three types of disposition are possible: one oval contains inside 1, 5, or 9 ovals not including each other, and outside it there lie 9, 5, or 1 oval, respectively, none including another.

Also certain general inequalities and congruences are known to which the numbers of dispositions of ovals of one kind or another are subject. For example, from one general result of Petrovskii it follows that the number of ovals of a curve of degree $2n$ not containing each other does not exceed $\frac{3}{2}n(n-1)+1$, from which it follows, in particular, that a curve of degree 6 cannot split into eleven ovals not contained in one another. A classification of all possible types of disposition of ovals is not known at present.

In the same lecture and in connection with the same problem Hilbert points to an analogy between the problem on ovals of a real algebraic curve and the limit cycles of a differential equation

$$\frac{dy}{dx}=\frac{f(x,y)}{g(x,y)},$$

where f and g are polynomials. In this problem not even an analogue to Harnack's theorem has been found, that is, no estimate for the number of limit cycles for such a differential equation in terms of the maximum degree of the polynomials f and g is known. No such estimate is known even when the maximum is equal to 2.

Exercises

1. Let k be an algebraically closed field of characteristic 0, $k\{t\}$ the field of fractions of the ring of power series in one variable. Show that this field has a unique extension of a given degree n, namely that which is obtained by adjoining $\sqrt[n]{t}$. *Hint:* Use the arguments in the proof of Theorem 1.

2. Let X and Y be smooth projective curves, $f:X\to Y$ a morphism, and $f(X)=Y$. Derive a formula that expresses the genus of X in terms of the genus of Y and the multiplicity of the ramification points of the morphism f. *Hint:* Consider triangulations of X and Y that are compatible with f.

3. Let X be a projective smooth model of the curve with the equation

$$y^2=(x-a)(x-b)(x-c)(x-d),$$

and $f:X(\mathbb{C})\to\mathbb{P}^1(\mathbb{C})$ the continuous mapping that corresponds to the morphism specifying the function x. We denote by α and β disjoint segments on the sphere $\mathbb{P}^1(\mathbb{C})$ joining a to b and c to d. Show that $f^{-1}(\mathbb{P}^1(\mathbb{C})-\alpha-\beta)$ splits into two connected components X_i each of which is mapped by f homeomorphically onto $\mathbb{P}^1(\mathbb{C})-\alpha-\beta$, and that $\mathbb{P}^1(\mathbb{C})-\alpha-\beta$ is homeomorphic to the sphere with two discs removed.

4. In the notation of Exercise 3, show that the boundary of the closure $\bar{X}_i$ is homeomorphic to $\alpha \cup \beta$, $i=1,2$, that X is obtained by identifying these boundaries and thus is homeomorphic to a torus in accordance with Theorem 5.

5. Show that if a real curve of degree 4 splits into three ovals, then none of them lies inside another. *Hint:* Otherwise there would exist a line intersecting the curve in six points.

6. Show that the normalization of a real curve is a real curve.

7. Consider the normalization of the projective closure of the curve

$$y^2 = -(x-e_1)\dots(x-e_{2g+2}),\ e_i \neq e_j,\ e_i \in \mathbb{R}.$$

Show that the number of ovals for it is the same as the bound given by Harnack's theorem.

Chapter VIII. Complex Analytic Manifolds

§ 1. Definitions and Examples

1. Definition. In the preceding chapter we have investigated the topological space $X(\mathbb{C})$ connected with an arbitrary algebraic manifold X defined over the field of complex numbers $\mathbb{C}$. The example of smooth projective curves shows to what extent this space characterizes the variety X. We have shown that in this case the only invariant of $X(\mathbb{C})$ is the genus of X. We can say, therefore, that the genus is the only topological invariant of a projective curve. Undoubtedly, the genus is a most important invariant of an algebraic curve, however, it does not determine the curve by any means. We have seen in Ch. III, § 5.6, that there exist very many non-isomorphic curves of one and the same genus. The connection between a variety X and the space $X(\mathbb{C})$ for varieties of higher dimensions bears a similar character.

When we focus our attention on the way in which the topology was defined in $X(\mathbb{C})$ (Ch. VII, § 1.1), we observe that in the same manner we can connect with X another object, which reflects vastly more properties of this variety. We do this here under the assumption that X is a smooth variety; the general case will be treated in § 1.5.

We begin just as we have done in the preceding chapter: we consider a point $x \in X(\mathbb{C})$, some system of local parameters $t_1, \ldots, t_n$ at this point, and a homeomorphism determined by them

$$\varphi : U \Longrightarrow V \subset \mathbb{C}^n \tag{1}$$

of some neighbourhood U of x and a neighbourhood of zero $V \subset \mathbb{C}^n$. We have used this homeomorphism to define on $X(\mathbb{C})$ the structure of a topological $2n$-dimensional manifold. An essential feature here is the compatibility of the various mappings (1) that are defined in various neighbourhoods U and by means of various systems of parameters. It follows from the fact that if a function $f \in \mathbb{C}(X)$ is regular at x, then $g = f\varphi^{-1}$ is an analytic function of n complex coordinates $z_1, \ldots, z_n$ in a neighbourhood of zero in $\mathbb{C}^n$. Of this property we have only used a very small part: the continuity of g, from which we have derived that any other local parameters $u_1, \ldots, u_n$ are continuous functions of $t_1, \ldots, t_n$.

At the basis of this argument lies the fact that the concept of a continuous complex-valued (or real-valued) function in a neighbourhood V of $x \in X(\mathbb{C})$ can be defined in an invariant manner. It is natural to use this name for a function $h: U \to \mathbb{C}$ for which $h \cdot \varphi^{-1}$ is continuous in $V \subset \mathbb{C}^n$, and this property does not depend on the choice of φ. We now recall that if f is a regular function at x, then $g = f\varphi^{-1}$ is not only continuous but analytic. Hence it follows that if $u_1, \ldots, u_n$ is another system of local parameters at x, then in some neighbourhood $U' \subset U$ of this point $t_1, \ldots, t_n$ are analytic functions of $u_1, \ldots, u_n$. Therefore, if for some continuous function $h: U \to \mathbb{C}$ the function $h\varphi^{-1}$ is analytic in a neighbourhood of zero, then the same property holds for every mapping (1) given by another system of local parameters at x.

Thus, the following concept is well-defined.

A complex-valued function h defined in some neighbourhood of a point $x \in X(\mathbb{C})$ is said to be *analytic at* x if $g(z_1, \ldots, z_n) = h\varphi^{-1}$, defined by means of (1), is an analytic function of the variables $z_1, \ldots, z_n$ in a neighbourhood of zero in $\mathbb{C}^n$.

The functions that are analytic at all points of an open set U form a ring, which is denoted by $\mathcal{O}_{\mathrm{an}}(U)$. Since the definition of being analytic has local character, the mapping $U \to \mathcal{O}_{\mathrm{an}}(U)$ determines a sheaf $\mathcal{O}_{\mathrm{an}}$, a so-called sheaf of analytic functions. Evidently it is a subsheaf of the sheaf of continuous function on $X(\mathbb{C})$, and on the other hand, the sheaf of regular functions $\mathcal{O}$ is a subsheaf of it.

In preceding parts of the book we have determined an algebraic variety by its topological space (in the spectral topology) and the sheaf of regular functions. Similarly a topological space with a sheaf of analytic functions given on it leads to a new concept which we wish to define.

To begin with we consider a domain W in the space $\mathbb{C}^n$ of complex variables. For every open set $U \subset W$ the collection of all functions that are analytic at all points of U form an algebra $\mathcal{O}_{\mathrm{an}}(U)$ over $\mathbb{C}$, and the assignation $U \to \mathcal{O}_{\mathrm{an}}(U)$ determines a sheaf of algebras over $\mathbb{C}$, which is a subsheaf of the sheaf of continuous functions on W. We call $\mathcal{O}_{\mathrm{an}}$ the sheaf of analytic functions on W.

These ringed spaces play in our theory the role of the simplest objects analogous to that of affine schemes in the definition of the general concept of a scheme.

Definition. A ringed space $(X, \mathcal{O}_X)$ consisting of a topological Hausdorff space X and a sheaf $\mathcal{O}_X$ of algebras over $\mathbb{C}$ given on it, which is a subsheaf of the sheaf of continuous complex-valued functions, is called a *complex analytic manifold* if it satisfies the following condition: every point $x \in X$ has a neighbourhood U such that the ringed space defined by the restriction of $\mathcal{O}_X$ to U is isomorphic to $(W, \mathcal{O}_{an})$, where W is a domain in $\mathbb{C}^n$ and $\mathcal{O}_{an}$ the sheaf of analytic functions on W.

Continuous functions that are sections of $\mathcal{O}_X$ over an open set $U \subset X$ are called *analytic functions* on U.

A mapping $f: X \to Y$ of two analytic manifolds is called *holomorphic* if it is continuous and determines a morphism of the ringed spaces. The latter is equivalent to the fact that the mapping f^* carries analytic functions into analytic functions.

If x is a point of an analytic manifold, U a neighbourhood of x, and $f: U \to W \subset \mathbb{C}^n$ an isomorphism onto an open set in $\mathbb{C}^n$, then the number n is called the complex dimension at x. From the definition it follows that as a topological space X is a $2n$-dimensional manifold at x. Therefore the complex dimension is one and the same for two points of a single connected component of an analytic manifold. If X is connected, this number is called its *complex dimension.*

Above we have constructed the sheaf $\mathcal{O}_{an}$ on any smooth algebraic manifold X defined over the field of complex numbers. Evidently $(X(\mathbb{C}), \mathcal{O}_{an})$ determines an analytic manifold, which we denote by X_{an}. Every morphism $f: X \to Y$ of algebraic manifolds determines a holomorphic mapping $X_{an} \to Y_{an}$ of analytic manifolds, which we denote by f_{an}. The remaining part of this chapter is mainly devoted to a study of connections between algebraic varieties and the analytic manifolds corresponding to them. For example, the following problem arises:

Is every analytic manifold of the form X_{an}, where X is some algebraic variety?

Is every homomorphic mapping $X_{an} \to Y_{an}$ of the form f_{an}, where $f: X \to Y$ is a regular mapping of algebraic varieties?

Does an isomorphism of analytic manifolds X_{an} and Y_{an} imply an isomorphism of the algebraic varieties X and Y?

The answers to all these questions are negative, and relevant examples are not hard to construct. On the first two, see Exercises 1 and 3; on the third see § 3.2. But if we restrict ourselves to compact manifolds, then the same questions become much deeper and the answers to them less trivial. We consider them in the next few sections.

2. Factor Spaces. Here we describe a new construction of analytic manifolds. As a first application we obtain a number of examples, which enable us to answer some of the problems discussed at the end of the preceding subsection.

Let X be a topological space and G a group consisting of homeomorphisms of this space. We say that G acts on X freely and discretely if the following two conditions hold: 1) every point $x \in X$ has a neighbourhood U such that

$$gU \cap U = \emptyset \tag{1}$$

for every $g \in G$, other than the identity transformation, and 2) any two points $x, y \in X$ for which $y \neq gx$ for any g in G have neighbourhoods U of x and V of y such that

$$gU \cap V = \emptyset$$

for all $g \in G$.

We denote by X/G the set of equivalence classes of points, where two points x_1 and x_2 are called equivalent if there exists a $g \in G$ such that $gx_1 = x_2$. By assigning to every point the class containing it we determine a mapping

$$\pi : X \to X/G .$$

In X/G we introduce a topology, by calling a subset $U \subset X/G$ open if $\pi^{-1}(U)$ is open in X. Condition 2) guarantees that if X is a Hausdorff space, then so is X/G. If $x \in X$, $y = \pi(x)$, U a neighbourhood of x satisfying (1), then $V = \pi(U)$ is a neighbourhood of y. By virtue of (1)

$$\pi^{-1}(V) = \bigcup gU , \quad g_1 U \cap g_2 U = \emptyset \quad \text{for} \quad g_1 \neq g_2 , \tag{2}$$

and the mapping

$$\pi : gU \to V$$

is a homeomorphism. A mapping of topological spaces $\pi : X \to Y$ having the analogous property is called an *unramified covering*.

Now we assume that $(X, \mathcal{O}_X)$ is an analytic manifold and that G, acting freely and discretely on X, consists of automorphisms of this analytic manifold. In that case we define on X/G a sheaf $\mathcal{O}_{X/G}$, taking for $\mathcal{O}_{X/G}(V)$ the collection of those continuous functions f on V for which $\pi^* f \in \mathcal{O}_X(\pi^{-1}(V))$.

Let us show that $(X/G, \mathcal{O}_{X/G})$ is an analytic manifold. For this purpose we consider a neighbourhood U of $x \in X$ that satisfies simultaneously the condition in the definition of an analytic manifold and (1), and let

$$\varphi : U \to W$$

be an isomorphism of it with a domain in $\mathbb{C}^n$.

We set $\pi(U) = V$ and consider a continuous function f on V. By (2) $f \in \mathcal{O}_{X/G}(V)$ if and only if $\pi^* f \in \mathcal{O}_X(gU)$ for all $g \in G$. On the other hand, since g is an automorphism of the analytic manifold X, it determines an isomorphism of U and gU under which the restrictions of π to U and gU go over into each other. Hence it follows that if

$$\pi_1 : U \to V$$

is the restriction of π to U, then f belongs to $\mathcal{O}_{X/G}(V)$ if and only if $\pi_1^* f$ belongs to $\mathcal{O}_X(U)$. Since π_1 is a homeomorphism, it follows that π_1 is an isomorphism of the analytic manifolds U and V. Therefore, we have the isomorphism

$$\varphi \pi_1^{-1}: V \to W \subset \mathbb{C}^n,$$

whose existence shows that X/G is an analytic manifold.

Example 1. We regard the n-dimensional vector space $\mathbb{C}^n$ over the field of complex numbers as a $2n$-dimensional space over $\mathbb{R}$, and we choose in it m linearly independent vectors (over $\mathbb{R}$) $a_1, \ldots, a_m$. The set of vectors of the form

$$a = l_1 a_1 + \cdots + l_m a_m, \quad l_i \in \mathbb{Z},$$

is called an *m-dimensional lattice*, which we denote by Ω. The transformations

$$g_a(z) = z + a, \, z \in \mathbb{C}^n, \, a \in \Omega,$$

are automorphisms of the analytic manifold $\mathbb{C}^n$. Since

$$g_{a_1 + a_2} = g_{a_1} \cdot g_{a_2},$$

all the g_a, $a \in \Omega$, form a group G. Clearly, it acts on $\mathbb{C}^n$ freely and discretely. For let us supplement $a_1, \ldots, a_m$ to a basis $a_1, \ldots, a_{2n}$ and denote by U the open set consisting of the vectors of the form

$$a = x_1 a_1 + \cdots + x_{2n} a_{2n}, \, x_i \in \mathbb{R},$$
$$-\tfrac{1}{2} < x_i < \tfrac{1}{2} \quad \text{for} \quad i = 1, \ldots, m.$$

Then the set $z + U$ consisting of the vectors $z + u$, $u \in U$, is a neighbourhood of z and

$$g_a(z + U) \cap (z + U) = \emptyset$$

for $a \in \Omega$, $a \neq 0$.

Thus, $\mathbb{C}^n/G$ is an analytic manifold, which is easily seen to be compact if and only if $m = 2n$.

Suppose then that $m = 2n$. In that case $\mathbb{C}^n/G$ has a very simple topological structure. Since

$$\mathbb{C}^n = \mathbb{R} a_1 + \cdots + \mathbb{R} a_{2n}, \, \Omega = \mathbb{Z} a_1 + \cdots + \mathbb{Z} a_{2n},$$

we see that $\mathbb{C}^n/G$ is homeomorphic to the product of $2n$ copies of $\mathbb{R}/\Gamma$, where Γ consists of the translations $t \to t + n$, $t \in \mathbb{R}$, $n \in \mathbb{Z}$. Evidently R/Γ is homeomorphic to a circle, and $\mathbb{C}^n/G$ homeomorphic to a $2n$-dimensional torus. The varieties $\mathbb{C}^n/G$ (with $m = 2n$) are therefore called *complex tori*. Later we shall see that as analytic manifolds they are by no means always isomorphic.

Example 2. Let $X = \mathbb{C}^n - 0$, c a positive number, $c \neq 1$, and G the group consisting of the transformations

$$(z_1, \ldots, z_n) \to (c^l z_1, \ldots, c^l z_n), \quad l \in \mathbb{Z}.$$

The fact that G acts on X freely and discretely can easily be verified directly, but the subsequent arguments make it perfectly obvious.

We write any point $z \in X$ in the form

$$z = r \cdot u\,,$$

where r is a positive number and $u = (u_1, \ldots, u_n)$ is such that

$$|u_1|^2 + \cdots + |u_n|^2 = 1\,.$$

This representation is obviously unique and determines a homeomorphism of X onto

$$\mathbb{R}_+ \times S^{2n-1}\,,$$

where $\mathbb{R}_+$ is the set of positive real numbers and S^{2n-1} the sphere of dimension $2n-1$. In this representation the transformations in G act trivially on S^{2n-1}, and on $\mathbb{R}_+$ as multiplication by a power of c. If we map $\mathbb{R}_+$ onto $\mathbb{R}$ by means of the log-function, the latter action turns into the translation by the lattice vector $\mathbb{Z}\log c$. Hence it is clear that G acts freely and discretely and that X/G is homeomorphic to $(\mathbb{R}/\mathbb{Z}\log c) \times S^{2n-1}$, that is, to $S^1 \times S^{2n-1}$. The compact analytic manifold we have constructed is called a *Hopf manifold*.

3. Commutative Algebraic Groups as Factor Spaces. We return to Example 1 of the preceding subsection: the factor space $\mathbb{C}^n/G$, where G consists of translations by vectors of a certain lattice Ω. This lattice is a subgroup of $\mathbb{C}^n$, and the factor space $\mathbb{C}^n/G$ is homeomorphic to the factor group $\mathbb{C}^n/\Omega$, therefore is a group. It is very easy to check that the mapping

$$m: \mathbb{C}^n/\Omega \times \mathbb{C}^n/\Omega \to \mathbb{C}^n/\Omega\,,$$

defined by the group law is holomorphic. Thus, $\mathbb{C}^n/\Omega$ is a commutative complex analytic Lie group.

Let us assume that the manifold $\mathbb{C}^n/\Omega$ originates in some algebraic variety X, in other words, is of the form X_{an}. In that case it can be shown that $m = \mu_{\text{an}}$, where

$$\mu: X \times X \to X$$

is the morphism that defines in this way on X the structure of an algebraic group. In the most interesting case when X is compact this follows from a theorem of the next section. Thus, in this case X is an Abelian variety.

We show presently that, conversely, every commutative algebraic group over the field of complex numbers can be represented in the form $\mathbb{C}^n/\Omega$, where Ω is a lattice. To do this we need one auxiliary result.

Lemma. *An invariant one-dimensional differential form on a commutative algebraic group is closed.*

Proof. Let φ be an invariant differential form on a group G. Then $d\varphi$, as is easy to verify, is also invariant. Therefore we need only show that $(d\varphi)(e)=0$, from which it follows that $d\varphi=0$.

Writing φ in the form

$$\varphi=\Sigma\,\psi_m du_m\,,$$

we make use of formula (5) in Ch. III, § 5.2:

$$\sum_l c_{ml}\psi_l=\psi_m(e)\in\mathbb{C}\,.$$

From this it follows that

$$\sum_l \frac{\partial\psi_l}{\partial u_i}c_{ml}+\sum_l \psi_l\frac{\partial c_{ml}}{\partial u_i}=0\,.$$

We consider this equality at e. Since $c_{ml}(e)=\delta_{m,l}$, it follows from this equality that

$$\frac{\partial\psi_m}{\partial u_i}(e)+\sum_l \psi_l(e)\frac{\partial c_{ml}}{\partial u_i}(e)=0\,.$$

To prove the equality

$$\left(\frac{\partial\psi_m}{\partial u_i}\right)(e)=\left(\frac{\partial\psi_i}{\partial u_m}\right)(e)\,, \tag{1}$$

which expresses the fact that φ is closed, it is sufficient for us to verify that

$$\frac{\partial c_{ml}}{\partial u_i}(e)=\frac{\partial c_{il}}{\partial u_m}(e)\,.$$

It is here that we use the commutativity of G. By the formulae (4) in Ch. III, § 5.2:

$$c_{ml}=\sum_j v_{lj}(g)\frac{\partial w_{lj}}{\partial u_m}(e)\,.$$

Since the group is commutative,

$$\mu^*(u_m)(g_1,g_2)=\sum_j v_{mj}(g_1)w_{mj}(g_2)=\sum_j w_{mj}(g_1)\,v_{mj}(g_2)\,.$$

Therefore,

$$c_{ml}(g)=\sum_j v_{lj}(g)\,\frac{\partial w_{lj}}{\partial u_m}(e)=\sum_j w_{lj}(g)\frac{\partial v_{lj}}{\partial u_m}(e)$$

and consequently,

$$\frac{\partial c_{ml}}{\partial u_i}(e)=\sum_j\frac{\partial v_{l,j}}{\partial u_i}(e)\frac{\partial w_{lj}}{\partial u_m}(e)=\sum_j\frac{\partial w_{lj}}{\partial u_i}(e)\frac{\partial v_{l,j}}{\partial u_m}(e)=\frac{\partial c_{il}}{\partial u_m}(e)\,.$$

This proves formula (1) and the lemma.

Now we consider an arbitrary n-dimensional commutative algebraic group A defined over the field of complex numbers. By the proposition in Ch. III, § 5.2, the space of invariant one-dimensional differential forms on A is n-dimensional. We denote by $\omega_1, \ldots, \omega_n$ a basis of it. According to the lemma, the differential forms ω_i are closed, hence there exist holomorphic functions $f_1, \ldots, f_n$ defined in some complex neighbourhood U of the zero element $o \in A$ such that

$$\omega_i = d f_i, \quad f_i(o) = 0.$$

From the invariance of the forms ω_i it follows that

$$d t_g^* f_i = d f_i$$

in the domain $U \cap t_g^* U$. In other words,

$$t_g^* f_i = f_i + \alpha_i, \quad \alpha_i \in \mathbb{C}. \tag{2}$$

But $(t_g^* f_i)(g_1) = f_i(g + g_1)$ (we write the group law on A additively). Therefore (2) indicates that $f_i(g + g_1) = f_i(g) + \alpha_i$ if g, $g + g_1 \in U$. In particular, setting $g = 0$ we see that

$$f_i(g + g_1) = f_i(g) + f_i(g_1) \tag{3}$$

for g, $g + g_1 \in U$.

Thus, the f_i determine a "local homomorphism" of a neighbourhood of o in A into a neighbourhood of 0 in $\mathbb{C}^n$. From the proposition in Ch. III, § 5.2, we see that the $d_0 f_i$ form a basis in Θ_o^*, and this means that the Jacobian $\det\left(\frac{\partial f_i}{\partial t_j}(e)\right) \neq 0$ for every system of local parameters $t_1, \ldots, t_n$ at o. By virtue of this, the mapping φ

$$\varphi(g) = (f_1(g), \ldots, f_n(g))$$

is an analytic isomorphism between the neighbourhood U of o and the neighbourhood V of 0 in $\mathbb{C}^n$, and by (3) a "local isomorphism" of the groups.

Now we construct a homomorphism $\psi: \mathbb{C}^n \to A$, by setting for $z \in \mathbb{C}^n$

$$\psi(z) = l \varphi^{-1}\left(\frac{z}{l}\right),$$

where the integer l is chosen sufficiently large so that $z/l \in V$. That this is well-defined (independent of l) follows at once from (3). So we have constructed the homomorphism

$$\psi: \mathbb{C}^n \to A,$$

which on $V \subset \mathbb{C}^n$ is the same as φ^{-1}. Hence it is easy to deduce that ψ is holomorphic. From the fact that it maps isomorphically onto V it follows that $\psi(\mathbb{C}^n)=A$. We denote by Ω the kernel of ψ. Then $\Omega \cap V=0$, that is, Ω is a discrete subgroup of $\mathbb{C}^n$. By a standard argument it follows that Ω is a lattice. Thus, $A \simeq \mathbb{C}^n/\Omega$, where Ω is a lattice.

4. Examples of Compact Analytic Manifolds that are not Isomorphic to Algebraic Varieties. Here we are concerned with the first of the problems stated in § 1.2 in connection with the concept of an analytic manifold: does every analytic manifold originate in some algebraic variety, in other words, is it of the form X_{an}? Since the problem only becomes really interesting when we restrict ourselves to compact analytic manifolds, we make this assumption in the statement of the problem.

The fact that with this restriction the problem becomes far more subtle is clear if only because for one-dimensional varieties the answer is in the affirmative: every one-dimensional compact manifold is isomorphic to one of the form X_{an}, where X is a smooth projective curve. This proposition is called Riemann's existence theorem. We do not prove it here—in every form a proof requires certain arguments of an analytic character connected with the theory of harmonic functions. A proof can be found, for example, in [31], Ch. 8.

It is all the more interesting that for dimensions greater than 1 the answer to our problem is in the negative. Here we come across a familiar phenomenon in algebraic geometry: many difficulties do not arise in the one-dimensional case. In this subsection we give some examples of compact manifolds that are not algebraic, and for simplicity we confine ourselves to the two-dimensional case.

Owing to the fact that this problem is connected with the most central concepts, we analyse two construction principles for such examples: in Example 1 we use almost exclusively algebraic arguments, and in Examples 2 and 3 more geometrical ones.

Example 1. Our manifold is a complex torus, that is, of the form $\mathbb{C}^2/\Omega$, where Ω is a four-dimensional lattice. If it were an algebraic variety, then by what has been shown in § 1.3 it would be Abelian. We now establish a property of Abelian varieties which turns out to fail in some part of Ω. This property is called Poincaré's theorem on complete reducibility and reads as follows.

Proposition. *If A and B are Abelian varieties and $\varphi: A \to B$ an epimorphism, then there exists an Abelian variety $C \subset A$ such that* $\dim C = \dim B$ *and $\varphi: C \to B$ is an epimorphism.*

For brevity we assume that $\dim A = 2$, $\dim B = 1$, and that the ground field is of characteristic 0. We shall apply the proposition only with these restrictions.

We consider a point $a \in A$ and the subvariety $Y = \varphi^{-1}(\varphi(a))$ containing it. By the theorem on the dimension of fibres, Y is a curve. There exists another irreducible curve X passing through a, but not contained in Y. This follows because A is algebraic; it is sufficient to consider an affine neighbourhood of a and to take for X the closure of a suitable hyperplane section or a component of it. The morphism $\psi : X \to B$, the restriction of φ, evidently has finite degree and the fibres $\psi^{-1}(b)$ are finite.

For a divisor D on X, $D = \sum_{i=1}^{r} l_i x_i$, we denote by $S(D)$ the point $l_1 x_1 \oplus \cdots \oplus l_r x_r \in A$, where $\oplus$ is the group operation on A, and we set

$$f(b) = S(\psi^*(b)).$$

We claim that f is a morphism of B into A. To begin with we verify that it is a rational morphism. Let θ be a primitive element of the extension $k(X)/k(B)$,

$$k(X) = k(B)(\theta).$$

Consequently, the coordinates t_l of some point $x \in X$ have the form $F_l(\theta)$, $F_l \in k(B)(T)$. If $\theta_1, \ldots, \theta_n$ are all the conjugates of θ over $k(B)$, then the points x_i with coordinates $F_l(\theta_i)$ have a common image in B:

$$\psi(x_i) = \psi(x) = b.$$

Obviously, the coordinates of $f(b)$ can be expressed in terms of the coordinates of the x_i rationally and symmetrically, that is, they are symmetric functions of the θ_i and are therefore contained in $k(B)$. This shows that f is a rational morphism.

From the fact that A is complete and B is a smooth curve it follows that f is a morphism, and from Theorem 4 in Ch. III, § 3, that it is a homomorphism. We set $C = f(B)$. To prove the proposition we have to verify that $\varphi(C) \neq 0$. But by definition

$$\varphi C = \varphi f B = \nu B,$$

where ν is the endomorphism of multiplication by n in B:

$$\nu(b) = b \oplus \cdots \oplus b.$$

Since the ground field is of characteristic 0, $\operatorname{Ker} \nu$ is finite (Exercise 2 to Ch. III, § 3), $\nu B = B \neq 0$. This proves the proposition.

Now we can complete the construction of the example. The idea is to construct a complex torus for which this proposition does not hold. Suppose that a lattice Ω in $\mathbb{C}^2$ has a basis of four vectors

$$(1, 0), (i, 0), (0, 1), (\alpha, \beta).$$

It is easy to verify that they are independent over $\mathbb{R}$, provided that β is not real. We set $A=\mathbb{C}^2/\Omega$, $B=\mathbb{C}^1/\Omega'$, where Ω' is the lattice with the basis $(0, 1)$ and (α, β). The mapping $(z_1, z_2)\to z_2$ induces, as is easy to see, a holomorphic homomorphism $\varphi: A\to B$. We assume that A is an Abelian variety. From Riemann's existence theorem it follows that B is an algebraic curve (we shall verify this directly in Ch. IX, § 2). As we shall see a little later (Theorem 2 of §3), it follows that φ is a morphism. Now we may apply Poincaré's theorem on complete reducibility and obtain that there exists a one-dimensional Abelian variety $C\subset A$ such that $\varphi C=B$.

We denote by Λ the inverse image of C in $\mathbb{C}^2$. This is a closed subgroup of $\mathbb{C}^2$, and all the closed subgroups in any $\mathbb{R}^n$ are easy to determine. A simple argument (see [25], Pontryagin) shows that they are of the form $\mathbb{Z}e_1+\cdots+\mathbb{Z}e_s+\mathbb{R}e_{s+1}+\cdots+\mathbb{R}e_{s+r}$, where $e_1,\dots, e_{s+r}$ are independent over $\mathbb{R}$. In our case $\Lambda\supset\Omega$, hence contains four independent vectors over $\mathbb{R}$. We denote by Λ_0 the connected component of zero in Λ. This is an $\mathbb{R}$-admissible linear subspace. Since C is defined in A by a single local equation and is smooth, Λ_0 is defined in $\mathbb{C}^2$ by a local equation $f(z_1, z_2)=0$, in which the linear part does not vanish. Let

$$f=f_1+f_2+\cdots$$

be the decomposition of the Taylor series for f into its homogeneous constituents. Then for sufficiently small $\alpha\in\mathbb{R}$

$$f(\alpha z_1, \alpha z_2)=\alpha f_1+\alpha^2 f_2+\cdots,$$

and since Λ_0 is a linear subspace over $\mathbb{R}$, we have $f_2=0, \dots$. So we see that Λ_0 is given by the linear equation $f_1=0$. As a result we obtain that

$$\Lambda=\mathbb{Z}e_1+\mathbb{Z}e_2+\mathbb{R}e_3+\mathbb{R}e_4=\mathbb{Z}e_1+\mathbb{Z}e_2+\Lambda_0,$$

where $\Lambda_0=\mathbb{R}e_3+\mathbb{R}e_4$ is a $\mathbb{C}$-linear subspace of $\mathbb{C}^2$. In other words, $e_4=\lambda e_3$, $\lambda\in\mathbb{C}$. In conclusion we recall that $\Lambda\supset\Omega$. Hence it follows that under the projection onto $\mathbb{Z}e_1+\mathbb{Z}e_2$ a two-dimensional sublattice of Ω goes into 0. Therefore $\Lambda_0\cap\Omega$ is a two-dimensional sublattice of Λ_0. Consequently, we have reached the conclusion that Poincaré's theorem in our case simply indicates the existence of a complex line

$$\Lambda_0=\mathbb{R}e_3+\mathbb{R}e_4, \quad e_4=\lambda e_3, \quad \lambda\in\mathbb{C},$$

for which $\Lambda_0\cap\Omega$ is a two-dimensional lattice and which projects onto the whole line z_1, that is, does not coincide with the line $z_2=0$.

In other words, to verify that the theorem holds we have to find in Ω a vector e for which $z_1\neq 0$ and a vector $\lambda e\in\Omega$ for a non-real complex number λ. Let us find out whether this is always possible. Let

$$e = a(1,0) + b(i,0) + c(0,1) + d(\alpha,\beta),$$
$$\lambda e = a(\lambda,0) + b(i\lambda,0) + c(0,\lambda) + d(\lambda\alpha,\lambda\beta),$$
$$a,b,c,d \in \mathbb{Z}.$$

All the z_2-coordinates of the vectors in Ω are contained in $\mathbb{Z}+\mathbb{Z}\beta$. In particular, $(c+d\beta)\lambda \in \mathbb{Z}+\mathbb{Z}\beta$, therefore λ must be contained in the field $\mathbb{Q}(\beta)$ (we recall that $c+d\beta \neq 0$, by hypothesis). Similarly, from a discussion of the z_1-coordinates we obtain that $\alpha \in \mathbb{Q}(\beta,\lambda,i) = \mathbb{Q}(\beta,i)$. Evidently, this condition does not always hold: for example, we can set $\beta = i$, $\alpha = \sqrt{2}$.

The construction of the example is now complete. It is interesting to look once more at our arguments, in order to understand at what place we have made essential use of the fact that A is an algebraic variety. It is easy to convince ourselves that all the arguments except a single one can be carried through equally for analytic manifolds. This only essential argument occurs in the proof of Poincaré's theorem, at the place where we have drawn the curve X that does not coincide with $\varphi^{-1}(b)$. From this we can conclude that this property is not true for the torus constructed in our example. Thus, this torus has a mapping φ to a curve B such that the only compact analytic manifolds in A are the fibres $\varphi^{-1}(b)$. (More details on the notion of a subvariety will be given in § 1.5, here we can interpret this as the image of a projective curve under a holomorphic mapping in A.) So we see that A is very barren in one-dimensional subvarieties, and this is the main difference to algebraic surfaces, which are loaded with curves all over.

Example 2. This also refers to a two-dimensional torus, but utilizes some topological arguments. Again, let

$$\Omega = \mathbb{Z}e_1 + \mathbb{Z}e_2 + \mathbb{Z}e_3 + \mathbb{Z}e_4.$$

Then A is homeomorphic to the torus $(\mathbb{R}/\mathbb{Z})^4$. Therefore $H_2(A,\mathbb{Z}) = \mathbb{Z}^6$, and their generators are the six cycles $S_{i,j}$, $1 \leqslant i < j \leqslant 4$, the images of the planes $\mathbb{R}e_i + \mathbb{R}e_j$, $i<j$, in A.

In this example we begin with the same argument as at the end of the preceding example. If A were algebraic, we could find in it an algebraic curve $C \subset A$. If $v: C^v \to C$ is the normalization mapping, then by triangulating C^v on the basis of Theorem 3 in Ch. VII, § 3, we would make C into a singular cycle. In particular,

$$C \sim \sum_{1 \leq i < j \leq 4} a_{i,j} S_{i,j}, \quad a_{i,j} \in \mathbb{Z}.$$

We show that the cycle C is not homologous to 0, consequently that not all the a_{ij} vanish. To see this we observe that the differential form $\frac{1}{2i}(dz_1 \wedge d\bar{z}_1 + dz_2 \wedge d\bar{z}_2)$ on $\mathbb{C}^2$ is invariant under Ω, therefore determines a differential form ω on A. We show that $\int_C \omega > 0$, from which it follows that C is not homologous to 0. If we regard z_1 and z_2 as functions in a neighbourhood of a point $x \in C$, then in a neighbourhood of $y = v^{-1}(x)$ our form is equal to

$$\frac{1}{2i}\left(\left|\frac{dv^*(z_1)}{dt}\right|^2 + \left|\frac{dv^*(z_2)}{dt}\right|^2\right) dt \wedge d\bar{t} > 0, \tag{1}$$

where t is a local parameter at y.

Now we consider in a similar way the differential form η corresponding to $dz_1 \wedge dz_2$. On the one hand, by Stokes' theorem

$$\int_C \eta = \sum a_{i,j} \int_{S_{i,j}} \eta,$$

where $\int_{S_{i,j}} \eta$ is easy to compute: the reader is recommended to verify that if $e_i = (\alpha_i, \beta_i)$, then

$$\int_{S_{i,j}} \eta = \alpha_i \beta_j - \alpha_j \beta_i.$$

On the other hand,

$$\int_C \eta = 0.$$

For by analogy with (1) $v^*(\eta)$ is on C^v equal to

$$\frac{dv^*(z_1)}{dt} \frac{dv^*(z_2)}{dt} dt \wedge dt = 0.$$

So we see that under the assumption that A is algebraic we have a relation

$$\sum a_{i,j}(\alpha_i \beta_j - \alpha_j \beta_i) = 0, \quad a_{i,j} \in \mathbb{Z}, \tag{2}$$

in which not all the $a_{i,j}$ vanish. Of course, it is not hard to choose $\alpha_1, \ldots, \alpha_4$, $\beta_1, \ldots, \beta_4$ so that the numbers $\alpha_i \beta_j, \ldots, \alpha_j \beta_i$ are independent over $\mathbb{Z}$. The corresponding torus is not algebraic.

It is easy to verify that this torus has even fewer one-dimensional subvarieties than the one constructed in Example 1: in fact, it has no compact one-dimensional analytic manifolds at all.

Notes. 1. We write the coordinates of the vectors of a basis of the lattice Ω in the form of a matrix of type

$$\Omega = \begin{pmatrix} \alpha_1 \alpha_2 \alpha_3 \alpha_4 \\ \beta_1 \beta_2 \beta_3 \beta_4 \end{pmatrix}$$

and consider the skew-symmetric 4×4 matrix A in which the elements $a_{i,j}$ with $i<j$ are the same as in (2). Then (2) can be written in the form of an equation

$$\Omega A \Omega' = 0\,, \tag{3}$$

where Ω' is the transposed matrix. The existence of an integral matrix A satisfying this relation is evidently necessary for the torus corresponding to the matrix Ω to be projective or even algebraic.

Other conditions are given by inequalities of the type (1). To obtain conditions as general as possible, we consider the form

$$\omega = \frac{1}{2i}(\lambda_1\bar\lambda_1 dz_1\wedge d\bar z_1 + \lambda_1\bar\lambda_2 dz_1\wedge d\bar z_2 + \bar\lambda_1\lambda_2 dz_2\wedge dz_1 + \lambda_2\bar\lambda_2 dz_2\wedge dz_2)\,.$$

Arguing exactly as in the proof of (1) we see that $\int_C \omega \geqslant 0$. Furthermore, it is easy to verify that if the torus $\mathbb{C}^2/\Omega$ is projective and C corresponds to a hyperplane section in one of its embeddings, then $\int_C \omega = 0$ only for $\lambda_1=\lambda_2=0$ (see Exercise 9). The last condition can be written differently. Let Ω^* be the Hermitian conjugate matrix to Ω. Then the 2×2 matrix $\Omega A \Omega^*$ is Hermitian, as is easy to see, that is, it corresponds to a Hermitian form $F(x)$. A simple substitution shows that $\int_C \omega = F(\lambda)$, where $\lambda=(\lambda_1,\lambda_2)$. Therefore the relation we have derived indicates that F is a positive-definite form. We write this as follows:

$$\Omega A \Omega^* > 0\,. \tag{4}$$

Thus, the relations (3) and (4) are necessary for the torus corresponding to the period matrix Ω to be projective. Precisely the same relations are necessary for an n-dimensional torus with the $n\times 2n$ period matrix Ω to be projective. They are called the *Frobenius relations.* It can be shown that they are also sufficient for the torus to be projective. Some hint on the idea of its proof is contained in Ch. IX, § 2.

2. In discussing the last example we could replace the formula (1) by a reference to the proposition in Ch. VII, § 1.3. Namely, a word-for-word repetition of the arguments given these shows that if $\omega_{C^\vee}$ is an orienting cycle of the curve $C^\vee$, then $v_*(\omega_{C^\vee})$ is not homologous to 0 on $A(\mathbb{C})$. The reference to the triangulation of $C^\vee$ can also be replaced by integration over the cycle. This is a convenient way to proceed in the following example.

Example 3. Let X be a Hopf manifold (Example 2 of § 1.2). Since X is homeomorphic to $S^1\times S^{2n-1}$, for $n>1$ its two-dimensional Betti number is equal to 0. The proposition in Ch. VII, § 1.3, shows us that X is not a projective variety. It is easy to prove that it is not algebraic.

5. Complex Spaces. Analytic manifolds are analytic analogues to smooth algebraic varieties. To restrict ourselves only to this concept would be very inconvenient: varieties with singular points can arise even in the study of smooth algebraic varieties as subvarieties or images under regular mappings. Furthermore, the majority of the arguments by which we have attempted to show in Ch. V the necessity of introducing the concept of a scheme is applicable in the analytic situation. The corresponding analytic concept is not used in the remainder of the book. However, it would be a pity not to mention it at all. Therefore we give its definition and indicate without proofs some of its basic properties.

We begin with a special case. Let $W \subset \mathbb{C}^n$ be a domain in the space of n complex variables, and $f_1, \ldots, f_l$ functions holomorphic in W. We denote by Y the set of common zeros of the functions $f_1, \ldots, f_l$. We define a sheaf $\mathcal{O}_Y$ on Y, by setting

$$\mathcal{O}_Y(V) = \mathcal{O}_W(\bar{V})/(f_1, \ldots, f_l),$$

where V is an open set on Y, $V = Y \cap \bar{V}$, $\bar{V}$ open in W (all open sets on Y can be represented in this form), $\mathcal{O}_W$ the sheaf of holomorphic functions on W, and $(f_1, \ldots, f_l)$ the ideal generated by these functions. Since Y is the set of common zeros of $f_1, \ldots, f_l$, the right-hand side does not depend on the choice of the open set $\bar{V}$. Topological spaces Y with the so defined sheaves are called *local models*.

Now we come to the global definitions. We define a *complex ringed space* as a topological space X equipped with a sheaf $\mathcal{O}$, which is a sheaf of algebras over $\mathbb{C}$. Any open set $U \subset X$ is itself a ringed space if it is equipped with the restriction of $\mathcal{O}$ to U.

Definition. *A complex analytic space* is a complex ringed space $(X, \mathcal{O})$ such that every point $x \in X$ has a neighbourhood U that is isomorphic, as a ringed space, to some local model.

As in the case of schemes, the stalks of the structure sheaf of an analytic space are local rings. If they do not contain nilpotent elements, then the space is said to be *reduced*. In this case the sheaf $\mathcal{O}$ is a subsheaf of the sheaf of continuous functions on X, and on a local model Y the stalk $\mathcal{O}_y$ consists of those functions that are induced on Y by functions on W holomorphic at y. In this case continuous functions $f \in \mathcal{O}_y$ are called functions on Y holomorphic at y. Henceforth we only deal with reduced analytic spaces without saying so. In this context morphisms of analytic spaces are called *holomorphic mappings*.

Suppose that a closed subspace $X' \subset X$ has the following property: for every point $x \in X'$ there is a neighbourhood $U \subset X$ and functions $f_1, \ldots, f_l$ holomorphic in it such that X' coincides with the collection of their common zeros in U. We equip X' with the sheaf obtained by restricting to X' the functions holomorphic on X. It is easy to verify that in this way we arrive at an analytic space. It is called a *subspace* of X.

An analytic space X is said to be *reducible* if $X = X' \cup X''$, where X' and X'' are subspaces of it, other than X. It is not hard to show that any analytic space X is the union of a family of irreducible subspaces:

$$X = \bigcup X_\alpha ,$$

and that only finitely many X_α pass through every point $x \in X$. From now on we only consider irreducible spaces.

A point $x \in X$ is called *simple* if it has a neighbourhood isomorphic to an analytic manifold. It can be shown that the set of simple points of an irreducible analytic space X is connected, hence as a connected analytic manifold has a well-defined dimension. This number is called the *dimension* of X. Every proper subspace of X has smaller dimension. In particular, it can be shown that the set of singular (that is, non-simple) points is a subspace. In view of this an analytic space X is the union of finitely many analytic varieties (not closed in X): the set of simple points, the set of simple points of the subspace of singular points, etc.

Proofs of these properties can be found, for example, in [17], Ch. I–V.

Complex spaces are analytic analogues of algebraic varieties and even of schemes, at least in the sense that with every scheme X of finite type over the field of complex numbers we can associate a certain complex space X_{an} (here again we consider schemes and spaces that are not necessarily reduced). Let us describe the construction of the space X_{an}.

With a scheme X we associate the topological space $\tilde{X} = X_{\mathrm{red}}(\mathbb{C})$ of complex points of its reduced subscheme (in the complex topology).

An affine scheme X of finite type over $\mathbb{C}$ is obviously a local model, and the open set W is the whole of $\mathbb{C}^N$ in which X is contained. It is easy to verify that this model does not depend on the embedding $X \to \mathbb{A}^N$. The structure sheaf on this model is denoted by $\mathcal{O}_{\mathrm{an}}$.

If X is any scheme of finite type over $\mathbb{C}$ and $X = \bigcup U^{(i)}$ an affine covering of it, then the sheaves $\mathcal{O}_{\mathrm{an}}^{(i)}$ on $U^{(i)}$ just defined together determine a single sheaf $\mathcal{O}_{\mathrm{an}}$ on the space $\tilde{X}$. The pair $(\tilde{X}, \mathcal{O}_{\mathrm{an}})$ is the complex space X_{an} associated with X.

In § 1.4 we have been faced with questions on the connections between the concepts of a complex manifold and a smooth algebraic variety. Of course, similar problems arise in the context of complex spaces and their connections with arbitrary algebraic varieties.

The only positive result we have stated in § 1.4: Riemann's existence theorem, has an analogue in this general case. Namely, every compact reduced one-dimensional complex space is isomorphic to an algebraic curve. This result can be derived from Riemann's existence theorem by means of the normalization process of a complex space, about which we want to say a few words, omitting all proofs.

A reduced complex space is called normal if the local rings $\mathcal{O}_x$ of its structure sheaf are integrally closed. Following very closely the arguments we used in the case of algebraic varieties, we can construct for every reduced and irreducible complex space X its normalization, that is, a normal space X^ν and a holomorphic mapping $\nu: X^\nu \to X$ having the properties of Theorem 1 in Ch. VI, § 1. If X is compact, then so is X^ν. A detailed account of all the arguments is contained, for example, in [1], 447. In the case of one-dimensional spaces, with which we are concerned henceforth, the position is somewhat simplified, and the reader can try to provide these arguments himself by way of an exercise (by no means trivial).

Let X be a compact reduced one-dimensional complex space with the structure sheaf $\mathcal{O}$. According to Riemann's theorem X^ν is a projective algebraic curve. On X we introduce the spectral topology in which the finite subsets and X itself are closed, and we define the sheaf $\tilde{\mathcal{O}}$ by the property $\tilde{\mathcal{O}}(U) = \mathcal{O}(U) \cap \mathbb{C}(X^\nu)$. It is not hard to verify that in this way we define an algebraic curve $\tilde{X}$, and that $\tilde{X}_{\text{an}} = X$.

Exercises

1. Construct an example of a holomorphic mapping $g: \mathbb{C}^1 \to \mathbb{C}^1$ that is not of the form f_{an}, where f is a morphism $f: \mathbb{A}^1 \to \mathbb{A}^1$.

2. Let X be a smooth irreducible curve, and f a holomorphic function on X_{an}. Show that if f is bounded on the set $X(\mathbb{C})$, then $f \in \mathbb{C}$.

3. Show that the disc $|z| < 1$ in $\mathbb{C}^1$ is not isomorphic to X_{an} for any smooth curve X.

4. Show that if A is an elliptic curve, then the number of solutions of the equation $mx = o$, $x \in A$, $m > 0$ an integer, o the zero point on A, is equal to m^2. If A is an n-dimensional Abelian variety, then the number of solutions of this equation is equal to m^{2n}. Deduce that a plane smooth cubic curve has nine points of inflexion (see Exercise 8 to Ch. III, § 2).

5. Show that a one-dimensional Hopf manifold is isomorphic to a complex torus.

6. Let $X = (\mathbb{C}^2 - o)/G$ be a two-dimensional Hopf manifold. Show that the mapping $\mathbb{C}^2 - 0 \to \mathbb{P}^1: (z_1, z_2) \to (z_1 : z_2)$ determines a holomorphic mapping $X \to \mathbb{P}^1$ whose fibres are one-dimensional complex tori.

7. In the notation of Exercise 6, show that on X there are no one-dimensional complex analytic submanifolds exept the fibres of a mapping $X \to \mathbb{P}^1$.

8. Let X be the complex space $\mathbb{C}^2$, g the automorphism $g(z_1, z_2) = (-z_1, -z_2)$, $G = \{1, g\}$ a group of order 2. Show that the factor space X/G (see Exercise 1 to Ch. V, § 3) is a complex space and isomorphic to the cone in $\mathbb{A}^3$ with the equation $xy = z^2$.

9. Let $X = \mathbb{C}^2/\Omega$ be a complex torus,

$$\omega = \frac{1}{2i}(|\lambda_1|^2 \, dz_1 \wedge d\bar{z}_1 + \lambda_1 \bar{\lambda}_2 \, dz_1 \wedge d\bar{z}_2 + \bar{\lambda}_1 \lambda_2 \, d\bar{z}_1 \wedge dz_2 + |\lambda_2|^2 \, dz_2 \wedge d\bar{z}_2),$$

C an analytic curve on X, $x \in C$. Identify the tangent plane to X at x with $\mathbb{C}^2$ by means of the mapping $\mathbb{C}^2 \to X$ and introduce coordinates in this way. Show that if x is a simple point on C and if the coordinates of the tangent vector to C at x are μ_1, μ_2, and if $\lambda_1 \bar{\mu}_1 + \bar{\lambda}_2 \mu_2 \neq 0$, then $\int_C \omega > 0$ (that is, $\neq 0$). Deduce that $\int_C \omega > 0$ if X is a projective torus and C the class of a hyperplane section.

§ 2. Divisors and Meromorphic Functions

1. Divisors. Now we return to the theory of analytic manifolds. The problem we consider is to construct for them an analogue to the theory of divisors. We must begin with an account of some simple properties of the stalk $\mathcal{O}_x$ of the structure sheaf of an analytic manifold. By definition, the ring $\mathcal{O}_x$ is isomorphic to the ring $\mathbb{C}\{z_1,\dots,z_n\}$ of power series in $z_1,\dots,z_n$ that are convergent in some neighbourhood of x (the neighbourhood depending on the series). This ring is very similar to the ring of formal power series. In particular, it is a regular local ring, and the analogue to Weierstrass' preparation theorem holds for it, which is stated exactly and proved almost exactly as for formal power series. A proof can be found in [30], Ch. I, § 2. From this theorem it follows word-for-word as for formal power series that in the ring $\mathbb{C}\{z_1,\dots,z_n\}$ decomposition into prime factors is possible and unique. In particular, this ring has no divisors of zero.

Let U be a connected analytic manifold and $\mathcal{O}(U)$ the ring of functions holomorphic on the whole of U. This ring has no divisors of zero. For if $f, g \in \mathcal{O}(U)$ and $fg=0$, then the set of points where $f\neq 0$ is open, and $g=0$ on this set. But then $g=0$ on the whole of U, by the uniqueness property of analytic functions. The elements of the field of fractions of $\mathcal{O}(U)$ are called *meromorphic fractions* on U. If $V\subset U$ is a connected open subset, then the restriction $\mathcal{O}(U)\to\mathcal{O}(V)$ extends to an isomorphic embedding of the field of meromorphic fractions on U into the corresponding field on V. Frequently we shall identify two corresponding meromorphic fractions.

Definition. A *divisor on an analytic manifold X* is a covering $X=\bigcup U_\alpha$ by connected open sets and a meromorphic fraction φ_α on every U_α, which must satisfy the condition: $\varphi_\alpha^{-1}\varphi_\beta$ is holomorphic and does not vanish on $U_\alpha\cap U_\beta$.

Equality of divisors and their addition are defined exactly as for locally principal divisors on algebraic varieties. A divisor is said to be *effective* if all the meromorphic fractions φ_α are holomorphic in their open sets.

Theorem 1. *Every divisor is the difference of two effective divisors.*

Lemma. *If two functions f and g are holomorphic at a point $x\in\mathbb{C}^n$ and are relatively prime as elements of the ring $\mathcal{O}_x=\mathbb{C}\{z_1,\dots,z_n\}$, then there exists a neighbourhood U of x such that f and g are holomorphic in U and relatively prime as elements of every ring $\mathcal{O}_y$, $y\in U$.*

Proof of the Lemma. When we multiply f and g by invertible elements of $\mathcal{O}_x$ and apply Weierstrass' theorem, we can achieve that f and g become polynomials in z_1 with coefficients in $\mathbb{C}\{z_2,\dots,z_n\}$ with the highest coefficient 1. Since they are relatively prime, there exist $u, v \in \mathbb{C}\{z_1, \ldots, z_n\}$ such that

$$fu + gv = r, \qquad r \in \mathbb{C}\{z_2, \ldots, z_n\}, \tag{1}$$

and this equations holds in some neighbourhood U of x. Suppose that f and g have a common factor $h \in \mathcal{O}_y$, $y \in U$. Then $h|r$, and again by applying Weierstrass' theorem we see that h differs by an invertible element of $\mathcal{O}_y$ from an element $h_1 \in \mathbb{C}\{z_2, \ldots, z_n\}$. But $h_1 | f$, and since the polynomial f in z_1 over the ring $\mathbb{C}\{z_2, \ldots, z_n\}$ has the highest coefficient 1, we see that h_1 is invertible in $\mathbb{C}\{z_2, \ldots, z_n\}$. This proves the lemma.

Proof of the Theorem. We may assume that the divisor D is given by a covering U_α and a collection of meromorphic fractions φ_α, such that

$$\varphi_\alpha = f_\alpha / g_\alpha \quad \text{in} \quad U_\alpha,$$

f_α and g_α are holomorphic on U_α and relatively prime at every point $y \in U_\alpha$. Then it follows from the uniqueness of the decomposition into prime factors in $\mathcal{O}_y$ that the f_α determine a divisor D' and the g_α a divisor D'', both these divisors being effective and $D = D' - D''$. This proves the theorem.

Evidently every effective divisor D determines some complex subspace of X, namely the one that is given by the equation $\varphi_\alpha = 0$ in the open set U_α. It is called the *support* of D and is denoted by Supp D. If $D = D' - D''$ is a representation in the form of a difference of effective divisors, then by definition Supp D = Supp $D' \cup$ Supp D''. Making use of the concept of dimension of a complex space that was introduced at the end of the last section we can state the following result.

Proposition. *The support of a divisor is of codimension* 1.

First of all, we have to specify this subspace by a more economical system of equations. For this purpose we decompose at every point $x \in U_\alpha$ the function φ_α into irreducible factors in $\mathcal{O}_x$, and we denote by ψ_x the product of these factors to the first power. The function ψ_x is holomorphic in some neighbourhood U_x of x, and all these functions determine the same subspace Supp D as φ_x (although possibly another divisor). Thus, we may assume that from the very beginning the divisor is given by functions φ_x having no multiple factors in $\mathcal{O}_x$, $x \in U_\alpha$.

According to Weierstrass' preparation theorem we may assume that for some point $x \in U_\alpha$ the function φ_α is given in the form

$$\varphi_\alpha = z_1^m + a_1 z_1^{m-1} + \cdots + a_m,$$

where $a_i \in \mathbb{C}\{z_2, \ldots, z_n\}$, and $z_1, \ldots, z_n$ are local parameters at x. By virtue of the assumption about the functions φ_α made above we may take it that $\partial\varphi_\alpha/\partial z_1$ is coprime to φ_α in $\mathcal{O}_x$, hence $\partial\varphi_\alpha/\partial z_1$ is not identically 0 on Supp D in a neighbourhood of x. Now we divide the points $y \in \operatorname{Supp} D$ into two types: those at which all the $\partial\varphi_\alpha/\partial z_i = 0$ for $i = 1, \ldots, n$, and the remaining points. Evidently the first points form a subspace $S \subset \operatorname{Supp} D$, and as we have just seen, $S \neq \operatorname{Supp} D$.

The proposition is an obvious consequence of two assertions: a) the points of the first type are singular points of the subspace Supp D, that is, in their neighbourhood this subspace is not isomorphic to an analytic manifold, and b) in a neighbourhood of the points of the second type Supp D is isomorphic to a manifold of dimension $n-1$.

The assertion a) follows from the representation of the local ring of $y \in \operatorname{Supp} D$:

$$\mathcal{O}_{y, \operatorname{Supp} D} = \mathcal{O}_x/(\varphi_\alpha) \tag{2}$$

(the verification of (2) is left to the reader). If y is a point of the first type, then $\varphi_\alpha \in \mathfrak{m}_x^2$, where $\mathfrak{m}_x$ is the maximal ideal of the local ring $\mathcal{O}_x$. From this it follows at once that $\mathcal{O}_{y, \operatorname{Supp} D}$ is not a regular local ring, hence that Supp D is not a manifold.

The assertion b) is a direct consequence of the implicit function theorem. If, for example, $\partial\varphi_\alpha/\partial z_1(y) \neq 0$, then z_1 is a holomorphic function of $z_2, \ldots, z_n$ on Supp D in a neighbourhood of y, hence $z_2, \ldots, z_n$ determine an isomorphism of this neighbourhood with a domain in $\mathbb{C}^{n-1}$.

We do not develop the theory of divisors on analytic manifolds any further. It can be carried through to results completely analogous to those we have obtained for algebraic varieties. Namely, every divisor has a unique representation in the form of a linear combination of irreducible effective divisors, and irreducible divisors correspond one to one to complex subspaces of codimension 1. Proofs of these facts are contained in the book [36], Appendix to the Russian edition, 185–202. They are quite elementary and do not depend on other parts of that book.

2. Meromorphic Functions. Now we consider meromorphic functions on analytic manifolds, which are an analogue to rational functions on an algebraic variety. The basic auxiliary tool is the concept of a meromorphic fraction, which was introduced in § 2.1.

Definition. A *meromorphic function* on an analytic manifold X is given by a covering $X = \bigcup U_\alpha$ by connected open sets and a system of meromorphic fractions φ_α on U_α such that the restrictions of φ_α and φ_β to the set $U_\alpha \cap U_\beta$ are identical for any α and β. Such systems of functions are called *compatible*.

A covering $X = \bigcup U_\alpha$ and a compatible system of functions φ_α determine the same meromorphic function as the covering $X = \bigcup V_\beta$ and system ψ_β if the restrictions of φ_α and ψ_β agree on $U_\alpha \cap V_\beta$.

If φ_α is a meromorphic fraction on U_α, $\varphi_\alpha = f/g$, f and g holomorphic in U_α, and $g(x) \neq 0$ at some point $x \in U_\alpha$, then φ_α coincides with the holomorphic function f/g in a neighbourhood of x. This concept can be carried over naturally to meromorphic functions. Thus, for every meromorphic function φ on X there exists an open set $U \subset X$ and a function f holomorphic in U such that the restriction of φ to U coincides with f. We say that φ is *holomorphic* at the points of U.

Algebraic operations on meromorphic functions are defined in terms of the corresponding meromorphic fractions. Evidently all meromorphic functions on a manifold X form a ring. If X is connected, then the ring of meromorphic functions on it is a field. For let φ be given by a covering $\{U_\alpha\}$ and a compatible system of functions φ_α. If $\varphi \neq 0$, then at least one $\varphi_\alpha \neq 0$. But from the compatibility of these functions it follows that then $\varphi_\beta \neq 0$ for all β for which $U_\alpha \cap U_\beta$ is not empty. Since X is connected, it then follows that all the $\varphi_\gamma \neq 0$, hence the function φ^{-1} given by the system φ_γ^{-1} exists. Henceforth we only consider connected manifolds. The field of meromorphic functions on such a manifold is denoted by $\mathscr{M}(X)$.

If the manifold is of the form X_{an}, where X is an irreducible smooth algebraic variety, then clearly the rational functions on X determine meromorphic functions on X_{an}. In other words, $\mathbb{C}(X) \subset \mathscr{M}(X_{\mathrm{an}})$. Of course, in general, equality does not hold. However, if X is complete, then the two fields are the same, as we shall show in § 3.

A comparison of the definitions of the two concepts shows that every meromorphic function φ determines a divisor, which we denote by (φ). From the definition it follows that (φ) is an effective divisor if and only if φ is holomorphic on the whole manifold X. For compact connected manifolds this is possible only if φ is a constant, just as in Corollary 1 to Theorem 2 in Ch. I, § 5.

Theorem 2. *A function φ that is holomorphic at all points of a compact connected manifold X is constant.*

The function $|\varphi|$ obviously is continuous on X and therefore attains a maximum at some point x_0. We consider a neighbourhood U of x_0, isomorphic to an open set $V \subset \mathbb{C}^n$.

We may assume that V consists of the points $(z_1, \ldots, z_n)$, $\Sigma |z_i|^2 < 1$, and under the isomorphism $f: U \to V$ we have $f(x_0) = 0 = (0, \ldots, 0)$. The function $\psi = (f^{-1})^* (\varphi)$ is holomorphic on V and its modulus attains a maximum at 0. For every point $(\alpha_1, \ldots, \alpha_n) \in V$ we consider the one-dimensional complex subspace $z_i = \alpha_i t$, $i = 1, \ldots, n$. On it the function ψ determines a holomorphic function of a single argument t, which by the maximum modulus principle is a constant. Hence it follows that ψ is constant on V and therefore φ is constant on U. Since X is connected, by the uniqueness theorem φ is a constant on the whole of X.

Since for divisors of meromorphic functions the following identical relation holds:

$$(\varphi\cdot\psi^{\pm 1})=(\varphi)\pm(\psi),$$

the theorem has the following corollary.

Corollary. *On a compact manifold a meromorphic function is uniquely determined by its divisor, to within a constant factor.*

Having introduced the concepts of meromorphic functions and their divisors, we can throw new light on the examples worked out in § 1.4 of compact analytic manifolds that are not algebraic varieties.

We begin with Example 2: the two-dimensional torus A, which is not algebraic because there is not a single algebraic curve on it.

As was shown in § 1.5, one-dimensional complex subspaces are algebraic curves. Therefore on the torus A there is not a single one-dimensional complex subspace, that is, no divisor different from 0. Hence it follows that the divisor of every meromorphic function on A is 0, hence that all these functions are constants, by virtue of Theorem 2. In other words, $\mathscr{M}(A)=\mathbb{C}$. So we have a new characterization of the non-algebraic torus A: on it there are far fewer meromorphic functions than on an algebraic variety on which necessarily all rational functions are meromorphic.

We now look at Example 1. There we have constructed a two-dimensional torus A and a homomorphism $f: A\to B$ of it onto an elliptic curve. This torus is non-algebraic because the only irreducible curves lying on it are the fibres $f^{-1}(b)$.

The divisor of an arbitrary meromorphic function φ on A can be represented, on the basis of Theorem 1, in the form

$$(\varphi)=D'-D'',$$

where D' and D'' are effective divisors. From the proof of this theorem it is easy to see that the set $\operatorname{Supp} D'\cap\operatorname{Supp} D''$ may consist only of isolated points, and since distinct fibres $f^{-1}(b)$ do not intersect, in general, we have

$$\operatorname{Supp} D'=\bigcup f^{-1}(b_i'),\qquad \operatorname{Supp} D''=\bigcup f^{-1}(b_j''),$$
$$b_i'\neq b_j''.$$

Applying this reasoning to the functions $\varphi-c$, $c\in\mathbb{C}$, we verify that every meromorphic function on A is constant on the fibres of f. At a point $a\in A$ we choose local parameters z_1, z_2 so that $z_1=f^*(t)$ and t is a local parameter at $b=f(a)$. In this coordinate system φ can be represented as a meromorphic fraction $F(z_1,z_2)$ that does not depend on z_1, in other words, is locally of the form $f^*(\psi)$, where ψ is meromorphic on B. Hence it follows that also on the whole of A the equality $\varphi = f^*(\psi)$, $\psi \in \mathscr{M}(B)$, holds. But B is an algebraic curve, and according to a theorem we shall quote and prove in § 2.3, $\mathscr{M}(B) = \mathbb{C}(B)$. So we have shown that for the torus A

$$\mathscr{M}(A) = f^*\mathbb{C}(B).$$

We see that the non-algebraic character of A is again is expressed in this relation. For an algebraic surface X the field $\mathscr{M}(X)$ contains $\mathbb{C}(X)$ and is therefore of transcendence degree at least 2 over $\mathbb{C}$, whereas in our case the transcendence degree is 1.

The same arguments are applicable to Example 3 in § 1.4 (see Exercises 6 and 7 to § 1). On a Hopf surface X the equality $\mathscr{M}(X) = \mathbb{C}(\mathbb{P}^1)$ holds.

3. Siegel's Theorem. The examples at the end of the preceding subsection show that on a compact analytic manifold X there can be "few" meromorphic functions compared with an algebraic variety of the same dimension: more precisely, the transcendence degree of $\mathscr{M}(X)$ can be less than the dimension of X. Quite a number of important properties of compact analytic manifolds follow from the fact that there cannot be too "many" meromorphic functions on them. This is what we are going to prove next.

Theorem 3. *The transcendence degree of the field of meromorphic functions on a compact analytic manifold does not exceed the dimension of the manifold.*

The proof of this theorem is completely elementary. We preced it by a simple remark.

Schwarz's Lemma. *Let $f(z_1, \ldots, z_n) = f(z)$ be a holomorphic function in the domain $|z_i| \leqslant 1$, $i = 1, \ldots, n$, and $M = \max_{|z_i| \leqslant 1} |f(z)|$. If $f \in \mathfrak{m}_0^h$, where $\mathfrak{m}_0$ is the maximal ideal of the local ring of functions analytic at the origin of coordinates (that is, if all the derivatives of f of order $\leqslant h$ vanish at the origin of coordinates), then*

$$|f(z)| < M \max_i |z_i|^h \quad \text{for} \quad |z_i| < 1, \quad i = 1, \ldots, n. \tag{1}$$

Proof. For a point $z = (z_1, \ldots, z_n) \in \mathbb{C}^n$ we set $|z| = \max_i |z_i|$. For a fixed z with $|z| < 1$ we set $g(t) = f(tz)$, $t \in \mathbb{C}$. This function is holomorphic for $|t| \leqslant |z|^{-1}$, and the first h coefficients of its Taylor series vanish at 0. Therefore $g(t)/t^h$ is holomorphic for $|t| \leqslant |z|^{-1}$. By the maximum modulus principle, in this closed disc we have $|g(t)/t^h| \leqslant M/|z|^{-h} = M|z|^h$. Setting $t = 1$ we obtain (1).

In the proof of the theorem we use the fact that the dimension of the space of polynomials in ν variables of degree $\leqslant l$ is given by

$$\frac{(l+1)(l+2)\dots(l+\nu)}{\nu!}.$$

We denote this number by $H_\nu(l)$. It is a polynomial in l of degree ν.

Proof of the Theorem. Let $f_1,\dots,f_{n+1}$ be $n+1$ meromorphic functions on an n-dimensional compact manifold X. Our aim is to construct a polynomial $F(T_1,\dots,T_{n+1})$ for which

$$F(f_1,\dots,f_{n+1})=0. \tag{2}$$

For every point $x\in X$ we choose three neighbourhoods $U_x\supset V_x\supset W_x$. The neighbourhood U_x is such that in it

$$f_i=\frac{P_{i,x}}{Q_{i,x}}, \qquad i=1,\dots,n+1, \tag{3}$$

with $P_{i,x}$ and $Q_{i,x}$ being holomorphic in U_x and relatively prime in every point of this neighbourhood. The existence of such neighbourhoods follows from the lemma in § 2.1. The neighbourhood V_x together with its closure is contained in U_x. In it there exists a local system of coordinates $(z_1,\dots,z_n)$ such that $|z|<1$. The neighbourhood W_x is given by the condition $|z|<\frac{1}{2}$.

From the fact that for distinct points x and y the expressions (3) give representations of one and the same function f_i, and that $P_{i,x}$ and $Q_{i,x}$ are relatively prime, it follows that

$$Q_{i,x}=Q_{i,y}\,\varphi_{i,x,y},$$

where $\varphi_{i,x,y}$ is holomorphic and non-zero in $U_x\cap U_y$.

From the system W_x we select a finite covering of X (here we use that it is compact):

$$X=\bigcup W_\xi.$$

We denote the number of sets W_ξ (and of points $\xi\in X$) by r and we set

$$\varphi_{\xi,\eta}=\prod_{i=1}^{n+1}\varphi_{i,\xi,\eta},\quad C=\max_{\xi,\eta}\ \max_{V_\xi\cap V_\eta}|\varphi_{\xi,\eta}|.$$

Observe that $|\varphi_{\xi,\eta}|$ is bounded in $V_\xi\cap V_\eta$, because the closure of this set is contained in $U_\xi\cap U_\eta$, where $\varphi_{\xi,\eta}$ is holomorphic. Furthermore, $C>1$ since $\varphi_{\xi,\eta}\varphi_{\eta,\xi}=1$.

For the polynomial $F(T_1,\dots,T_{n+1})$ of degree l in $T_1,\dots,T_{n+1}$, which is not yet defined, we set

$$F(f_1,\dots,f_{n+1})=\frac{R_x}{Q_x^l}\quad\text{in}\quad V_x,$$

where

$$Q_x = \prod_{i=1}^{n+1} Q_{i,x} .$$

Clearly $R_\xi = \varphi^l_{\xi,\eta} R_\eta$ in $V_\xi \cap V_\eta$.

Having introduced the notation we can now go over to the business part of the proof. As a first approximation to (2) we show that for every given h the polynomial F can be chosen so that $F \neq 0$ and

$$R_\xi \in \mathfrak{m}_\xi^h \tag{4}$$

for all r points ξ.

These conditions can be written in the form of relations

$$(D^s R_\xi)(\xi) = 0 ,$$

where D^s is a partial derivative of order $< h$. Therefore, there are linear relations on the coefficients of F. The number of relations is $rH_n(h-1)$. If we choose the degree l of F so that

$$H_{n+1}(l) > rH_n(h-1) , \tag{5}$$

then we can find a non-zero polynomial F for which (4) holds.

By Schwarz's lemma, with this choice of F the functions R_ξ are small in the neighbourhoods W_ξ: if

$$M = \max_\xi \max_{x \in V_\xi} |R_\xi(x)| ,$$

then

$$|R_\xi(x)| \leqslant \frac{M}{2^h} \quad \text{for} \quad x \in W_\xi . \tag{6}$$

This tells us that $M = 0$, in other words, that (2) holds, for sufficiently large l and h. For suppose that the maximum M is attained at a point $x_0 \in V_\eta$. Then $x_0 \in W_\xi$ for some point ξ. Therefore

$$M = |R_\eta(x_0)| = |R_\xi(x_0)| \, |\varphi_{\xi,\eta}(x_0)|^l .$$

If l and h are already such that (5) holds, then (6) is also true, and hence

$$M \leqslant \frac{M}{2^h} C^l .$$

It remains for us to choose l and h so that apart from (5) we also have

$$C^l / 2^h < 1 ,$$

and then we find that $M = 0$. The required choice is possible: if $C = 2^\lambda$, $\lambda > 0$ (because $C > 1$), then it is sufficient to take $l < \lambda^{-1} h$ and satisfying (5). For example, for $l = h/m$, where is m is any integer greater than λ, we obtain in (5) on the left-hand side a polynomial of much higher degree in h than on the right-hand side, hence the left-hand side for sufficiently large h divisible by m is, in fact, larger than the right-hand side. The proof of the theorem is now complete.

By means of similar arguments it can be shown that if the transcendence degree of the field $\mathcal{M}(X)$ is l and if $f_1, \ldots, f_l$ are algebraically independent meromorphic functions on X, then the degree of an irreducible relation

$$F(f, f_1, \ldots, f_l) = 0,$$

satisfied by an arbitrary meromorphic function f is bounded above. Therefore the field $\mathcal{M}(X)$ not only has finite transcendence degree, but is even finitely generated.

Exercises

1. Define an analytic vector bundle by analogy to what we have done in Ch. VI, § 1.2, with this difference that E and X are analytic manifolds and $p: E \to X$ a holomorphic mapping. Show that the correspondence between bundles and transition matrices that was established in Ch. VI, § 1.2, remains valid for analytic vector bundles.

2. Show that the association of a divisor with a linear bundle, as described in Ch. VI, § 1.4, carries over to analytic bundles. Here we have to state the definition of this association in Ch. VI, § 1.4, in terms of transition matrices [Formula (2) of Ch. VI, § 1.4]. Show that also for analytic bundles equivalent divisors determine isomorphic bundles.

3. Let X be an analytic manifold, $U_\alpha \subset X$ an open set for which there exists an isomorphic mapping onto an open set in $\mathbb{C}^n$, and $z_1^{(\alpha)}, \ldots, z_n^{(\alpha)}$, the inverse images of coordinates in $\mathbb{C}^n$ relative to this isomorphism. If U_β is another such open set, then in $U_\alpha \cap U_\beta$ we set

$$\varphi_{\alpha\beta} = \frac{\partial(z_1^{(\alpha)}, \ldots, z_n^{(\alpha)})}{\partial(z_1^{(\beta)}, \ldots, z_n^{(\beta)})}.$$

Show that $\varphi_{\alpha\beta}$ are the transition functions of some linear bundle $\mathcal{K}$. Show that if $X = Y_{an}$, where Y is an algebraic variety, then $\mathcal{K} = K_{an}$, where K is the bundle corresponding to the canonical class on Y. In the general case $\mathcal{K}$ is called the *canonical bundle*.

4. Let X be an analytic manifold, $X = \bigcup U_\alpha$ a covering for which there are isomorphisms $\varphi\alpha: U_\alpha \to \mathbb{C}^n$ onto open sets of $\mathbb{C}^n$. Suppose that in $\mathbb{C}^n$ functions f_α are given that are holomorphic on $\varphi_\alpha(U_\alpha)$ such that under the isomorphisms $\varphi_\beta \varphi_\alpha^{-1}: \varphi_\alpha(U_\alpha \cap U_\beta) \to \varphi_\beta(U_\alpha \cap U_\beta)$ the forms $f_\alpha\, dz_1 \wedge \cdots \wedge dz_n$ and $f_\beta\, dz_1 \wedge \cdots \wedge dz_n$ are carried into each other. By definition such a collection determines a holomorphic form ω on X. Show that the functions $\varphi_\alpha^*(f_\alpha)$ determine a divisor on X, the so-called divisor of the form ω. Show that the divisors of any two holomorphic forms are equivalent. Show that if a holomorphic differential form exists on X, then the linear bundle defined by its divisor is isomorphic to the canonical bundle.

5. Show that the canonical bundle of a complex torus is trivial.

6. Let Ω be a $2n$-dimensional lattice in $\mathbb{C}^n$, $X = \mathbb{C}^n/\Omega$ an n-dimensional torus (Example 1 of § 1.2), and $\chi: \Omega \to \mathbb{C}^*$ a homomorphism of the group Ω into the multiplicative group of non-zero complex numbers. Define an action of Ω on the space $\mathbb{C}^n \times \mathbb{C}^1$ by the condition

$$a(x, z) = (x + a, \chi(a)\, z),\ x \in \mathbb{C}^n,\ z \in \mathbb{C}^1,\ a \in \Omega.$$

Show that Ω acts on $\mathbb{C}^n \times \mathbb{C}^1$ freely and discretely. The projection $\mathbb{C}^n \times \mathbb{C}^1 \to \mathbb{C}^n$ is permutable with the action of Ω and determines a mapping $p: (\mathbb{C}^n \times \mathbb{C}^1)/\Omega \to \mathbb{C}^n/\Omega = X$. Show that p is holomorphic and determines in $E_\chi = (\mathbb{C}^n \times \mathbb{C})/\Omega$ the structure of a linear bundle over X (see Exercise 1).

7. In the notation of Exercise 6, show that two bundles E_χ and $E_{\chi'}$ are isomorphic if and only if there exists a holomorphic function g on $\mathbb{C}^n$, vanishing nowhere, such that $g(x+a)\,g(x)^{-1} = \chi'(a)\,\chi(a)^{-1}$ for all $x \in \mathbb{C}^n$, $a \in \Omega$.

8. In the notation of Exercises 6 and 7 assume, in addition, that $|\chi(a)| = 1$ for all $a \in \Omega$. Show that if χ and χ' have this property, then E_χ and $E_{\chi'}$ are isomorphic only if $\chi = \chi'$.

9. Show that Theorem 3 of Ch. VI, § 1 has no analogue in the theory of analytic linear bundles: not every linear bundle is determined by some divisor.

§ 3. Algebraic Varieties and Analytic Manifolds

1. Comparison Theorem. Now we are in a position to prove some fundamental facts, which show that for complete and projective algebraic varieties X over the field of complex numbers many properties of the corresponding analytic manifold X_{an} can be reduced to algebraic properties of X.

Theorem 1. *If an algebraic variety X is complete, then a meromorphic function on the analytic manifold X_{an} is a rational function on X.*

Let f be a meromorphic function on X_{an}. According to Theorem 3 of § 2, it is algebraic over $\mathbb{C}(X)$ (because X_{an} is compact). Therefore we need only show that a function f that is meromorphic on X_{an} and algebraic over $\mathbb{C}(X)$ is rational on X. In the proof of this fact the completeness of X does not play any role.

Let

$$F(f) = f^m + a_1 f^{m-1} + \cdots + a_m = 0$$

be the irreducible equation over $\mathbb{C}(X)$ whose root f is. By removing from X the poles of the rational functions a_i, we may assume that the a_i are regular on X. Then f is holomorphic on X. This follows from the fact that in the rings $\mathcal{O}_{x,\text{an}}$ the decomposition into prime factors is unique, so that they are integrally closed in their fields of fractions.

In the product $X \times \mathbb{A}^1$ we consider the set X' of points (x, z) satisfying the relation

$$z^m + a_1(x)\, z^{m-1} + \cdots + a_m(x) = 0\,.$$

The algebraic variety X' is irreducible, and $\mathbb{C}(X') = \mathbb{C}(X)(f)$. We denote by $p: X' \to X$ the natural projection. Once more we diminish X and X', by removing from X the points at which the discriminant of the polynomial $F(T)$ vanishes, and from X' the inverse image (relative to p) of this set. We denote the so obtained irreducible varieties by X and X', as before. In this way we achieve that $p^{-1}(x)$ for every point $x \in X$ consists of m distinct points, and at every such point (x, z) we have $F'_T(x, z) \neq 0$.

If $z_1, \ldots, z_n$ are local parameters at $x \in X$, then it follows that $p^*(z_1), \ldots, p^*(z_n)$ are local parameters at any point $p^{-1}(x)$. Therefore there exists a sufficiently small complex neighbourhood U of x such that $p^{-1}(U)$ splits into m disjoint sets $U_1, \ldots, U_m$ and the projection $p: U_i \to N$ is an isomorphism of the corresponding analytic manifolds. It is sufficient for us to prove that p is a homeomorphism. This assertion means that $X'(\mathbb{C})$ is an unramified covering of $X(\mathbb{C})$.

The function f determines a section of this unramified covering, that is, a continuous mapping

$$\varphi(x) = (x, f(x)),$$

for which $p \cdot \varphi = 1$.

The information we have obtained is now enough to show that $m = 1$, hence $f \in \mathbb{C}(X)$. For if $m > 1$, then $\varphi(X) \neq X'$, because $\varphi(x)$ is a single point, whereas $p^{-1}(x)$ consists of m points. We show that the sets $\varphi(X(\mathbb{C}))$ and $X'(\mathbb{C}) - \varphi(X(\mathbb{C}))$ are closed and disjoint, from which it follows that $X'(\mathbb{C})$ is disconnected. This contradicts the theorem in Ch. VII, § 2, because X' is an irreducible algebraic variety.

All the assertions that remain to be verified are of local character, so that we need only verify them for the sets U and $p^{-1}(U)$ instead of X and X', where U is any neighbourhood of $x \in X$. In particular, we may take U to be connected and such that

$$p^{-1}(U) = U_1 \cup \cdots \cup U_m, \quad U_i \cap U_j = \emptyset \quad \text{for} \quad i \neq j.$$

Then $\varphi(U)$ must coincide with one of the U_i, and all we need then clearly follows.

Theorem 2. *If X and Y are complete varieties, then any holomorphic mapping $f: X_{\text{an}} \to Y_{\text{an}}$ is of the form g_{an}, where $g: X \to Y$ is a morphism.*

Let $x \in X$, $y = f(x)$, and U be an affine neighbourhood of y. We assume that $U \subset \mathbb{A}^N$ and denote by $t_1, \ldots, t_N$ coordinates in this space. According to Theorem 1 the $f^*(t_i)$ are rational functions on X.

If we can show that they are regular at x, then we have $f = g_{\text{an}}$ in some neighbourhood of x. In this way we construct a system of morphisms $g_i: V_i \to Y$ on open sets V_i covering X. Clearly they determine a single morphism $g: X \to Y$ for which $f = g_{\text{an}}$.

Thus, everything is reduced to a local proposition.

Lemma. *If the rational function g is holomorphic at x, then it is regular there.*

We set $g = u/v$, $u, v \in \mathcal{O}_x$. We denote by $\mathcal{O}_{x,\text{an}}$ the ring of all functions that are holomorphic in some neighbourhood of x, and by $\hat{\mathcal{O}}_x$ the ring of formal power series. By assigning to a holomorphic function its power series we define an embedding $\mathcal{O}_{x,\mathrm{an}} \subset \hat{\mathcal{O}}_x$ such that

$$\mathcal{O}_x \subset \mathcal{O}_{x,\mathrm{an}} \subset \hat{\mathcal{O}}_x .$$

The fact that g is holomorphic at x means that $v|u$ in $\mathcal{O}_{x,\mathrm{an}}$. Then a fortiori $v|u$ in $\hat{\mathcal{O}}_x$. But according to 1 and 2 at the end of Ch. II, § 3.3, it then follows that $v|u$ in $\mathcal{O}_x$, hence that g is regular at x.

Corollary. *If for two complete algebraic varieties X and Y the analytic manifolds X_{an} and Y_{an} are isomorphic then X and Y are isomorphic.*

Theorem 3. *If X is a projective variety, then any compact analytic submanifold V of X_{an} is of the form Y_{an}, where Y is a closed algebraic subvariety of X.*

Since X is contained in a projective space, it is sufficient to prove the theorem for the case $X = \mathbb{P}^N$. Furthermore, it is sufficient to prove the theorem for connected submanifolds $V \subset \mathbb{P}^N_{\mathrm{an}}$, because from the compactness of V it follows that it consists of finitely many connected components. Therefore, in what follows, we assume V to be connected.

We denote by Y the closure of V in the spectral topology of $\mathbb{P}^N$, that is, the intersection of all algebraic subvarieties of this space containing V. Let us show that the projective variety Y is irreducible. To see this we need only show that the homogeneous ideal determined by it is prime, that is, if P and Q are homogeneous polynomials such that $P \cdot Q = 0$ on V, then $P = 0$ or $Q = 0$ on V. If P does not vanish identically on V, then the set $U \subset V$ of those points where $P \neq 0$ is open in V. Suppose that the homogeneous coordinate x_0 does not vanish identically on V. Then the function Qx_0^{-l}, $l = \deg Q$, vanishes on U and is holomorphic. By the uniqueness theorem it vanishes on the whole connected component V containing U, that is, on the whole of V. But this means that $Q = 0$ on V.

From the definition of Y it follows that every rational function $\varphi \in \mathbb{C}(Y)$ determines a certain meromorphic function on V. In other words,

$$\mathbb{C}(Y) \subset \mathcal{M}(V) . \tag{1}$$

We set $\dim V = n$, $\dim Y = m$. Since in a neighbourhood of each of its simple points Y is a $2m$-dimensional manifold and $Y \supset V$, we have $m \geqslant n$. But according to Theorem 3, § 2.3 the transcendence degree of the field $\mathcal{M}(V)$ does not exceed n, so that from (1) we obtain.

$$\dim Y = \dim V = n . \tag{2}$$

From (2) it is easy to derive that $Y_{\mathrm{an}} = V$, which is what we have to show. For we denote by S the set of singular points of Y. The algebraic variety $Y - S$ is irreducible, hence by Theorem 2 of Ch. VI, § 1, the manifold $(Y - S)_{\mathrm{an}}$ is connected. The set $V - (V \cap S)$ is closed in $(Y - S)_{\mathrm{an}}$, because is V is closed in Y_{an}. On the other hand, from (2) it follows that

$V-(V\cap S)$ is open in $(Y-S)_{\mathrm{an}}$. Therefore $V-(V\cap S)=(Y-S)_{\mathrm{an}}$, that is, $(Y-S)_{\mathrm{an}}\subset V$. Since V is closed and since by Lemma 1 of Ch. VII, § 2.1, $(Y-S)_{\mathrm{an}}$ is everywhere dense in Y_{an}, we have $Y_{\mathrm{an}}\subset V$, that is, $Y_{\mathrm{an}}=V$. This proves the theorem.

2. An Example of Non-Isomorphic Algebraic Varieties that are Isomorphic as Analytic Manifolds. We construct two algebraic varieties X and Y that are not isomorphic, but for which the analytic manifolds X_{an} and Y_{an} are isomorphic. According to the corollary to Theorem 1 in this situation X and Y cannot be complete.

First we give a description of the relevant example. Let C be a smooth projective curve of degree 3, o a point of it, B the incomplete curve $C-o$, and p one of its points. In Ch. VI, § 1.4, we have seen that to every divisor on B there corresponds a linear bundle $E\to B$. We take for X the bundle corresponding to the divisor p, and for Y the direct product $B\times\mathbb{A}^1$ corresponding to the zero divisor. We have to prove two facts: 1) the algebraic varieties X and Y are non-isomorphic, and 2) the analytic manifolds X_{an} and Y_{an} are isomorphic.

1) First of all we observe that X and Y are non-isomorphic as linear bundles. According to Theorem 3 in Ch. VI, § 1, to prove this we need only show that the divisors corresponding to them are non-equivalent, that is, that p is not equivalent to zero on B. If this were so, then there would exist a function f regular on B having a simple zero at p. The divisor of this function on C must be of the form $p-lo$. By the corollary to Theorem 1 in Ch. III, § 2, $l=1$, and by Theorem 1 at the same place this contradicts the fact that C is a non-rational curve.

We now assume that there exists an isomorphism $\varphi: X\to Y$ of algebraic varieties. We denote by p_X and p_Y the projections of X and Y onto B defined by specifying the structure bundles on them. For every point $b\in B$ the curve $p_X^{-1}(b)$, hence also $\varphi p_X^{-1}(b)$, is isomorphic to $\mathbb{A}^1$. If $p_Y\varphi p_X^{-1}(b)$ is not a point, then p_Y determines an embedding of $\mathbb{C}(B)$ in $\mathbb{C}(\varphi p_X^{-1}(b))$, which contradicts Lüroth's theorem, because the curve $\varphi p_X^{-1}(b)$ is rational, but B is not. Thus, φ carries a fibre of X into a fibre of Y.

We see that there exists a mapping $\psi: B\to B$ such that the diagram

$$\begin{array}{ccc} X & \xrightarrow{\varphi} & Y \\ {\scriptstyle p_X}\downarrow & & \downarrow{\scriptstyle p_Y} \\ B & \xrightarrow{\psi} & B \end{array}$$

commutes. If s_X is the zero section of X, then $\psi=p_Y\varphi s_X$, from which it follows that ψ is a morphism, hence an automorphism of B. We denote by $\psi\times 1$ the automorphism of the bundle $Y=B\times\mathbb{A}^1$ acting as ψ on B and trivially on $\mathbb{A}^1$. Then $\varphi' = (\psi \times 1)^{-1}\varphi$ is also an isomorphism of X and Y, but now $p_Y\varphi' p_X^{-1}(b) = b$ for $b \in B$ and $\psi' = p_Y\varphi' s_X = 1$.

We set $\varphi' s_X = t : B \to Y$. This is a section of the bundle Y. We now recall that Y is a vector bundle, so that it makes sense to talk of subtraction of vectors in one of its fibres. We set

$$\varphi''(x) = \varphi'(x) - t p_X(x) .$$

Obviously, this is again an isomorphism of the varieties X and Y, however, now not only does every fibre go into itself, but the zero point is preserved. But the only automorphisms of the line $\mathbb{A}^1$ preserving the zero point are the linear transformations $\alpha \to \lambda\alpha$. Thus, φ'' must be an isomorphism of the vector bundles X and Y, but as we have seen, they are non-isomorphic.

2) We make use of the fact that the association $D \to L_D$ in Ch. VI, § 1.4, carries over verbatim to analytic manifolds and meromorphic functions (see Exercises 1 and 2 to § 2). In particular, if we can show that the point p is the divisor of some meromorphic function, then this shows that X_{an} and Y_{an} are isomorphic manifolds (even as "analytic vector bundles").

Thus, our task reduces to constructing a holomorphic function on B having a single zero of order 1 at p.

To make everything quite specific, we assume that C is given by the equation

$$y^2 = x^3 + ax + b , \tag{1}$$

and that o is the point at infinity on it. Then B is given by the equation (1) in the affine plane.

We consider on C three rational differential forms:

$$\omega_1 = \frac{dx}{y}, \omega_2 = x\frac{dx}{y}, \omega_3 = \frac{1}{2}\frac{y - y_0}{x - x_0}\cdot\frac{dx}{y},$$

where $p = (x_0, y_0)$. Let us investigate their behaviour at o. A local parameter at this point is $t = x/y$, and

$$x = \frac{u}{t^2}, y = \frac{v}{t^3}, u, v \in \mathcal{O}_o, u(o) = v(o) = 1 .$$

Hence it follows that ω_1 is a regular form at o, ω_2 has there a pole of order 2, and ω_3 a simple pole. From (1) (divided by y^2) it is easy to derive that $xt^2 \equiv 1(t^4)$, $yt^3 \equiv 1(t^4)$. Hence it follows that

$$\omega_2 = \left(-\frac{2}{t^2} + f\right)dt, \quad \omega_3 = \left(-\frac{1}{t} + g\right)dt, \ f, g \in \mathcal{O} . \tag{2}$$

Since the genus of C is 1, according to the results of Ch. VII, § 3.3, the topological space $C(\mathbb{C})$ is homeomorphic to a torus. We denote by α

and β a basis of its two-dimensional homology group, for example, its parallel and meridian.

It is easy to see that ω_1 is a regular form on C, and that ω_2 has only a single pole of order 2, namely at o. The integral of ω_1 over a one-dimensional cycle σ depends only on its homology class: if σ is homologous to $a\alpha + b\beta$, then

$$\int_\sigma \omega_1 = a\int_\alpha \omega_1 + b\int_\beta \omega_1 .$$

Although the form ω_2 is not regular in the neighbourhood of o, its integral over a small contour around o is 0, because its expansion (2) has no term with $1/t$. Therefore the analogous formula

$$\int_\sigma \omega_2 = a\int_\alpha \omega_2 + b\int_\beta \omega_2$$

holds if σ does not contain o.

Finally, for the form ω_3 we obtain similarly

$$\int_\sigma \omega_3 = a\int_\alpha \omega_3 + b\int_\beta \omega_3 + 2\pi i n, \quad n \in \mathbb{Z}, \tag{3}$$

because ω_3 has simple poles at p and o, and by analogy to the expansion (2) $\omega_3 = \frac{1}{u} + h\,du$, $h \in \mathcal{O}_p$, where u is a local parameter at p.

The vectors $\left(\int_\alpha \omega_1, \int_\beta \omega_1\right)$ and $\left(\int_\alpha \omega_2, \int_\beta \omega_2\right)$ are not proportional. For if a combination of them with coefficients λ and μ would vanish, then we would have the relation

$$\int_\sigma (\lambda\omega_1 + \mu\omega_2) = 0$$

for every cycle σ. This means that the function $\varphi(x) = \int_q^z (\lambda\omega_1 + \mu\omega_2)$ (for some fixed point q) is single-valued and meromorphic on C_{an}. By Theorem 1 it must be a rational function on C. If $\mu \neq 0$, then it has only one simple pole at o, which is impossible, because C is not a rational curve. And if $\mu = 0$, then it is regular everywhere, which is also impossible.

Utilizing the independence of these vectors we can find numbers λ and μ such that

$$\left(\int_\alpha \omega_3, \int_\beta \omega_3\right) = \lambda\left(\int_\alpha \omega_1, \int_\beta \omega_1\right) + \mu\left(\int_\alpha \omega_2, \int_\beta \omega_2\right).$$

We set $\eta = \omega_3 = \lambda\omega_1 - \mu\omega_2$. The equation (3) shows that

$$\int_\sigma \eta = 2\pi i n, \quad n \in \mathbb{Z},$$

for every cycle σ. Therefore the function $\varphi(x) = \exp \int_q^x \eta$ is single-valued on C_{an}. It is meromorphic on B_{an} and regular everywhere except possibly at p. In a neighbourhood of this point $\omega_3 = \left(\frac{1}{u} + h\right) du$, $h \in \mathcal{O}_u$, and η has the same expansion, hence $\varphi = u \cdot \psi$, where ψ is holomorphic and different from 0 at p. This shows that the divisor of φ on B_{an} consists of the point p with coefficient 1.

3. Example of a Non-Algebraic Compact Manifold with the Maximal Number of Independent Meromorphic Functions. The transcendence degree of $\mathcal{M}(X)$ (which is finite by Theorem 3) is the basic invariant by means of which we might try to classify compact analytic manifolds. Here we report, omitting all proofs, about what is known in this direction.

From this point of view the manifolds closest to algebraic varieties are those for which the transcendence degree of $\mathcal{M}(X)$ is equal to the dimension of X. We begin by constructing an example of a non-algebraic manifold with this property.

We use a construction very similar to the one we have used in Ch. VI, § 2.3, to construct an example of a non-projective algebraic variety. Here we apply the concept of a σ-process along a smooth subvariety to the case when the enveloping manifold is analytic. The reader can easily verify that the definitions and simplest properties deduced in Ch. VI, § 2.2, carry over word for word to this case.

We consider the projective space $\mathbb{P}^3$ and in it a curve C having a double point x_0 with distinct tangents, for example, the curve with the equation (1) in Ch. I, § 1.1. There exists a neighbourhood U of x_0 (in the complex topology of $\mathbb{P}^3$) such that the analytic manifold $U \cap C$ is reducible and splits into two one-dimensional irreducible smooth submanifolds C' and C'' intersecting transversally, namely, the two branches of C at x_0.

We consider the σ-process of the variety U, $\sigma_1 : U_1 \to U$ with centre in the subvariety C'. The inverse image $C'_1 = \sigma_1^{-1}(C')$ is a smooth surface, with fibres isomorphic to $\mathbb{P}^1$. We set $\sigma^{-1}(x_0) = L_1$. The inverse image $\sigma^{-1}(C'')$ is reducible and consists of two one-dimensional components: L_1 and a smooth subvariety C''_1, which is mapped by σ, isomorphically onto C''. Both subvarieties intersect transversally at the point $x_1 = L_1 \cap C''_1$. We now consider the σ-process $\sigma_2 : \bar{U} \to U_1$ of the variety U_1 with centre in C''_1. Again the inverse image $\sigma_2^{-1}(L_1)$ consists of two one-dimensional components: $\sigma_2^{-1}(L_1) = \bar{L} \cup \bar{L}_1$, where $\bar{L} = \sigma_2^{-1}(x_1)$, and where $\bar{L}_1$ is mapped isomorphically by σ_2 onto L_1. We set $\bar{\sigma} = \sigma_2 . \sigma_1 : \bar{U} \to U$. On the other hand, we consider the σ-process $\sigma : V \to \mathbb{P}^3 - x_0$ of the variety $\mathbb{P}^3 - x_0$ with centre in the subvariety $C - x_0$. Since σ coincides in $U - x_0$ with the σ-process in $C - x_0$, the two varieties and mappings we have constructed can be combined into a single one:

$$\sigma : X \to \mathbb{P}^3 .$$

Evidently $\mathbb{C}(\mathbb{P}^3) \subset \mathcal{M}(X)$, hence the transcendence degree of the field $\mathcal{M}(X)$ is 3. We show that X is not an algebraic variety. To do this we assume that it is algebraic, and for the curve situated on it we use the concept of numerical equivalence introduced in connection with the similar example in Ch. VI, § 2.3. We also use the fact that on an algebraic variety an irreducible smooth curve is not equivalent to zero. For as we have seen in Ch. VI, § 2.3, for this purpose it is sufficient to construct an effective divisor intersecting our curve in a non-empty finite set of points. Let $E \subset X$ be our curve and $U \subset X$ an affine open set (it is here that we assume X to be algebraic) having a non-empty intersection with E. In U we can find a divisor intersecting $U \cap E$ in a finite and non-empty set of points, for example, a hyperplane section F in the enveloping affine space, passing through some point $x \in U \cap E$ but not through another point $x' \in U \cap E$. The closure $\bar{F}$ of the divisor F in the whole of X then has all the properties we need.

Now it is sufficient to find on X an irreducible curve equivalent to zero so as to obtain a contradiction to the fact that X is an algebraic variety. For this purpose we use the result that under a σ-process with centre on a curve the inverse images of all points of this curve are equivalent to each other. We take the points $x \in C - x_0$, $x' \in C' - x_0$, $x'' \in C'' - x_0$, and let $L = \sigma^{-1}(x)$, $L' = \sigma^{-1}(x')$, $L'' = \sigma^{-1}(x'')$. Considering L'' as inverse image of $\sigma_1^{-1}(x'')$ under σ_2 we find that

$$L \sim L'' \sim \bar{L}. \qquad (1)$$

On the other hand, on U_1

$$\sigma_1^{-1}(x') \sim L,$$

but on $\bar{U}$

$$L' \sim \sigma_0^{-1}(L_1) = L + L_1.$$

Thus,

$$L \sim \bar{L} + \bar{L}_1.$$

In conjunction with (1) this shows that $\bar{L}_1 \sim 0$.

Observe that in all these arguments we could have considered instead of equivalence of curves on X homology of the corresponding cycles, using the results of Ch. VII, § 1.3.

The dimension 3 in our example is the smallest possible, because it can be shown that a compact analytic manifold of dimension 2 on which there exist two algebraically independent meromorphic functions is an algebraic variety, hence even projective, as was indicated in Ch. VI, § 2.3.

Analytic manifolds X for which the transcendence degree of $\mathcal{M}(X)$ is equal to the dimension of X are very close to algebraic varieties. In this case $\mathcal{M}(X)$ is isomorphic to the field $\mathbb{C}(X')$ of rational functions on some algebraic variety X', $\dim X' = \dim X$, so that X is "bimeromorphically isomorphic" to an algebraic variety. This fact can be made more precise by proving for such varieties an analogue to Chow's lemma in Ch. VI, § 2.1. All this suggests that for such manifolds there exists a purely algebraic description and that the analogous objects can be defined for an arbitrary field. In fact, such a concept, a so-called "algebraic space", has recently been introduced. The reader can get acquainted with them in the papers [5] and [22].

4. Classification of Compact Analytic Surfaces. In our classification we now go over to the type of manifolds for which the transcendence degree of $\mathcal{M}(X)$ is $\dim X - 1$. Owing to Riemann's existence theorem this case is impossible for $\dim X = 1$, and we can expect to meet it first for $\dim X = 2$, that is, for analytic surfaces. Examples of such surfaces are known to us. They are the complex tori in Example 1 of § 1.4 and the Hopf manifolds (Example 3 there). A general classification of them is given by the following theorem of Kodaira:

A compact analytic surface X for which the field $\mathcal{M}(X)$ is of transcendence degree 1 has a holomorphic mapping $p: X \to Y$ onto an algebraic curve Y such that $\mathcal{M}(X) = p^*\mathbb{C}(Y)$ and all its fibres $p^{-1}(y)$, except finitely many, are elliptic curves.

A similar fact can be proved for manifolds of arbitrary dimension, but in a weaker form:

If X is a compact n-dimensional analytic manifold and the transcendence degree of $\mathcal{M}(X)$ is $n-1$, then X is bimeromorphically isomorphic to a manifold X' having a holomorphic mapping $p: X' \to Y$ onto an $(n-1)$-dimensional algebraic variety Y such that $\mathcal{M}(X) = \mathcal{M}(X') = p^*\mathbb{C}(Y)$, and $p^{-1}(y)$ is an elliptic curve for all points y in some set that is open in Y in the spectral topology.

Other types of analytic manifolds have been investigated almost exclusively in the case of analytic surfaces. The only remaining type for them is $\mathcal{M}(X) = \mathbb{C}$. Here we describe a classification of this type of surface due to Kodaira.

First of all, we note that the concept of an exceptional subvariety naturally carries over to analytic manifolds. It can be shown that every analytic surface can be obtained by finitely many σ-processes from a surface not having exceptional curves. In this context we are going to talk simply of surfaces without exceptional curves.

Kodaira has shown that for a compact surface X without exceptional curves for which $\mathcal{M}(X) = \mathbb{C}$ the one-dimensional Betti number b_1 can only assume the three values: 4, 1, and 0.

If $b_1=4$, then X is a complex torus. We already know an example of a complex torus on which all meromorphic functions are constants (Example 2 to § 1.4).

If $b_1=0$, then the canonical bundle of the surface is trivial (the canonical bundle is defined by analogy with the case of algebraic varieties and is a substitute for the canonical class when we cannot use rational or meromorphic functions, see Exercises 3 and 4 to § 2). All surfaces of this type are homeomorphic to each other and to algebraic surfaces of type $K3$ (see Ch. III, § 5.7). They are called *analytic surfaces* of type $K3$.

The case $b_1=1$ has not yet been investigated completely. Examples of such surfaces can be constructed by generalizing the construction of a Hopf variety. Namely, compact surfaces having the form $(\mathbb{C}^2-0)/G$, where G is a discretely and freely acting group of automorphisms of $\mathbb{C}^2-0$ are called *generalized Hopf manifolds*. For example, for G we can take the cyclic group generated by the automorphism $(z_1,z_2)\to(\alpha_1 z_1, \alpha_2 z_2)$, $|\alpha_1|<1, |\alpha_2|<1$. It can be shown that if there do not exist integers n_1 and n_2, not both 0, for which $\alpha_1^{n_1}=\alpha_2^{n_2}$, then on such a surface all meromorphic functions are constant.*

Thus, with respect to the invariant quantity l, the transcendence degree of $\mathcal{M}(X)$, compact analytic surfaces can be classified as follows:

$l=2$ – algebraic surfaces;

$l=1$ – surfaces with a sheaf of elliptic curves;

$l=0$ – the surfaces are tori or are of type $K3$ or have $b_1=1$ (the classification of all such surfaces is an interesting problem).

In this classification a striking resemblance with the classification of algebraic surfaces, as explained in Ch. III, § 5.7, hits the eye. An understanding of this analogy is probably possible only in connection with a generalization of the two theories to manifolds of arbitrary dimension. This is one of the most interesting tasks of the theory of algebraic varieties and analytic manifolds.

Exercises

1. Let $A=\mathbb{C}^2/\Omega$ be a two-dimensional complex torus, g the automorphism $gx=-x$, $G=\{1,g\}$. Show that the ringed space $\tilde{X}=A/G$ (see Exercise 1 to Ch. V, § 3) is a complex space having 16 singular points $z_1,\dots,z_{16}$ corresponding to the points $x\in A$ for which $2x=0$.

* Recently Inue has constructed a series of such surfaces, different from generalized Hopf manifolds. Their second Betti number is 0. Bogomolov has proved that all surfaces with $b_1=1$, $b_2=0$ are covered by Inue's construction. Finally, Hirzebruch constructed surfaces with $b_1=1$, $b_2>0$. At present there are no conjectures about their structure.

See E. Bombieri, D. Husemaller: "Classification and embeddings of surfaces". Proceedings of Symposia in Pure Math., Vol. XXIX and F. Bogomolov, Izv. Acad. Nauk USSR, Ser. Math., Vol. 40, 1976, N2. (Footnote to corrected printing, 1977).

2. In the notation of Exercise 1, show that every singular point $z_i \in \tilde{X}$ has a neighbourhood isomorphic to a neighbourhood of the singular point of a quadric cone (see Exercise 8 to § 2).

3. In the notation of Exercises 1 and 2, show that there exists a complex manifold X and a holomorphic mapping $\varphi: X \to \tilde{X}$ such that on X there are 16 pairwise disjoint curves $C_1, \ldots, C_{16}$ each of which is isomorphic to $\mathbb{P}^1_{\mathrm{an}}$, $\varphi(C_i) = z_i$, and $\varphi: X - \bigcup C_i \to \tilde{X} - \bigcup z_i$ is an isomorphism. (*Hint*: Use Exercise 10 to Ch. II, § 4.)

4. In the notation of the preceding exercises, show that the differential form $dz_1 \wedge dz_2$, where z_1 and z_2 are coordinates in $\mathbb{C}^2$, determine a differential form on A that is holomorphic and vanishes nowhere (see Exercises 3 and 4 to § 2). Show that it also determines a holomorphic form on X vanishing nowhere. Deduce that the canonical bundle on X is trivial.

5. In the notation of the preceding exercises, show that if all the meromorphic functions on the torus A are constant, then the same is true for X. Show that X is not isomorphic to a complex torus (for example, verify that on X there are no one-dimensional holomorphic differential forms). Thus, X is an example of a non-algebraic surface of type $K3$.

6. Show that for any smooth projective variety X of dimension $n \geqslant 3$ there exists a non-algebraic complex compact n-dimensional manifold X such that $\mathcal{M}(X') = \mathbb{C}(X)$.

Chapter IX. Uniformization

§ 1. The Universal Covering

1. The Universal Covering of a Complex Manifold. In the preceding sections we have made use of the concept of factor space to construct many important examples of analytic manifolds. Now we show how this concept leads to a general method of studying such manifolds.

We begin by recalling some simple topological facts (see, for example, [25], §§ 49–50). Let X be an arcwise connected, locally connected, and locally simply-connected space. Later X will be a connected manifold, and all these conditions hold. The universal covering space $\tilde{X}$ of X has a projection

$$p:\tilde{X}\to X,$$

which turns $\tilde{X}$ into an unramified covering. The homeomorphisms g of $\tilde{X}$ into itself for which $p\cdot g=p$ form a group G, which is isomorphic to the fundamental group $\pi_1(X)$ of X. The group G acts on X discretely and freely, and

$$X=\tilde{X}/G. \tag{1}$$

Now we assume that X is an analytic manifold, and we denote by $\mathcal{O}_X$ its structure sheaf. The manifold $\tilde{X}$ (and more generally, any unramified covering) can also be turned into an analytic manifold, in fact, so that the projection p is a holomorphic mapping. To see this we consider the presheaf $\tilde{\mathcal{O}}$ on $\tilde{X}$ defined by the condition

$$\tilde{\mathcal{O}}(\tilde{U})=\mathcal{O}_X(p(\tilde{U}))$$

for $\tilde{U}$ open in $\tilde{X}$ [then $p(\tilde{U})$ is also open in X, because p determines an unramified covering]. The sheaf associated with $\tilde{\mathcal{O}}$ is denoted by $\mathcal{O}_{\tilde{X}}$. Every point $\tilde{x}\in\tilde{X}$ has a neighbourhood $\tilde{U}$ that is mapped by means of p homeomorphically onto $p(\tilde{U})$. Therefore the sheaf $\mathcal{O}_{\tilde{X}}$ is given uniquely by its restrictions to these sets $\tilde{U}$. It is easy to see that for them $\mathcal{O}_{\tilde{X}}$ is obtained simply by transferring the sheaf $\mathcal{O}_X$ by means of the homeomorphism p.

From what we have said it follows that the pair $(\tilde{X},\mathcal{O}_{\tilde{X}})$ determines an analytic manifold. For if $p:\tilde{U}\to p(\tilde{U})$ is a homeomorphism, then the projection p determines an isomorphism of the ringed spaces $(\tilde{U},\mathcal{O}_{\tilde{X}}|_{\tilde{U}})$

and $(p(\tilde{U}), \mathcal{O}_X|_{p(\tilde{U})})$. Therefore, if $p(\tilde{U})$ is isomorphic to a domain in $\mathbb{C}^n$, then the same is true for $\tilde{U}$. It is also evident that p is a holomorphic mapping. Furthermore, since the complex structure on $\tilde{X}$ is defined by means of the projection p, and since the homeomorphisms $g \in G$ do not change under this projection, they are automorphisms of the analytic manifold $\tilde{X}$. Hence it follows that (1) is an isomorphism of analytic manifolds.

Suppose that two manifolds X and X' have a common universal covering $\tilde{X}$. Then

$$X = \tilde{X}/G, X' = \tilde{X}/G'$$

and there are two unramified coverings $p: \tilde{X} \to X$ and $p': \tilde{X} \to X'$. Let us find out when X and X' are isomorphic. Here we use an elementary topological fact (which justifies the term "universal" covering): if $p: \tilde{X} \to X$ is a universal covering and $q: X_1 \to X$ any connected unramified covering, then there exists a continuous mapping $\varphi: \tilde{X} \to X_1$ such that $q\varphi = p$. Let $f: X' \to X$ be an isomorphism. Then $q = fp'$ determines an unramified covering $q: \tilde{X} \to X$. By applying the results stated above we construct a continuous mapping $\varphi: \tilde{X} \to \tilde{X}$ such that the diagram

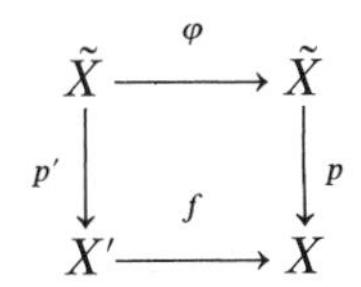

commutes. Hence it follows that φ is a holomorphic mapping. For it follows from the commutativity of the diagram that $p\varphi$ is holomorphic, that is, for a function $u \in \mathcal{O}_{X,x}$ the function $(p\varphi)^* u = \varphi^*(p^* u)$ is holomorphic at the points $\tilde{x} \in (p\varphi)^{-1}(x) = \varphi^{-1}(p^{-1}(x))$. But all functions holomorphic in a neighbourhood of $\tilde{x} \in p^{-1}(x)$ have the form $p^*(u)$ locally, and this implies that φ is holomorphic. By changing the roles of X' and X we see that φ is an automorphism of the analytic manifold $\tilde{X}$.

We recall that the groups G and G' consist of the automorphisms of $\tilde{X}$ for which

$$p\gamma = p \quad \text{for} \quad \gamma \in G$$

and

$$p'\gamma' = p' \quad \text{for} \quad \gamma' \in G'.$$

Multiplying the second equality by f and using the commutativity of the diagram (2), we see that $G = \varphi G' \varphi^{-1}$.

So we have proved the following result:

Theorem 1. *Any connected analytic manifold can be represented in the form* (1), *where* $\tilde{X}$ *is a simply-connected analytic manifold and* G *is a discretely and freely acting group of automorphisms of it. In all such* *representations of one and the same manifold X the groups G are conjugate in the group of all automorphisms of $\tilde{X}$.*

2. Universal Coverings of Algebraic Curves. Theorem 1 enables us to reduce the study of arbitrary analytic manifolds to that of simply-connected manifolds and their groups of automorphisms. Of course, the problem is only shifted in this way—everything depends on how much we know about simply-connected analytic manifolds. In general this is very little; more details on this will be in § 4. An exception are one-dimensional manifolds, and from now on we confine our attention mainly to them.

Classification of connected simply-connected one-dimensional analytic manifolds is very simple. There are altogether three:

1) the projective line $\mathbb{P}^1_{\mathrm{an}}$;

2) the affine line $\mathbb{A}^1_{\mathrm{an}} = \mathbb{C}^1$;

3) the interior of the unit disc D, defined in $\mathbb{C}^1$ by the condition $|z| < 1$.

[The manifolds 1) and 2) are known in the theory of analytic functions as the Riemann sphere and the finite plane.] This theorem can be proved by the same methods as Riemann's existence theorem. A proof can be found, for example, in the book [31], Ch. 9, § 1.

It is easy to verify that the three manifolds in the theorem are not isomorphic. The first of them is not isomorphic to the second or the third, because it is compact and they are not. The third is not isomorphic to the second because on it there exist bounded holomorphic non-constant functions, whereas by Liouville's theorem on the second there are none.

Thus, all connected one-dimensional analytic manifolds fall into three classes depending on which of the three types their universal coverings belong to. The types 1), 2), and 3) are called *elliptic*, *parabolic*, and *hyperbolic* manifolds; this terminology also applies to non-compact one-dimensional analytic manifolds.

In order to investigate manifolds of these three types we have to know the discretely and freely acting groups of automorphisms of their universal coverings. The answer follows easily from simple facts of the theory of analytic functions of a single complex variable.

Proposition. *Every automorphism of the manifold $\mathbb{P}^1_{\mathrm{an}}$ has a fixed point. A discretely and freely acting group G of automorphisms of $\mathbb{C}^1$ for which $\mathbb{C}^1/G$ is compact consists of the translations $z \to z + a$, where a ranges over the vectors of a two-dimensional lattice on $\mathbb{C}^1$. All the automorphisms of the unit disk have the form*

$$z \to \theta \frac{z - \alpha}{1 - \bar{\alpha} z}, \ |\theta| = 1, |\alpha| < 1 . \tag{1}$$

According to Theorem 2 of Ch. VIII, § 3, every automorphism of $\mathbb{P}^1_{\mathrm{an}}$ is of the form g_{an}, where g is an automorphism of the algebraic variety $\mathbb{P}^1$, hence is a fractional linear transformation. Since every fractional linear transformation has a fixed point, the first assertion of the proposition follows from this.

An automorphism of $\mathbb{C}^1$ is given by an entire function $f(z)$. If this function had an essential singularity at infinity, then it would assume in any neighbourhood values arbitrarily close to every given number (Weierstrass' theorem). This contradicts the fact that f determines an automorphism. For if $f(a)=b$, then f assume all values sufficiently close to b in some neighbourhood of a and cannot assume them in a neighbourhood of ∞. Thus, f is a polynomial. If its degree is n, then it takes every value n times. Therefore f determines an automorphism only for $n=1$. In other words, every automorphism of $\mathbb{C}^1$ is of the form

$$f(z)=az+b, a\neq 0\,. \tag{2}$$

The automorphisms occuring in a freely acting group G do not have fixed points. Hence for them $a=1$ in (2). So we see that G must consist of the translations $f(z)=z+b$. If we make use of the group structure in $\mathbb{C}^1$, we can restate our result by saying that G is a subgroup of $\mathbb{C}^1$, and X the factor group $\mathbb{C}^1/G$.

In Ch. VIII, § 1.4, we have already used the simple theorem that determines all discrete subgroups $G\subset\mathbb{C}^1$ with compact factor group. In our case it shows that G must coincide with a two-dimensional lattice $\mathbb{Z}\omega_1+\mathbb{Z}\omega_2$, where $\omega_1, \omega_2\in\mathbb{C}^1$ are independent over $\mathbb{R}$.

Finally, let D be the interior of the unit disc. Substitution shows that the transformations (1) form a group and that this group acts transitively on D. Therefore, when we multiply any automorphism by some automorphism (1) we can obtain an automorphism γ leaving the point 0 fixed. It is therefore sufficient to show that these automorphisms are of the form (1). If $\gamma(0)=0$, then by Schwarz's lemma (Ch. VIII, § 2.3) in D

$$|\gamma(z)/z|\leqslant 1\,,$$

and since $\gamma(z)$ and z are symmetrical, we also have $|z/\gamma(z)|\leqslant 1$, hence $|\gamma(z)/z|=1$. From this it follows that the function $\gamma(z)/z$ is constant:

$$\gamma(z)=\theta z, |\theta|=1\,.$$

This proves the proposition.

Thus, the classification of manifolds of elliptic type is trivial: they are all isomorphic. For in the representation (1) of § 1.1 for them $\tilde{X}=\mathbb{P}^1_{\mathrm{an}}$, also $G=e$ and $X=\tilde{X}=\mathbb{P}^1_{\mathrm{an}}$.

Compact manifolds of parabolic and hyperbolic types are worked out in the next two subsections. We show that for every discretely and freely acting group G for which $\tilde{X}/G$ is compact this manifold is a projective algebraic curve, and we construct an explicit projective embedding of these manifolds. In this way we give a proof of Riemann's existence theorem, starting out from the classification of simply-connected one-dimensional manifolds. Furthermore, we show that compact manifolds of parabolic type coincide with algebraic curves of genus 1 (elliptic curves), and manifolds of hyperbolic type with curves of genus greater than 1.*

3. Projective Embeddings of Factor Spaces. In several special cases we have had to investigate the following general situation. Let $\tilde{X}$ be a one-dimensional analytic manifold, and G a freely and discretely acting group of automorphisms of it. We assume that the factor space $X=\tilde{X}/G$ is compact; how can we construct an immersion of it in a projective space $\mathbb{P}^n$?

We shall give such an immersion by $n+1$ functions $f_0, \ldots, f_n$ that are holomorphic on the whole of $\tilde{X}$. We assume that they do not vanish simultaneously at any point $\tilde{x} \in \tilde{X}$. Then

$$\tilde{f}: \tilde{X} \to \mathbb{P}^n, \quad \tilde{f}(\tilde{x}) = (f_0(\tilde{x}), \ldots, f_n(\tilde{x})) \tag{1}$$

is a holomorphic mapping.

For $\tilde{f}$ to be a mapping of X into $\mathbb{P}^n$ we could require invariance of the functions f_i under all $g \in G$. But then these functions would be holomorphic on X, hence constant by Theorem 2 of Ch. VIII, § 2. However, this condition can be weakened by requiring only that for every $g \in G$ there should exist a function φ_g on $\tilde{X}$ such that

$$g^* f_i = f_i \varphi_g, \quad i = 0, \ldots, n. \tag{2}$$

It then follows that $\tilde{f} \cdot g = \tilde{f}$ for all $g \in G$, hence $\tilde{f}$ can be decomposed: $\tilde{f} = f \cdot \pi$, where π is the projection $\tilde{X} \to X$ and f some holomorphic mapping $X \to \mathbb{P}^n$. We say that $\tilde{f}$ determines the mapping $f: X \to \mathbb{P}^n$. From the fact that the functions f_i are holomorphic and do not vanish simultaneously on $\tilde{X}$ it follows that the functions φ_g are holomorphic and do not vanish anywhere on $\tilde{X}$.

Let us clarify when this system of functions determines an isomorphic embedding $f: X \to \mathbb{P}^n$.

Proposition. *Let $\tilde{X}$ be a one-dimensional analytic manifold, G a discretely and freely acting group of automorphisms of it, $f_0, \ldots, f_n$ functions*

* The terminology is clearly very bad: elliptic curves belong to the parabolic type, and the projective line to the elliptic type! But it has been universally accepted for such a long time that we dare not change it.

that are holomorphic on $\tilde{X}$ and satisfy (2), *where φ_g is a holomorphic function without zeros on $\tilde{X}$.*

We assume the following conditions to be satisfied:

$$\text{A)}\quad \operatorname{rg}\begin{pmatrix} f_0(x'),\dots, f_n(x') \\ f_0(x''),\dots, f_n(x'') \end{pmatrix} = 2$$

for all points $x', x'' \in \tilde{X}$, provided that $x'' \neq gx'$ for all $g \in G$, and

$$\text{B)}\quad \operatorname{rg}\begin{pmatrix} f_0(x),\dots, f_n(x) \\ f'_0(x),\dots, f'_n(x) \end{pmatrix} = 2$$

for all $x \in \tilde{X}$ [$f'(x)$ denotes the derivative of f as a function of the local parameter at x, and condition B) *does not depend on the choice of this parameter]. Then the mapping* (1) *determines an isomorphic embedding of the manifold $X = \tilde{X}/G$ in $\mathbb{P}^n$.*

The proof reduces to a simple verification. Condition A) guarantees that all the functions f_i do not vanish simultaneously at any point $x \in \tilde{X}$, so that (1) in fact gives a point of the projective space. Condition (2) shows that f determines a mapping

$$f: X \to \mathbb{P}^n,$$

which is holomorphic by the preceding remark. Condition A) guarantees that it is one-to-one.

Suppose that $f_0(x_0) \neq 0$ for some $x_0 \in \tilde{X}$. The corresponding point $x_0 \in X$ has a neighbourhood U in which f is given by equations

$$y_i = g_i(x) = f_i(x)/f_0(x),\ i = 1,\dots, n,$$

where $y_1,\dots, y_n$ are coordinates in the affine space $\mathbb{A}^n$ into which U is mapped. From condition B) it follows that for some $i > 0$ we have $g'_i(x_0) \neq 0$. We assume that $i = 1$, that is, $g'_1(x_0) \neq 0$. By virtue of this we may express the local parameter z at x_0 as an analytic function of $y_1 = g_1(z)$:

$$z = h(y_1).$$

So we see that $f(X)$ in a neighbourhood of $f(x_0)$ is given by analytic equations

$$y_i - g_i(h(y_1)) = 0,\ i = 2,\dots, n,$$

where the functions $u_1 = y_1$, $u_i = g_i(h(y_1)) - y_i$, $i = 2,\dots, n$, form a system of local coordinates in a neighbourhood of $f(x_0)$ in $\mathbb{P}^n$. This shows that $f(X)$ is an analytic submanifold in $\mathbb{P}^n$.

Finally, the mapping inverse to f is given in a neighbourhood of $f(z_0)$ by the function $z=h(y_1)$ (we recall that z can be regarded as a local coordinate on X). Therefore f is an isomorphic embedding. This proves the proposition.

Exercises

1. Show that the universal covering of an n-dimensional Abelian manifold over $\mathbb{C}$ is isomorphic to $\mathbb{C}^n$.

2. Show that birationally isomorphic smooth projective surfaces have isomorphic fundamental groups.

3. Let X be a compact analytic manifold. Show that there exists only finitely many non-isomorphic manifolds Y having a holomorphic mapping $f: Y \to X$ that turn Y into an unramified covering of X of given degree m.

4. Determine the fundamental group and the universal covering $\tilde{X}$ for the manifold $X=\mathbb{P}^1(\mathbb{C})-(0)-(\infty)$, and find a representation $X=\tilde{X}/G$, where G is a discrete group of automorphisms of $\tilde{X}$.

5. The same as in Exercise 4 for $X=D-(0)$, where $D=\{z, |z|<1\}$.

6. Show that the universal covering of the manifold $X=\mathbb{P}^1(\mathbb{C})-\alpha-\beta-\gamma$, where $\alpha, \beta, \gamma \in \mathbb{P}^1(\mathbb{C})$ are three distinct points, is isomorphic to D. Use the classification of simply-connected one-dimensional manifolds given at the beginning of § 1.2.

7. Deduce from the result of Exercise 6 Picard's theorem: if an entire function f does not assume two values α and β, $\alpha \neq \beta$, then it is a constant. *Hint.* Interpret f as a mapping $\mathbb{C}^1 \to \mathbb{P}^1(\mathbb{C})-\alpha-\beta-(\infty)$.

§ 2. Curves of Parabolic Type

1. θ-Functions. From the proposition in § 1.3 it follows that any compact manifold of parabolic type is a one-dimensional torus, that is, has the form $\mathbb{C}^1/\Omega$, where Ω is a two-dimensional lattice. According to Theorem 1 of § 1, two lattices Ω and Ω' lead to isomorphic factor spaces if and only if the groups of translations corresponding to them are conjugate under some automorphism f of $\mathbb{C}^1$. Evidently this is equivalent to the fact that $\Omega'=f\Omega$. Since f must be expressible in the form $f(z)=az+b$, it follows that the lattices Ω' and Ω are similar.

Our aim is to show that every one-dimensional torus X is of the form Y_{an}, where Y is a projective curve. For this purpose we use the method that was described at the end of the preceding section. First of all, since we can replace the lattice by a similar one without changing the torus $\mathbb{C}^1/\Omega$, we may assume that it has the basis $1, \tau$ with $\operatorname{Im}\tau>0$. We try to construct an embedding of $\mathbb{C}^1/\Omega$ in $\mathbb{P}^n$ by means of functions $f_0, \ldots, f_n$ satisfying the following special form of the relations (2) of § 1.3:

$$\begin{gathered} f_i(z+1)=f_i(z), \\ f_i(z+\tau)=e^{-2\pi i l z} f_i(z) \qquad (1) \\ i=0, \ldots, n, \end{gathered}$$

where l is a positive integer. Formally speaking, this choice does not require a justification, if only we can show that for some l we can find linearly independent functions satisfying the relations (1) and the conditions of the proposition in § 1.3. However, it can be shown that, in fact, the functions giving any mapping of X in $\mathbb{P}^n$ reduce to this form. The fact of the matter is that we do not change the mapping when we multiply all the functions $f_i(z)$ by $e^{u(z)}$, where $u(z)$ is an entire function. It is not hard to show that by making use of this we can always put the relations (2) of § 1.3 into the special form (1).

Definition. Entire functions satisfying the condition (1) are called *θ-functions* of weight l.

Clearly all θ-functions of equal weight form a linear space, which we denote by $\mathscr{L}_l$.

Theorem 1. *The dimension of the space $\mathscr{L}_l$ is l.*

Proof. The first of the conditions (1) shows that $f(z)=\varphi(t)$, where φ is a function holomorphic in $\mathbb{C}^1-0$ and $t=e^{2\pi i z}$. For $\varphi(t)=f\left(\frac{1}{2\pi i}\log t\right)$ is a single-valued analytic function in $\mathbb{C}^1-0$.

Let

$$\varphi(t)=\sum_{m=-\infty}^{\infty} c_m t^m$$

be the expansion of this function in a Laurent series. Setting $e^{2\pi i\tau}=\lambda$ we can rewrite the second condition (1) in the form

$$\sum_{m\in\mathbb{Z}} c_m t^m \lambda^m = \sum_{m\in\mathbb{Z}} c_m t^{m-l} = \sum_{m\in\mathbb{Z}} c_{m+l}\, t^m,$$

or

$$c_{m+l}=c_m\lambda^m, m\in\mathbb{Z}. \tag{2}$$

We set

$$m=l\cdot r+a,\quad 0\leqslant a<l. \tag{3}$$

From the relations (2) we find that

$$c_m=c_a\lambda^{ra+l\frac{r(r-1)}{2}}.$$

Thus, the function φ is given uniquely by the numbers $c_0,\ldots,c_{l-1}$, from which it follows that $\dim\mathscr{L}_l\leqslant l$.

To complete the proof of the theorem it is sufficient to show the convergence of the series corresponding to arbitrary sequences satisfying the relations (2). We may restrict ourselves to the single arithmetic progression (3). Then we obtain the series

$$c_a t^a \sum_{r\in\mathbb{Z}} u^r \mu^{\frac{r(r-1)}{2}},$$

where $u = t^l\lambda^a$, $\mu = \lambda^l$. From the condition $\operatorname{Im}\tau > 0$, $l > 0$, it follows that $|\mu| < 1$. Therefore the convergence of the series

$$\sum |u|^r |\mu|^{\frac{r(r-1)}{2}}$$

is obvious, and the theorem is proved.

Note. From the theorem it follows that to within a factor a θ-function of weight 1 is unique. If in (2) we set $c_0 = 1$, this function is uniquely determined. It is denoted by $\theta(z)$.

2. Projective Embedding. Now we can prove the main result of this section.

Theorem 2. *The θ-functions of weight $l \geqslant 3$ determine an isomorphic embedding of the manifold $X = \mathbb{C}^1/\Omega$.*

We give the proof for $l = 3$; the general case is completely similar. We make use of the following obvious remark. If $f(z)$ is a θ-function of weight l, and $a_1, \ldots, a_m$ complex numbers with $a_1 + \cdots + a_m = 0$, then

$$g(z) = \prod_{i=1}^{m} f(z + a_i)$$

is a θ-function of weight lm. In particular, for arbitrary a and b

$$f(z) = \theta(z + a)\,\theta(z + b)\,\theta(z - a - b)$$

is a θ-function of weight 3.

We have to verify that the conditions A) and B) of the proposition in § 1.3 hold for three linearly independent θ-functions of weight 3. If A) does not hold for three basis functions of $\mathscr{L}_3$, then there exist α and β, not both zero, and z' and z'' such that $z' - z'' \notin \Omega$, and $\alpha f(z') = \beta f(z'')$ for any function $f \in \mathscr{L}_3$. In particular,

$$\alpha\theta(z' + u)\,\theta(z' + a)\,\theta(z' - u - a) = \beta\theta(z'' + u)\,\theta(z'' + a)\,\theta(z'' - u - a)$$

for arbitrary u and a. We set $z' + u = z$, $z'' - z' = \zeta$, and we regard z as variable and the remaining quantities as fixed. So we see that

$$\alpha\theta(z)\,\theta(z' + a)\,\theta(2z' - a - z) = \beta\theta(z + \zeta)\,\theta(z'' + a)\,\theta(z' + z'' - a - z).$$

We choose a so that the functions $\theta(z)$ and $\theta(z'+z''-a-z)$ do not have common zeros. Then the functions $\theta(z+\zeta)$ and $\theta(2z'-a-z)$ have the same property. Therefore $\theta(z)/\theta(z+\zeta)$ has no zeros nor poles, from which it follows that $\theta(z+\zeta)=e^{g(z)}\,\theta(z)$, where g is an entire function. From the definition of $\theta(z)$ it follows that

$$g(z+1)=g(z)+2l\pi i\,, \tag{1}$$

$$g(z+\tau)=g(z)-2\pi i\zeta+2\pi i l',\ \ l,l'\in\mathbb{Z}\,. \tag{2}$$

Thus, the function $g'(z)$ has the periods 1 and τ, therefore, it is bounded on $\mathbb{C}^1$, and since it is entire, it is a constant. So we see that $g(z)=\alpha z+\beta$, and from (1) it follows that $\alpha=2l\pi i$, and from (2)

$$2l\pi i\tau=-2\pi i\zeta+2\pi i l'\,,\zeta=l'-l\tau\in\Omega\,.$$

This shows that condition A) in the proposition of § 1.3 holds. Condition B) is verified similarly. Namely, if it does not hold, then there exists a $z_0\in\mathbb{C}^1$ such that $(f'/f)(z_0)=0$ for all $f\in\mathscr{L}_3$. In particular, we can set $f=\theta(z+u)\,\theta(z+a)\,\theta(z-u-a)$. We find that

$$\frac{\theta'(z_0+u)}{\theta(z_0+u)}+\frac{\theta'(z_0+a)}{\theta(z_0+a)}+\frac{\theta'(z_0-u-a)}{\theta(z_0-u-a)}=0\,. \tag{3}$$

Again we regard u as variable and we choose a so that the functions $\theta(z_0+u)$ and $\theta(z_0-u-a)$ do not have common zeros. Here (3) is possible only if all three terms on the left-hand side are entire functions of u. This, in its turn, is possible only if $\theta(z)$ has no zeros, that is, $\theta(z)=e^{g(z)}$, where $g(z)$ is an entire function. This representation at once leads to a contradiction to the definition of θ. For it follows from this definition that

$$g(z+1)=g(z)+2m\pi i\,, \tag{4}$$

$$g(z+\tau)=g(z)-2l\,\pi iz+2l'\,\pi i\,. \tag{5}$$

Hence we find, as above, that $g''(z)$ is constant. Therefore $g(z)=\alpha z^2+\beta z+\gamma$, and from (4) we see that $\alpha=0$ and from (5) that $l=0$ against the condition $l>0$. This proves the theorem.

Note. In a similar but more complicated manner it can be shown that a many-dimensional torus is projective if its period matrix satisfies the Frobenius relations of which we have talked in Ch. VIII, § 1.4.

3. Elliptic Functions, Elliptic Curves, and Elliptic Integrals. Having constructed the mapping

$$f:X\to\mathbb{P}^n\,,$$

we are interested in studying it in more detail. We have seen that $f(X)$ is a smooth algebraic curve. The addition of points on the torus determines a group structure on Y. Here the addition mapping

$$\mu: Y \times Y \to Y$$

determines a holomorphic mapping of the corresponding complex manifolds. According to Theorem 2 in Ch. VIII, § 3, it follows that μ is a morphism. Thus, Y is a one-dimensional Abelian variety. In Ch. III we have seen that in this case the canonical class of Y is 0, hence the genus is 1. So we have shown that compact manifolds of parabolic type are smooth projective curves of genus 1 (elliptic curves) and only they.

Note that it makes sense to talk of the zeros of θ-functions on X: although the value of $\theta(z)$ changes when z is replaced by $z+a$, $a \in \Omega$, if $\theta(z) = 0$, then also $\theta(z+a)=0$. The usual definition allows us to talk of the divisor of a θ-function on X.

θ-functions of weight 3 lead to an isomorphic embedding of X in $\mathbb{P}^2$. In this case Y is a smooth plane curve of genus 1. From the formula for the genus of a plane curve it follows that the degree of Y is 3. In particular, every θ-function of weight 3 determines on X a divisor, and the mapping $f: X \to \mathbb{P}^2$ assigns to it the divisor of the section of $Y = f(X)$ with a line in $\mathbb{P}^2$, which is of degree 3. Applying this remark to the function θ^3, where θ is of weight 1, we see that the divisor of θ on X consists of a single point with multiplicity 1. If this point is x_0, then $\theta(z-a+x_0)$ has the divisor a. This sheds new light on the role of the θ-functions. If we make use of the θ-functions (which, of course, are neither meromorphic nor functions on X, in general), then every divisor becomes principal.

The embedding f we have constructed determines an isomorphism of the fields $\mathbb{C}(Y)$ and $\mathcal{M}(X)$. On the other hand, $\mathcal{M}(X)$ can be described as the field of meromorphic functions on $\mathbb{C}^1$ having the periods 1 and τ. Such functions are called *elliptic*. Thus, the field $\mathbb{C}(Y)$ is isomorphic to the field of elliptic functions. In particular, if

$$F(x, y) = 0$$

is the equation of an affine model of the curve Y, then there exists a parametrization

$$x = \varphi(z), \ y = \psi(z)$$

by elliptic functions. This parametrization is called a *uniformization* of Y. In this way a connection is established between elliptic functions and elliptic curves: the former uniformize the latter.

Let Y be an elliptic curve. How can we find a lattice Ω corresponding to it for which $Y = \mathbb{C}^1/\Omega$? Let ω be a regular differential form on Y. Since the genus of Y is 1, it is uniquely determined to within a constant factor. If $f:\mathbb{C}^1 \to Y$ is the holomorphic mapping we wish to find, then $f^*\omega$ is a holomorphic differential form on $\mathbb{C}^1$, which must be invariant under translations by vectors of Ω. This means that $f^*\omega = u(z)\,dz$, where $u(z)$ is an entire function that is invariant under translations in Ω. Hence it follows that u is constant and, using the fact that ω is arbitrary, we may assume that

$$f^*\omega = dz\,.$$

Let $z_0 \in \Omega$, that is, $f(z_0) = f(0)$, and let s be a path joining 0 to z_0 in $\mathbb{C}^1$. Then

$$z_0 = \int_0^{z_0} dz = \int_s f^*\omega = \int_{f(s)} \omega\,. \tag{1}$$

The path $f(s)$ is closed on $Y(\mathbb{C})$, so that it determines an element of the group $H^1(Y(\mathbb{C}), \mathbb{Z})$. Clearly all the elements of this group are obtained in this manner. Formula (1) shows that the lattice Ω coincides with the set of complex numbers

$$\int_\sigma \omega, \sigma \in H^1(Y(\mathbb{C}), \mathbb{Z}),.$$

In particular, a basis of it consists of the numbers

$$\int_{\sigma_1} \omega, \int_{\sigma_2} \omega\,,$$

where σ_1 and σ_2 form a basis of $H^1(Y(\mathbb{C}), \mathbb{Z})$. For example, if the curve Y is given by an equation $v^2 = u^3 + Au + B$, we may set $\omega = du/v$, and the basis of Ω consists of the numbers

$$\int_{\sigma_1} \frac{du}{(u^3 + Au + B)^{\frac{1}{2}}}, \qquad \int_{\sigma_2} \frac{du}{(u^3 + Au + B)^{\frac{1}{2}}}\,.$$

The integrals $\int_\sigma \omega$ are called *elliptic integrals*, and $\int_\sigma \omega$, where $\sigma \in H^1(Y(\mathbb{C}), \mathbb{Z})$, are called their periods. Thus, the lattice Ω, which determines the torus $\mathbb{C}^1/\Omega$ isomorphic to the elliptic curve Y, consists of periods of the elliptic integral connected with this curve.

In conclusion we mention that the uniformization of elliptic curves puts us in a position to understand from a new point of view the fundamental fact of which we talked in Ch. III, § 5.6: not all curves of genus 1 are isomorphic to each other. We can even form some idea of the structure of the set of equivalence classes of elliptic curves.

To do this we represent any elliptic curve in the form $\mathbb{C}^1/\Omega$, and replacing, if necessary, the lattice Ω by a similar one we choose in it a basis 1, τ, $\operatorname{Im}\tau > 0$. It is easy to see that two such bases 1, τ and 1, τ'

determine similar lattices if and only if there exist integers a, b, c, d such that

$$\tau' = \frac{a\tau + b}{c\tau + d}, \quad ad - bc = 1 . \tag{2}$$

The set of all transformations of the form (2) is a group G, the so-called *modular group*. We denote by H the upper half-plane $\operatorname{Im} \tau > 0$. Since elliptic curves are isomorphic if and only if the lattices corresponding to them are similar, the set of isomorphism classes of elliptic curves is in one to one correspondence with the points of the factor space H/G.

It can be shown that the group G acts on H discretely, but not freely (it has fixed points). Nevertheless, the factor space H/G is a one-dimensional complex manifold. Moreover, it is isomorphic to $\mathbb{C}^1$. The function $j: H/G \to \mathbb{C}^1$ that realizes this isomorphism establishes a one-to-one correspondence between classes of isomorphic elliptic curves and complex numbers. An algebraic description of it can be extracted from Exercises 12 and 13 to Ch. III, § 5.

Exercises

1. Show that if an elliptic curve X is determined by an equation with real coefficients, then it is isomorphic to $\mathbb{C}^1/\Omega$, where $\Omega = \mathbb{Z} + i\mathbb{Z}$ or $\Omega = \mathbb{Z} + \frac{1+i}{2}\mathbb{Z}$, and in the first case $X(\mathbb{R})$ consists of a single oval, in the second of two.

2. Show that if a real elliptic curve X consists of a single oval, then it is not homologous to 0 on $X(\mathbb{C})$.

3. Show that if a real elliptic curve X consists of two ovals T_1 and T_2, then for a suitable orientation T_1 and T_2 are homologous in $X(\mathbb{C})$.

4. Show that all θ-functions of weight 0, 1,... with given periods 1, τ form a ring and that this ring can be generated by the θ-functions of weight $\leqslant 3$.

5. Let f be an elliptic function with the period lattice Ω. Show that the number of zeros of f that are inequivalent modulo Ω is equal to the number of its poles that are inequivalent modulo Ω.

6. In the notation of Exercise 5, let $\alpha_1, \ldots, \alpha_m$ and $\beta_1, \ldots, \beta_m$, respectively, be inequivalent zeros and poles of f. Show that $\alpha_1 + \cdots + \alpha_m - \beta_1 - \cdots - \beta_m \in \Omega$. Show that any numbers with this property are the collection of zeros and poles of some elliptic function.

7. Let $X = \mathbb{C}^1/\Omega$, $X' = \mathbb{C}^1/\Omega'$ be two elliptic curves. Show that the group $\operatorname{Hom}(X, X')$ of homomorphisms (of algebraic groups) of X into X' is isomorphic to the group of those complex numbers $\alpha \in \mathbb{C}$ for which $\alpha\Omega \subset \Omega'$.

8. Show that one can choose forms ω and ω', regular on X and X', such that the number $\alpha \in \mathbb{C}$ corresponding by virtue of Exercise 7 to a homomorphism $f \in \operatorname{Hom}(X, X')$ is defined by the condition $f^*\omega' = \alpha\omega$.

9. Show that for an elliptic curve X defined over $\mathbb{C}$ the ring $\operatorname{End} X = \operatorname{Hom}(X, X)$ is isomorphic either to $\mathbb{Z}$ or to $\mathbb{Z} + \mathbb{Z}\gamma$, where γ satisfies an equation $\gamma^2 + a\gamma + b = 0$, $a, b \in \mathbb{Z}$, without real roots. In the second case $X = \mathbb{C}^1/\Omega$, where the lattice Ω is similar to an ideal of the ring of numbers $\mathbb{Z} + \mathbb{Z}\gamma$.

§ 3. Curves of Hyperbolic Type

1. Poincaré Series. Let us consider the interior of the unit disc D and a group G consisting of automorphisms of D, of which we assume that it acts on D discretely and freely and that the factor space $X=D/G$ is compact. We construct an embedding of X into a projective space. Just as in the preceding section, the construction of this embedding relies on a study of functions f holomorphic in D and satisfying the condition

$$g^*(f)=f\cdot\varphi_g,\ g\in G, \tag{1}$$

where φ_g is a holomorphic function that vanishes nowhere on D. From (1) it follows at once that

$$\varphi_{g_1g_2}=g_2^*(\varphi_{g_1})\,\varphi_{g_2}\,. \tag{2}$$

For an automorphism g we set

$$j_g=\frac{dg}{dz}\,.$$

The rule for the differentiation of a compound function shows that equation (2) holds for $\varphi_g=j_g$, hence for $\varphi_g=j_g^l$ with any positive integer l.

It can be shown that, in a sense, all solutions of equations (2) can be reduced to $\varphi_g=j_g^l$. We shall consider only this case.

Definition. A function f holomorphic in D and satisfying the relation

$$g^*(f)=f\cdot j_g^l,\ g\in G\,,$$

is called an *automorphic form of weight* l relative to the group G.

Our next aim is the construction of automorphic forms. For this purpose we take an arbitrary function h holomorphic and bounded in D and consider the series

$$\sum_{g\in G} g^*(h)\,j_g^l\,. \tag{3}$$

A series of this kind is called a *Poincaré series.* If it defines an analytic function, then a formal verification shows that this function is an automorphic form. Thus, it remains to prove the following result:

Proposition. *A Poincaré series converges for* $l\geqslant 2$ *absolutely and uniformly on any compact set* $K\subset D$.

We use the following simple property of analytic functions.

Lemma. *If a function $f(z)$ is analytic in the disc $|z| \leqslant r$, then*

$$|f(0)|^2 \leqslant \frac{1}{\pi r^2} \int_{|z| \leqq r} |f(z)|^2 \, dx \wedge dy ,$$

$$z = x + iy .$$

Proof. For any ϱ, $0 < \varrho < r$,

$$f(0)^2 = \frac{1}{2\pi i} \int_0^{2\pi} f(\varrho e^{i\varphi})^2 \, d\varphi .$$

Multiplying this equality by $\varrho d\varrho$ and integrating with respect to ϱ from 0 to r we obtain

$$\frac{r^2}{2} f(0)^2 = \frac{1}{2\pi i} \int_0^r \int_0^{2\pi} f(\varrho e^{i\varphi})^2 \, \varrho d\varrho \wedge d\varphi = \frac{1}{2\pi i} \int_{|z| \leqq r} f(z)^2 \, dx \wedge dy .$$

Therefore,

$$|f(0)|^2 = \frac{1}{\pi r^2} \left| \int_{|z| \leqq r} f(z)^2 \, dx \wedge dy \right| \leqslant \frac{1}{\pi r^2} \int_{|z| \leqq r} |f(z)|^2 \, dx \wedge dy .$$

Proof of the proposition. Since h is bounded on D, it is sufficient for us to show the convergence of the series $\sum_{g \in G} |j_g|^l$ or even the series

$$\sum_{g \in G} |j_g|^2 . \tag{4}$$

The proof uses the fact that $|j_g|^2$ is the Jacobian of g. We denote by $s(U)$ the area of a domain U determined by the Euclidean metric of the plane $\mathbb{C}^1$ in which D is contained. Then

$$s(g(U)) = \int_{gU} dx \wedge dy = \int_U |j_g(z)|^2 \, dx \wedge dy . \tag{5}$$

The convergence of the series (4) follows at once from this remark. For let U be a circle with centre at z_0 and sufficiently small radius r so that $g(U) \cap U = \emptyset$ for $g \in G$, $g \neq e$. According to Lemma 1 and the remark made above

$$\sum_{g \in G} |j_g(z_0)|^2 \leqslant \frac{1}{\pi r^2} \sum_{g \in G} \int_U |j_g(z)|^2 \, dx \wedge dy = \frac{1}{\pi r^2} \sum_{g \in G} s(g(U)) \leqslant \frac{s(D)}{\pi r^2} = \frac{1}{r^2} , \tag{6}$$

which proves the convergence.

To prove uniform convergence we observe that gK_1 and K_2 for any two compact sets K_1 and K_2 intersect only for finitely many elements

$g \in G$. For according to the definition of a discretely and freely acting group (Ch. VIII, § 1.2) any two points x and y have neighbourhoods U and V such that $gU \cap V = \emptyset$ for all $g \in G$ except possibly one. We take an arbitrary point $x \in K_1$, and for any point $y \in K_2$ we choose neighbourhoods U_y of x and V_y of y such that $gU_y \cap V_y = \emptyset$ for all $g \in G$ except possibly one. From the compactness of K_2 it follows that there exists a neighbourhood U of x such that $gU \cap K_2 = \emptyset$ for all $g \in G$ except finitely many. Our assertion now follows at once from the compactness of K_1.

Now we take a sufficiently small $r > 0$ so that the circle of radius r with centre at an arbitrary point of K is contained in a compact set $K' \subset D$. For every $\varepsilon > 0$ let $C \subset D$ be a circle, sufficiently close to D, such that $K' \subset C$ and $s(D - C) < \varepsilon$. We denote by q the number of those elements $g \in G$ for which $gK' \cap K' \neq \emptyset$. For all $g \in G$ except finitely many $gK' \subset D - C$. Denoting by Σ' the sum extended over these g we find that as in the derivation of (5):

$$\sum' |j_g(z)|^2 \leqslant \frac{1}{\pi r^2} \sum' s(g(U)) < \frac{q}{\pi r^2} s(D - C) < \frac{q}{\pi r^2},$$

from which the uniform convergence of the series (4) follows.

Note. We denote by M the multiplicative group of functions holomorphic on D and vanishing nowhere. It is a module over G with respect to the action $f \to g^*(f)$. Condition (2) agrees with the definition of a one-dimensional cocycle. In the construction of a modular form by means of a Poincaré series we can recognize the idea of a proof of the so-called Hilbert's Theorem 90 in homological algebra.

2. Projective Embedding. Now we can proceed to the main result.

Theorem. *Let G be a group of automorphisms acting discretely and freely on D such that the factor space $X = D/G$ is compact. Then there exist finitely many automorphic forms of one and the same weight which determine an isomorphic embedding of X in $\mathbb{P}^n$.*

Of course, the matter concerns a verification of conditions A) and B) of the proposition in § 1.3 for these forms. To begin with we aim at satisfying them locally.

Lemma. *Let $z', z'' \in D$ be arbitrary points such that $z'' \neq gz'$ for all $g \in G$. Then there exist automorphic forms f_0 and f_1 satisfying condition A) of the proposition in § 1.3 for these points. For every point $z_0 \in D$ there exist automorphic forms f_0 and f_1 satisfying condition B) of the same proposition at this point. In both cases we may assume that*

$$f_0(z) \neq 0, \; f_1(z) \neq 0, \quad \text{for} \quad z = z', z'', z_0 .$$

We are looking for forms satisfying condition A) for the points z' and z'', in the form of Poincaré series

$$f_i = \Sigma g^*(h_i)\, j_g^l, \quad i=0,1. \tag{1}$$

From the convergence of the Poincaré series it follows that $|j_g(z')|<1$ and $|j_g(z'')|<1$ for all $g \in G$ except finitely many. Let $g=e, g_1, \dots, g_N$ be these excluded elements. Then

$$\sum_{g \neq e, g_1, \dots, g_N} |j_g(z)|^l \to 0 \quad \text{for} \quad z = z', z'', \quad l \to \infty. \tag{2}$$

We take functions h_i, $i=0,1$, such that they satisfy the conditions

$$h_i(g_m(z'))=0, \quad m=1,\dots,N$$
$$h_i(g_m(z''))=0,$$
$$h_0(z')\,h_1(z')\,h_0(z'')\,h_1(z'') \neq 0,$$
$$h_0(z')\,h_1(z'') - h_0(z'')\,h_1(z') \neq 0.$$

Such functions can be found, for example, among the polynomials.

Then

$$f_i(z') = h_i(z') + \sum_{g \neq e, g_1, \dots, g_N} h_i(g(z'))\, j_g(z')^l = h_i(z') + u_i^{(l)}(z'),$$

where $u_i^{(l)}(z') \to 0$ as $l \to \infty$, by virtue of (2). A similar relation holds for z''. Hence it follows that

$$f_0(z')\,f_1(z')\,f_0(z'')\,f_1(z'') \neq 0,$$
$$f_0(z')\,f_1(z'') - f_0(z'')\,f_1(z') \neq 0$$

for sufficiently large l.

Now we construct functions satisfying condition B) of the proposition in § 1.3. Again we are looking for them in the form (1). Let $e, g_1, \dots, g_N$ be the elements $g \in G$ for which $|j_g(z_0)| \geqslant 1$. We take h_i, $i=0,1$, so that

$$h_i(g_m(z_0)) = h_i'(g_m(z_0)) = 0, \, i=0,1, \, m=1,\dots,N,$$
$$h_0(z_0)\,h_1(z_0) \neq 0,$$
$$h_0(z_0)\,h_1'(z_0) - h_1(z_0)\,h_0'(z_0) \neq 0.$$

As before,

$$f_i(z_0) = h_i(z_0) + u_i^{(l)}(z_0),$$
$$f_i'(z_0) = h_i'(z_0) + v_i^{(l)}(z_0), \quad i = 0, 1,$$

where $u_i^{(l)}(z_0) \to 0$, $v_i^{(l)}(z_0) \to 0$ as $l \to \infty$. Therefore

$$f_0(z_0)\, f_1'(z_0) - f_1(z_0)\, f_0'(z_0) \neq 0$$

for sufficiently large l. This proves the lemma.

Proof of the Theorem. We observe, first of all, that if two functions f_0 and f_1 satisfy condition B) of the proposition in § 1.3 for the point z_0, then they satisfy condition A) for arbitrary points $z', z'', z' \neq z''$, in a sufficiently small neighbourhood of z_0.

For the function

$$F(z_1, z_2) = \frac{f_1(z_1)\, f_0(z_2) - f_1(z_2)\, f_0(z_1)}{z_1 - z_2}$$
$$= \frac{f_0(z_1)\,(f_1(z_1) - f_1(z_2)) - f_1(z_1)\,(f_0(z_1) - f_0(z_2))}{z_1 - z_2}$$

is analytic and

$$F(z, z) = f_0(z)\, f_1'(z) - f_1(z)\, f_0'(z)\,.$$

Therefore $F(z_0, z_0) \neq 0$, hence $F(z_1, z_2) \neq 0$ for points z_1 and z_2 sufficiently close to z_0, from which our assertion follows.

Obviously, if condition A) or B) holds for certain functions and points z', z'' or z_0, then it also holds for sufficiently close points. Using the lemma we choose a finite covering of the compact manifold $X = D/G$ by open sets U_i such that in U_i condition B) is satisfied for functions $\{f_{0,i}, f_{1,i}\}$, $i = 1, \ldots, N$. According to the remark made above there exists a neighbourhood U of the diagonal in $X \times X$ such that at every point of this set some pair $f_{0,i}, f_{1,i}$ satisfies condition A). Since $X \times X - U$ is a compact set, we can find a finite set of pairs of functions $\{f_{0,i}, f_{1,i}\}$, $i = 1, \ldots, N$, such that at any point of the space $X \times X$ some pair of functions satisfies condition A).

Let the weight of the functions $f_{0,i}$ and $f_{1,i}$ be m_i, and $M = \Pi\, m_i$, $l_i = M/m_i$. We consider the system consisting of all products of the form $f_{0,i}^{2l_i}$, $f_{1,i}^{2l_i}$, $(f_{0,i}, f_{1,i})^{l_i}$, $f_{0,i}^{l_i-1}\, f_{1,i}^{l_i+1}$, $i = 1, \ldots, N$. Clearly all these are automorphic forms of weight $2M$. We show that for them conditions A) and B) in the proposition of § 1.3 hold. Indeed, if the functions $f_{0,i}$, and

$f_{1,i}$ satisfy condition A) at the points z', z'', then the minor

$$(f_{0,i}^{l_i} f_{1,i}^{l_i})(z')(f_{0,i}^{l_i-1} f_{1,i}^{l_i+1})(z'') - (f_{0,i}^{l_i} f_{1,i}^{l_i})(z'')(f_{0,i}^{l_i-1} f_{1,i}^{l_i+1}(z')$$
$$= f_{0,i}^{l_i-1}(z') f_{0,i}^{l_i-1}(z'') f_{1,i}^{l_i}(z') f_{1,i}^{l_i}(z'')(f_{0,i}(z') f_{1,i}(z'') - f_{0,i}(z'') f_{1,i}(z'))$$

is different from 0. Condition B) is verified similarly, and this proves the theorem.

Note. In the proof of the theorem we have made very little use of specific properties of the interior of the unit disc D. Even the fact that it is one-dimensional does not play an essential role in it. It can be carried over almost without change to the case when D is any bounded domain in $\mathbb{C}^n$ and G a group acting discretely and freely on it for which the factor space D/G is compact. Only we must understand by j_g in the definition of an automorphic form the Jacobian of the transformation $g \in G$.

3. Algebraic Curves and Automorphic Functions. In § 1 and § 2 we have proved that algebraic curves of elliptic and parabolic type coincide with the curves of genus 0 and 1. Therefore curves of hyperbolic type must be those of genus $g > 1$. The theorem shows that these curves coincide with compact manifolds of the form D/G, where D is the interior of the unit disc and G a group acting discretely and freely on it.

Now we shall describe algebraically an embedding of a curve $X \simeq D/G$ in a projective space, which is defined by automorphic forms. Let $f(z)$ be an automorphic form of weight l'. The expression $\eta = f(z)(dz)^{l'}$ determines a holomorphic differential form of weight l' on D (the definition of a holomorphic differential form is given in Exercise 4 to Ch. VIII, § 2, that of a differential form of weight > 1 in Exercise 7 to Ch. III, § 5). From the definition of an automorphic form it follows that η is invariant under the automorphisms of G. Indeed,

$$g^*\eta = g^*(f)(dg(z))^{l'} = f \cdot j_g^{l'} \cdot j_g^{-l'}(dz)^{l'} = \eta .$$

Therefore $\eta = \pi^*\omega$, where π is the projection $D \to D/G$, and ω is a holomorphic differential form of weight l' on D/G. Finally, if $\varphi: D/G \to X$ is an isomorphism with the algebraic curve, then $\omega' = (\varphi^{-1})^*\omega$ is a holomorphic differential form of weight l' on X. Hence it follows that ω' is a rational form of X. It is sufficient to take its quotient with any rational form of the same weight on X; by Theorem 1 of Ch. VIII, § 3, this is a rational function on X. The lemma in Ch. VIII, § 3, shows that ω' is a regular form of weight l' on X.

It is easy to show that, conversely, every regular differential form of weight l' on X is obtained in this way.

So we see that the space of automorphic forms of weight l' is isomorphic to the space of regular differential forms of weight l'.

Thus, the mapping by means of all automorphic forms of weight $l'>2$ coincides with the mapping corresponding to the class $l'\cdot K_X$. In Ch. III, § 5.6, we have deduced from the Riemann-Roch theorem that this mapping is an embedding for $l'\geqslant 3$. Hence this is also true for the mapping defined by the automorphic forms of weight l'. In addition, we obtain an interesting analytic application of the Riemann-Roch theorem: the dimension of the space of automorphic forms of weight l' is finite and (by the Riemann-Roch theorem) equal to

$$l(l'\cdot K_X)=(2l'-1)(g_X-1).$$

As in § 2.3, it follows from Theorem 1 that the field $\mathbb{C}(X)$ is isomorphic to the field of meromorphic functions on D that are invariant under the group G. Such functions are called *automorphic*. Thus, every curve of genus $g>1$ can be uniformized by automorphic functions.

Let us compare the picture we have obtained with that which holds in the parabolic case. In both cases the description of the curves reduces to a description of certain discrete groups. In the parabolic case the corresponding discrete groups are very simple: they are lattices in $\mathbb{C}^1$. What happens in the hyperbolic case?

Poincaré found a general method of constructing groups that act discretely and freely on the interior of the unit disc. His method is based on the fact that a metric can be defined in D in which analytic motions coincide with analytic automorphisms of D, and as a metric space D is isomorphic to the Lobachevskii plane. In this metric isomorphic lines of the Lobachevskii geometry correspond to circular arcs contained in D and orthogonal to the unit disc, the boundary of D. We do not need the definition of this metric. We only mention that in it the magnitude of an angle is the same is that between circles (that is, between their tangents at the point of intersection) in the Euclidean metric of the plane $\mathbb{C}^1$ of a complex variable containing D.

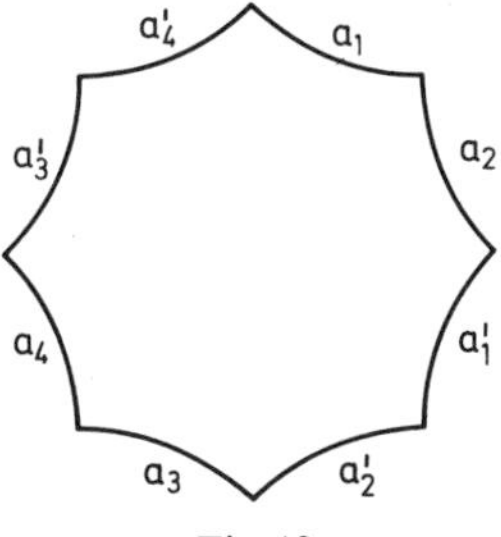

Fig. 19

Poincaré showed that every group G acting freely and discretely in D for which the factor space D/G is compact is defined by a certain polygon in the geometry described above. This polygon plays in the hyperbolic case the same role as the fundamental parallelogram of the lattice Ω in the parabolic case and is called the *fundamental polygon* of G.

If the genus of an algebraic curve D/G is g, then the fundamental polygon of G has $4g$ sides. We denote its sides in some direction of going around the polygon by $a_1, a_2, a'_1, a'_2, a_3, a_4, a'_3, a'_4, \ldots, a_{2g-1}, a'_{2g-1}, a'_{2g}$, in the order of the chosen direction of going around the polygon. Fig. 19 illustrates the case $g=2$. Then the following relations hold:

1) the sides a_i and a'_i are equal to each other.

2) the sum of the (interior) angles of the polygon is 2π.

The group G is determined by its fundamental polygon in the following way. We denote by $g_i (i=1, \ldots, 2g)$ the motions (without change of orientation) that carry the side a_i into a'_i with the opposite direction. Then the motions g_i generate G.

Conversely, if a polygon Φ is given satisfying the conditions 1) and 2), then the group G generated by the transformations g_i acts on D discretely and freely, and Φ is its fundamental polygon. Geometrically this is expressed by the fact that if F is the interior of Φ, then by applying the transformations g_i first to F and then to the domains $g_i F$ etc. we cover the whole of D by domains that intersect only in sides of the boundary.

It would be natural to try and obtain from this picture a description of classes of isomorphic curves of genus $g>1$, by analogy with the explanations at the end of § 2.3. However, the situation here is vastly more complicated and less well investigated. The corresponding complex space and even the algebraic variety can be defined precisely (in [4] the reader can find an analytic, and in [23] an algebraic definition). It is called the variety of moduli of curves of genus g. However, very little is known of its properties. Of the problems stated in precise terms the most interesting is the following: is the variety of moduli rational or perhaps unirational? The rationality has been proved only for $g=2$. The problem of unirationality is apparently easier. For small values of the genus ($g=3, 4, 5$) it is not hard to prove. Severi has proved the unirationality of the variety of moduli of curves of genus g for $g<11$. Nothing more is known.

Exercises

1. Show that the automorphic forms of weight 3 determine a projective embedding of a compact manifold D/G.

2. Show that for a fixed group G of automorphisms of the interior of the unit disc D for which D/G is compact the equation $\sum_{g \in G} g^*(h)\, j_g^l = 0$ has infinitely many linearly independent solutions among functions h that are holomorphic and bounded in D.

3. Show that the genus g of a curve D/G and the area (in the sence of Lobachevskii geometry) S of the fundamental polygon Φ of a group G are connected by the relation $g-1=\frac{1}{4\pi}S$. *Hint*: Use the theorem of Lobachevskii geometry according to which the sum of the angles of a triangle is less than π by the value of its area, and use also the connection of the genus with the Euler characteristic.

§ 4. On the Uniformization of Manifolds of Large Dimension

1. Simple Connectivity of Complete Intersections. Hardly anything is known on universal coverings and fundamental groups of varieties of dimension greater than 1. We give some simple examples and make a few remarks, with the object of throwing light on the nature of the problems arising.

The basic new phenomenon that we encounter here is the following. Among the smooth complete algebraic curves only one, the projective line, is simply-connected, so that the transition to the universal covering almost always reduces the study of the curve to that of another variety, which one might hope, and which turns out, to be simpler. No such thing happens for varieties of dimension $\geqslant 2$: among them very many are simply-connected, and for them the transition to the universal covering does not yield anything new. To make the expression "very many" a little more precise we prove for one wide class of varieties of dimension $\geqslant 2$ that they are simply-connected; among them, in particular, all smooth projective hypersurfaces.

Definition. A projective variety $X\subset\mathbb{P}^N$ of codimension n is called a *complete intersection* if it is the intersection of n hypersurfaces that are transversal at each of their points of intersection.
By our definition complete intersections are smooth algebraic varieties. From now on we consider them over the field of complex numbers.

We show that if the dimension of a complete intersection X is greater than 1, then the topological manifold $X(\mathbb{C})$ is simply-connected.

This is a consequence of a general result, which we use repeatedly later:

Proposition. *If V is an n-dimensional projective variety over the field of complex numbers and W a hyperplane section of it for which $V-W$ is smooth, then the embedding $W(\mathbb{C})\to V(\mathbb{C})$ determines an isomorphism of their homotopy groups*

$$\pi_r(W(\mathbb{C}))\simeq\pi_r(V(\mathbb{C}))$$

in dimensions $r<n-1$.

The proposition is a simple application of Morse theory (see, for example, [21]). We have to make use of Theorem 7.4 in [21] and the exact homotopy sequence of the pair $(V(\mathbb{C}), W(\mathbb{C}))$.

We prove simple connectivity of complete intersections by induction on their codimension in projective space. At the first step of the induction we must use the fact that the space $\mathbb{P}^n(\mathbb{C})$ is simply-connected.

Let $X \subset \mathbb{P}^n$ be the intersection of n transversal hypersurfaces $E_1, \ldots, E_n$. We set $Y = E_1 \cap \cdots \cap E_{n-1}$. Clearly Y is a complete intersection, and by the inductive hypothesis the space $Y(\mathbb{C})$ is simply-connected. We consider the Veronese mapping

$$v_m : \mathbb{P}^n \to \mathbb{P}^M, \quad M = \binom{n+m}{m} - 1,$$

where $m = \deg E_n$. Let $V = v_m(Y)$, $W = v_m(X)$. Evidently W is the section of V by the hyperplane $v_m(E_n)$. Since the space $V(\mathbb{C})$ is homeomorphic to $Y(\mathbb{C})$, it is simply-connected. We may apply the proposition and find that for $\dim V > 2$, that is, $\dim X \geqslant 2$

$$\pi_1(W(\mathbb{C})) = \pi_1(V(\mathbb{C})) = 0.$$

Since the space $X(\mathbb{C})$ is homeomorphic to $W(\mathbb{C})$, it is also simply-connected.

2. Example of a Variety with a Preassigned Finite Fundamental Group. However, there exist many non simply-connected algebraic varieties of any given dimension. Presently we shall give an illustration of this phenomenon, which in a certain sense is opposite to what we talked about in § 4.1. Namely, we show that for every finite group Γ and every integer $n \geqslant 2$ there exists an n-dimensional complete algebraic variety whose fundamental group is isomorphic to Γ.

To begin with, we construct the example for the case when $\Gamma = S_m$ is the symmetric group of degree m. To do this we consider the product $\Pi = \mathbb{P}^s \times \cdots \times \mathbb{P}^s$ of m copies of the s-dimensional projective space. We write a point $x \in \Pi$ in the form $x = (x_1, \ldots, x_m)$, $x_i \in \mathbb{P}^s$.

On this variety the group S_m acts by permuting the individual points

$$g(x_1, \ldots, x_m) = (x_{i_1}, \ldots, x_{i_m}), \quad g = \begin{pmatrix} 1 \ldots m \\ i_1 \ldots i_m \end{pmatrix} \in S_m.$$

The main step in the construction of the example is the construction of the factor space $\Pi' = \Pi / S_m$, that is, of the normal variety Π' and a finite morphism $\varphi : \Pi \to \Pi'$, such that $\varphi(x) = \varphi(x')$ if and only if $x' = g \cdot x$ for some $g \in S_m$.

Suppose that in the j-th copy of $\mathbb{P}^s$ homogeneous coordinates are denoted by $x_{0,j},\ldots,x_{s,j}$. We introduce $s+1$ auxiliary variables $t_0,\ldots,t_s$, and we consider the form

$$F(x,t)=\prod_{j=1}^{m}\left(\sum_{i=0}^{s}x_{i,j}t_i\right)=\prod_{j=1}^{m}L_j(x_j,t)\,, \qquad (1)$$

$$L_j(x_j,t)=\sum_{i=0}^{s}x_{i,j}t_i,\ x_j=(x_{0,j},\ldots,x_{s,j})\,.$$

We denote a monomial of degree m in the variables $t_0,\ldots,t_s$ by $T^{(\alpha)}$, and the number of them by $N+1$. Then

$$F(x,t)=\sum_{\alpha=0}^{N}F_\alpha(x)\,T^{(\alpha)}\,,$$

where the $F_\alpha(x)$ are forms in the variables $x_{i,j}$, linear in each system $x_{0,j},\ldots,x_{s,j}$. We consider the rational mapping

$$\varphi(x)=\{F_\alpha(x),\alpha=0,\ldots,N\}\,,$$

defined by these forms. This mapping is regular: if all the $F_\alpha(\bar{x})=0$ for some point $\bar{x}\in\Pi$, then $F(\bar{x},t)=0$, and this means that $L_j(\bar{x}_j,t)=0$ for some j, that is, all the coordinates of x_j are 0. There is a simple connection between φ and that embedding of Π as a closed subset $\bar{\Pi}$ of some projective space which was constructed in Ch. I, § 5.1. Namely, it is easy to verify that φ is a projection of $\bar{\Pi}$ and that for this projection the conditions of Theorem 8 in Ch. I, § 5, hold, so that the mapping $\varphi:\Pi\to\Pi'\subset\mathbb{P}^N$, $\Pi'=\varphi(\Pi)$ is finite.

Under the action of S_m on Π the factors in (1) are permuted, from which it follows that $\varphi\cdot g=\varphi$, that is, if $x=g(y)$, then $\varphi(x)=\varphi(y)$. Conversely, if $\varphi(x)=\varphi(y)$, then $x=g(y)$ for some $g\in G$. For if $\varphi(x)=\varphi(y)$, then $F(x,t)=cF(y,t)$, $c\neq 0$, and from the unique factorization of polynomials it follows that the points y_j, $j=1,\ldots,m$, are obtained by a permutation of $x_1,\ldots,x_m$:

$$\varphi^{-1}(x')=\{g(x),g\in S_m\},\ x'\in\Pi'\,. \qquad (2)$$

Now we show that Π' is a normal variety. For this purpose we observe that the polynomials $F_\alpha(x_1,\ldots,x_m)$ do not change when the points $x_1,\ldots,x_m$ are permuted. The converse is also true: every polynomial in homogeneous coordinates of $x_1,\ldots,x_m$ that is homogeneous in the coordinates of each of these points and invariant under arbitrary permutations of the points is a form in the polynomials F_α. This is an analogue to the fundamental theorem on symmetric functions and is proved in exactly the same way. A proof can be found in older textbooks of algebra, for example, [6], Ch. XIX, § 89.

Let H be a form in homogeneous coordinates of points on Π', and $Y \subset \Pi'$ the affine open set defined by the condition $H \neq 0$, $X = \varphi^{-1}(Y)$; it is also affine and is defined by the condition $\varphi^* H \neq 0$. We verify that the ring $\varphi^* k[Y]$ consists precisely of the elements of $k[X]$ that are invariant under S_m:

$$\varphi^* k[Y] = k[X]^{S_m}. \tag{3}$$

For a function $f \in k[X]$ has the form

$$f = \frac{H_1}{(\varphi^* H)^l},$$

where H_1 is a form whose degree is equal to that of $(\varphi^* H)^l$. If f is invariant under S_m, then so is H_1 (the form $\varphi^* H$ is necessarily invariant). Therefore it follows from the generalized theorem on symmetric functions quoted above that H_1 is a form in the polynomials F_α, and this means that $f \in \varphi^* k[Y]$.

From (3) it follows that $k[Y]$ is integrally closed. For if a function $f \in k(Y)$ is integral over $k[Y]$, then $\varphi^*(f)$ is a fortiori integral over $k[X]$. But X is smooth and hence normal. Therefore $\varphi^*(f) \in k[X]$. When we now apply (3), we find that $\varphi^*(f) \in \varphi^* k[Y]$, that is, $f \in k[Y]$. So we have shown that Y is a normal variety. Since the hyperplane $H = 0$ was chosen arbitrarily, it now follows that Π' is normal.

Clearly from (3) an analogous equality follows for the field of fractions $\varphi^* k(\Pi) = k(\Pi)^{S_m}$. From the simplest results of Galois theory it now follows that $k(\Pi)/k(\Pi')$ is a Galois extension with Galois group S_m. In particular,

$$\deg \varphi = m! \tag{4}$$

We denote by $\Delta \subset \Pi$ the closed set consisting of those points $(x_1, \ldots, x_m)$ for which $x_i = x_j$ for some $i \neq j$, and we set $\Delta' = \varphi(\Delta) \subset \Pi'$, $W = \Pi - \Delta$, $W' = \Pi' - \Delta'$. If $x' \in W'$, then by (2), $\varphi^{-1}(x')$ consists of $m!$ distinct points. Comparing this with (4) we see that $\varphi : W \to W'$ is an unramified mapping. Since W is smooth, it follows from Corollary 3 to Theorem 8 in Ch. II, § 5, that W' is also smooth.

We have now constructed two smooth varieties W and W' and an unramified covering $\varphi : W \to W'$ with automorphism group S_m. However, this is not what we need, because both our varieties are incomplete. To avoid this difficulty we intersect Π' with a linear subspace $L \subset \mathbb{P}^n$ such that L does not intersect Δ' and that the variety $Y = L \cap \Pi'$ is smooth.

Such a subspace exists and can be given by d independent linear equations provided that

$$d > \operatorname{codim}_{\Pi'} \Delta' . \tag{5}$$

The variety consisting of the points $(x_1, \ldots, x_m) \subset \Pi$ for which $x_p = x_q$ is of codimension s in Π, hence

$$\operatorname{codim}_{\Pi} \Delta = \operatorname{codim}_{\Pi'} \Delta' = s ,$$

so that condition (5) takes the form $d > s$. We can choose the linear space so that the dimension of Y is determined by the theorem on the dimension of an intersection:

$$\dim Y = \dim \Pi' - d = m \cdot s - d .$$

Obviously, by choosing s sufficiently large we can achieve that the relations

$$\dim Y = m \cdot s - d = n, \quad d > s ,$$

hold. For this it is sufficient that

$$s(m-1) > n .$$

Since $Y \cap \Delta' = \emptyset$, that is, $Y \subset W'$, we see that $X \subset \varphi^{-1}(Y)$ is an unramified covering of Y with automorphism group S_m.

All the preceding arguments were purely algebraic. Let us suppose now that all the varieties are defined over the field of complex numbers $\mathbb{C}$. As we have seen in Ch. II, § 5.3, the mapping of topological spaces $\varphi : X(\mathbb{C}) \to Y(\mathbb{C})$ is an unramified covering. The proposition in § 4.1 applied for $r = 0$ shows that $X(\mathbb{C})$ is connected. For X is obtained from Π by intersections with hypersurfaces that can be regarded as hyperplanes under the standard embedding of Π in a projective space. The same proposition can be applied for $r = 1$. Since $\Pi(\mathbb{C})$ is simply-connected, the proposition shows that $X(\mathbb{C})$ is also simply-connected.

So we see that $X(\mathbb{C})$ is the universal covering of the manifold $Y(\mathbb{C})$ and that $S_m = \pi_1(Y(\mathbb{C}))$. Observe that by construction $Y(\mathbb{C})$ is projective.

Starting out from the unramified covering

$$\varphi : X \to Y$$

with the group S_m it is easy to construct a covering with an arbitrary finite group Γ. To do this we assume that $\Gamma \subset S_m$. We have seen that the extension $\mathbb{C}(X)/\mathbb{C}(Y)$ has the Galois group S_m, and by Galois theory, to the subgroup Γ there corresponds a subfield K such that $\mathbb{C}(X) \supset K \supset \mathbb{C}(Y)$ and $\mathbb{C}(X)/K$ has the Galois group Γ. We denote by $\bar{Y}$ the normalization of Y in K. According to the general properties of normalizations, we have morphisms

$$X \xrightarrow{\bar{\varphi}} \bar{Y} \xrightarrow{\psi} Y,$$

with

$$\psi \cdot \bar{\varphi} = \varphi .$$

From the general properties of finite morphisms it follows that $\bar{\varphi}$ and ψ are finite. We show that $\bar{Y}$ is smooth and $\bar{\varphi}$ unramified. For since $\deg \varphi = \deg \bar{\varphi} \deg \psi$ and since the number of inverse images $\varphi^{-1}(y)$ for a closed point $y \in Y$ is equal to $\deg \varphi$, the number of inverse images $\psi^{-1}(y)$ is equal to $\deg \psi$, and for $\bar{y} \in \bar{Y}$ the number of inverse images $(\bar{\varphi})^{-1}(y)$ is equal to $\deg \bar{\varphi}$. Thus, φ and ψ are unramified, and since Y is smooth, by Corollary to Theorem 8 of Ch. II, § 5, $\bar{Y}$ is also smooth. So we see that $X(C)$ is a universal covering for $\bar{Y}(C)$ and that $\pi_1(Y(C)) = \Gamma$. This completes our construction. Observe that from the unproved theorem on the normalization of a projective variety being projective it follows that the variety $\bar{Y}$ we have constructed is projective.

3. Notes. As a supplement to the example constructed above it should be noted that we can construct examples of a large number of projective varieties whose fundamental groups are infinite. Thus, if X is an n-dimensional Abelian variety, then by the results of Ch. VII, § 1.3, the manifold $X(\mathbb{C})$ is homeomorphic to a $2n$-dimensional torus and $\pi_1(X(\mathbb{C})) = \mathbb{Z}^{2n}$. If $n \geqslant 3$, then according to the proposition in § 4.1 a smooth hyperplane section Y of the torus X has the same fundamental group. Its universal covering is a subvariety of $\mathbb{C}^n$, but apparently nothing is known about its structure.

In these examples we have encountered two types of fundamental groups and of universal coverings of complete algebraic varieties.

Type I. The universal group $\pi_1(X)$ is finite. In that case it can be shown that the fundamental covering is a complete algebraic variety, and if X is projective, also a projective variety. In the case of dimension 1 this type is represented only by the line $\mathbb{P}^1$, the only variety of elliptic type.

To define *Type II* in precise terms is difficult. However, in this case the fundamental group is necessarily infinite, and the universal covering is a "very large" analytic manifold, far removed from being projective or complete algebraic. In the case of dimension 1 these are curves of parabolic and hyperbolic type. For dimension $\geqslant 2$ they are Abelian varieties or (by the remark at the end of § 3.2) varieties of the form D/G, where D is a bounded domain in $\mathbb{C}^n$ and G a group acting on D discretely and freely such that D/G is compact. Hyperplane sections of these varieties also belong to this type.

In trying to characterize more accurately the second type of variety we are led to the definitions of complex spaces of two types that play a fundamental role in the general theory of complex spaces and manifolds.

A complex space X is said to be *holomorphically convex* if for every sequence of points $x_n \in X$ having no limit points on X there exists a function f holomorphic on X such that $|f(x_n)| \to \infty$ as $n \to \infty$.

Every compact space is trivially holomorphically convex. Another example are the spaces X_{an}, where X is an algebraic affine variety. If $X \subset \mathbb{A}^n$, then even among the coordinates in $\mathbb{A}^n$ we can find a function such as required in the definition.

A holomorphically convex analytic space X is said to be *holomorphically complete* if the holomorphic functions on it separate its points, that is, for any two points $x', x'' \in X$, $x' \neq x''$, there exists a holomorphic function f on X such that $f(x') \neq f(x'')$.

From Theorem 2 in Ch. VIII, § 2, it follows that a compact analytic manifold is holomorphically complete only if it consists of a single point. The same is true for compact analytic spaces.

Analytic spaces of the form X_{an}, where X is an affine algebraic variety, are obviously holomorphically complete. Altogether, holomorphically complete analytic spaces occupy in the general theory of analytic spaces a place analogous to affine varieties in algebraic geometry. For example, they too are "antipodal" to compact spaces, like affine varieties are to projective or complete varieties.

Now we can give a more accurate description of the two types of universal coverings of algebraic varieties that occur in our examples. In Type I the universal coverings are compact, and in Type II they are holomorphically complete. It is natural to expect that the general case is in some sense "a mixture" of these two extremes. In the theory of analytic spaces there is a fundamental result that can be regarded as a more precise definition of the term "mixture". This is the so-called reduction theorem of Remmert, which asserts that every holomorphically convex normal analytic space X has a proper holomorphic mapping $f: X \to Y$ onto a holomorphically complete space Y (a mapping f is said to be proper if the inverse image of every compact set is compact; in particular, its fibres are compact).

In view of this theorem an interesting question arises: is perhaps the universal covering of a complete algebraic variety holomorphically convex? (Perhaps it would be more cautious to confine ourselves to projective varieties.)

Clearly all compact varieties are holomorphically convex. A typical example of a variety that is not holomorphically convex is $\mathbb{C}^2 - 0$. For it can be shown that a function holomorphic on $\mathbb{C}^2 - 0$ is also holomorphic at 0 (analogous to the fact that a rational function on a smooth algebraic variety is non-regular at the points of an integral divisor). Therefore, if $x_n \to 0$, then for every function f holomorphic on $\mathbb{C}^2 - 0$ we have $f(x_n) \to f(0)$. In fact, the variety $\mathbb{C}^2 - 0$ is the universal covering of non-algebraic varieties, for example, of Hopf surfaces (Ch. VIII, § 1.2). Kodaira has shown that no compact manifold whose universal covering coincides with $\mathbb{C}^2 - 0$ is algebraic.

In § 3.2 we have remarked that if a bounded domain $D \subset \mathbb{C}^n$ is the universal covering of a compact manifold, then the latter is an algebraic variety. On the other hand, it has been shown that every such domain D is holomorphically convex. These and some other examples give reason to hope that the answer to the question raised above is in the affirmative. A proof of this would be a great advance in the problem of the structure of universal coverings of algebraic varieties. At present almost nothing is known on universal covering varieties of arbitrary algebraic varieties. Apparently the only result in this direction is due to Griffiths, who has shown that any projective variety contains an affine open subset whose universal covering is isomorphic to a bounded domain in $\mathbb{C}^n$.

Exercises

1. Let $C \subset \mathbb{P}^2$ be a smooth projective plane curve given by the equation $F(x_0, x_1, x_2) = 0$ of degree n, $V \subset \mathbb{P}^3$ a projective surface given by the equation $F(x_0, x_1, x_2) = x_3^n$, and $f: V \to \mathbb{P}^2$ the projection with centre at $(0:0:0:1)$. Show that $f: V - f^{-1}(C) \to \mathbb{P}^2 - C$ is an unramified covering and that $V - f^{-1}(C)$ is the universal covering for $\mathbb{P}^2 - C$. Deduce that $\pi_1(\mathbb{P}^2 - C) \simeq \mathbb{Z}/n$.
2. Show that $(\mathbb{P}^1)^n / S_n = \mathbb{P}^n$.
3. Let X be a smooth projective curve, $G = \{1, g\}$, $g(x, x') = (x', x) \in X \times X$. Show that the ringed space $Y = (X \times X)/G$ is a complex space and even a variety (notwithstanding the fact that g has fixed points). Show that $\pi_1(Y) \simeq H_1(X)$.

Bibliography

[1] Abhyankar, S. S.: Local analytic geometry. New York-London: Academic Press, 1964.
[2] Aleksandrov, P. S., Efremovich, V. A.: A survey of the fundamental concepts of topology. Moscow: ONTI 1936.
[3] Algebraic surfaces. Trudy Mat. Inst. Steklov **75** (1965). English translation: Algebraic Surfaces (by the members of the seminar of I. R. Šafarevič). Amer. Math. Soc. Translations **75** (1967).
[4] Ahlfors, L. V., Bers, L.: Spaces of Riemann surfaces and quasiconformal mappings. Moscow: Izdat. Inost. Lit., 1961. (Translation into Russian of two papers by Ahlfors and four by Bers, see MR **24** A 229.)
[5] Artin, M.: Algebraic spaces. Mimeographed notes. New Haven, Conn.; Yale University, 1969.
[6] Bôcher, M.: Introduction to higher algebra. Cambridge Mass.: Harvard Univ., 1909.
[7] Borevich, Z. I., Shafarevich, I. R.: Number theory. Moscow: Nauka, 1964. Translation, New York and London: Academic Press, 1966.
[8] Bourbaki, N.: Elements of mathematics, Book III, General topology, 2 vols. Paris, Reading, Mass.: Hermann Cie. – Addison-Wesley Publ. Co., 1966.
[9] Cartan, H.: Elementary theory of analytiç functions of one or several complex variables. Paris – Reading, Mass.: Hermann & Cie. – Addison Wesley Publ. Co., 1963.
[10] Cartan, H., Eilenberg, S.: Homological algebra. Princeton, N. J.: Univ. Press, 1956.
[11] Chern, S. S.: Complex manifolds. Univ. Recife 1959.
[12] Eilenberg, S., Steenrod, N. E.: Foundations of algebraic topology. Princeton, N. J.; Univ. Press, 1962.
[13] Fam, F.: Introduction to the topological investigation of Landau singularieties. Moscow: Mir, 1970.
[14] Grothendieck, A.: Cohomologie locale des faisceaux cohérents et théorèmes de Lefschetz locaux et globaux. 2 fasc., third rev. ed., Paris: Inst. Hautes Ét. Sci., 1968.
[15] Grothendieck, A., Dieudonné, J.: Éléments de géometrie algébrique, vol. 1. Berlin-Heidelberg-New York: Springer-Verlag, 1971.
[16] Goursat, E.: A course in mathematical analysis, vol. 1. New York: Dover reprint, 1959.
[17] Gunning, R. C., Rossi, H.: Analytic functions of several complex variables. Englewood Cliffs, N. J.: Prentice-Hall, 1965.
[18] Hodge, W. V. D., Pedoe, D.: Methods of algebraic geometry, vol. II. Cambridge: Univ. Press, 1952.
[19] Husemoller, D.: Fibre bundles. New York-London: McGraw-Hill, 1966.
[20] Lang, S.: Introduction to algebraic geometry. New York-London: Interscience, 1958.
[21] Milnor, J. W.: Morse theory. Princeton, N. J.: Univ. Press, 1963.
[22] Moishezon, B. G.: An algebraic analogue to compact complex spaces with a sufficiently large field of meromorphic functions. Izv. Akad. Nauk SSSR Ser. Mat. **33**, 174–238, 323–367, 506–548 (1969).
[23] Mumford, D.: Geometric invariant theory. Berlin-Heidelberg-New York: Springer-Verlag, 1965.

[24] Mumford, D.: Lectures on curves on an algebraic surface. Princeton, N. J.; Univ. Press, 1966.

[25] Pontryagin, L. S.: Topologische Gruppen, second ed. 2 vols. Leipzig: Teubner, 1957.

[26] de Rham, G.: Variétés différentiables. Paris: Hermann & Cie. 1955.

[27] Samuel, P.: Méthodes d'algèbre abstraite en géométrie algébrique. Berlin-Heidelberg-New York: Springer-Verlag, 1967.

[28] Seifert, H., Threllfall, W.: Lehrbuch der Topologie. Leipzig: Teubner, 1934.

[29] Serre, J. P.: Lecture Notes in Math. Vol. II, Algèbre Locale multiplicités. Third ed. Berlin-Heidelberg-New York: Springer-Verlag, 1975.

[30] Siegel, C. L.: Analytic functions of several complex variables. Princeton, N. J.: Inst. for Adv. Studies, 1950.

[31] Springer, G.: Introduction to Riemann surfaces. Reading, Mass.: Addison-Wesley Publ. Co., 1957.

[32] Vinogradov, I. M.: Elements of number theory, seventh ed. Moscow: Nauka, 1965. Translations: New York: Dover, 1954 or Oxford: Pergamon 1952.

[33] van der Waerden, B. L.: Algebra, 2 vols., fifth ed. Berlin-Göttingen-Heidelberg: Springer-Verlag, 1960.

[34] Walker, R. J.: Algebraic curves, Princeton, N. J.; Univ. Press., 1950.

[35] Wallace, A. H.: Differential topology: First steps, New York-Amsterdam: W. A. Benjamin, 1968.

[36] Weil, A.: Introduction à l'étude des variétés Kähleriennes. Paris: Hermann & Cie., 1958.

[37] Zariski, O., Samuel, P.: Commutative algebra, 2 vols. Princeton, N. J.; Van Nostrand and Co. 1958/60.

Historical Sketch

This sketch does not pretend to give a systematic account of the history of algebraic geometry. Its aim is to describe in very general terms how the ideas and concepts with which the reader has become acquainted in this book were created. In explaining the research of one mathematician or another we often omit important work of his (sometimes even the most important) if it has no bearing on the contents of our book.

We shall try to formulate the results as closely as possible to the way the authors have done it, using only occasionally contemporary notation and terminology. In cases where this is not immediately obvious we shall explain these investigations from the point of view of the concepts and results of our book. Such places are marked by an asterisk * (at the beginning and the end).

* Naturally algebraic geometry arose first as the theory of algebraic curves. Only by going beyond the frame of rational curves do we encounter properties of algebraic curves that are characteristic for algebraic geometry. Therefore we leave aside the theory of conics, which are all rational. Next in complexity and hence the first non-trivial example are curves of genus 1, that is, elliptic curves and, in particular, non-singular curves of the third degree. And historically the first step in the development of the theory of algebraic curves consisted in a clarification of the basic concepts and ideas of this theory in the example of elliptic curves.

Thus, it would seem that these ideas developed in the same sequence in which they are now set forth (for example, in Ch. I, § 1). However, in one respect this is by no means the case. The complex of concepts and results that we now call the theory of elliptic curves arose as part of analysis, and not of geometry: as the theory of integrals of rational functions on an elliptic curve. It was precisely these integrals that originally were called by the name elliptic (they occur in connection with the computation of the arc length of an ellipse), and later the name was transferred from them to functions and to curves. *

1. Elliptic Integrals. They were an object of study as early as the XVII century, as an example of integrals that cannot be expressed in terms of elementary functions and lead to new transcendental functions.

At the very end of the XVII century Jacob, and later Johann, Bernoulli came up against a new interesting property of these integrals (see J. Bernoulli [1], Vol. 1, p. 252). In their investigations they considered integrals expressing the arc length of certain curves. They found certain transformations of one curve into another that preserve the arc length of the curve, although the corresponding arcs cannot be superposed to one another. It is clear that analytically this leads to the transformation of one integral into another. In some cases there arise transformations of an integral into itself. In the first half of the XVIII century many examples of such transformations were found by Fagnano.

In general form the problem was raised and solved by Euler. He communicated his first results in this direction in a letter to Goldbach in 1752. His investigations on elliptic integrals were published from 1756 to 1781 (see Euler [1]).

Euler considers an arbitrary polynomial $f(x)$ of degree 4 and asks for the relations between x and y if

$$\frac{dx}{\sqrt{f(x)}} = \frac{dy}{\sqrt{f(y)}}. \tag{1}$$

He regards this as a differential equation connecting x and y. The required relation is the general integral of this equation. He finds this relation: it turns out to be algebraic of degree 2 both in x and in y. Its coefficients depend on the coefficients of the polynomial $f(x)$ and on one independent parameter c.

Euler formulates this result in another form: the sum of the integrals $\int_0^\alpha \frac{dx}{\sqrt{f(x)}}$ and $\int_0^\beta \frac{dx}{\sqrt{f(x)}}$ is equal to a single integral:

$$\int_0^\alpha \frac{dx}{\sqrt{f(x)}} + \int_0^\beta \frac{dx}{\sqrt{f(x)}} = \int_0^\gamma \frac{dx}{\sqrt{f(x)}}, \tag{2}$$

and γ can be expressed rationally in terms of α and β. Euler also brings forward arguments why such a relation cannot hold if the degree of the polynomial $f(x)$ is greater than 4.

For arbitrary elliptic integrals of the form $\int \frac{r(x)\,dx}{\sqrt{f(x)}}$ Euler proves a relation that generalizes (2):

$$\int_0^\alpha \frac{r(x)\,dx}{\sqrt{f(x)}} + \int_0^\beta \frac{r(x)\,dx}{\sqrt{f(x)}} - \int_0^\gamma \frac{r(x)\,dx}{\sqrt{f(x)}} = \int_0^\delta V(y)\,dy, \tag{3}$$

where γ is the same rational function of α and β as in (2), and where δ and V are also rational functions.

* The reason for the existence of an integral of the equation (1) and of all its special cases discovered by Fagnano and Bernoulli is the presence of a group law on an elliptic curve with the equation $s^2 = f(t)$ and the invariance of the everywhere regular differential form $s^{-1}dt$ under translations by elements of the group. The relations found by Euler that connect x and y in (1) can be written in the form

$$(x, \sqrt{f(x)}) \oplus (c, \sqrt{f(c)}) = (y, \sqrt{f(y)}),$$

where $\oplus$ denotes addition of points on the elliptic curve. Thus, these results contain at once the group law on an elliptic curve and the existence of an invariant differential form on this curve.

The relation (2) is also an immediate consequence of the invariance of the form $\varphi = \dfrac{dx}{\sqrt{f(x)}}$. In it

$$(\gamma, \sqrt{f(\gamma)}) = (\alpha, \sqrt{f(\alpha)}) \oplus (\beta, \sqrt{f(\beta)})$$

and

$$\int_0^\alpha \varphi + \int_0^\beta \varphi = \int_0^\alpha \varphi + \int_\alpha^\gamma t_g^* \varphi = \int_0^\alpha \varphi + \int_\alpha^\gamma \varphi = \int_0^\gamma \varphi,$$

where t_g is the translation by $g = (\alpha, \sqrt{f(\alpha)})$. Observe that we write here the equation between integrals formally, without indicating the paths of integration. Essentially this is an equation "to within a constant of integration", that is, an equation between the corresponding differential forms. This is how Euler understood them.

Finally, the meaning of the relation (3) will become clear later, in connection with Abel's theorem (see **3.**). *

2. Elliptic Functions. After Euler the theory of elliptic integrals was developed mainly by Legendre. His investigations, beginning in 1786, are collected in the three-volume "Traité des fonctions elliptiques et des intégrales Eulériennes (Legendre [1]).* In his preface to the first supplement published in 1828 Legendre writes: "So far the geometers have hardly taken part in investigations of this kind. But no sooner had this book seen the light of day, no sooner had it become known to scholars abroad, than I learned with astonishment as well as joy that two young geometers, Herr Jacobi in Königsberg and Herr Abel in Christiania, have achieved in their works substantial progress in the highest branches of this theory".

* Legendre called elliptic functions what we now call elliptic integrals. The contemporary terminology became accepted after Jacobi.

Abel's papers on the theory of elliptic functions appeared in 1827–1829. He starts out (see Abel [1], Vol. I, No. XVI, No. XXIV) from the elliptic integral

$$\theta = \int_0^{\lambda} \frac{dx}{\sqrt{(1-c^2x^2)(1-e^2x^2)}},$$

where c and e are complex numbers; he regards it as a function of the upper limit and introduces the inverse function $\lambda(\theta)$ and the function $\Delta(\theta) = \sqrt{(1-c^2\lambda^2)(1-e^2\lambda^2)}$. From the properties of elliptic integrals known at that time [essentially, from Euler's relations (2) in **1**.] he deduces that the functions $\lambda(\theta \pm \theta')$ and $\Delta(\theta \pm \theta')$ can be simply expressed in the form of rational functions of $\lambda(\theta)$, $\lambda(\theta')$, $\Delta(\theta)$, and $\Delta(\theta')$. Abel shows that both these functions have in the complex domain two periods 2ω and $2\tilde{\omega}$:*

$$\omega = 2\int_0^{1/c} \frac{dx}{\sqrt{(1-c^2x^2)(1-e^2x^2)}},$$

$$\tilde{\omega} = 2\int_0^{1/e} \frac{dx}{\sqrt{(1-c^2x^2)(1-e^2x^2)}}.$$

He finds representations of the functions introduced by him in the form of infinite products extended over their zeros.

As an immediate generalization of the problem with which Euler had been occupied, Abel [1] (Vol. I, No. XIX) raises the question: "To list all the cases in which the differential equation

$$\frac{dy}{\sqrt{(1-c_1^2y^2)(1-e_1^2y^2)}} = \pm a \frac{dx}{\sqrt{(1-c^2x^2)(1-e^2x^2)}} \tag{1}$$

can be satisfied by taking for y an algebraic function of x, rational or irrational".

This problem became known as the transformation problem for elliptic functions. Abel showed that if the relation (1) can be satisfied by means of an algebraic function y, then it can also be done by means of a rational function. He showed that if $c_1 = c$, $e_1 = e$, then a must either be rational or a number of the form $\mu' + \sqrt{-\mu}$, where μ' and μ are rational numbers and $\mu > 0$. In the general case he showed that the periods ω_1 and $\tilde{\omega}_1$ of the integral of the left-hand side of (1), multiplied by a common factor, must be expressible in the form of an integral linear combination of the periods ω and $\tilde{\omega}$ of the integral of the right-hand side.

* As E. I. Slavutin has remarked, already Euler [2] drew attention to the fact that the function $\int_0^y \frac{dx}{\sqrt{1-x^4}}$ has in the real domain a "modulus of multi-valuedness" similar to the inverse trigonometric functions.

Somewhat later than Abel, but independently, Jacobi [1] (Vol. I, Nos. 3 and 4) also investigated the function inverse to the elliptic integral, proved that it has two independent periods, and obtained a number of results on the transformation problem. Transforming into series the expressions for elliptic functions that Abel had found in the form of products, Jacobi arrived at the concept of θ-functions* and found numerous applications for them, not only in the theory of elliptic functions but also in number theory and in mechanics.

Finally, after Gauss's posthumous works were published, especially his diaries, it became clear that long before Abel and Jacobi he had mastered some of these ideas to a certain extent.

$*$ The first part of Abel's results requires hardly any comment. The mapping $x=\lambda(\theta)$, $y=\Delta(\theta)$ determines a uniformization of the elliptic curve $y^2=(1-c^2x^2)(1-e^2x^2)$ by elliptic functions. Under the corresponding mapping $f:\mathbb{C}^1\to X$ the regular differential form $\varphi=\frac{dx}{y}$ goes over into a regular differential form on $\mathbb{C}^1$ that is invariant under translations by the vectors of the lattice $2\omega\mathbb{Z}+2\tilde{\omega}\mathbb{Z}$. This form differs by a constant factor from $d\theta$, and we may assume that $d\theta=f^*\frac{dx}{y}$, that is, $\theta=\int\frac{dx}{y}$.

The integration of the equation (1) has the following geometric meaning. Let X and X_1 be elliptic curves with the equations $u^2=(1-c^2x^2)\cdot(1-e^2x^2)$ and $v^2=(1-c_1^2y^2)(1-e_1^2y^2)$. The point is to investigate curves $C\subset X\times X_1$ (which corresponds to an algebraic relation between x and y). Since an elliptic curve is its own Picard variety (see Ch. III, § 3.5), C determines a morphism $f: X\to X_1$. This makes it clear why the problem reduces to the case when y is a rational function of x. According to Theorem 4 in Ch. III, § 3, f can be regarded as a homomorphism of the algebraic groups X and X_1. Thus, Abel studied the group Hom (X, X_1) and for $X=X_1$ the ring End X. A homomorphism $f\in\mathrm{Hom}(X,X_1)$ determines a linear transformation of the one-dimensional spaces $f^*:\Omega^1[X_1]\to\Omega^1[X]$, which is given by a single number, the factor $\pm a$ in (1). See also Exercises 7, 8, and 9 to Ch. IX, § 2. $*$

3. Abelian Integrals. The transition to arbitrary algebraic curves proceeded entirely within the framework of analysis: Abel showed that the basic properties of elliptic integrals can be generalized to integrals of arbitrary algebraic functions. These integrals later became known as Abelian integrals.

* θ-functions occured first in 1826 in a book by Fourier on the theory of heat.

In 1826 Abel wrote a paper (see Abel [1], Vol. I, No. XII), which was the beginning of the general theory of algebraic curves. He considers in it an algebraic function y determined by two equations

$$\chi(x, y) = 0, \tag{1}$$

and

$$\theta(x, y) = 0, \tag{2}$$

where $\theta(x, y)$ is a polynomial that depends, apart from x and y, linearly on some parameters $a, a', \ldots$, the number of which is denoted by α. When these parameters are changed, some simultaneous solutions of (1) and (2) may not change. Let $(x_1, y_1), \ldots, (x_\mu, y_\mu)$ be variable solutions, and $f(x, y)$ an arbitrary rational function. Abel shows that

$$\int_0^{x_1} f(x, y)\, dx + \cdots + \int_0^{x_\mu} f(x, y)\, dx = \int V(g)\, dg, \tag{3}$$

where $V(t)$ and $g(x, y)$ are rational functions depending also on the parameters $a, a', \ldots$. Abel interpreted this result by saying that the left-hand side of (3) is an elementary function.

Using the freedom in choosing the parameters $a, a', \ldots$ Abel shows that the sum of any number of integrals $\int_0^{x_i} f(x, y)\, dx$ can be expressed in terms of $\mu - \alpha$ such integrals and a term of the same type as that on the right-hand side of (3). He establishes that the number $\mu - \alpha$ depends only on (1). For example, for $y^2 + p(x)$, where the polynomial p is of degree $2m$, we have $\mu - \alpha = m - 1$.

Next Abel investigates for what functions f the right-hand side of (1) does not depend on the parameters $a, a', \ldots$. He expresses f in the form $\dfrac{f_1(x, y)}{f_2(x, y)\, \chi'_y}$, and he shows that $f_2 = 1$, and f_1 satisfies a number of restrictions as a consequence of which the number γ of linearly independent ones among the required functions f is finite. Abel shows that $\gamma \geqslant \mu - \alpha$ and that $\gamma = \mu - \alpha$, for example, if (using a much later terminology) the curve $\chi(x, y) = 0$ has no singular points.

* The discussion of the solutions $(x_1, y_1), \ldots, (x_\mu, y_\mu)$ of the system consisting of (1) and (2) leads us at once to the contemporary concept of equivalence of divisors. Namely: let X be the curve with the equation (1) and D_λ the divisor cut out on it by the form θ_λ (in homogeneous coordinates), where λ is the system of parameters $a, a', \ldots$. By hypothesis, $D_\lambda = \bar{D}_\lambda + D_0$, where D_0 does not depend on λ. Therefore all the $\bar{D}_\lambda = (x_1, y_1) + \cdots + (x_\mu, y_\mu)$ are equivalent to each other. The problem with which Abel was concerned reduces to the investigation of the sum $\int_{\alpha_1}^{\beta_1} \varphi + \cdots + \int_{\alpha_\mu}^{\beta_\mu} \varphi$, where φ is a differential form on X, α_i and β_i are

points on $X, (\alpha_1)+\dots+(\alpha_\mu)\sim(\beta_1)+\dots+(\beta_\mu)$. We give a sketch of a proof of Abel's theorem that is close in spirit to the original proof. We may assume that

$$(\alpha_1)+\dots+(\alpha_\mu)-(\beta_1)-\dots-(\beta_\mu)=(g)\,, \qquad g\in\mathbb{C}(X)\,,$$
$$(\alpha_1)+\dots+(\alpha_\mu)=(g)_0\,, \qquad (\beta_1)+\dots+(\beta_\mu)=(g)_\infty\,.$$

We consider a morphism $g\colon X\to\mathbb{P}^1$ and the corresponding extension $\mathbb{C}(X)/\mathbb{C}(g)$. For simplicity we assume that this is a Galois extension (the general case easily reduces to this), and we denote its Galois group by G. The automorphisms $\sigma\in G$ act on the curve X and the field $\mathbb{C}(X)$ and carry the points $\alpha_1, \dots, \alpha_\mu$ into each other, because $\{\alpha_1, \dots, \alpha_\mu\}=g^{-1}(0)$. Therefore $\{\alpha_1, \dots, \alpha_\mu\}=\{\sigma\alpha, \sigma\in G\}$, where α is one of the points α_i. Similarly $\{\beta_1, \dots, \beta_\mu\}=\{\sigma\beta, \sigma\in G\}$. Representing φ in the form $u\,dg$, we see that

$$\sum_{i=1}^{\mu}\int_{\alpha_i}^{\beta_i}\varphi=\sum_{\sigma\in G}\int_{\sigma\alpha}^{\sigma\beta}u\,dg=\int_{\alpha}^{\beta}\Big(\sum_{\sigma\in G}\sigma u\Big)\,dg\,. \tag{4}$$

The function $v=\sum_{\sigma\in G}\sigma u$ is contained in $\mathbb{C}(g)$, and Abel's theorem follows from this.

We see that every sum of integrals $\sum_i\int_0^{x_i}f(x, y)\,dx$ can be expressed as a sum of l integrals $\sum_{j=1}^{l}\int_0^{x'_j}f(x, y)\,dx+\int V(g)\,dg$ if the equivalence

$$\sum_i((\alpha_i)-o)\sim\sum_{j=1}^{l}((\alpha'_j)-o) \tag{5}$$

holds, where $\alpha_i=(x_i, y_i)$, $\alpha'_j=(x'_j, y'_j)$, and o is the point with $x=0$. From the Riemann-Roch theorem it follows at once that the equivalence (5) holds (for arbitrary α_i and certain α'_j corresponding to them), with $l=g$ (see Exercise 19 to Ch. III, § 5). Thus, the constant $\mu-\alpha$ introduced by Abel is the same as the genus.

If $\varphi\in\Omega^1[X]$, then also $v\,dg\in\Omega^1[\mathbb{P}^1]$, where $v=\sum_{\sigma\in G}\sigma u$ in (4). Since $\Omega^1[\mathbb{P}^1]=0$, in this case the term on the right-hand side of (3) disappears. Hence it follows that $\gamma\geqslant g$. In some cases arising naturally the two numbers coincide.

We see that this paper of Abel's contains the concept of the genus of an algebraic curve and the equivalence of divisors and gives a criterion for equivalence in terms of integrals. In the last relation it leads to the theory of Jacobian varieties of algebraic curves (see § 5). *

4. Riemann Surfaces. In his dissertation published in 1851 Riemann [1] (No. 1) applied a completely new principle of investigating functions of a complex variable. He assumes that the function is given not on the plane of a complex variable but on some surface that "extends in many sheets" over this surface. The real and imaginary parts of this function satisfy the Laplace equation. This function is uniquely determined if the points are known at which it becomes infinite, the curves along which cuts make it single-valued, the character of its singularities at these points, and the many-valuedness in passing through these curves. Riemann also works out a method of constructing a function from these data that is based an a variational principle which Riemann called "the Dirichlet principle".

In the first part of the paper "Theory of Abelian functions", which appeared in 1857, Riemann [1] (No. II) applied these ideas to the theory of algebraic functions and their integrals. The paper begins with the investigation of properties of the corresponding surfaces that belong, as Riemann says, to Analysis Situs. By means of an even number $2p$ of cuts the surface becomes a simply-connected domain. By arguments taken from Analysis Situs he shows that $p = w/2 - n + 1$ where n is the number of sheets and w the number of branch points of the surface over the plane of the complex variable (taken with the appropriate multiplicities).

Riemann investigates functions that, speaking generally, are many-valued on the surface, but single-valued in the domain obtained after making the cuts, and on passing through the cuts their values change by constants, the so-called moduli of periodicity of the function. The Dirichlet principle gives a method of constructing such functions. In particular, there are p linearly independent everywhere finite such functions: the "integrals of the first kind". Similarly functions are constructed that become infinite at given points. In order to form from them functions that are single-valued on the surface one has to equate to zero their moduli of periodicity. Hence it follows that among the single-valued functions that become infinite only at m given points not fewer than $m - p + 1$ are linearly independent and, if $m > p$, this inequality becomes an equality for points in "general" position.

Riemann shows that all the functions that are single-valued on a given surface are rational functions of two of them: s and z, which are connected by a relation $F(s, z) = 0$. He calls two such relations belonging to one "class" if they can be rationally transformed into one another. In that case the corresponding surfaces have one and the same number p. But the converse is not true. Studying the possible dispositions of the branch points of surfaces, Riemann shows that the set of classes depends for $p > 1$ on $3p - 3$ independent parameters, which he calls "moduli".

∗ The surfaces introduced by Riemann closely correspond to the contemporary concept of a one-dimensional complex analytic manifold; these are the sets on which analytic functions are defined. Riemann raises and solves the problem of the connection of this concept with that of an algebraic curve. (The appropriate result is called Riemann's existence theorem.)

This circle of Riemann's ideas did not by any means become clear at once. An important role in their clarification is played by Klein's lectures [2], in which he emphasizes that a Riemann surface a priori is not connected with an algebraic curve or an algebraic function. A definition of a Riemann surface that differs only terminologically from the presently accepted definition of a one-dimensional analytic manifold was given by H. Weyl [1].

Riemann's paper marks the beginning of the topology of algebraic curves. The topological meaning of the dimension p of the space $\Omega^1[X]$ is explained in it: it is half the dimension of the first homology group of the space $X(\mathbb{C})$.

Analytically Riemann proves the inequality $l(D) \geqslant \deg D - p + 1$. The Riemann-Roch equality was then proved by his pupil Roch.

Finally, in this paper the field $k(X)$ emerges for the first time as an original object connected with a curve X, and the concept of a birational isomorphism appears. *

5. The Inversion Problem. Already Abel had raised the question of the inversion of integrals of arbitrary algebraic functions. He observed, in particular, that the function inverse to the hyperelliptic integral connected with $\sqrt{\psi(x)}$ has periods equal to half the value of this integral taken between two roots of the polynomial ψ (see Abel [1], Vol. II, No. VII).

Jacobi drew attention to the fact that we are concerned here with a function of a single complex variable having more that two periods if the integral is not elliptic, and that this is impossible for a reasonable function. If X is a polynomial of degree 5 or 6, Jacobi proposes to consider the pair of functions

$$u = \int_0^x \frac{dx}{\sqrt{X}} + \int_0^y \frac{dx}{\sqrt{X}}, \qquad v = \int_0^x \frac{x\,dx}{\sqrt{X}} + \int_0^y \frac{x\,dx}{\sqrt{X}}.$$

He suggests expressing $x - y$ and xy as analytic functions of the two variables u and v and conjectures that this expression is possible by means of a generalization of θ-functions (see Jacobi [1], Vol. II, Nos. 2 and 4). This conjecture was verified in a paper by Göpel [1] published in 1847.

The second part of Riemann's paper [1] (No. II) on Abelian functions is concerned with the connection between θ-functions and the inversion problem in the general case. He considers a series in p variables

$$\theta(v)=\sum_m e^{F(m)+2(m,v)}, \tag{1}$$

where $m=(m_1, \ldots, m_p)$ ranges over all integral p-dimensional vectors, $v=(v_1, \ldots, v_p)$, $(m, v)=\Sigma m_i v_i$, $F(m)=\Sigma \alpha_{jl} m_j m_l$, $\alpha_{jl}=\alpha_{lj}$. This series converges for all values of v if the real part of the quadratic form F is negative definite. The main property of the function θ is the equality

$$\theta(v+\pi i r)=\theta(v), \qquad \theta(v+\alpha_j)=e^{L_j(v)}\theta(v), \tag{2}$$

where r is an integral vector, α_j a column of the matrix (α_{jl}), and $L_j(v)$ a linear form.

Riemann shows that one can choose cuts $a_1, \ldots, a_p, b_1, \ldots, b_p$ which make his surface simply-connected, and a basis $u_1, \ldots, u_p$ of everywhere finite integrals on this surface such that the integrals u_j over a_l are equal to 0 for $j \neq l$ and to πi for $j=l$, and the same u_j over b_l form a symmetric matrix (α_{jl}), satisfying conditions under which the series (1) converges. He considers the function θ corresponding these coefficients α_{jl} and the function $\theta(u-e)$, where $u=(u_1, \ldots, u_p)$ (the u_i are everywhere finite integrals) and e is an arbitrary vector.

Riemann shows that $\theta(u-e)$ has on the surface p zeros $\eta_1, \ldots, \eta_p$ or vanishes identically. For a suitable choice of the lower limits in the integrals u_i in the first case

$$e \equiv u(\eta_1)+\cdots+u(\eta_p), \tag{3}$$

where the congruence is taken modulo integral linear combinations of the periods of the integrals u_i. In this way the points $\eta_1, \ldots, \eta_p$ are uniquely determined. In the second case there also exist points $\eta_1, \ldots, \eta_{p-2}$ such that

$$e \equiv -\left(u(\eta_1)+\cdots+u(\eta_{p-2})\right). \tag{4}$$

Riemann knew that the periods of any $2n$-periodic function of n variables satisfy relations similar to those that are necessary for the convergence of the series defining the θ-function. These relations between the periods were described explicitly by Frobenius [1], who showed that they are necessary and sufficient for the existence of non-trivial functions satisfying the functional equation (2). Hence it follows that these relations are necessary and sufficient for the existence of a meromorphic function with given periods that cannot be reduced by a linear change of variables to a function of a smaller number of variables. One only has to apply the theorem that every $2n$-periodic analytic function can be represented as a quotient of entire functions satisfying the functional equation of the θ-function. This theorem, stated by Weierstrass, was proved by Poincaré [2]. In 1921 Lefschetz [1] proved that when the Frobenius relations hold, the θ-functions determine an embedding of the manifold $\mathbb{C}^n/\Omega$ (where Ω is the lattice corresponding to the given period matrix) into a projective space.

* The inversion problem is connected with questions which in this book we only touched upon incidentally, often without proofs. The matter concerns the construction of the Jacobian variety of an analytic curve and properties of arbitrary Abelian varieties (see Ch. III, § 3.5, and Ch. VIII, § 1.3).

If o is a fixed point on a curve X, then $f(x)=x-o$ is evidently an algebraic family of divisors of degree zero on X. The basis of this family coincides with X. By the definition of the Jacobian variety J_X of X (we recall that this is the name for the Picard variety if X is a curve) there exists a morphism $\varphi: X \to J_X$, which is an embedding if the genus p of X is different from 0. It can be shown that $\varphi^*: \Omega^1[J_X] \to \Omega^1[X]$ is an isomorphism. Therefore in the representation

$$J_X = \mathbb{C}^p/\Omega \tag{5}$$

the $2p$-dimensional lattice $\Omega \subset C^p$ consists of the periods of p independent differential forms $\omega \in \Omega^1[X]$. Riemann also starts out from this analytic specification of the Jacobian variety and then develops an algebraic method of investigating it.

If D_0 is an arbitrary effective divisor of degree p, then $g(x_1, \ldots, x_p) = x_1 + \cdots + x_p - D_0$ determines a family of divisors of degree 0 on X. A basis of this family can be taken to be the factor space X^p/S_p of the product of p copies of X with respect to the symmetric group acting by permutations of the factors. By the definition of the Jacobian variety there exists a morphism $\psi: X^p/S_p \to J_X$. It follows easily from the Riemann-Roch theorem that it is an epimorphism and one-to-one on an open set in J_X. Therefore it is a birational isomorphism. In the analytical representation ψ takes the form (3), and by definition it is not one-to-one at those points $(x_1, \ldots, x_p)$ for which $l(x_1 + \cdots + x_p) > 1$. It follows from the Riemann-Roch theorem that this is equivalent to the condition $l(K - x_1 - \cdots - x_p) > 0$, that is (because $\deg K = 2g-2$), to

$$x_1 + \cdots + x_p \sim K - y_1 - \cdots - y_{p-2}$$

for certain points $y_1, \ldots, y_{p-2}$. The latter relation is the same as (4), to within the additional term K, hence to a shift by a point in J_X.

The Frobenius relations are the condition for the analytic manifold $\mathbb{C}^p/\Omega$ to be projective. They are written down in Ch. VIII, § 1.4 [formulae (3) and (4)]. *

6. Geometry of Algebraic Curves. So far we have seen how the concepts and results that nowadays form the foundation of the theory of algebraic curves have been created under the influence and within the framework of the analytic theory of algebraic functions and their integrals. A purely geometric theory of algebraic curves developed independently of this trend of research. For example, in a book published in 1834 Plücker found formulae connecting the class, the degree of a curve, and the number of its double points (see Exercise 13 to Ch. IV, § 3). There he also proved the existence of nine points of inflexion on a plane cubic curve (see Exercise 4 to Ch. VIII, § 1). But research of a similar kind took second place in the mathematics of the time – no deeper ideas were linked with it.

Only in the period following the era of Riemann did the geometry of algebraic curves occupy a central place in the contemporary mathematics, alongside the theory of Abelian integrals and Abelian functions. Basically this change of view point was connected with the name of Clebsch. Whereas for Riemann the foundation is the function, Clebsch takes as the fundamental object the algebraic curve. One can say that Riemann considered a finite morphism $f: X \to \mathbb{P}^1$, and Clebsch the algebraic curve X itself. In the book by Clebsch and Gordan [1] a formula is deduced for the number p of linearly independent integrals of the first kind (that is, for the genus of the curve X), which expresses it in terms of the degree of the curve and the number of singular points (see Exercise 12 to Ch. IV, § 3). There it is also shown that for $p=0$ the curve has a rational parameterization, and for $p=1$ becomes a plane cubic curve.

An error Riemann had made turned out to be exceptionally useful for the development of the algebraic-geometrical aspect of the theory of algebraic curves. In the proof of his existence theorems he had regarded as obvious the solubility of a certain variational problem: the "Dirichlet principle". Before long Weierstrass showed that not every variational problem has a solution. Therefore Riemann's results remained unfounded for some time. One of the ways out was an algebraic proof of these theorems: they were stated essentially in algebraic form. These investigations, which were undertaken by Clebsch (see Clebsch and Gordan [1]), furthered considerably the clarification of the essentially algebraic-geometrical character of the results of Abel and Riemann, hidden under an analytic cloak.

The trend of research begun by Clebsch achieved its bloom in the work of his pupil M. Noether. Noether's ideas are particularly clearly outlined in his joint paper with Brill [1]. In it the problem is raised of developing the geometry on an algebraic curve lying in a projective plane, as the collection of results that are invariant under biunique (that is, birational) transformations. The foundation is the concept of the group of (coincident or distinct) points of the curve. They consider systems of groups of points that cut out on the original curve linear systems of curves (that is, systems whose equations form a linear space). It can happen that all groups of such a system contain a common group G, that is, consist of G and another group G'. The system of groups G' obtained in this way is called linear. If the dimension of the linear (projective) space of equations of the cutting curves is equal to q and the groups G' consists of Q points, then the system is denoted by $g_Q^{(q)}$. Two groups of one and the same system are called corresidual. Clearly this corresponds to the modern concept of equivalence of effective divisors, and if G is contained in a linear system $g_Q^{(q)}$, then in the modern notation $\deg G = Q$, $l(G) \geqslant q+1$ (we recall that $l(G)$ is the dimension of the vector space and q that of the corresponding projective space).

Every group of points determines a largest linear system $g_Q^{(q)}$ containing all the groups corresidual to the given one. The numbers q and Q are connected by the Riemann-Roch theorem, which is proved purely algebraically.

Of course, the Riemann-Roch theorem presupposes a definition analogous to the canonical class. It is given without appeal to the concept of a differential form, but the connection with this concept is very easily established. For if a curve of degree n has the equation $F=0$ and is smooth, then the differential forms $\omega \in \Omega^1[X]$ can be written in the form $\omega = \frac{\varphi}{F'_v} dx$, where φ is a homogeneous polynomial of degree $n-3$ (Ch. III, § 5.4). It can be shown that if the curve has only the simplest singularities, then this expression remains valid if it is required that φ vanishes at all singular points. These polynomials are said to be associated. Associated polynomials of degree $n-3$ determine the linear system that is an analogue to the canonical class.

In their paper Brill and Noether consider a mapping of a curve into the $(p-1)$-dimensional projective space defined by the associated polynomials of degree $n-3$. Its image is called a normal curve. They show that a single-valued (in present-day terminology, birational) correspondence of curves reduces to a projective transformation of normal curves (provided that the curves are not hyperelliptic).

Noether [1] applied these ideas to the investigation of space curves. In modern language we can say that his paper is concerned with the study of the irreducible components of the Chow variety of curves in three-dimensional space.

7. Many-Dimensional Geometry. At the beginning of the second half of the nineteenth century many special properties of algebraic varieties of dimension greater than 1, mainly surfaces, had been found. For example, cubic surfaces had been investigated in detail, in particular, Salmon and Cayley had proved in 1849 that on any cubic surface without singular points there are 27 distinct lines. However, for a long time these results were not combined by any general principles and were not connected with the deep ideas that had been worked out at the time in the theory of algebraic curves.

The decisive step in this direction was taken, apparently, by Clebsch. In 1868 he published a small note [1] in which he considers algebraic surfaces from the point of view (using modern terminology) of birational isomorphism. He considers everywhere finite double integrals on the surface and mentions that the maximal number of linearly independent among them is invariant under birational isomorphism.

These ideas were developed in Noether's paper [2], which consists of two parts. As is clear from the very title, in it he considers varieties of an arbitrary number of dimensions. However, the major part of the results refers to surfaces. This is typical for the whole subsequent period of algebraic geometry: although very many results were in fact true for varieties of arbitrary dimension, they were stated and proved only for surfaces.

In the first part Noether considers "differential expressions" on a variety of arbitrary dimension, and it is interesting that he writes down an integral sign only once. Thus, here the algebraic character of the concept of a differential form already becomes formally obvious. Noether considers only forms of maximal degree. He shows that they make up a finite-dimensional space whose dimension is invariant under single-valued (that is, birational) transformations.

In the second part he considers curves on surfaces. (Only the last section contains some interesting remarks on three-dimensional varieties.) Noether gives a description of the canonical class (in modern terminology) by means of associated surfaces, analogous to the way in which this was done earlier for curves. He raises the question of the surfaces V that cut out on a curve C lying on V its canonical class (again in modern terminology). He calls the curves cutting them out on V associated with C and gives an explicit description for them which leads him to a formula for the genus of a curve on a surface. This formula essentially is the same as (1) in Ch. IV, § 2.3; however, an understanding of the fact that an associated curve is of the form $K+C$ was achieved only 20 years later in the work of Enriques.

In the same paper Noether investigates the concept of an exceptional curve, which contracts to a point under a birational isomorphism.

The most brilliant development of the ideas of Clebsch and Noether came not in Germany, but in Italy. The Italian school of algebraic geometry exerted an immense influence on the development of this branch of mathematics. Undoubtedly many ideas created by this school have so far not been fully understood and developed. The founders of the Italian geometrical school are Cremona, C. Segre, Bertini. Its most important representatives are Castelnuovo, Enriques, and Severi. Castelnuovo's papers began to appear at the end of the 1880's. Enriques was a pupil (and relative) of Castelnuovo. His papers appeared at the beginning of the 1890's. Severi began to work about ten years after Castelnuovo and Enriques.

One of the main achievements of the Italian school is the classification of algebraic surfaces. As a first result we can quote here a paper of Bertini [1] in which a classification of involutory transformations of the plane is given. The matter concerns (in present-day terminology) the classification, to within conjugacy in the group of birational automorphisms of the plane, of all group elements of order 2. The classification turns out to be very simple, in particular, it is easy to derive from it that the factor space of the plane with respect to a group of order 2 is a rational surface. In other words, if a surface X is unirational and a morphism $f: \mathbb{P}^2 \to X$ is of degree 2, then X is rational.

The general case of Lüroth's problem for algebraic surfaces was solved (affirmatively) by Castelnuovo [1]. After this he raised the problem of characterizing rational surfaces by numerical invariants and solved it in [2]. The classification of surfaces that we have explained in Ch. III, § 5.7, was obtained by Enriques in a series of papers, which was completed in the first decade of our century (see Enriques [2]).

In the context of Lüroth's problem for three-dimensional varieties Fano investigated certain types of these varieties, suggesting a proof of the fact that they are not rational. Enriques had shown that many of them are unirational. This would give a negative solution of Lüroth's problem, but Fano pointed out many obscure places in the proof. Some intermediate propositions turned out to be not true. The problem was solved finally when the last pages of this book were already written. V.A. Iskovskii and Yu. I. Manin have shown that Fano's basic idea can be made to work. They have shown that smooth hypersurfaces of degree 4 in $\mathbb{P}^4$ are non-rational (the fact that some of them are unirational had been proved by B. Segre). Simultaneously Griffiths and Clemens have found a new analytical method of proving that certain varieties are non-rational, for example, smooth hypersurfaces of degree 3 in $\mathbb{P}^4$ (see Exercise 18 to Ch. III, § 5). Of course, these results are only the first steps on the way to a classification of unirational varieties.*

The main tool of the Italian school was the investigation of families of curves on surfaces – linear and algebraic (the latter were called continuous). This led to the concept of linear and algebraic equivalence (in our book linear equivalence is simply called equivalence). The connection between these two concepts was first investigated by Castelnuovo [3]. He discovered a link of this problem with an important invariant of

* A third method to construct non-rational but unirational threefolds was discovered by Artin and Mumford. See v. A. Iskovskii and Yu. I. Manin, Mat. Sborn. 86 (1971), 140—166, C. H. Clemens and P. Griffiths, Ann. of Math. 95 (1972) 281—358; M. Artin and D. Mumford Proc. of Lond. Math. Soc. XXV (1972) 75—95. (Footnote to corrected printing, 1977).

the surface, the so-called irregularity. We do not give here the definition of irregularity, which was used by Castelnuovo; it is connected with ideas close to the cohomology theory of sheaves. Formula (1) below gives another interpretation of this concept.

Castelnuovo [3] proved that if not every continuous system of curves is contained in a linear system (that is, if algebraic and linear equivalence are not one and the same thing), then the irregularity of the surface is different from zero. Enriques [1] proves the converse proposition. Furthermore, he shows that every sufficiently general curve (in an exactly defined sense) lying on a surface of irregularity q is contained in an algebraically complete (that is, maximal) continuous family, which is stratified into linear families of the same dimension, and the basis of the stratification is a variety of dimension q. Castelnuovo [1] showed that the fibering of linear systems (that is, classes of divisors) determines on a q-dimensional basis of the fibering constructed by Enriques a group law by virtue of which this basis is an Abelian variety and is therefore uniformized by Abelian ($2q$-periodic) functions. This Abelian variety does not depend on the curve from which we have started out and is determined by the surface itself. It is called the Picard variety of this surface.

The irregularity turned out to be connected with the theory of one-dimensional differential forms on the surface, the beginning of which goes back to Picard [1]; in this paper it is proved that the space of everywhere regular forms is finite-dimensional. In 1905 Severi and Castelnuovo proved that this dimension is the same as the irregularity; in our notation

$$q = h^1 = \dim \Omega^1[X]. \tag{1}$$

Severi [1] investigated the group of classes relative to algebraic equivalence and proved that it is finitely generated. His proof is based on a connection of the concept of algebraic equivalence with the theory of one-dimensional differential forms. Namely, an algebraic equivalence $n_1 C_1 + \cdots + n_r C_r \approx 0$ is equivalent to the fact that for some one-dimensional differential form the set of its "logarithmic singularities" coincides exactly with the curves $C_1, \ldots, C_r$, taken with the multiplicities $n_1, \ldots, n_r$. (A curve C is a logarithmic singularity of multiplicity n for a form ω if locally $\omega = n f^{-1} df$, where f is a local equation of C.) Picard had already proved earlies that the so-defined relation of equivalence by means of differential forms gives rise to a finitely generated group of classes (see Picard and Simart [1]).

8. The Analytic Theory of Manifolds. Although a considerable part of the concepts of algebraic geometry arose in analytic form, their algebraic meaning cleared up in time. Now we pass on to concepts and results which are essentially connected with analysis (at least from the present point of view).

At the beginning of the 1880's there appeared papers by Klein and Poincaré devoted to the problem of uniformization of algebraic curves by automorphic functions. The aim was to uniformize arbitrary curves by functions that are now called automorphic, just as elliptic functions uniformize curves of genus 1. (The term "automorphic" was proposed by Klein, previously these functions were called by various names.) Klein [1] (No. 84) started out from the theory of modular functions. The field of modular functions is isomorphic to the field of rational functions, but one can consider functions that are invariant under various subgroups of the modular group and so obtain more complicated fields. In particular, Klein considered functions that are automorphic relative to the group consisting of all transformations $z \to \dfrac{az+b}{cz+d}$ in which a, b, c, and d are integers, $ad-bc=1$, and

$$\begin{pmatrix} a & b \\ c & d \end{pmatrix} \equiv \begin{pmatrix} 1 & 0 \\ 0 & 1 \end{pmatrix} (\bmod 7) .$$

He proved that these functions uniformize the curve of genus 3 with the equation $x_0^3 x_1 + x_1^3 x_2 + x_2^3 x_0 = 0$. The fundamental polygon of this group can be deformed so as to obtain new groups uniformizing new curves of genus 3.

A similar train of ideas lay at the basis of the papers by Klein [1] (No. 101–103) and Poincaré [1], (p. 92, 108, 169), but Poincaré used for the construction of automorphic functions the series that now bear his name. They both conjectured correctly that every algebraic curve admits a uniformization by the corresponding group and made substantial progress towards a proof of this result. However, a complete proof was not achieved at the time, but only in 1907 by Poincaré (and independently by Koebe). An important role was played by the fact that by this time Poincaré had investigated the concept of fundamental group and universal covering.

The topology of algebraic curves is very simple and was completely studied by Riemann. In the investigation of the topology of algebraic surfaces Picard developed a method that is based on a study of the fibres of a morphism $f: X \to \mathbb{P}^1$. The point is to find out how the topology of the fibre $f^{-1}(a)$ changes when the point $a \in \mathbb{P}^1$ changes, in particular, when this fibre acquires a singular point. He proved, for example, that smooth surfaces in $\mathbb{P}^3$ are simply-connected (see Picard and Simart [1], Vol. I). By this method Lefschetz [2], [3] obtained many deep results on the topology of algebraic surfaces, and also on varieties of arbitrary dimension.

The study of global properties of analytic manifolds began fairly recently (Hopf [1], A. Weil [2]). This domain developed vigorously in the 1950's, in connection with the creation and application by Cartan and Serre of the theory of analytic coherent sheaves (see Cartan [1] and Serre [1]). We do not give a definition of this concept – it is an exact analogue of the concept of an algebraic coherent sheaf (but we must emphasize that the analytic concept was introduced before the algebraic one). One of the basic results of this theory was the proof that the cohomology groups (and, in particular, the groups of sections) of an analytic coherent sheaf over a compact manifold are finite-dimensional. In this context Cartan gave a definition of an analytic manifold that is based on the concept of a sheaf and expressed the idea that the definition of various types of manifolds is linked with the specification of sheaves of rings on them.

9. Algebraic Varieties over an Arbitrary Field. Schemes. Formally the study of varieties over an arbitrary field began only in the twentieth century, but the foundations for this were laid earlier. An important role was played here by two papers printed in one and the same issue of Crelle's Journal in 1882. Kronecker [1] investigates problems that would nowadays be referred to the theory of rings of finite type without divisors of zero and of characteristic 0. In particular, for integrally closed rings he constructs a theory of divisors.

The paper by Dedekind and Weber [1] is devoted to the theory of algebraic curves. Its aim is to give a purely algebraic account of a considerable part of this theory. The authors emphasize that they do not use the concept of continuity anywhere, and their results remain true if the field of complex numbers is replaced by the field of all algebraic numbers.

The principal significance of the paper by Dedekind and Weber lies in the fact that in it the basic object of study is the field of rational functions on an algebraic curve. Concrete (affine) models are employed only as a technical tool, and the authors use the term "invariance" to denote concepts and results that do not depend on the choice of model. In this paper the whole account becomes to a significant degree parallel to the theory of fields of algebraic numbers. In particular, the analogy between prime ideals of a field of algebraic numbers and points of the Riemann surface of a field of algebraic functions is emphasized (we could say that in both cases we are concerned with the maximal spectrum of a one-dimensional scheme).

Interest in algebraic geometry over "non-classical" fields arose first in connection with the theory of congruences, which can be interpreted as equations over a finite field. In his lecture at the International Congress of Mathematicians in 1908 Poincaré says that the methods of the theory of algebraic curves can be applied to the study of congruences in two unknowns.

The ground for a systematic construction of algebraic geometry was prepared by the general development of the theory of fields and rings in the 1910's and 20's.

In 1924 Artin published a paper (see Artin [1], No. 1) in which he studied quadratic extensions of the field of rational functions of one variable over a finite field of constants, based on their analogy to quadratic extensions of the field of rational numbers. Of particular importance for the subsequent development of algebraic geometry was his introduction of the concept of the ζ-function of this field and the formulation of an analogue to the Riemann hypothesis for the ζ-function. We introduce (which Artin did not do) a hyperelliptic curve defined over a finite field k for which the field in question is of the form $k(X)$. Then the Riemann hypothesis gives a best possible estimate for the number N of points $x \in X$ that are defined over a given finite extension K/k, that is, for which $k(x) \subset K$ (just like the Riemann hypothesis for the field of rational numbers gives a best possible estimate for the asymptotic distribution of prime numbers). More accurately, the Riemann hypothesis is equivalent to the inequality $|N-(q+1)| < 2g\sqrt{q}$, where q is the number of elements of the field K and g the genus of the curve X.

Attempts to prove the Riemann hypothesis (which, as becomes clear at once, can be formulated for any algebraic curve over a finite field) led in the 1930's to work by Hasse and his pupils on the theory of algebraic curves over an arbitrary field. Here the hypothesis itself was proved by Hasse [1] for elliptic curves.

Strictly speaking, this theory concerns not curves but the corresponding fields of functions, and the authors nowhere use geometric terminology. With this style one can become acquainted in the book by Hasse [2] (see the sections devoted to function fields). The possibility of this birationally invariant theory of algebraic curves is connected with the uniqueness of a smooth projective model of an algebraic curve. Therefore great difficulties arise in applying this approach to the many-dimensional case.

On the other hand, in a sequence of papers published under the general title „Zur algebraischen Geometrie" in the Mathematische Annalen between the end of the 1920's and the beginning of the 1930's, van der Waerden made progress in the construction of algebraic geometry over an arbitrary field. In particular, he set up a theory of intersections (as we would say nowadays, he defined the ring of classes of cycles) over a smooth projective variety.

In 1940 A. Weil succeeded in proving the Riemann hypothesis for an arbitrary algebraic curve over a finite field. He found two ways of proving it. One of them is based on the theory of correspondences of the curve X (that is, divisors on the surface $X \times X$), and the other on an analysis of its Jacobian variety. Thus, in both cases many-dimensional varieties are brought into play. The book by Weil [1] contains the construction of algebraic geometry over an arbitrary field: the theory of divisors, cycles, intersections. Here "abstract" (not necessarily quasiprojective) varieties are defined for the first time by the process of pasting together affine pieces (similar to Ch. V, § 3.2).

A definition of a variety based on the concept of a sheaf is contained in the paper by Serre [2], where the theory of coherent algebraic sheaves is constructed, for which the recently created theory of coherent analytic sheaves served as a prototype (see § 8).

Generalizations of the concept of an algebraic variety, close in spirit to the later concept of a scheme, were proposed at the beginning of the 1950's. Apparently the first and for the time very systematic working out of these ideas is due to Kähler [1], [2]. The concept of a scheme, as well as the majority of results in the general theory of schemes, is due to Grothendieck. The first systematic account of these ideas is contained in a lecture by Grothendieck [1].

Bibliography for the Historical Sketch

Abel, N. H.: [1] Oeuvres complètes. Christiania 1881.

Artin, E.: [1] Collected papers. New York-London: Addison-Wesley, 1965.

Bernoulli, J.: [1] Opera omnia, vol. I–IV. Lausannae et Genevae: Bosquet 1742.

Bertini, E.: [1] Ricerche sulle transformazioni univoche involutorie nel piano. Ann. Mat. Pura Appl. (2) **8** (1877).

Brill, A., Noether, M.: [1] Über die algebraischen Funktionen und ihre Anwendung in der Geometrie. Math. Ann. **7** (1873).

Cartan, H.: [1] Variétés analytiques complexes et cohomologie, Coll. sur les fonctions de plusieurs variables. Bruxelles, March 1953.

Castelnuovo, G.: [1] Sulla razionalità delle involuzioni piane. Rend. Accad. Lincei **2** (1893).

[2] Sulle superficie di genere zero. Mem. Soc. Ital. Sci. **10** (1896).

[3] Alcuni proprietà fondamentali dei sistemi lineari di curve tracciati sopra una superficie, ibid.

[4] Sugli integrali semplici appartenenti and una superficie irregolars. Rend. Accad. Lincei **14** (1905).

Clebsch, A.: [1] Sur les surfaces algébriques. C. R. Acad. Sci. Paris **67**, 1238–1239 (1868).

Clebsch, A., Gordan, P.: [1] Theorie der Abelschen Funktionen. Leipzig: Teubner, 1866.

Dedekind, R., Weber, H.: [1] Theorie der algebraischen Funktionen einer Veränderlichen. J. Reine Angew. Math. **92** (1882).

Enriques, F.: [1] Sulla proprietà caratheristica delle superficie irregolary. Rend. Accad. Bologna **9** (1904).

[2] Superficie algebriche. Bologna 1949.

Euler, L.: [1] Integral calculus, Vol. I, Ch. VI.

[2] Opera omnia, Ser. I, Vol. XXI, 91–118.

Frobenius, G.: [1] Über die Grundlagen der Theorie der Jakobischen Funktionen. J. Reine Angew. Math. **97**, 16–48, 188–223 (1884).

Göpel: [1] Theoriae transcendentium Abelianarum primi ordinis adumbrato levis. J. Reine Angew. Math. **35** (1847).

Grothendieck, A.: [1] The cohomology theory of abstract algebraic varieties. Internat. Congr. Math. Edinburgh 1958.

Hasse, H.: [1] Zur Theorie der abstrakten elliptischen Funktionenkörper. J. Reine Angew. Math. **175** (1936).

[2] Zahlentheorie. Berlin: Akademie-Verlag, 1950.

Hopf, H.: [1] Zur Topologie der komplexen Mannigfaltigkeiten. Studies and essays presented to R. Courant. New York 1958, 167–187.

Jacobi, C. G. J.: [1] Gesammelte Werke. Berlin 1881.

Kähler, E.: [1] Algebra und Differentialrechnung. Berichte Math. Tagung. Berlin 1953.

[2] Geometria arithmetica. Ann. Mat. Pura Appl. (4), **45** (1958).

Klein, F.: [1] Gesammelte Mathematische Abhandlungen, Vol. III. Berlin: Springer-Verlag, 1923.
[2] Riemannsche Flächen. Berlin 1891–1892.
Kronecker, L.: [1] Grundzüge einer arithmetischen Theorie der algebraischen Größen. J. Reine Angew. Math. **92** (1882).
Lefschetz, S.: [1] Numerical invariants of algebraic varieties. Trans. Amer. Math. Soc. **22** (1921).
[2] L'Analysis situs et la géométrie algébrique. Paris 1924.
[3] Géométrie sur les surfaces et les variétés algébriques. Mem. Sci. Math. **40** (1929).
Legendre, A. M.: [1] Traité des fonctions elliptiques et des intégrales Eulériennes, 3 vols. Paris 1825–1828.
Noether, M.: [1] Zur Grundlegung der Theorie der algebraischen Raumkurven. J. Reine Angew. Math. **93** (1882).
[2] Zur Theorie des eindeutigen Entsprechens algebraischer Gebilde von beliebig vielen Dimensionen. Math. Ann. **2** (1870), **8** (1875).
Picard, E.: [1] Sur les intégrales des différentielles totales algébriques de première espèce. C. R. Acad. Sci. Paris **99** (1884).
Picard, E., Simart, G.: [1] Théorie des fonctions algébriques de deux variables indépendentes. Paris 1897–1906.
Poincaré, H.: [1] Oeuvres, Vol. II. Paris 1916.
[2] Sur les propriétés du potentiel et sur les fonctions Abéliennes. Acta Math. **22**, 89–178 (1899).
Riemann, B.: [1] Gesammelte Werke.
Serre, J. P.: [1] Quelques problemes globaux relatifs aux variétés de Stein. Coll. sur les fonctions de plusieurs variables. Bruxelles, March 1953.
[2] Faisceaux algébriques cohérents. Ann. of Math. (2) **61** (1955).
Severi, F.: [1] La base minima pour la totalité des courbes algébriques tracées sur une surface algébrique. Ann. École Norm. Sup. **25** (1908).
Weil, A.: [1] Foundations of algebraic geometry. New York 1946.
[2] Sur la théorie des formes différentielles attachées à une variété analytique complexe. Comm. Math. Helv. **24** (1947).
Weyl, H.: [1] Die Idee der Riemannschen Fläche. Berlin 1923.

Subject Index

List of Notation